A Short Textbook of Radiotherapy

A Short Textbook of Radiotherapy

Radiation Physics, Therapy, Oncology

J. Walter
MA BM FRCP FRCR DMRE
Former Consultant Radiotherapist,
Weston Park Hospital, Sheffield
Honorary Clinical Lecturer in Radiotherapy,
University of Sheffield

H. Miller
OBE MA PhD FInstP
Former Chief Physicist, Trent Regional Health Authority
Professor Associate Emeritus, University of Sheffield

C. K. Bomford
BSc MPhil MInstP
Principal Physicist, Weston Park Hospital, Sheffield
Honorary Lecturer, University of Sheffield

FOURTH EDITION

CHURCHILL LIVINGSTONE
EDINBURGH LONDON AND NEW YORK 1979

CHURCHILL LIVINGSTONE
Medical Division of Longman Group Limited

Distributed in the United States of America by
Longman Inc., 19 West 44th Street, New York,
N.Y. 10036, and by associated companies,
branches and representatives throughout
the world.

First, second and third
editions by
J. Walter and H. Miller

First edition 1950
Second edition 1959
Third edition 1969
Fourth edition 1979

ISBN 0 443 01389 6

British Library Cataloguing in Publication Data
Walter, Joseph
A short textbook of radiotherapy. – 4th ed.
1. Radiotherapy
I. Title II. Miller, Harold III. Bomford, C K
615'.842 RM847 78–40776

Printed in Singapore by
Kua Co., Book Manufacturer, Pte Ltd.

To Marianne
and to the memory
of my parents (JW)

Preface

This volume is more than a revision of the third edition. It is virtually a new book and very little of the older editions remains.

The original basic purpose is the same a text for the qualifying diploma examination of the College of Radiographers (DCR). But we have not hesitated to go further in some sections and include material appropriate for the higher diploma (HDCR).

Major changes in Part 1 include the elimination of an elementary treatment of the structure of matter, on the assumption that this will have been covered in the first stage of radiographers' training. This has allowed room for considerable extension of the details of multifield planning techniques including simulator methods and of the principles and practice of the use of unsealed radioactive sources. A brief account is included of new developments in technique which have an influence on radiotherapy practice. These include ultrasonics and computerised transverse axial tomography as diagnostic tools and the use of neutron beams for therapy.

Medical physics technicians will also find much of relevance to their work in radiotherapy physics and of help towards their qualifying examinations for the Ordinary and Higher National Certificates (ONC and HNC) or in those courses proposed and run under the auspices of the Technician Education Council.

In Part 2 the medical coverage has been considerably extended to include the whole range of therapy carried out in most departments of radiotherapy and oncology. This systematic survey makes it approximate to a text for trainee radiotherapists and medical postgraduates. We hope they will find it even more useful than previous editions as a first approach to the specialty and that medical students and clinicians, as well as paramedical workers (physicists, biochemists etc.) will also continue to find it helpful. Some material, especially in the general chapters, has been incorporated from the companion volume for nurses (J. Walter, *Cancer and Radiotherapy* 2nd edition).

While all sections have been updated, there are major alterations and additions in epidemiology, central nervous system, lymphomas, bone and soft tissues, cancer in children. Medical oncology has been included, not only because it is in the examination syllabus, but also as it is so intimately connected with the work of radiotherapy departments. Therapy radiographers will (and should) be interested in this aspect of their patients' treatment, though they will not require technical detail such as drug dosages.

Our grateful thanks go out to many helpers—Mrs Susan Sharman, Miss Carol Cheetham and Mrs Audrey Rawson for typing; Mrs Marianne Walter for a number of diagrams; colleagues in the Medical Physics Department for helpful discussion and criticism; Dr F. E. Neal, Professor D. S. Munro, Dr I. G. Emmanuel, Dr J. S. L. Lilleyman.

Words in the text marked with an asterisk are defined in the glossary.

Sheffield, 1978

JW
HM
CKB

Contents

PART I

Radiation physics

1. The emission of X-rays

Production of X-rays

X-rays are produced whenever fast electrons are slowed down in passing through matter. In a conventional X-ray tube in particular X-rays arise when the electron beam, which forms the current through the tube, is stopped at the metal target or anode of the tube. The electrons arriving at the target with an energy determined by the voltage applied to the tube penetrate the target metal and interact with the atoms of the target. The interaction is mainly with the electrons in the outer structure of the target atoms though less frequently an interaction with the electrical field around the nucleus also occurs. The electrical forces between the impinging electrons of the beam and the charged components of the atoms of the target cause the impinging electrons to lose energy so that after penetrating the target material for a distance representing the electron range in the material the impinging electrons come to 'rest'; that is they then form part of the population of the target's conduction electrons and contribute to the electron current passing into the external circuit.

The interaction between the impinging electrons and the target atoms occurs in three main ways. Most of the electrons suffer many minor deflections before coming to rest and each deflection transfers a little of the electron energy to the atom which has caused the deflection. This increased energy of the target atoms appears as heat. For normal voltages applied to the tube most of the energy of the impinging electrons appears therefore as heat in the target and as a result the cooling of the target becomes a major part of X-ray tube design (see page 19).

A second type of interaction between the impinging electron and the atoms of the target, however, is a more significant type of collision, in which the energy lost by the impinging electron appears directly as a photon of electromagnetic radiation. A fast electron may sometimes lose all its energy in one collision with an atom in this way. It is more likely however that it will lose only a part of its energy in the collision and then proceed further, interacting with other target atoms before coming to rest. A beam of electrons interacting with the target in this way therefore produces photons of electromagnetic radiation with energies spread over a complete range from very small values to the maximum energy of the electrons in the beam. This gives rise to the *Continuous spectrum* of X-rays and it is this process which accounts for most of the emission of X-rays from the target.

A third type of interaction occurs however, also leading to the emission of X-rays. Some of the impinging electrons interact with electrons in the orbital electron structure of the target atoms and cause them to be ejected from the atoms. The vacant spaces so created in the electron orbits are then filled by electrons transferring from outer orbits. The transfer of an electron to an inner orbit in any atom results in an energy loss and this energy is emitted as a photon of electromagnetic radiation. The photon so emitted has an energy which is determined by the electron structure of the atom concerned and is characteristic of the atom. If the photon energy is high enough the radiation is in the X-ray region of the electromagnetic spectrum. Photons emitted in this way contribute to a line spectrum of X-rays characteristic of the target element. This third type of interaction therefore gives rise to *characteristic radiation*.

Efficiency of X-ray production

The efficiency of X-ray production, that is the proportion of the total energy of the incident electron beam that appears as X-ray emission, is normally very small. A low efficiency of X-ray production means that a correspondingly high proportion of the electron beam energy appears as heat. The efficiency of X-ray production is proportional, approximately, to both the tube voltage and the atomic number of the target. A target of high atomic number material therefore is very advantageous for an X-ray tube and tungsten with an atomic number of 74 is generally used as the target material. It is also a suitable material because of its high melting point, reasonable heat conductivity and good mechanical properties. For a tungsten target in an X-ray tube at 100 kV the efficiency of X-ray production is approximately 0.5 per cent—that is 99.5 per cent of the electron beam energy goes into heat. The X-ray production efficiency at 200 kV is approximately 1 per cent. Accelerating voltages of 20–40 million volts are required before the X-ray production becomes very efficient, say 60 or 70 per cent.

The energy of the individual electrons bombarding the tube target is determined by the voltage applied to the tube. It is convenient to express this energy in terms of the unit, the *electron volt*. One electron volt is the energy acquired by an electron when accelerated through a potential difference of one volt. When a voltage V volts is applied to the tube the electrons arrive at the target with an energy V electron volts. The efficiency of conversion of this electron energy to X-ray energy is approximately proportional to the applied tube voltage and the energy of the electrons is also proportional to this voltage. The total X-ray emission therefore will increase approximately proportionally to the square of the applied tube voltage if the tube current is kept constant, that is, if the number of electrons striking the target per second is constant. There is thus a rapid change of X-ray output with tube voltage and much attention must be given in the design of X-ray equipment to ensure the constancy of the voltage applied to the X-ray tube.

The distribution of X-rays round the target

When electrons strike a metal target which is just thick enough to bring them to rest, X-rays are found to be emitted from the target in all directions. The distribution of the X-radiation depends however very much on the tube voltage. For tube voltages of about 50 to 100 kV the maximum intensity of emission is approximately at right angles to the electron beam but as the beam voltage increases the direction of maximum intensity moves more and more towards that of the electron beam and for beams in the megavoltage range most of the X-ray emission is confined to a narrow cone near the forward direction of the electron beam.

Figure 1.1 illustrates by polar diagrams the distribution of X-ray energy from a thin target for three representative electron energies. The curves are such that the distance of the curve from the origin O in any direction is proportional to the intensity of the beam in that direction. It can be seen that for electrons of 30 keV energy the X-ray intensity at right angles to the electron stream is greater than in the forward direction whereas at 2 MeV the forward intensity is 10 times that at right angles.

The actual distribution of X-ray intensity round the thick targets of practical X-ray tubes is different from that shown in Figure 1.1 because of the attenuation of the X-ray beams in the target material especially for beams generated at low voltages. The design of targets for X-ray tubes is however much influenced by the distribution of X-rays illustrated in this figure. At low voltages it is advantageous to use the X-ray beam coming from the target at right angles to the electron stream and most diagnostic and therapy X-ray tubes up to 300 kV use therefore a *reflection target* in which the useful beam is taken from the front of the target (see p. 18). At tube voltages of 1 million or more however it is advantageous to use the X-ray beam in the same direction as the incident electron stream in spite of the attenuation in the target. Such targets which are made just thick enough to stop all the bombarding electrons are known as *transmission targets* (see p. 24).

The variation of X-ray beam intensity with direction gives rise to problems in the production of uniform irradiation of a very large field. The most important example of this is the need to introduce arrangements for field flattening in megavoltage radiotherapy because of the concentration of the X-ray emission in a very narrow angle round the direction of the electron beam (see p. 24).

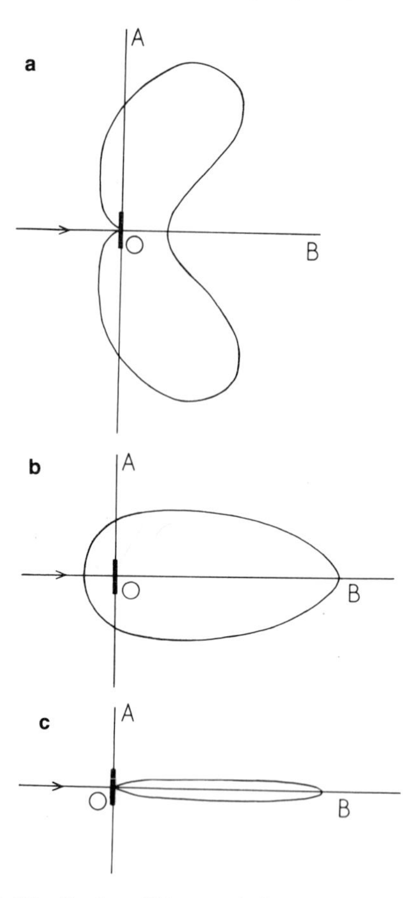

Fig. 1.1 Distribution of X-ray emission.
(a) 30 keV electrons
(b) 2 MeV electrons
(c) 20 MeV electrons

The continuous spectrum

As seen above the X-ray beam consists mainly of photons arising from multiple interactions of many electrons with the target atoms. Photons are present therefore of all energies up to a maximum value corresponding to the energy of the electrons accelerated by the maximum voltage applied to the tube. This forms the continuous spectrum. The continuous spectrum has a maximum photon energy equal to the tube voltage when the energy is expressed in electron volts. Photons of maximum photon energy have the *minimum wavelength* in the spectrum. This minimum wavelength may be calculated thus:

Energy of photon $= V$ electron volts $= hn = \frac{hc}{\lambda}$

Minimum wavelength $= \lambda_{min} = \frac{hc}{V_{max}}$

where h is Planck's constant, n the frequency of the radiation and λ its wavelength and c the velocity of light.

If appropriate units are put in for the physical constants and if appropriate energy units are used this formula becomes

$$\text{Minimum wavelength} = \frac{1.24}{kV} \text{ nanometres}$$

where kV is the tube voltage expressed in kilovolts. (1 nanometre $= 10^{-9}$ metres)

This relationship is sometimes referred to as *Duane-Hunt's Law.*

The distribution of the relative intensity of the radiations in the continuous spectrum as a function of photon energy expressed in kilo-electron volts is shown in Figure 1.2 for three representative tube voltages in the superficial X-ray range. In this diagram the ordinate represents the energy flow per unit interval of energy as expressed by the number of photons with that particular energy multiplied by the energy of those photons. The three curves show the continuous spectrum for tube voltages of 60 kV, 90 kV and 120 kV. Note that in these curves any characteristic radiations have not been shown, and the beam current is assumed constant for the three curves.

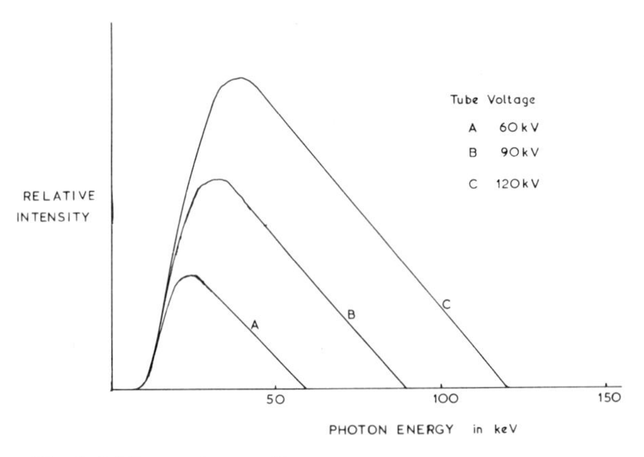

Fig. 1.2 The continuous X-ray spectrum.

Note the following properties of these curves:

1. The existence of a maximum photon energy (minimum wavelength) for each tube voltage.
2. The existence of a peak in the relative intensity distribution at a photon energy about one third of the maximum value.
3. The low intensity of low energy photons which arises because of the decrease in penetrating power of the photons as the photon energy decreases so that they escape less readily from the target.
4. An increase in tube voltage increases the intensity of emission at all photon energies and introduces higher energy photons not present at lower voltages. The total intensity of the radiation emitted from the target is easily seen to be proportional to the area under the appropriate curve and the family of curves illustrates that the total emission is approximately proportional to the square of the voltage applied to the tube.
5. With increase in applied voltage the proportion of high energy photons increases and the peak of the curve moves to higher photon energies. Since normally higher energy photons have greater penetrating power (see p. 35) the figure also illustrates the fact that increases of voltage on the tube gives increased penetration power of the beam.

Characteristic radiation

The curves of Figure 1.2 are drawn for a fixed beam current with an arbitrary intensity scale. They represent the continuous spectrum from a target of any element arising from multiple interactions of the bombarding electrons with the target atoms. We have seen however that a third type of interaction occurs between the electrons of the tube current and the target atoms. In this case the electrons of the target atoms may be ejected from their shell and the vacancy is filled by electrons falling from outer shells or from outside the atom altogether. Energy is released in this process and this appears as photons having an energy equal to the energy difference between the level from which the electrons fall and the one into which they fall. This radiation is a *line spectrum*, that is it consists of a few quite precise and discrete photon energies entirely determined by the electron structure of the target material. It is thus radiation which is characteristic of the target element and of the particular shell from which an electron was dislodged to initiate the process. Such photon emission is referred to as characteristic radiation.

For a tungsten target electrons can be ejected from the K shell if the bombarding electrons have an energy of 69 keV or more. The resultant characteristic radiation, consisting of photons with energies in one of a small group of discrete energies of nearly the same value, appears as a line spectrum superposed on the continuous spectrum for all tube voltages above 69 kV. If the target material were of some other element the voltage needed to dislodge the K electron would be different and the characteristic radiation would have a different wavelength. Figure 1.3. shows the two spectra including both continuous and characteristic radiation for a tungsten (W) target and a tin (Sn) target for the same tube voltage (120 kV) and the same tube current.

Note the following points:

1. The continuous spectrum from tin (atomic number 50) is less intense than from tungsten (atomic

number 74)—the total emission being approximately proportional to the atomic number of the target.
2. The characteristic radiation from tungsten appears between 57 keV and 69 keV photon energy whenever the tube voltage exceeds 69 keV, while that from the tin target appears between approximately 25 keV and 29 keV whenever the tube voltage is above 29 kV.
3. Note that although the intensity of the lines of the characteristic radiation may be high compared to that of the continuous spectrum its contribution to the total emission of X-ray energy from the tube is relatively small.

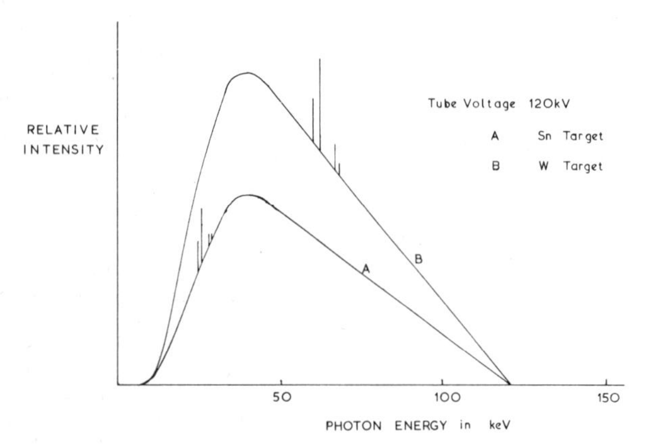

Fig. 1.3 X-ray spectra for tungsten and tin targets.

Filtration of the X-ray beam

Any X-ray beam is of practical use only after it has emerged from the target and passed through the glass wall of the tube, the tube shield, and any materials inevitably or deliberately placed in its path. Since the attenuation of the photons increases with decrease in photon energy, the spectral distribution of the X-ray beam changes in passing through any material between the target and the point of use. This important effect is discussed in detail later in its practical applications (see p. 49). Figure 1.4 illustrates the effect however by plotting the spectral distribution of a beam generated at 140 kV from a tungsten target for three conditions, (A) with no intervening attenuating material (B) after passing through 0.3 mm of copper and (C) after passing through 0.6 mm of copper. The conditions of curve B correspond approximately to one practical case used in some low Kilovoltage tubes for superficial therapy (see p. 21). It will be noted that the total intensity as represented by the area under the curve is reduced by about 50 per cent for the addition of the first filter but by a smaller fraction by the addition of a second 0.3 mm copper. It will also be noted that the beam, after passing through the copper, loses progressively an increasing proportion of low energy photons and therefore becomes more penetrating. This process is known as *Filtration* and is dealt with in detail later (see p. 48).

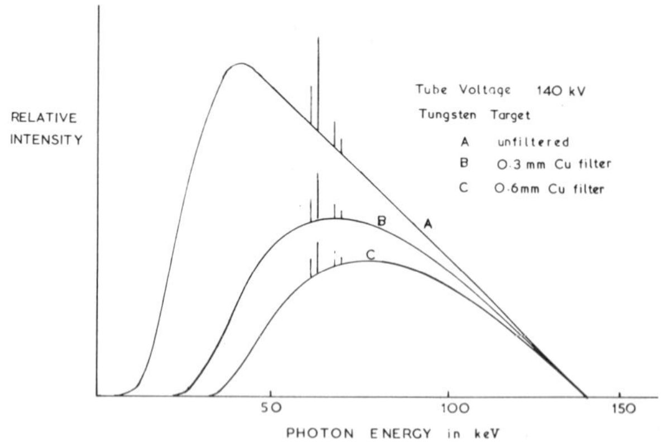

Fig. 1.4 X-ray spectra showing effect of filtration.

2. Radioactivity

Radioactivity

Radioactivity is the property of some categories of atoms to emit ionising radiations spontaneously. The radiations may be in the form of charged particles or photons of electromagnetic radiation and are emitted from the nuclei of radioactive atoms because of the unstable structure of these nuclei. The emission of these radiations causes a change in the nuclear structure and results in the production of a more stable nucleus. If, as is generally the case, charged particles are emitted from the nucleus the atomic number of the atom is changed and an atom of a different element is produced. This new atomic nucleus may be a stable form of the element or it may itself be unstable and radioactive.

Two terms are in common use in describing types of atoms. These are 'nuclide' and 'isotope'. The word *nuclide* refers to a species of atom characterised by a particular atomic number and a particular mass number, and also a particular value of its nuclear energy (see p. 9). It will be recalled that the atomic number Z is the number of positively charged protons in the nucleus of an atom and that the mass number A is the total number of protons and neutrons in the nucleus, neutrons being the uncharged particles of similar mass. A nuclide is specified therefore by a particular value of Z and a particular value of A. For example an atom having $Z=11$ and $A=23$ is the nuclide forming the stable element sodium.

The word *isotope* refers to one of a number of atomic species sharing the same atomic number. Isotopes therefore are nuclides of the same element i.e. the same atomic number with different mass numbers.

For most elements both stable and unstable isotopes are known, that is, the elements exist in both non-radioactive and radioactive forms. The radioactive nuclides are those in which the number of neutrons present in the nucleus falls short of, or is in excess of, the number present in the stable nuclides. For example the element carbon of atomic number $Z = 6$ is known to have isotopes with the following mass numbers 10, 11, 12, 13, 14, 15. Of these the isotopes of $A = 12$ and 13 are stable and exist in nature with 6 and 7 neutrons respectively in the nuclei. The isotopes $A = 10, 11$ are neutron deficient and those with $A = 14, 15$ have a neutron excess. All four are radioactive. For most elements, radioactive isotopes can be produced by various techniques of nuclear bombardment to be described later (p. 13).

Some radioactive nuclides do occur in nature however, either because their rate of disintegration is extremely low or because they are being produced continuously by the disintegration of other very slowly disintegrating nuclides. An example of the former is the radioactive isotope of potassium ($Z = 19$, $A = 40$) which occurs as part of the normal naturally occurring element. An example of the latter is the radioactive element radium ($Z = 86$, $A = 226$) which occurs in nature as a radioactive by-product of the slow disintegration of the radioactive element uranium.

Radiations from radioactive nuclides

Radioactive nuclei may give off one of four different types of particulate radiation and may at the same time give off photons of electromagnetic radiation. The particulate radiations–alpha particles, beta particles, positrons or neutrons are described below in more detail. Electromagnetic radiation from the nuclei of radioactive atoms is called gamma radiation. In a small number of radioactive nuclides only gamma radiation is emitted and in this case since no charged particle leaves the nucleus the atomic number of the disintegrating atom does not change. The resulting atom in this case is of the same chemical element as the initial one and the radioactive process has merely changed the internal energy of the nucleus.

Alpha particles

Certain types of radionuclides emit positively charged particles called alpha particles, (α-particles). These particles have a mass of 4 on the atomic scale, taking the mass of the proton as 1 and a positive charge of 2 units, taking the electron as the unit negative charge. They are the nuclei of helium atoms. Alpha particle emitters are relatively unimportant in normal medical work for two reasons. First the nuclides concerned are all those of elements with high atomic numbers greater than $Z = 82$. Several of the naturally occurring radionuclides such as radium 226 or polonium 210 are alpha emitters as are also many of the artificially prepared radionuclides of atomic numbers greater than 92, such as plutonium 239. Radioactive heavy elements are not in general so useful medically as light elements though radium 226 has special characteristics which has led to very important clinical applications (see p. 83). Secondly, the range of the emitted alpha particles in tissue is very small. The maximum range of these particles is always less than 0.1 mm and those from most radionuclide alpha emitters much smaller than this. They do not therefore penetrate the superficial layers of dead cells forming the outer coating of the skin. On the other hand if alpha emitting nuclides are ingested and deposited in tissues the alpha particles will cause ionisation along their paths. Since the concentration of ionisation along the track of the alpha particles is very high these particles are very effective in producing biological damage in tissues (see p. 183). Radiation hazard produced by ingestion of alpha particle emitters is therefore important (see p. 151). Alpha particles from a particular nuclide have a specific energy of emission and therefore a range characteristic of the nuclide and travel in straight lines until they have lost most of their energy.

Beta particles

Most of the commonly available radionuclides emit beta particles, (β-particles). These are negatively charged particles with the same charge and mass as an electron. They are therefore electrons of high energy emitted from the nucleus of the atom. The radionuclides of the lighter elements in which the nucleus has a higher neutron content than the normal stable isotopes are all negative beta emitters as are some of the naturally occurring radionuclides. Examples of the former are sodium-24, phosphorus-32 and of the latter potassium-40 and lead-210 (radium D). Beta particles are easily deflected in passing through tissue and lose their energy in producing ionisation along a tortuous path but those from some nuclides have ranges much greater than alpha particles. The maximum range in tissue of the beta particles from phosphorus-32 for example is 8 mm. On the other hand the maximum range from some very important beta emitters is very small. For example the maximum range of the β-particle from carbon-14 is only 0.3 mm in tissue.

The reference in the last two sentences to 'maximum' range refers to the fact that, unlike alpha particles, the beta particle emission from a particular radionuclide is a continuous spectrum with a maximum energy characteristic of the nuclide and a peak intensity at an energy something like one-third of the maximum. For a particular beta emitting nuclide therefore it is possible to specify both a maximum energy of emission (E_{max}) and an average energy of emission ($\bar{E}_\beta$). These are normally set out in tables of physical constants in units of kilo-electron volts (see p. 3). The first (E_{max}) is important in considering the thickness of shielding necessary to provide adequate protection from radiation hazard (see p. 106) and the second, ($\bar{E}_\beta$), is important in calculations of the radiation dose produced in tissue by locally deposited beta emitting nuclides (see p. 98). The ion density along the track of a beta particle in tissue is about 10 ion pairs per micrometre, some hundred times smaller than that produced by the normal alpha particle.

Positrons

A number of radionuclides emit particles of the same mass as electrons but positively charged. These are the *positive electrons* or positrons, and are the same kind of particles as are produced in the pair production process when very high energy photons interact with matter (see p. 37). Although the positrons have a very short life they behave in most respects exactly like negative electrons during their passage through matter when they lose energy by production of ions. When the energy has been completely lost in this way the positron combines with a negative electron and the rest mass of the two particles is converted into two photons of electromagnetic radiation each of energy 0.51 MeV. This is called *annihilation radiation* (see p. 37). The nuclides which emit positrons during radioactive decay are those with nuclei in which there is a neutron deficiency compared to the stable isotopes. A common example is the isotope sodium-22.

Neutrons

A rare type of emission during radioactive decay, but one with some possible importance in radiotherapy, is the emission of neutrons. The artificially prepared radionuclide californium-252 for example emits neutrons of considerable energy during its disintegration (see Ch. 12). Neutrons are strictly not themselves ionising radiations. They have no charge and can pass freely through the charged electron structure of the atoms of a medium without producing ionisation. However, the collision of fast neutrons with the nuclei of these atoms produces an efficient transfer of energy to these nuclei. Since these nuclei are charged and heavy they produce intense ionisation along their tracks. Ionisation produced in this way by neutron beams has some favourable advantages for therapy and californium-252 is being tried out for this purpose (see p. 136).

Radioactive transformations

The emission of charged particles from the nucleus of an atom causes a change in the atomic number Z of the atom and hence of the chemical nature of the atom concerned. The three main types of emission mentioned above may be illustrated by the following typical radioactive processes. In these equations the chemical symbol of the element is preceded by a superscript giving the mass number and a subscript giving the atomic number.

1. Alpha emission:

$$^{210}_{84}\mathrm{Po} \rightarrow {}^{4}_{2}\mathrm{He} + {}^{206}_{82}\mathrm{Pb}$$

polonium-210 → alpha particle + lead-206
(radium F) (stable)

2. Negative beta emission:

$$^{24}_{11}\text{sodium} \rightarrow {}^{0}_{-1}\beta^{-} + {}^{24}_{12}\text{magnesium} + \gamma$$

sodium-24 → beta particle + magnesium-24 (stable) + gamma photon

3. Positive beta emission:

$$^{22}_{11}\text{sodium} \rightarrow \beta^{+} + {}^{22}_{10}\text{Ne (stable)} + \gamma$$

sodium-22 → positron + neon-21 (stable) + gamma photon

In each of these cases a stable atom formed by the disintegration is of a different element from the initial radioactive nuclide. The emission of a gamma ray photon does not produce such a change in the chemical nature of the radioactive nuclide. Gamma ray photon emission represents a change in the energy state of the nucleus of the radioactive nuclide. These photons very frequently accompany the emission of beta particles from the nucleus as illustrated in 2 and 3 above as a consequence of a loss of energy before the final nucleus achieves a stable state.

Decay schemes

The changes involved in a radioactive decay are often represented by *decay schemes* in which the energy of the transforming nucleus at various steps in the decay process is represented by the height above a base line. The decay scheme for sodium-24 is represented thus:

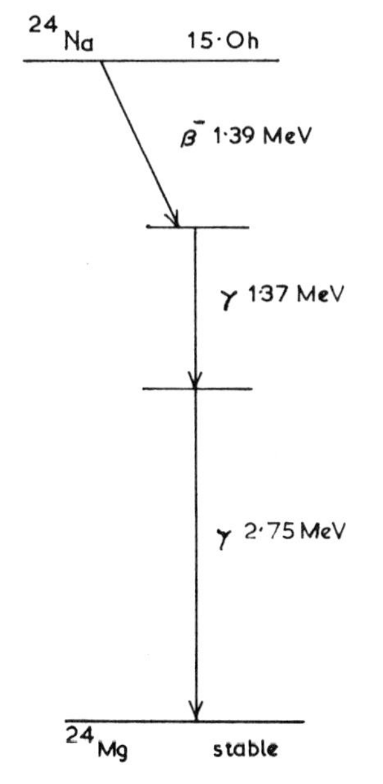

Fig. 2.1 Decay scheme for sodium-24.

The complete process of change of the sodium-24 nucleus to a stable magnesium-24 nucleus involves therefore three steps in succession, resulting in the emission of a beta particle with maximum energy 1.39 MeV followed by two photons of energies 2.76 MeV and 1.37 MeV respectively. Many radioactive decay processes are much more complicated than this and often involve alternative methods of disintegration (see p. 11). From the point of view of their use in medicine the information required is the energy and number of the particles or photons emitted per disintegration. This information is available in tables of physical constants of the radionuclides.

Other decay processes

A number of radionuclides disintegrate in ways which differ from the more common processes described above. Some decay, for example, by a process involving the capture of an orbital electron by the nucleus. This results in the emission of X-rays and sometimes of low energy electrons from the orbital electron structure of the atom. This process is known as *electron capture (EC)*. Other nuclides decay by the process of absorption of low energy nuclear gamma rays by the inner orbital electrons. This also results in the emission of low energy X-rays and electrons from the orbital electron structure. The process is known as *internal conversion (IC)*. Both electron capture and internal conversion give rise to the emission of electromagnetic photons from the orbital electron system of the disintegrating nucleus and these are the characteristic X-rays usually from the K shell of the atom concerned. Electromagnetic photons may also be emitted from the nucleus during the process of energy readjustment and these will be gamma rays.

Yet a third type of somewhat unusual disintegration process is known. Following the emission of a beta particle a nucleus may be left in an excited state and then change its energy by transition to a lower state with the emission of gamma radiation. In some cases the excited energy state has a finite lifetime of several minutes or hours and it is then said to be *metastable*. The excited state in this case is represented by m. For example, technetium-99m is the excited metastable state (half life 6 hours) of the nuclide technetium-99 and is an important isotope in clinical practice (see p. 112).

These three more unusual types of disentegration process are often of importance because they give rise to relatively low energy gamma and X-ray emission with no beta emission and are useful in clinical isotope procedures to reduce the deposition of energy at the disintegration site (see p. 95).

Radioactive decay

Disintegration of a radioactive nucleus is spontaneous and random. The break up of any particular nucleus is not predictable but the proportion of nuclei in any particular radionuclide which disintegrate in unit time depends on the instability of the nucleus. The fraction of the total number of atoms of a particular radionuclide which disintegrate per unit time is a constant for that nuclide and is known as the *decay constant* or *transformation constant* (λ). It can easily be demonstrated that the disappearance of a constant fraction of the atoms per unit time can be represented mathematically by the exponential law:

$$N = N_0 e^{-\lambda t}$$

where N is the number of atoms remaining at time t, N_0 is the initial number present at time $t = 0$ and λ is the transformation constant

This is the *law of exponential decay*. Since the intensity of emission from a radioactive sample is directly proportional to the number of radioactive atoms present the exponential law of decay can be observed readily by measuring under fixed geometrical conditions the variation with time of the number of beta particles emitted per unit time or the gamma ray intensity. Figure 2.2 shows a radioactive decay curve for sodium-24.

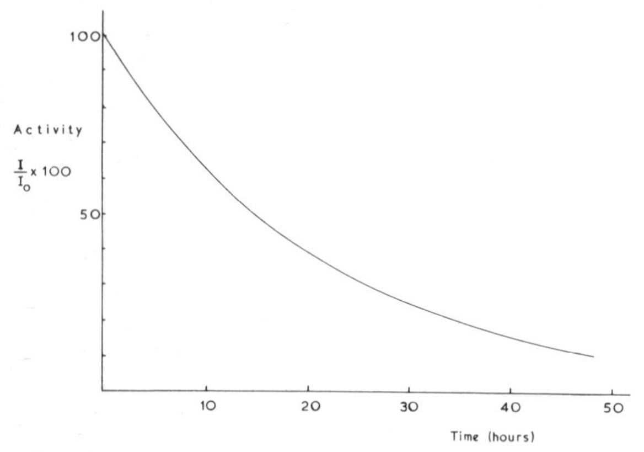

Fig. 2.2 Decay curve for sodium-24.

The exponential curve represented by the above equation has two important characteristics. First, if plotted logarithmically, that is, if log intensity is plotted against time a straight line results. This is an important aspect of the curve since if an experimentally determined logarithmic decay curve is not a straight line this is strong evidence that two or more radionuclides are present which are decaying at different rates.

The second important characteristic is that the time taken for the activity, i.e. the intensity of the emitted radiation or the number of radioactive atoms present, to be reduced to one half of the original value is a constant. Physically this means that any radionuclide has a characteristic transformation constant leading to a fixed time for the number of atoms of the nuclide to be reduced to half its original value. This time is the *half life*. It can be shown that the half life is related to the transformation constant by the simple relationship:

$$T\tfrac{1}{2} = \frac{0.693}{\lambda}$$

where $T\frac{1}{2}$ is the half life.

Half lives of radionuclides vary enormously corresponding to a wide range in stability of the radioactive nucleus. For example, the isotope polonium-210, quoted above, has a half life of 138 days, sodium-24 15.0 hours and sodium-22 2.6 years—but extreme values of half life are represented by potassium-40 1.3×10^9 years and polonium-214 (radium C^1) 1.6×10^{-4} seconds. It will be seen later (p. 95) that the half life of a radionuclide has an important bearing on its use in clinical practice.

The radioactivity of a particular sample of a radionuclide at a particular time is measured by its *disintegration rate*, that is the number of disintegrations per unit time (second). The unit of measurement of radioactivity is the *curie* (after the famous discoverer of radium). One curie (Ci) of radioactive material has a disintegration rate of 3.7×10^{10} disintegrations per second. The unit has sub-multiples of great practical use, the millicurie (1 mCi) equal to 10^{-3} curie and the microcurie (1μCi) equal to 10^{-6} curie. The multiple unit, the kilocurie equal to 10^3 curies, is used for the large amounts of isotopes involved in telecurie therapy equipment (see Ch. 3).

It will be appreciated that the curie or its sub-multiples are not a measure of the mass of the radioactive material present. In a radionuclide with a slow disintegration rate a large mass of substance will be necessary to provide 1 curie of activity. For radium-226 for example, with a half life of 1600 years, 1 curie of activity will be provided by 1 g of radium. In a rapidly decaying radionuclide however a very much smaller mass of material will be equivalent to 1 curie of activity. For example 1 curie of iodine-131 with a half life of 8 days will have a mass of 8×10^{-6}g.

A new unit of activity will be adopted in the future to fit with the S.I. system. This is the *becquerel* (Bq) which is a disintegration rate of 1 per second.

The physical properties of the available radionuclides have been very thoroughly investigated and are listed in appropriate tables of physical constants. The properties which are of special relevance to the clinical use of radioactive materials are the half life $T\frac{1}{2}$ (see p. 95), the mean energy E_β (see p. 98) and the maximum energy E_{max} of the beta particle emission and a further factor which gives an indication of the gamma ray photon emission per disintegration. This is normally specified by the 'k' factor—defined as the *specific gamma ray emission*. This is the exposure rate produced by the gamma rays from a unit point source of the nuclide at unit distance and is expressed in terms of Roentgens per hour per millicurie at 1 cm distance. Knowledge of the 'k' factor is important in design of telecurietherapy units (see p. 25) and in consideration of gamma ray protection requirements when working with radioactive materials (see p. 91).

Radioactive equilibrium

A radioactive disintegration process involving the emission of a charged particle generally results in the production of a new stable non-radioactive nuclide of a different element. Sometimes however the nuclide formed by the disintegration of a radionuclide is itself an unstable nuclide and undergoes a further radioactive transformation. The two nuclides are often referred to as *parent* and *daughter nuclei* respectively. An example of some importance is the radionuclide of strontium, strontium-90 (Z = 38, A = 90) which disintegrates with the emission of a negative beta particle forming an isotope of the element yttrium (Z = 39, A = 90). This isotope is itself radioactive and emits a further negative beta particle, the final product of this second disintegration process being a stable isotope of the element zirconium (Z = 40, A = 90). Strontium-90 has a relatively long half life of 28 years but its daughter product yttrium-90 is much more unstable energetically and has a half life of 64 hours. In a pure sample of strontium-90 the radionuclide yttrium-90 formed from the parent element will accumulate and the amount of yttrium-90 will grow until the number of yttrium-90 atoms breaking up per second equals the number being formed per second by disintegration of the parent element. A state of *radioactive equilibrium* is then said to exist between the parent and daughter radionuclides. In this case the parent nuclide (strontium-90) and the daughter nuclide (yttrium-90) then decay at the same rate—that is the decay rate of the parent element ($T_{\frac{1}{2}}$ = 28 years). It can be shown that the growth of the daughter product in the mixture is exponential and is governed by the decay rate of the daughter nuclide—that is, it will reach half its full equilibrium value in 64 hours and three-quarters of its equilibrium value in 126 hours and so on.

The practical importance of this particular example arises from the fact that, though the maximum range of the beta particle from strontium-90 is only 1.2 mm in water, the beta particles from yttrium-90 are much more energetic with a maximum range in water of 11 mm. An equilibrium mixture of the two radionuclides therefore supplies beta particles of the long range, but with an effective half life of 28 years. This proves to be a very useful therapeutic source of energetic beta radiation (see p. 91).

The phenomenon of radioactive equilibrium is very important in the production of short lived radionuclides for use in medical investigations. Several of these short lived products are available as daughter radioelements easily separated by elution from longer lived parent elements. The parent nuclide is generally absorbed on a suitable inert bed such as alumina contained in a robust plastic vessel and the daughter nuclide is separated by washing out by an appropriate eluting liquid. Such sources are referred to as *generator columns*. Elution produces a solution of the pure radioactive daughter product, and in the column regeneration of this short lived radionuclide occurs the activity rising to about 90% of the equilibrium value in a time equal to three half lives of the daughter nuclide. The whole activity of the generator column decays with the half life of the parent element. Examples of radioactive decay systems in common use in generators to produce short lived isotopes in hospitals are:

$$^{132}\text{Te (78 h)} \xrightarrow{\beta^-} {}^{132}\text{I (2.3 h)}$$

$$^{99}\text{Mo (67 h)} \xrightarrow{\beta^-} {}^{99m}\text{Te (6.0 h)}$$

$$^{113}\text{Sn (118 d)} \xrightarrow{\text{EC}} {}^{113m}\text{In (1.7 h)}$$

$$^{87}\text{Y (80 h)} \xrightarrow{\text{EC}} {}^{87m}\text{Sr (2.8 h)}$$

Successive disintegration

The naturally occurring radioactive elements arise from a series of decay processes producing several successive radioactive daughter products. Radioactive equilibrium is established between the primary long lived radioactive elements—uranium, thorium, actinium and several generations of daughter products.

The most important of these is the radium series. Radium itself is one of the radionuclides in the decay chain of the parent nuclide of uranium. It is isolated chemically from uranium ores. Radium (Z = 88, A = 226) is an alpha particle emitter with a half life of 1600 years.

Its immediate daughter product is an inert gas, radon, which is also an alpha emitting radionuclide of short life ($T_{\frac{1}{2}}$ = 3.8 days). Radon however is only one step in a series of *successive disintegrations* leading finally to a stable nucleus at the end of the chain. This nucleus is an isotope of the element lead (Z = 82, A = 206). Figure 2.3. gives a detailed account of the successive transformations of the radium series in which some ten different radionuclides are involved.

If the pure radium element as parent of this series is sealed in a container so that the successive radioactive products do not escape, all the radionuclides of the series accumulate until each is in equilibrium with the rest. At equilibrium each nuclide will decay at the same rate, each being produced by the decay of the radionuclide preceding it in the chain and giving rise to the nuclide following it. At radioactive equilibrium the weight of each nuclide present is inversely proportional to the half life of the nuclide. The importance of this *radioactive family*, as it is called, is that two of the radioactive nuclides in the series are emitters of penetrating gamma rays: lead-214 or radium-B, and bismuth-214 or radium-C. These two nuclides have short half lives of 26.8 min and 19.7 min respectively but in equilibrium with the rest of the chain of radioactive products they have an effective half life the same as the parent element radium. Radium in a sealed container therefore in equilibrium with its products will act as a long lived radioactive source ($T_{\frac{1}{2}}$ = 1600 years) emitting a mixture of alpha particles, beta particles and gamma rays. A container of suitable thickness (0.5 mm of platinum) therefore will act as a gamma emitter which is effectively constant in activity since the half life of radium is long enough to be considered infinite in clinical work. The use of sealed radium containers is dealt with in detail in Chapter 8.

days) (see p. 90). The presence of radon in the sealed radium container is important however since its chemical nature allows it to escape from a leaking container and this constitutes a hazard in the use of sealed radium containers which must be guarded against (see p. 94).

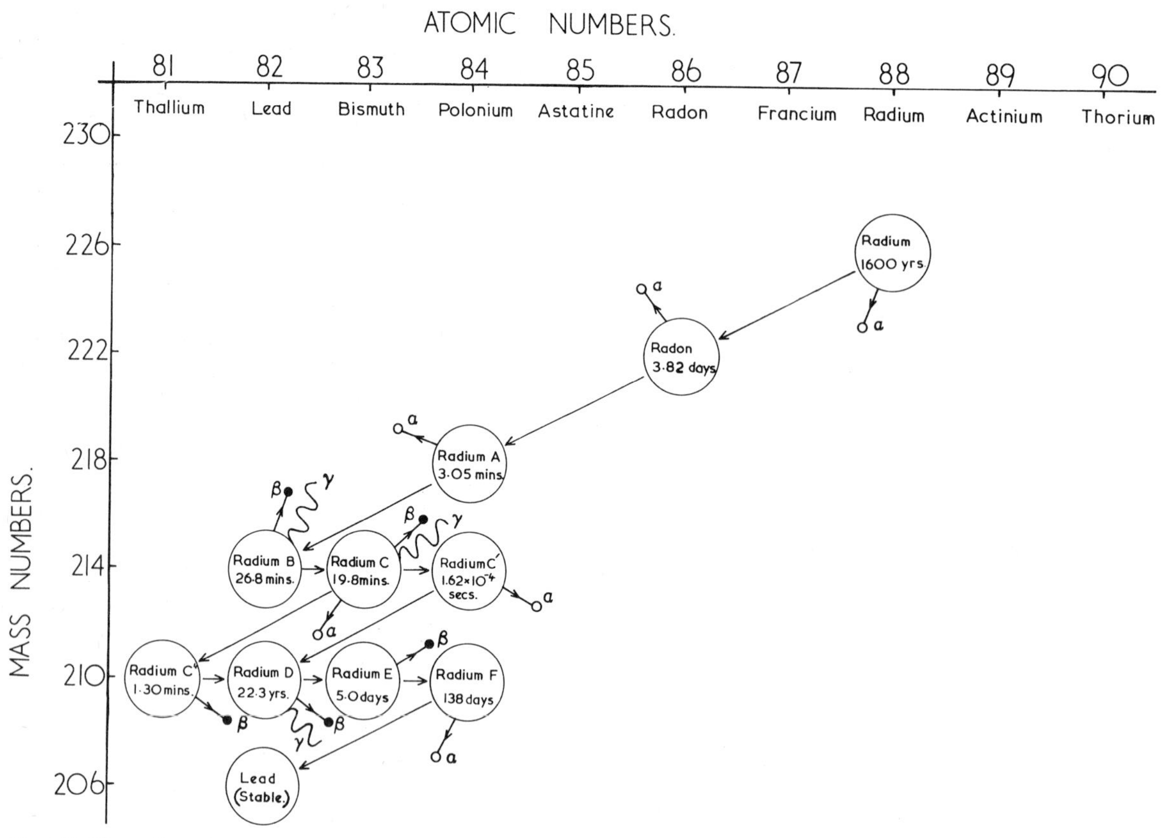

Fig. 2.3 The radium family.

It is to be noted that the first disintegration product of radium is the inert gas radon. Because it is very different chemically from the element radium it can be separated from its parent reasonably easily. If radon were itself to be sealed in a suitable container the successive radioactive products of the radium family would accumulate to equilibrium amounts, these products again including the gamma emitters lead-214 and bismuth-214. In this case the container would be an emitter of clinically useful penetrating gamma rays but with an effective half life equal to that of radon ($T_{\frac{1}{2}} = 3.8$ days). Radon sealed in thin walled gold capsules called *radon seeds* was used extensively until quite recent times for small sealed gamma ray sources of short life. Radon seeds have now been almost completely replaced by artificially produced radionuclides of similar half life and gamma emitting properties, particularly radioactive gold-198 ($T_{\frac{1}{2}} = 2.7$

Production of radioactive materials

The production of radioactive materials in forms suitable for use in medicine is an expensive and difficult matter. Amongst the naturally occurring radioactive elements, radium is the only one used medically in any major way. It is prepared by separation from uranium ores but since it is present in only about 1 part in 10 million by weight its separation is not easy. The radium salts, along with a large excess of the chemically similar barium salts, are separated from the ores by solution and chemical preparation. Separation of the radium from the barium salts is achieved by a laborious process of fractional crystallisation depending on the solubility of the chlorides or bromides of radium being slightly less than those of barium. Purified radium compounds, generally the insoluble sulphate, are finally prepared for packing into suitable containers for therapy (see p. 83).

The first decay product of radium—radon-222—can be separated relatively easily from radium salts in solution since it is an inert gas. The preparation for use as a clinical tool sealed in containers involves a somewhat elaborate purification process for separating the relatively small amount of radon gas from a much larger quantity of gaseous impurities, mainly hydrogen, oxygen and carbon dioxide. These arise from chemical decomposition of the solution caused by the intense ionisation of the emitted radiations. The radon seeds used in therapy enclosed the radon finally in a very narrow bore thin-walled gold capillary tube.

Radioactive isotopes of most of the normal elements are produced by processes in which the nuclear composition of an atom is modified by bombardment with a suitable agent to produce nuclear reactions. The bombarding agents used to produce nuclear reactions may be charged particles such as protons or alpha particles or uncharged ones such as neutrons. If charged particles are used they must in general be particles of nuclear dimension and positively charged. In order to make a collision with the positively charged nucleus of an atom these particles must have energies large enough to overcome the electrostatic repulsion between the two. Some form of instrument for accelerating these particles to a high energy is therefore required. A Van de Graaff accelerator (see p. 22) with a positive potential on the high voltage terminal is sometimes used for potentials up to a few million volts. A very successful alternative for producing particle beams of high energy for use in nuclear reactions is the cyclotron, described later. An important characteristic of the production of artificial radionuclides by charged particle bombardment is that the nuclide produced is often of a different chemical element from the target material. In this case a chemical separation is possible to obtain the pure radionuclide free from non-radioactive isotopes of the same element. These types of radioactive preparation, referred to as *carrier free* radioisotopes, are important in some diagnostic and therapeutic uses of these materials (see p. 95). An illustration of such a reaction is the production of the isotope of sodium, sodium-22 by the bombardment of magnesium by a high energy beam of heavy hydrogen nuclei or *deuterons*. The reaction may be represented:

$$^{24}_{12}\text{Mg} + ^{2}_{1}\text{H} \longrightarrow ^{22}_{11}\text{Na} + ^{4}_{2}\text{He}$$

The isotope sodium-22 with a half life of 2.6 years is produced together with the emission of an alpha particle and is separated chemically from the magnesium of the target.

The alternative bombarding particle, the uncharged neutron, can react with the nucleus of an atom without having a high initial energy. An apparatus for the production of a suitable cloud of bombarding neutrons is the nuclear reactor described on page 14. An example of the use of neutrons to produce an artificial radionuclide is the bombardment of sodium by neutrons. A neutron is captured into the nucleus with the emission of a gamma ray photon and an increase in one of the mass number

$$^{23}_{11}\text{Na} + ^{1}_{0}\text{n} \longrightarrow ^{24}_{11}\text{Na} + \gamma$$

The sodium-24 ($T_{\frac{1}{2}}$ = 15 hours is of the same chemical element as the target material and in this case cannot be separated chemically from the inactive sodium of the target.

The cyclotron

The cyclotron accelerates positively charged particles so that they acquire energies equivalent to those produced by acceleration through many million volts. It does this by a long series of steps each step giving an acceleration by a potential difference of a relatively small value. The arrangement for doing this is shown in Figure 2.4. Two hollow semicircular electrodes in an evacuated container are placed in a strong magnetic field perpendicular to the plane of the electrodes. An alternating electric field is applied between the two electrodes.

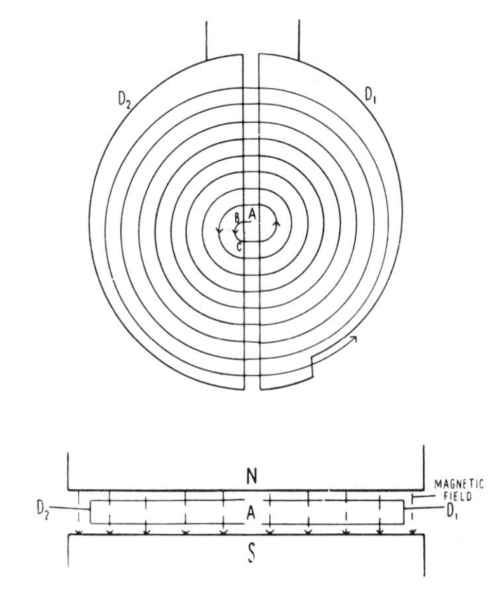

Fig. 2.4 The cyclotron.

Positive ions of the kind to be accelerated are produced at point A in the centre of the electrodes. With the electric field between the two electrodes D_1 and D_2 in the right direction the ions are accelerated to point B. Under the influence of the magnetic field they move in a circular path and arrive at the gap between the electrodes at C. By correct adjustment of the frequency of the electric field between D_1 and D_2 the ions are now accelerated across the gap into the opposite electrode. They then move again in a circular orbit arriving at the gap again in time to receive a further acceleration. This is possible because the time taken to travel round the semicircular paths under the influence of the magnetic field remains constant. With increasing velocity of the ions only the radius of the path increases. The ions therefore travel on a spiral path of gradually increasing radius and are accelerated by the peak potential difference between the electrodes each time they cross the gap. In a typical cyclotron the peak potential difference between the electrodes might be 20000 volts but ions are accelerated across the gap 800 times in electrodes of 30 inch radius, thus producing an energy equivalent to an acceleration by 16 million volts. At the edge of the electrodes the beam is no longer under the influence of the magnetic field and emerges tangentially.

The high energy positive ion beams produced in the cyclotron can be used to bombard directly a suitable target in which nuclear reactions then result in the production of radionuclides as described in the previous paragraph. Beams of protons, deuterons, or alpha particles are used in these reactions and a large number of radionuclides are produced, particularly many which are positive beta emitters.

If energetic alpha particles bombard a beryllium target intense beams of high energy neutrons are produced. These beams can be used directly as ionising radiation for radiotherapy and there are some circumstances in which neutron beams have advantages over X-ray beams for this purpose (see p. 135). Similar high energy neutron beams for therapeutic use can also be produced from a nuclear reaction involving hydrogen isotopes accelerated in a sealed off tube at relatively low voltages. (See Chapter 12).

The nuclear reactor

The nuclear reactor is an important device for releasing energy from atomic nuclei and is the basis of present methods of producing electrical power in this way. It is based on the *nuclear fission* produced in the nucleus of some uranium isotopes by neutron bombardment. The uranium nucleus involved is unstable in a manner different from almost all other nuclei. Under neutron bombardment capture of a neutron takes place and the nucleus breaks up into two approximately equal parts. The fission is accompanied by the emission of several further neutrons from the broken nucleus. Other nuclei of the same type in the vicinity can be stimulated to undergo fission under suitable conditions by capturing these secondary neutrons. A *chain reaction* is thus produced if a suitable mass of uranium surrounds the initiating uranium nucleus. A very large amount of energy is released at each disintegration, corresponding to the conversion of a small part of the mass of the fissionable uranium nucleus into kinetic energy of the two bits of the broken nucleus or the emitted neutrons. Uranium fission and the chain reaction are the fundamental processes underlying the release of energy in the atom bomb and in the nuclear reactor, sometimes referred to as the *chain reacting pile*.

The nuclear reactor consists essentially of a mass of uranium often in the form of a pile of bars of uranium metal in which the chain reaction is controlled by the introduction of interleaved rods of a material which slows down the neutrons (the moderator) and the other rods of a material which absorbs the neutrons completely (control rods). The arrangement of these three components varies a great deal from one pile to another depending on the purposes for which it is designed. The uranium nuclei which undergo fission are those of mass number 235 which form one part in 140 of the total uranium nuclei in natural uranium. The initial fission is produced by a few neutrons always present in natural radiation and fission reactions in the surrounding uranium-235 nuclei are induced by the released neutrons slowed down by the moderator to a speed at which they produce the effect most efficiently. The chain reaction is kept to an equilibrium controllable level by the neutron absorbing control rods often of cadmium which prevent excessive numbers of neutrons being produced in the uranium bulk.

The form of the pile is largely determined by the form of the moderator. This may consist of a large block of graphite in which the uranium bars are embedded. For the production of artificial radioactive isotopes holes are made through the very massive concrete radiation shield round the pile and then into the uranium graphite pile itself. Samples of material to be irradiated are then introduced along these holes into the reactor core so that they may then be exposed for a suitable time to a very high flux of neutrons.

The fission process releases a large amount of energy in the form of heat and the pile can be used as a source of power by using this heat to generate steam to feed conventional turbo-alternators. In the reactors of the U.K. Central Electricity Generating Board the heat is often removed from the pile by high pressure carbon-dioxide pumped through the units of the pile.

Figure 2.5 shows a picture of the DIDO reactor at the U.K. Atomic Energy Establishment at Harwell which is used for the production of radioactive isotopes. From outside the protective concrete screen dominates the appearance and this is seen in the lower portion of the picture. The protective screen is necessary to give protection not only from the large neutron emission in the reactor core but also from the beta and gamma rays arising from the radioactive materials produced in the core. The picture of Figure 2.5 illustrates the complexity of the control and experimental equipment surrounding the shielded reactor and shows also the loading face on the top of the shield through which materials may be introduced into the core of the reactor for activation and subsequently withdrawn safely by a special shielded transfer device.

Radioactive isotopes can be produced in the pile in a

further way in addition to the induction of radioactivity in neutron bombarded stable elements. The products of the fission of uranium 235 nuclei are nuclei having mass numbers and atomic numbers about half that of the original nuclei. These *fission products* are all radioactive isotopes of elements near the middle of the periodic table. They accumulate in the rods of uranium fuel as the fission process proceeds. Radioactive isotopes of elements between $Z = 34$ and $Z = 62$ are all present in the fission material but not all in the same abundance. They have a wide range of radioactive properties and some of them have great value in the medical field, both in therapeutic and diagnostic uses. The spent uranium bars are a valuable source of some isotopes. Three typical isotopes present in the fission products, for example, are caesium-137 a beta-gamma emitter of half-life 30 years, strontium-90, a pure beta emitter of half-life 28 years and iodine-131, the beta-gamma emitter of 8 day half-life commonly used in the diagnosis of thyroid diseases. The separation of the useful isotopes from the spent uranium bars is a difficult and expensive process but some radioactive materials are prepared in this way.

The spent uranium bars are valuable for a second reason. The uranium isotope which forms the majority of atoms in natural uranium, i.e. uranium-238, captures a neutron and produces a new nucleus which is itself radioactive and a beta-particle emitter. The decay of this material results in the production of other radioactive atoms which have atomic numbers greater than 92, the greatest naturally occurring atomic number. One of these products is the element plutonium ($Z = 94, A = 239$). This, though having important radioactive properties of normal type also undergoes fission under neutron bombardment. Plutonium therefore represents an important second fissionable material prepared entirely artificially which can act as the basis for a chain reaction.

If in the pile the controlling and absorbing materials were withdrawn the number of neutrons available to cause fission in the uranium would become uncontrollable, and tremendous heat would be developed by the energy released from the disintegrating nuclei. The energy produced in the fission process can be released suddenly if a large mass of atoms of uranium-235 or plutonium can be suddenly assembled. The mass must be great enough so that the neutrons produced in any fission event have little chance of escaping, but are instead absorbed to produce further fission in neighbouring nuclei. This can be achieved by bringing together suddenly masses of fissile material which are themselves not large enough to produce a chain reaction in this way, but are large enough when joined together. This is, in effect, the method by which atom bombs are made. When the total mass is great enough to produce a chain reaction the energy release is immense, violent and very sudden. The energy release represents a real conversion of a small proportion of the uranium nuclear mass into heat in a small time. This gives rise to tremendous explosive force as well as to a large release of radioactive materials arising from the fission involved in the chain reaction.

It is useful to note that in nuclear technology involving the separation of some fraction of the fissile nuclide uranium-235 from natural uranium one by-product is a

Fig. 2.5 The DIDO reactor Harwell. (By courtesy of U.K. Atomic Energy Authority.)

uranium metal in which the proportion of this nuclide has been substantially reduced. This material is referred to as *depleted uranium*. Since its density is very high ($\rho = 18.7$) it is used as a shielding material where small quantities of very dense material are advantageous. Such sites include portions of the inner shield around the source in a cobalt therapy unit or the collimetor blades of a megavoltage therapy equipment.

3. The production of X- and gamma ray beams

Introduction

We have seen that X-rays are produced whenever an electron stream travelling at high speed is brought to rest in a solid target. To produce an X-ray beam for therapeutic purposes, it is necessary to have an evacuated tube in which the electron stream can be produced and accelerated on to a suitable target by a voltage generator. Thus the basic components of an X-ray set are:

1. An evacuated tube
2. An electron stream
3. A target
4. A voltage generator.

Gamma-rays are in nature identical to X-rays except that they are spontaneously emitted from certain radioactive isotopes. A gamma-ray beam may be produced by suitably encapsulating such an isotope in a substantial shield and permitting the intense gamma radiation to emerge through an appropriate aperture in the shield. Teleisotope beam units will be described later in this chapter.

The X-ray Tube and Housing

The X-ray tube is a highly evacuated glass envelope containing two electrodes and, as such, its electrical behaviour is the same as a diode valve. In construction, however, there are important differences. The accelerating voltage used to generate X-rays is very large and, therefore, great care has to be taken to insulate the anode from the cathode. For this purpose the electrodes are shaped to avoid sharp corners—the electrostatic field strength is very intense at sharp corners and would give rise to sparking—and the surfaces are as smooth as possible. The two electrodes are usually sealed into the glass envelope so that the insulating path along the glass is as long as possible.

The electron stream is produced by thermionic emission from a heated tungsten filament (the cathode). The filament is either a flat spiral or a helix of tungsten wire at the centre of a conical or semi-cylindrical *focussing cup* (Fig. 3.1). This cup is maintained at the same potential as the filament and so focusses the electrons on to the X-ray target (anode). As in the case of the thermionic diode under saturation conditions, the number of electrons emitted is controlled by the temperature of the filament and increases rapidly with that temperature. In X-ray generator circuits the filament is heated using a highly stabilised circuit and the tube current is controlled by varying the current through the filament (Fig. 3.2).

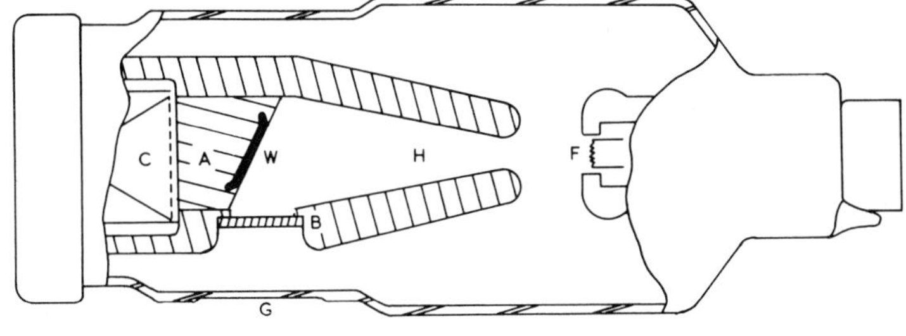

Fig. 3.1 A 250 kV X-ray tube showing the copper anode A, the tungsten target W, the cooling oil spray C, together with the copper anode hood H and the beryllium window B. The filament F and the thin window in the glass G are also shown. (Adapted from a diagram supplied by Philips Medical Systems Ltd)

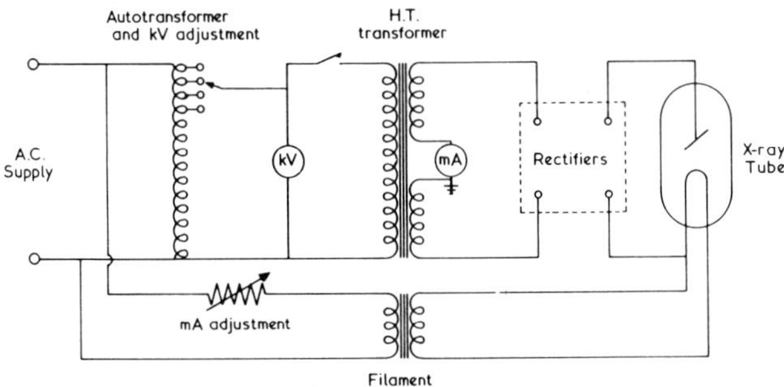

Fig. 3.2 A simple X-ray generator circuit.

So far the word *target* and the word *anode* have been used interchangeably, but it is important now to differentiate between the two. The *anode* is the positive electrode in the X-ray tube and the *target* is the piece of tungsten bonded to the surface of the anode (Fig. 3.1) on to which the electrons impinge and in which the X-rays are generated. Tungsten is chosen as a target material for its high atomic number ($Z = 74$) which leads to a relatively efficient conversion of electron energy to X-ray energy and for its high melting point which removes the chance of melting the target under the bombardment of the electrons. Furthermore, the target material must not evaporate readily when heated in a vacuum since, if it did, it would become deposited on the tube walls causing absorption of X-rays and a breakdown in the electrical insulation of the tube. In the hooded anode, this deposition is on the hood and so rendered harmless.

Target angle and heel effect

It has been seen in Chapter 1 (p. 4) that for conventional energies X-rays are produced in a *reflection target* and the useful beam is considered to be at right angles to the electron beam. No mention has been made about the target angle. The *target angle* is the angle between the axis of the X-ray beam and the face of the target, or, the angle between the face of the target and the normal to the electron beam (Fig. 3.3). The effect of this angle is threefold.

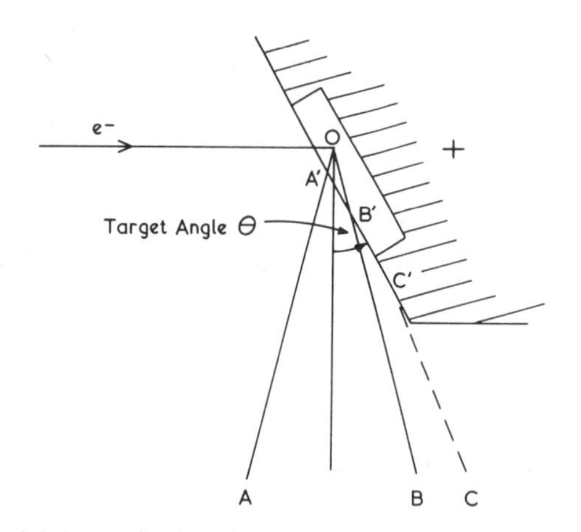

Fig. 3.3 The definition of target angle and the heel effect.

The thin target theory suggests that although the (low energy) X-rays are produced predominantly in the direction at right angles to the electron beam (Fig. 1.1), the X-ray intensity is greater at angles less than 90° than at angles of greater than 90° to the direction of the electrons. In order therefore to achieve some degree of symmetry to the X-ray beam it is necessary to attenuate the radiation on the anode side of the X-ray beam axis to a greater extent than that on the cathode side. Figure 3.3 shows the source of X-rays at a point O below the surface of the target and the two X-rays OA and OB. From this simple construction, it can be seen that OA is attenuated by a thickness of target material OA′ and OB by a thickness of OB′. In particular, OB′ is greater than OA′. Since the relative length of OA′ and OB′ is dependent on the target angle θ, an angle can be chosen to produce a symmetrical beam of radiation by attenuating the more intense beam OB by a suitably greater thickness OB′.

Attenuation is, however, dependent on photon energy as well as atomic number. It follows, therefore, that a particular target angle will only produce a symmetrical beam of radiation for one photon energy and to operate the X-ray tube at energies below the design energy will result in an asymmetric beam—with the anode edge of the beam being relatively less intense (Fig. 3.4). Over small changes in accelerating voltage this may not be clinically important for deep X-ray therapy, while the effect is less pronounced at superficial X-ray energies.

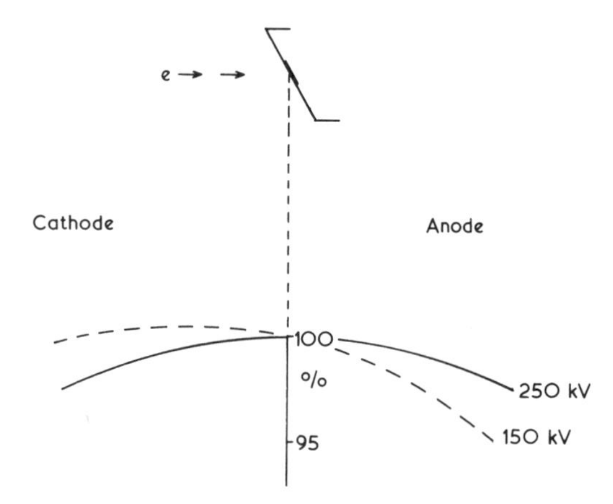

Fig. 3.4 The effect on beam symmetry of using a kilovoltage lower than that for which the target angle was designed.

At voltages between 200 and 300 kV, the target angle is usually about 30°.

The second important criterion which is dependent on target angle is the overall field size or maximum beam divergence. It can be seen from Figure 3.3 that as the beam divergence increases, OC′ increases rapidly and the intensity at C falls producing a beam edge. It follows that the half angle of beam divergence cannot exceed the target angle and, in practice, the useful beam is limited to a few degrees less than the target angle.

Thirdly, the geometry of the target is such that the area of the tungsten bombarded by the electrons is larger than either the cross sectional area of the electron beam or the area from which the X-rays are seen to emanate (Fig. 3.5). The latter is known as the *effective focal spot size* and decreases with target angle. In diagnostic X-ray tubes, where the effective focal spot size is of paramount importance, very small target angles are used—15° is common and some are as low as 7°. In therapy, however, large symmetrical beams are of greater importance, than the size of the focal spot. In either discipline, the larger the area of target bombarded by electrons the larger the area for the dissipation of heat—an important subject which will now be dealt with.

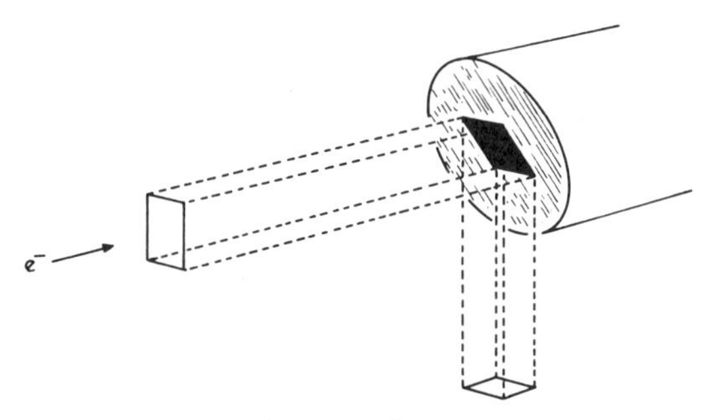

Fig. 3.5 The effective focal spot size.

Cooling the X-ray target

The generation of heat in the target by the bombardment of electrons leads to real problems of cooling in X-ray tubes. In the conventional X-ray tube only about 1 per cent of the energy dissipated in the target is converted into X-rays and 99 per cent into heat (p. 3). The method of cooling depends on the design of the target and the voltage at which it operates (Fig. 3.6). In X-ray tubes operating at less than 50 kV (grenz* rays) and those using resonant transformers (p. 22), the target is at earth potential and can be cooled directly by a flow of cold water through the anode behind the target (Fig. 3.6c). Where the X-ray tube is immersed in oil to provide additional electrical insulation around the glass envelope, the oil itself will be used to cool the target either by natural or forced convection. The oil is then passed through an oil/water heat exchanger thus transferring the heat to the water and so to waste. The heat exchanger may be incorporated in the X-ray head assembly (Fig. 3.7), or in a separate unit connected to the X-ray head using flexible oil pipes. At megavoltage energies where the generation of X-rays is a more efficient process (about 50 per cent), less heat is produced and cooling is not a major problem. The target is at earth potential and a simple flow of water through the target assembly is all that is required.

The X-ray tube housing

The X-ray tube itself is often referred to as *the insert* and is the main component contained in the X-ray head. The X-ray head is a metal tank in which the insert is rigidly mounted and connected to the HT cables. The head contains the insulating transformer oil and some temperature sensitive device which will prevent the use of the tube when its temperature rises above a predetermined safe level. In a sealed system, this may be an expansion diaphragm which operates a microswitch as it takes up the expansion of the heated oil (Fig. 3.7).

Other essential features of the X-ray head include

1. The X-ray window and primary collimator through which the useful X-ray beam emerges
2. Lead protection around the head to reduce the leakage radiation (i.e. unwanted radiation) to a safe level (p. 28)
3. An assembly will be fitted on the primary collimator to house and electrically interlock the added filtration (p.48)
4. An ionisation chamber (p. 44)
5. The secondary collimation will be mounted on this assembly in the form of an adjustable diaphragm or interchangeable applicators (p. 28)
6. A mounting for beam direction devices (p. 71).

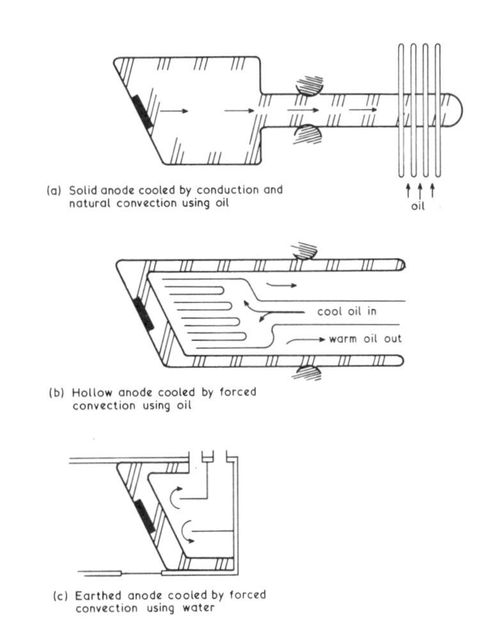

Fig. 3.6 The cooling of X-ray targets.

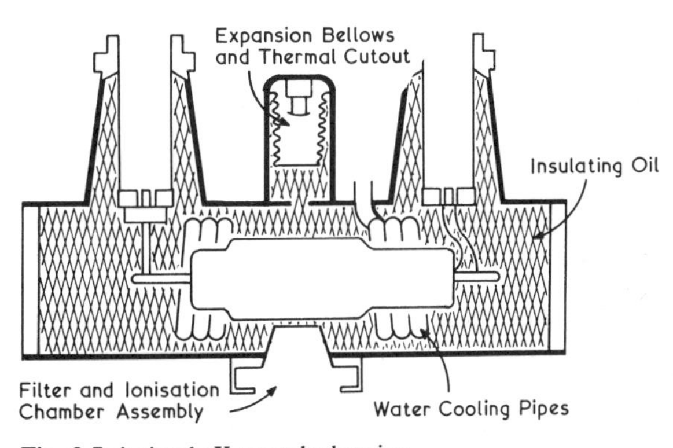

Fig. 3.7 A simple X-ray tube housing.

The X-ray generator circuits

The HT generator and the X-ray head may both be contained in the same assembly, but more frequently they are separate assemblies connected by shockproof HT cables. The cathode HT cable requires two inner conductors to carry the heating supply and the negative HT connection. Although the anode cable only requires one inner conductor to carry the positive HT connection, it is common practice to use a cable with two inner conductors connected in parallel. The two HT cables may, therefore, be identical—reducing the cost of manufacture and the cost of periodic replacement. The inner conductors are individually insulated and then surrounded by flexible solid rubber insulation covered by a strong earthed metal sheath connected to the X-ray head at one end and to the transformer tank at the other. Electrical or mechanical failure of the rubber insulation can result in an electrical breakdown between the inner HT conductors and the earthed sheath. It will be accompanied by the smell of burning and the sound of sparking.

A simple X-ray circuit is shown in Figure 3.2. Without rectifiers the circuit is of little practical value for two reasons (1) the X-ray tube must be made to withstand the full reverse voltage and (2) the hot anode can become a source of electrons—by thermionic emission—which will bombard and damage the filament when the potential is reversed during the negative half cycle of the applied voltage. The circuit will usually incorporate a rectifier network. This may be one or two rectifiers producing half-wave rectification, a bridge network of four rectifiers producing full-wave rectification, or a more complex circuit of rectifiers and capacitors (e.g. the Greinacher Circuit). These will now be briefly dealt with in turn.

The half-wave rectified circuit

This circuit is shown in Figure 3.8. During the half cycle of the applied voltage that makes the anode positive, current flows through the rectifier and the X-ray tube. (Remember the electron flow is in the opposite direction to the conventional current.) The voltage drop across the rectifier is small so the voltage applied across the X-ray tube rises and falls in the same way as the transformer voltage. During the reverse half cycle, the voltage is divided between the rectifier and the X-ray tube. The voltage developed across the tube is, therefore, considerably less than that produced by the transformer. It is quite common to find two rectifiers in this circuit—one in the anode and one in the cathode connection to the transformer—thus reducing the reverse voltage on the tube still further. The graphs in Figure 3.8 show that electrons are bombarding the target during practically the whole of the half cycle in which the anode is positive. The majority of these electrons will, however, be accelerated by voltages considerably less than the peak voltage. The radiation from a tube energised by such a circuit will, therefore, contain a larger proportion of low energy photons than if the tube was excited by a steady potential.

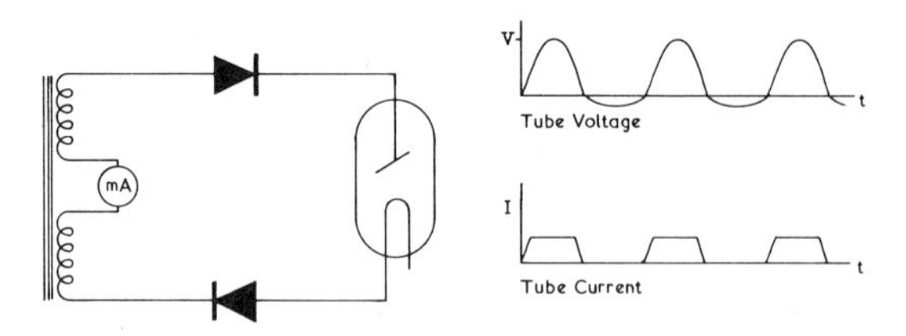

Fig. 3.8 The half-wave rectified circuit

The full-wave rectified circuit

The use of four rectifiers connected as a bridge network (Fig. 3.9) enables current to flow through the tube during both half cycles of the applied alternating potential and gives full-wave rectification. During each half cycle, one pair of rectifiers allows current to pass through the X-ray tube. The potential applied to the X-ray tube is thus unidirectional but pulsating at twice the frequency of the half-wave rectified circuit. If a capacitor is added in parallel to the X-ray tube, then the capacitor is charged to the full peak voltage during each half-cycle. As the transformer voltage falls, the capacitor begins to discharge through the tube. Providing the capacitor is large and the tube current small, the discharge will also be small. The potential of the anode will therefore not fall to zero between each pulse but it will be maintained at a value close to the peak. This slight drop in voltage every half cycle is called *ripple*. If the operating conditions of the X-ray tube are carefully chosen, the reduced anode voltage due to the ripple will not take the tube out of saturation and, therefore, the current through the tube will be steady. The output exposure rate from this *constant potential* tube will be considerably greater than that from a tube energised by the half-wave rectified circuit for the same nominal kilovoltage and, in addition, the number of low energy photons will be reduced.

In each circuit described so far, the peak voltage applied to the anode has been equal to, or a little less than, the voltage developed in the secondary winding of the transformer. In the circuit to be described now the anode voltage is twice that of the transformer and again almost constant. It is sometimes known as a voltage doubling circuit.

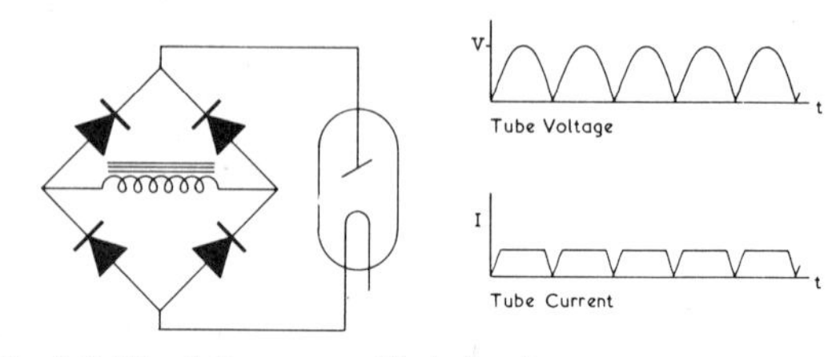

Fig. 3.9 The full-wave rectified circuit

The Greinacher circuit
The circuit is illustrated in Figure 3.10. Two rectifiers R_1 and R_2 are connected in series with the X-ray tube. Two series connected capacitors C_1 and C_2 are in parallel with the tube. The transformer secondary is connected to the mid-points of the two capacitors and of the two rectifiers. During the half cycle, in which end A of the transformer winding is positive, the capacitor C_1 is charged through the rectifier R_1 to the full peak voltage of the transformer, while rectifier R_2 is not conducting. During the next half cycle C_2 is charged to the same potential through R_2. Now C_1 and C_2 are series connected and their potentials are additive, thus, a voltage equal to twice that of the transformer is seen across the X-ray tube. As before, the capacitors discharge through the tube, but providing their capacitance is large and the tube current small, the fall in the voltage applied to the tube will be small. So once again there will be a ripple on the tube voltage (of say 5 per cent) but the tube current will be steady.

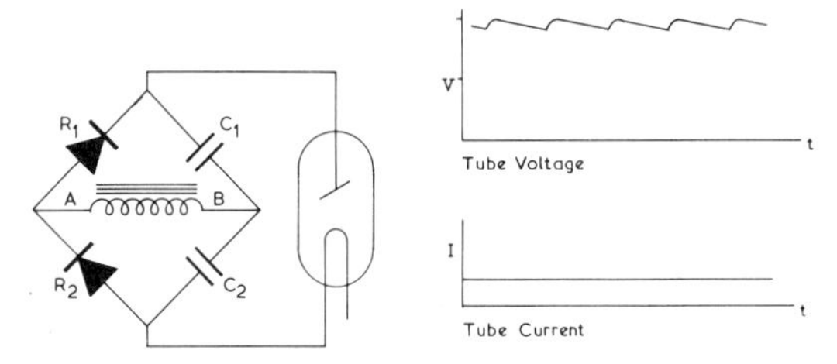

Fig. 3.10 The Greinacher circuit

The control desk

The basic design (Fig. 3.2) of an X-ray installation usually necessitates a control panel remote from the HT generator and the X-ray head. The control panel will usually include the following:

A mains voltage compensator. This control enables small fluctuations on the mains supply voltage to be corrected at the input to the HT generator. Some units will have an automatically controlled mains voltage compensator, while others will be controlled manually to a datum line on a meter.

Kilovoltage control and the kV meter. The HT transformer has a fixed turns ratio and the voltage applied to the tube is varied by adjusting the output voltage from the autotransformer. This may be a stepped control or a continuous smooth control. The kV meter is calibrated in kV but measures the primary voltage to the HT transformer.

Tube current control. It has already been explained that the tube current is determined by the temperature of the filament. The tube current control adjusts the input voltage to the filament transformer.

The X-ray switch. The X-ray beam is switched on by applying HT to the tube, the exposure is therefore controlled by switching the primary to the HT transformer. Owing to inductance, it is often necessary to switch on at a lower voltage and then to increase the voltage to the desired value. This increase may be manually or automatically controlled. On some sets, this may be done before the X-ray beam is allowed to emerge from the X-ray head by the opening of a shutter.

Treatment timer. The switching on of the X-rays starts an exposure timer, the purpose of which is to switch off the X-rays automatically after a pre-set exposure time. Recent legislation suggests that the timer should be one which counts up from zero to the time prescribed—continuing beyond that pre-set time if, for any reason, the beam is not switched off. This is clearly an improvement on the timers which have been used where the timer counts down from the pre-set value and switches off on reaching zero.

Tube current meter. This is generally a DC milliammeter measuring the current through the tube, but placed in that part of the high tension circuit which is at earth potential often at the midpoint of the secondary winding of the HT transformer where the voltage is divided or in the anode circuit where the anode is at earth potential.

Dose meter. Where possible, an ionisation chamber or solid state detector is fitted in the beam close to the target (but on the patient side of any additional filtration) to monitor the intensity of the emerging X-ray beam. The chamber should feed into an integrating dose meter (p. 43) or a doserate meter circuit, the latter being used in conjunction with a timer to determine the dose delivered to the patient.

Indicator lamps and interlocks. There are several safety devices which can be built in an X-ray unit to guard against overheating, overvoltage, overcurrent, etc., all of which may be indicated on the control desk. The essential indicator lamps are those to indicate and interlock the added filtration or wedge filter and to indicate the excitation of the tube. It is common practice now to interlock the filter, kilovoltage and the tube current so as to prevent a wrong combination from being used—this is particularly important at the grenz ray and superficial X-ray energies where units are frequently used under several different operating conditions.

The generators and control circuits described above have been used over the full range of X-ray energies from 10 kV to 500 kV. They therefore cover the ranges of energies commonly known as

Grenz rays	10–50 kV
Superficial X-rays	50–150 kV
Orthovoltage X-rays	200–500 kV

The need for X-ray energies greater than this has been recognised for many years, but the problems of the insulation, the weight of the transformer and of the thickness and flexibility of HT cables prevented the use of the circuits described above. Other techniques have to be used to generate *megavoltage* X-rays.

Megavoltage X-ray generators

The resonant transformer

One method of overcoming these problem is to dispense with the soft iron core from the transformer and the HT cables by making the transformer windings coaxial with the X-ray tube (Fig. 3.11). This system has been used to produce X-ray beams from about 300 kV up to two million volts (MV).

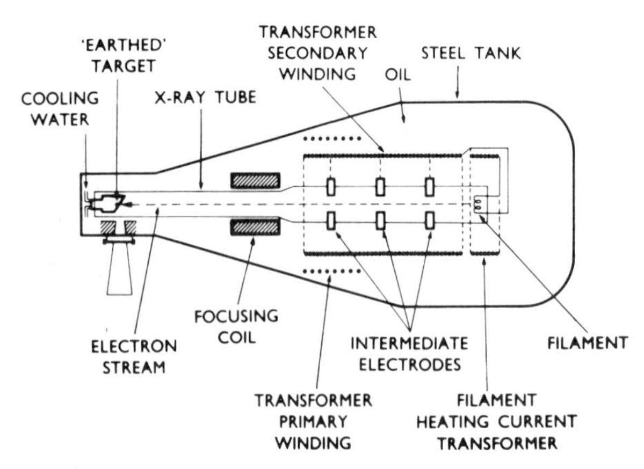

Fig. 3.11 A resonant transformer X-ray unit. (Meredith and Massey, *Fundamental Physics of Radiology*. Bristol: Wright)

Transformers which operate at the normal mains frequency require a soft iron core if they are to operate efficiently. If, however, the frequency is increased and the secondary circuit resonates at this same frequency, the transformer will function efficiently without the soft iron core. *A resonant transformer* can, therefore, be built around an X-ray tube with the secondary windings connected directly to the electrodes in the X-ray tube so dispensing with the HT cables.

A system of this kind requires a specially designed X-ray tube, the essential features of which are

1. The target is at earth potential
2. The cathode is at the peak (negative) voltage and connected to one end of the secondary winding
3. The X-ray tube is its own rectifier (a disadvantage)
4. The acceleration of the electrons is through a series of annular electrodes along the centre of the transformer and each connected to the appropriate point on the secondary winding.
5. These multiple electrodes and a focussing coil enable the electrons to travel the long distance from the cathode to anode without being scattered (by mutual repulsion) or lost to the walls (by attraction).

The disadvantages associated with self-rectification are more than offset by the higher voltages achieved and the more efficient cooling of the target permitting larger tube currents to be used.

At the higher energies where the transformers are large, the insulating properties of gas at high pressure are used in place of the more conventional transformer oil.

The Van de Graaf generator

An alternative approach to the production of high energy X-rays is to do away with the transformer and produce the high potentials by some other means—e.g. the Van de Graaf Generator—or to accelerate the electrons without the use of high potential differences—e.g. the linear accelerator and the betatron.

The Van de Graaf generator is an engineering development of an old simple device for the production of high potentials in the laboratory. It is used as an electron accelerator for the production of X-rays in the range of 1 to 5 million volts—2 MV being typical of the medical Van de Graaf generators.

The generator (Fig. 3.12) consists essentially of a rapidly moving continuous belt of an insulating material supported on an insulated column. Electric charge is sprayed on to the belt at one point and is carried along by the belt to another point where it is removed and accumulated. The charge is accumulated on a hollow metal hemisphere supported at the top of the insulating column. This insulating column is a stack of metal rings separated by glass insulators and is of sufficient height to withstand the high potential of the charged hemisphere. The purpose of such a complex insulating column in to achieve a uniform potential distribution along the column by arranging, for example, equal but very high resistances between each of the metal rings. With a high potential on the top terminal therefore, there is a small but equal potential drop between each ring because of the small current passing along the resistance chain.

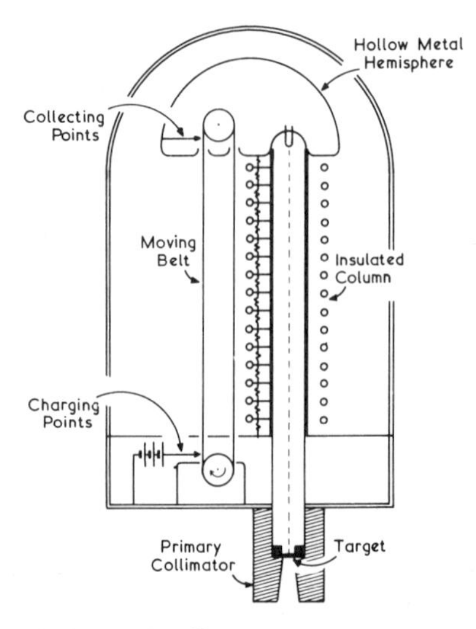

Fig. 3.12 A Van de Graaff generator.

The moving belt carrying the charge to the insulated terminal passes up the column between two pulleys, the one at the base is driven at high speed by an electric motor and the other is at the top inside the metal hemisphere. At the base of the column a row of very sharp metal points is arranged near the moving belt. These points are given a potential of several thousand volts and in the intense electrical fields thereby created at the points the air around them becomes ionised and ions of a similar sign to the charge on the points are repelled and driven onto the belt. At the top of the column a similar set of points effectively discharges the belt and transfers the charge to the hemispherical electrode. The ultimate potential reached by this electrode depends on the insulating properties of the column or the surrounding air and on the balance between the charging current on the belt and the current discharged through the X-ray tube or high resistance attached to the electrode.

a suitable support from the ceiling so that the beam can be directed vertically downwards or horizontally at any convenient height.

The X-ray beam emerges from the target in the same direction as the electron stream through a substantial lead cylinder. This acts as a primary collimator and carries at its end the beam defining system. A light beam is arranged to emerge through the beam defining system to indicate the size of the irradiated area (Fig. 3.23).

The linear accelerator

As the name implies, the electrons are accelerated in a straight line and this is accomplished by using *radio frequency* (rf) waves of approximately 10 cm wavelength. These radio waves are generated in a vacuum valve called a *magnetron* which operates in an intense magnetic field.

The electron acceleration is achieved in what is called the *corrugated waveguide* (Fig. 3.13). This is a hollow tube

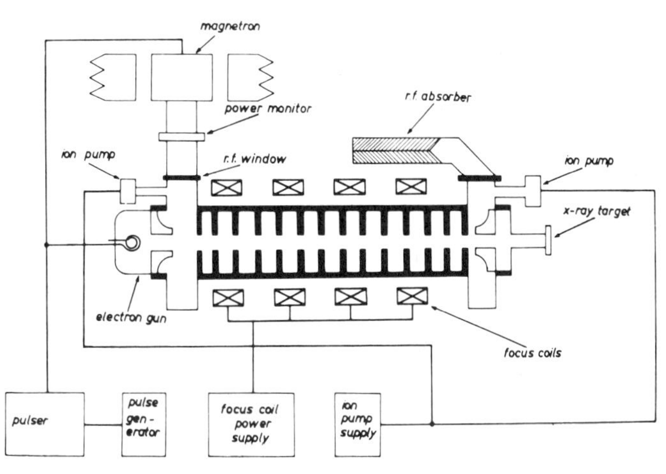

Fig. 3.13 A linear accelerator. (Courtesy of MEL Equipment Company Limited)

This electrostatic generator produces a steady potential and can be made either positive or negative by spraying charge of the appropriate sign on to the belt. For use as an X-ray generator the X-ray tube is inside the column alongside the moving belt. The tube itself is of the multiple electrode type—similar to that used in the resonant transformer generator—the electrodes being connected to the neighbouring metal rings of the column to ensure a uniform potential gradient down the tube. The cathode filament is heated by a current produced by a small dynamo inside the upper electrode and driven by the belt.

The target of the X-ray tube is a 3 mm thick gold disc at the lower end of the tube at earth potential. The target can therefore be readily cooled by a flow of water.

The generator is enclosed in a tank (approximately 2 metres long and 1 metre diameter) filled with an insulating mixture of carbon dioxide and nitrogen at a pressure 25 times that of the atmosphere. The tank can be mounted in

in which *iris diaphragms* are positioned with varying spacing and aperture diameter. These iris diaphragms limit the velocity at which the radio frequency wave can pass down the waveguide.

Electrons are produced by thermionic emission from a heated filament. They are injected into the waveguide in pulses so that they are in step with that part of the radio frequency wave—from the magnetron—in which the electric field will accelerate the electrons down the guide. The velocity of electromagnetic radiation is that of light in free space (3×10^8 ms^{-1}) but in the corrugated waveguide the iris diaphragms are arranged so as to slow the radio frequency wave at the input to the guide. The diaphragm spacing is then adjusted so that the rf wave is allowed to accelerate and carry with it the pulse of electrons. In most clinical linear accelerators the electrons can be accelerated to an energy corresponding to 4 MeV by a waveguide 1 metre in length—higher energies requiring longer waveguides. Machines are now becoming available which are more efficient and therefore produce higher energies with shorter waveguides. Linear accelerators of energies between 4 and 15 MV are in common use, but there is some preference for X-ray energies at the lower end of this range.

As in the resonant transformer, the electron stream is focussed by externally applied magnetic fields to prevent the beam spreading. The final focal spot should be only a few millimetres in diameter. The vacuum system is continuously pumped using ion pumps to maintain the pressure in the waveguide at a sufficiently low level.

At the end of the corrugated waveguide, the radio frequency wave is diverted and absorbed in a rf load while the high energy electrons continue in a straight line. They may bombard a *transmission target* directly but more frequently they are turned through 90° (or 270°) by an electromagnet before reaching the target, thereby producing an X-ray beam in a direction at right angles to the corrugated waveguide. This bending of the beam reduces the overall height of the installation and enables the linear accelerator to be mounted on an isocentric gantry (p. 72). Under certain circumstances the thick target is replaced by a thin window thus permitting the electrons themselves to emerge from the vacuum system as an electron beam. This has certain therapeutic uses which will be described later (p. 74). The X-ray beam is collimated in much the same way as in the Van de Graaf generator except that the thickness of the absorbing diaphragms will be increased appropriate to the photon energy.

Although the X-radiation from a linear accelerator is pulsed at between 100 and 500 pulses per second, the average doserate of the flattened beam may be several hundred rads/minute at 1 metre from the target. Each pulse lasts about 2 microseconds.

It has been seen (p. 4) that in the megavoltage range of energies the X-rays are predominantly produced in the same direction as the electrons bombarding the target. Although this theory is based on a thin target, the transmission target must be sufficiently thick to stop completely all the electrons bombarding it. There is, therefore, some modification of the spatial distribution. The theory, however, has two consequences particularly relevant to photon energies above 4 MV. Firstly, the photon energy sets a maximum practicable field size, in that the higher the energy the smaller the divergence of the X-rays produced. Secondly, the intensity along the central axis of the beam is greater than at the edges, that is the dose distribution across the beam is not uniform. Beam uniformity is achieved by the use of a beam *flattening filter* close to the target (Fig. 3.14). This is a metal filter essentially conical in section being thick at the centre to reduce the intensity of the central axis beam and tapering to zero at the edges so as not to reduce the intensity of the peripheral beam. The intensity distribution across the flattened beam should be such that it does not vary by more than 3 per cent from that at the central axis to within 3 cm of the edge of the beam (Fig. 3.15).

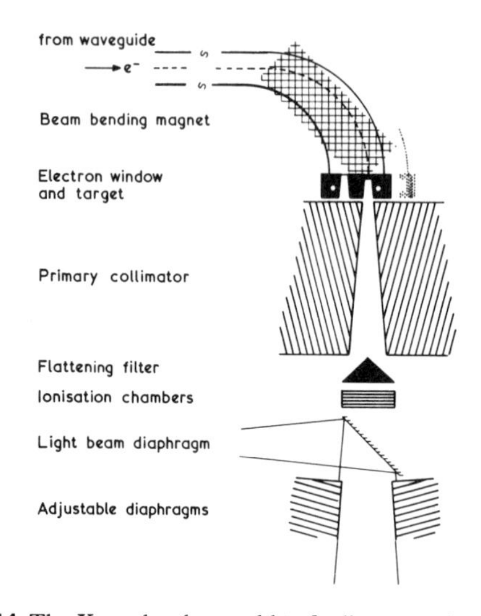

Fig. 3.14 The X-ray head assembly of a linear accelerator.

The betatron

The betatron, or induction accelerator, is able to cause electrons to travel round in a circular path with a continuously increasing energy. The betatron can produce X-rays of 15 MV to 50 MV photon energy although more frequently it is used as a source of high energy electrons (or beta particles) for electron therapy.

The name induction accelerator gives a clue to how the betatron works. If a changing magnetic field threads a coil of wire then an induced electromotive force (emf) is produced. In the betatron (Fig. 3.16) the wire is replaced by an evacuated annulus in which electrons are emitted by a hot wire filament. The evacuated annulus or *doughnut* is placed between the poles of an electromagnet, the coils of which are fed with an alternating current. If we consider

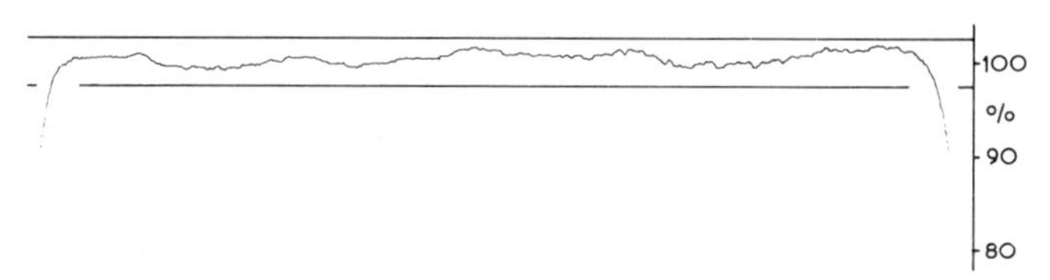

Fig. 3.15 A typical traverse across a flattened linear accelerator beam.

the conditions during the one-quarter of the cycle, BC, in which the current exciting the magnet is changing from zero to its peak, then the rising magnetic flux will produce a force on the electrons tending to move them in a direction at right angles to the magnetic field. An electron moving in the annulus will be deflected by the magnetic field in such a way that it moves in a circular path. By shaping the pole pieces correctly, the electrons can be kept moving in one orbit of a fixed radius whilst accelerating, i.e. whilst receiving more and more energy. When the magnetic flux reaches its peak, a subsidiary coil is energised and 'peels' the electron from its orbit so that it strikes the target where X-rays are produced or the window through which the electron beam can emerge.

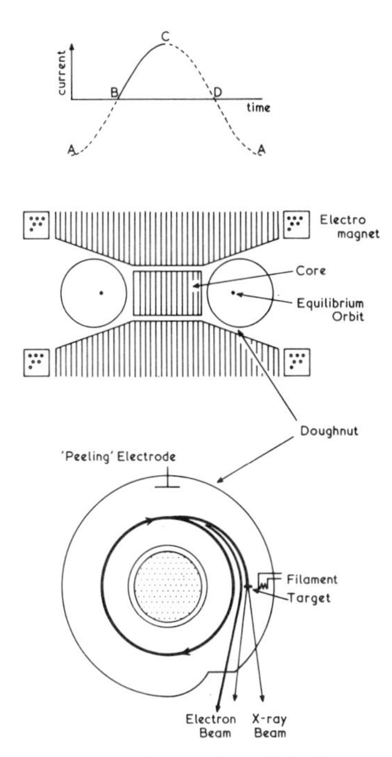

Fig. 3.16 The principle of operation of the betatron.

One advantage of the betatron over the linear accelerator is that the energy of the electrons arriving at the target (or electron window) can be varied relatively easily. As in the linear accelerator, the radiation is pulsed because only one quarter cycle of the alternating current in the electromagnet can be used. Again the high energy X-radiation is in the forward direction setting a limit on the maximum useful beam size even after passing through a beam flattening filter.

Gamma ray beam units

The choice of isotope

Gamma emitting radioactive nuclides may be used to produce a useful gamma ray beam for radiotherapy providing the photon energy is appropriately high and the specific activity is high enough to provide an adequate exposure rate at the chosen treating distance. Isotopes used in this way are summarised in Table 3.1. It must be remembered that owing to the decay of activity the source will need replacing from time to time. Any one source is normally used for a period a little longer than half the half life of the isotope, that is when the activity has decayed to about two-thirds of the initial activity and the treatment time is increased by about one-half. Owing to the discrete energies of gamma rays, the gamma beam is equivalent in many respects to X-ray beams generated at twice the photon energy.

Early units using radium had the advantage of a long half life and an effectively constant output doserate. There were however three disadvantages: the potential leakage of radon gas, the low specific activity of approximately one curie per gramme, the very high photon energy made the protection of the source housing difficult. These radium sources have been replaced with either caesium-137 or cobalt-60.

Caesium-137. Caesium-137 has a long half life (30 y) and is an attractive alternative to radium. Unfortunately, it has only a low specific activity (like radium) as well as a low k-factor, which together result in relatively low output exposure rates even from large sources. Caesium units, therefore, operate at source-skin-distances of less than 40 cm and the useful beam has a wide penumbra (p. 31). The decay scheme (Fig. 3.17) shows a gamma emission of 0.66 MeV photons which are high enough to overcome the differential absorption in body tissues and low enough to make source housing a manageable size. The disadvantages of caesium, however, outweigh its advantages and it

Table 3.1 Isotopes Used in Gamma Ray Beam Units

Element	Cobalt-60		Caesium-137	Radium-226 (in equilibrium)
Gamma Ray Energy	1.17, 1.33 MeV		0.662 MeV	0.19–2.43 MeV
Half Life	5.26 y		30 y	1620 y
Rate of decay (approx)	1% per month		1% per 6 months	1% per 25 years
k-Factor (R/h-mCi at 1 cm)	13.0		3.3	8.25
Typical Source Diameter	4 mm	17 mm	20–40 mm	40–70 mm
Typical SSD	5–20 cm	60–100 cm	5–40 cm	10 cm

is now rarely found in teletherapy units. It is widely used as a radium substitute in sealed sources (Ch. 8).

Cobalt-60. The effective photon energy of cobalt-60 is only slightly lower than that of radium and its half life is relatively short, but despite this, cobalt-60 is the most widely used isotope in teletherapy units. It can be made to a high specific activity which combined with a k-factor of 13 Rh^{-1} mCi^{-1} at 1 cm produce a source with a high useful beam exposure rate. The half-life is such that treatment times have to be increased by approximately 1 per cent per month and the useful life is approximately three and a half years.

Cobalt-60 is produced by neutron bombardment of cobalt-59 and decays to nickel-60 by emitting a beta particle and two gamma rays of 1.17 and 1.33 MeV.

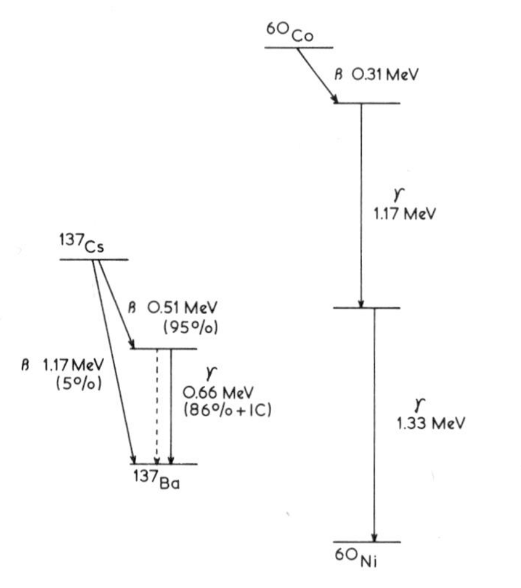

Fig. 3.17 The decay schemes of radioactive caesium-137 and radioactive cobalt-60.

The source design

To guard against the leakage of radioactive particles and to simplify the handling of the source, the radioactive cobalt (or caesium) is encapsulated in stainless steel (Fig. 3.18). The cobalt metal is in the form of discs approximately 2 mm thick and 17 mm diameter. These discs, probably 10 to 12 of them depending on the required activity are inserted into the inner capsule and any remaining space is taken up by similarly sized brass discs. The inner capsule is then closed with a screw cap which is sealed by brazing. The sealed inner capsule is then mounted in an outer case which is again closed with a screw cap and brazed.

The manufacturer of the cobalt unit provides the isotope laboratory with a source pencil or holder which will accept the double encapsulated source. The laboratory will load the source into the pencil and the pencil into the manufacturer's transit container. To facilitate the changing of a source on site, two such containers are brought to the hospital. The empty container is linked to the unit and the pencil with the 'dead' source is transferred from the treatment head to the transit container. The container with the new source pencil is then linked to the unit and the procedure reversed. The whole procedure may take a whole day, although the actual source transfer will only take a second or two. Since this procedure is only carried out every three or three and a half years, the student should watch it—preferably by closed circuit television—if the chance arises.

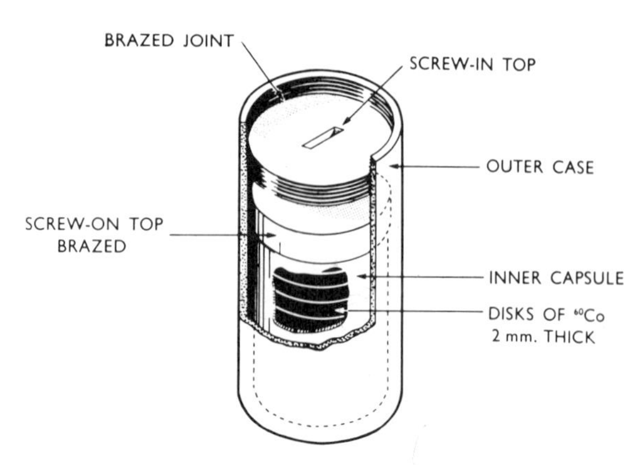

Fig. 3.18 A double encapsulated teletherapy cobalt-60 source. (Meredith and Massey, *Fundamental Physics of Radiology*. Bristol: Wright).

The alternative procedure is to dismantle the source housing from the unit and send it to the isotope laboratory to have the new source fitted. This results in a considerably longer 'down time' unless the isotope laboratory is only a short distance from the hospital.

Despite all the precautions, it is good practice to take a wipe test soon after a new source is fitted and periodically after that. To do this all readily accessible surfaces—especially the collimating system—are wiped with a damp swab. If any radioactive dust has leaked from the source capsule the damp swab will pick it up and it will be detected when the swab is monitored for free activity. The acceptable leakage levels are laid down along with recommendations on how the swab count is to be interpreted. If source is leaking, a simple wipe of the collimator and accessible surfaces is unlikely to pick up all the active dust particles and the activity measured on the swab will be only a fraction of that which has escaped from the source capsule.

The beam control

One fundamental difference between an X-ray unit and a gamma ray beam unit is that the gamma ray beam cannot be switched off. If the mains supply fails or is interrupted by a switch the X-ray beam is automatically safe. The gamma ray beam is not—unless it is designed to be so.

The gamma ray source, therefore, needs to be housed in

a protective shield such that the gamma ray exposure rates at the surface and at one metre from the source are below the accepted safety limits (Table 3.2). This means whatever else is required, the major part of the source housing will be either heavy alloy, depleted uranium or lead. The head of a 5000 Ci Cobalt unit may weigh 1000 kg. This protective material will surround the source in all directions except one—the direction of the useful beam. The protected treatment head will be mounted on a gantry similar to that of a linear accelerator or X-ray unit so as to enable the useful beam to be fired in directions appropriate to the treatment of the patient. It should be noted that although a unit may be designed to accept sources of a certain activity, say 5000 Ci, the source is often specified in terms of the roentgens-per-minute-at-a-metre (Rmm). This is directly related to the activity of a point source. In practice a long small diameter source may have the same activity (in curies) as a short large diameter source, but will have a lower output exposure rate owing to the inherent attenuation of the photons emitted from the nuclei at the back of the source. This *self absorption* within the source is taken into account in the specification of the Rmm in that it is a measure of the output rather than the content of the source.

The *moving source unit* (Fig. 3.19) has a mechanism whereby the source can be moved from the centre of the protective shield to the apex of the collimating system—that is the point from which the useful beam emanates. The mechanism moves the source between the 'Beam Off' position and the 'Beam On' position. In the *fixed source unit*, however, the source is fixed at the apex of the collimating system which either absorbs the useful beam or allows it to pass out of the head—the hole in the shutter through which the beam passes will form part of the collimating system. Whichever mechanism is employed, it is customary to provide a spring return and a manual return system in addition to the motor drive used to make the unit safe after an exposure, thus, failure of the

Table 3.2 Leakage Exposure Rates permitted from X-ray Tube Shields and Source Housings of Therapy Units (U.K. Code of Practice 1972)

		At 1 m from the source of radiation	At 5 cm from the readily accessible surfaces of the housing
Isotope Source Housings:			
short distance head and neck units[a]			
	'Beam Off'	2 mR/h	40 mR/h
	'Beam On'	1%[b]	—
conventional teletherapy units[a]			
	'Beam Off'	2 mR/h	20 mR/h
	'Beam On'	1 R/h or 0.1%[c]	—
Therapy X-ray Tube Housings:			
operating at < 500 kV maximum		1 R/h	30 R/h
operating at > 500 kV maximum		0.1%[b]	—

[a] It is assumed that sources in head and neck units give a useful beam exposure rate of less than 100 R/h whereas conventional units have an exposure rate in excess of 100 R/h at 1 metre.
[b] These percentages relate the leakage to the useful beam exposure rate at the same distance from the source.
[c] The permitted leakage is whichever is the greater.

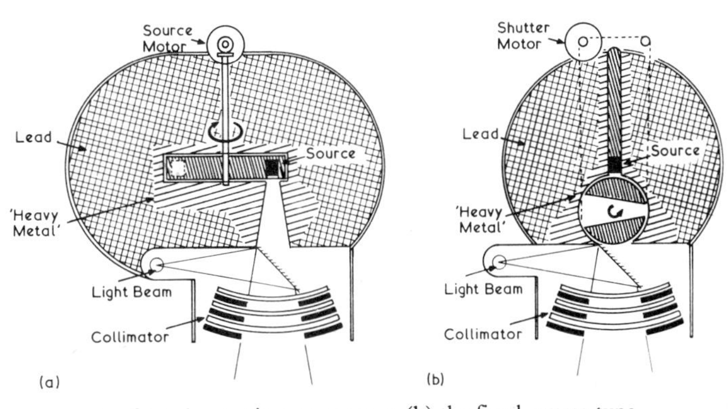

Fig. 3.19 Cobalt-60 teletherapy units of (a) the moving source type (b) the fixed source type

electrical supply or of the spring does not leave the source in the 'Beam On' position. The manual return system needs to be understood by everyone who may be required to operate the unit. Mechanical indicators are incorporated in the design of the unit to show that the shutter (or source) is in the safe 'Beam Off' position. Supplementary electrical and mechanical indicators may also be fitted.

Since the radiation is emitted spontaneously and the exposure rate decays with the half life of the isotope, the only controls required are those to operate the beam on/off mechanism. The control desk is, therefore, very simple. A treatment timer (preferably two) is used to terminate the exposure after a preset time. To correct for the decay of the source treatment times will need to be increased by approximately 1 per cent each month for cobalt-60 units (1 per cent every 6 months for caesium-137 units).

Leakage radiation

Reference has already been made to the lead protection incorporated in the tube shield of an X-ray unit and to the lead surrounding the source in a teleisotope unit. The function of this lead protection is primarily twofold. First to reduce the level of radiation reaching parts of the patient other than the treated area and second, to reduce the level of radiation reaching the walls of the treatment room, thereby reducing the wall thickness required. In the case of a teleisotope unit there is the additional need to make the unit safe in the 'Beam Off' condition to enable the radiographers to attend to the patient and other staff (cleaners, for example) to carry out their duties in the treatment room.

In any new installation or following any modifications which may affect the protection surrounding the head, a complete check is required to ensure the intensity of the leakage radiation is below the accepted limits and to ensure there are no weaknesses, such as cracks or pinholes, in the material used. The former is measured using an ionisation chamber and the latter by wrapping X-ray film around the surface of the housing.

Table 3.2 details the safety levels for the leakage through tube shields and source housings as specified in the United Kingdom Code of Practice.

Beam collimation

Beam collimation and collimating systems have been referred to above but will now be dealt with in greater detail. In general terms *beam collimation* means the limiting of the spread of the radiation to a predetermined direction and over a given solid angle. In practical radiotherapy it is necessary to be able to vary the size of the beam to suit the needs of each patient. This is referred to as field size or beam size.

The collimation system can be divided into primary collimation and secondary collimation.

Primary collimation

Primary collimation is achieved by a conical hole in a suitably thick metal block close to the source of radiation. The axis of the hole defines the beam axis and should pass through the centre of the source of radiation. The circular conical hole also defines the maximum divergence of the beam and, therefore, the maximum field size. It is this circular primary collimator that causes the larger square fields on some units to have 'rounded corners'. On a X-ray unit the primary collimator will be part of the X-ray tube housing or X-ray head (Fig. 3.7), on an accelerator or betatron it will be a heavy metal block close to or incorporating the X-ray target (Fig. 3.14) and on a fixed source cobalt unit (Fig. 3.19b) it may be part of the shutter mechanism. The maximum divergence is pre-determined in the lower energy X-ray units either by the target angle (the heel effect) in the insert or by the photon energy of the linear accelerator or betatron (p. 4). When an isotope source is used the maximum field size is only limited by the engineering problems associated with the secondary collimating system.

It needs to be remembered, however, that unless some compensation (e.g. a flattening filter) is introduced into the beam, the exposure rate at the corner of a large field will be lower than that at the centre owing to the effect of the inverse square law. Table 3.3 shows some typical values. In some centres, low atomic number flattening filters have been introduced into the superficial X-ray units for just this reason. At orthovoltage energies where scattered radiation plays an important role, the reduction of scatter further reduces the dose at the edges and in the corners.

Table 3.3 The Effect of Inverse Square Law on the Percentage Dose at the Corners of Large Fields

Typical Unit	SSD	Field Size	Dose at Corner Low By
Linear Accelerator	100	35 × 35	3%
Cobalt Unit	80	40 × 40	6%
Orthovoltage X-rays	50	25 × 25	6%
Superficial X-rays	25	30 cm dia	14%

Secondary collimation

Having defined the overall maximum field size and the beam direction, the secondary collimation provides a means of adjusting the field size to suit the particular requirements and the means of rotating square and rectangular fields about the beam axis. Secondary collimation may be achieved by either interchangeable applicators or adjustable diaphragms.

Interchangeable applicators are preferred where the SSD is short as the applicator provides an accurate means of setting up the source-skin-distance and, therefore, reducing errors due to the inverse square law. The closed ended applicators also simplify the use of bolus by presenting a surface normal to the beam axis against which the bolus can be packed (p. 62). On obese patients, the applicator can also be used as a means of applying compression, so reducing the distance between the skin surface and the treatment site (Fig. 6.16).

Adjustable diaphragms are essential where rotation techniques are to be used (as the SSD varies with the rotation) and for high energy radiation beams. In practice, applicators are used for low photon energies (below, say 500 kV) and where SSDs are short (below, say 20 cm); diaphragms are used on megavoltage units and for orthovoltage rotation techniques.

Interchangeable applicators

To provide an adequate choice of size and shape of field some twenty applicators may be required for each machine. Temporary shapes can be made using lead or lead rubber of adequate thickness over the end of a slightly over-large applicator, but these are not recommended. Table 3.4 shows the lead equivalence required to attenuate the incident beam to less than 2 per cent. The actual thickness of lead rubber is several times greater than its lead equivalence—i.e. the thickness of lead which will produce the same degree of attenuation.

Table 3.4 The lead equivalence required to reduce the transmitted primary beam to less than 2%

Therapy Beam	Lead Equivalence (approx)
50 kV 1 mm Al HVT	0.25 mm
100 kV 2 mm Al HVT	0.5 mm
150 kV 4 mm Al HVT	1.0 mm
250 kV 2 mm Cu HVT	2.0 mm
Caesium-137	31 mm
Cobalt-60	60 mm

Two types of applicator are shown in Figure 3.20. In both cases the applicator has a base which serves as a means of attaching the applicator to the X-ray head and incorporates a diaphragm which limits the beam to a size slightly larger than that required. The diaphragm is required to reduce the intensity outside the useful beam to less than 2 per cent of that of the useful beam. The lower edge of the applicator wall may be transparent to enable the coverage of the beam to be checked when the applicator is in contact with the patient. The final collimation is achieved at a point within approximately 5 cm from the skin surface. The face in contact with the skin indicates the actual area irradiated by the beam. In the design shown in Figure 3.20a the walls are not irradiated and can be made of steel, however, the straight sides make large applicators particularly bulky and difficult to handle. The 'Fulfield' applicator (Fig. 3.20b), on the other hand, has lead lined walls parallel to the rays at the edge of the field and the perspex end is taken across the face of the applicator to absorb secondary electrons from the lead. The perspex face is inscribed with the principal axes which, of course, intersect on the beam axis simplifying the setting-up of a patient and the use of 'bolus'.

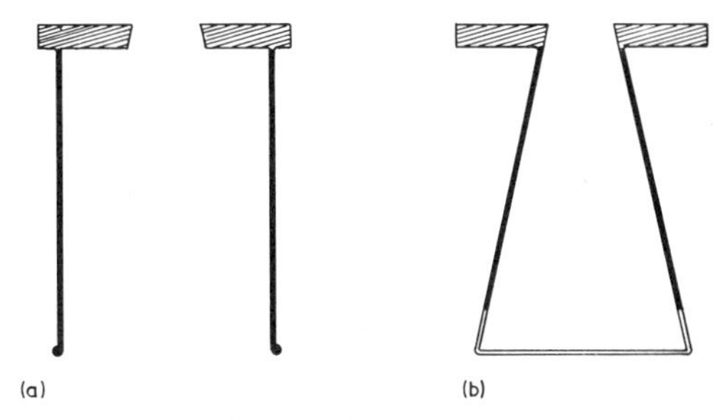

Fig. 3.20 Beam defining applicators
(a) Parallel-sided
(b) Fulfield

Applicators from time to time should be checked for accurate alignment (Fig. 3.21). The check is two-fold. First, set up a pointer to mark the beam axis on the surface of the applicator, then on rotating the applicator in its mounting, the pointer should remain at the same point on the applicator surface or describe a circle less than 2 mm diameter (Fig. 3.21a). Second, a film should be exposed (to a density of about 1.5) with markers indicating the corners of the applicator—the developed film should show coincidence of the markers and the corners of the blackened area. The blackened area should be checked carefully for a 'pin-cushion' effect as 'Fulfield' applicators tend to be handled in such a way that the lead lining is deformed inwards and becomes separated from the

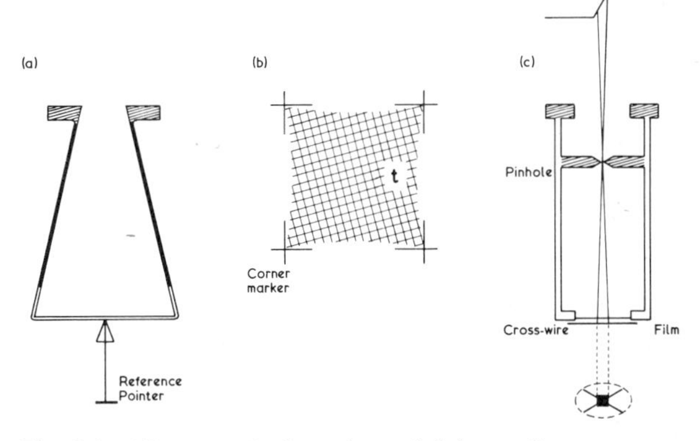

Fig. 3.21 Alignment checks on beam defining applicators
(a) Mechanical alignment checked on rotation about a reference pointer
(b) Beam coverage checked by corner markers on exposed film, the pin cushion effect is due to the lead lining becoming detached
(c) Alignment of focal spot checked with a pin-hole camera

applicator wall material. This effect is shown exaggerated in Figure 3.21b.

If these checks suggest the alignment on rotation is satisfactory, but the coincidence of the blackened area and markers is poor, then the focal spot may not be correctly aligned to the axis of rotation of the applicators. To check this requires a special device called a 'pin-hole camera' (Fig. 3.21c). This device is mounted in place of the applicator and checked for rotational accuracy. A film is then exposed and the focal spot is imaged on the film together with the shadow of the cross-wires, the intersection of which should be close to, or in the centre of, the image of the focal spot.

Adjustable diaphragms

Adjustable diaphragms are used for megavoltage units as the penetrating radiation requires a large thickness of heavy metal to reduce the transmitted intensity to less than the required 2 per cent of the useful beam intensity at the same distance. Applicators would be too heavy. In addition, the high energy radiation produces in the diaphragms secondary electrons which have a considerable range in air and so, if the diaphragms were close to the skin, these electrons would contribute to the skin dose negating the principal advantage of megavoltage radiation—the skin-sparing or build-up effect (p. 53). The optimum distance is based on the consideration of skin dose and penumbra width and is typically 20 cm, but the precise value depends on the beam energy and the design of the unit. In practice, the diaphragms are approximately halfway between the source of radiation and the patient.

As with applicators, there are several designs of adjustable diaphragms in common use, they may be conveniently referred to as (a) double plane, (b) single plane and (c) interleaved (Fig. 3.22). All of these limit the choice of field shape to squares and rectangles within the overall circle defined by the primary collimator. Two design features of adjustable diaphragms are based on the requirement to minimise the size of the *penumbra* (p. 31). This requires the collimating faces of the thick diaphragm blades to move in such a way that they always lie along the direction of propagation of the radiation, i.e. along a radius from the source and that the collimated length—the distance from the source to the most distant part of the diaphragm blade—should be as large as possible. A compromise has to be reached between the collimated length and the range of the secondary electrons from the diaphragm blades.

The double plane system has the clear disadvantage in that the collimated length of the two pairs of blades—which may be 10 to 12 cm thick—is different giving rise to different penumbra widths on the two principal axes of the beam. The other systems too have disadvantages. The single plane system requires a much larger housing to accommodate the blades when setting small field sizes. The interleaved system poses considerable engineering problems, particularly for the higher energy beams where the overall thickness of the blades is large.

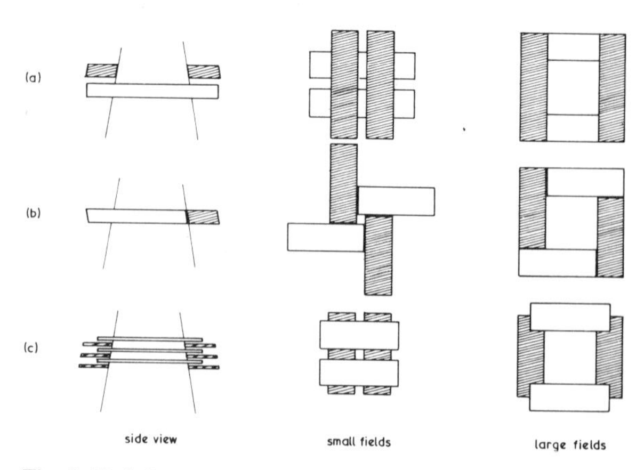

Fig. 3.22 Adjustable diaphragm systems
(a) Double plane
(b) Single plane
(c) Interleaved

Since the diaphragm system is some distance from the patient, a light beam is incorporated in the unit to indicate the area covered by the radiation. The light beam (Fig. 3.23) is provided by a small filament lamp positioned in such a way that its virtual image is coincident with the source of X- or gamma-rays. In this way, the visible light is seen to emanate from the same source as the radiation beam and, therefore, illuminating the area irradiated and defined by the diaphragm system. The thin surface-silvered mirror does not appreciably affect the photon beam. A specially constructed mirror is required if it is to

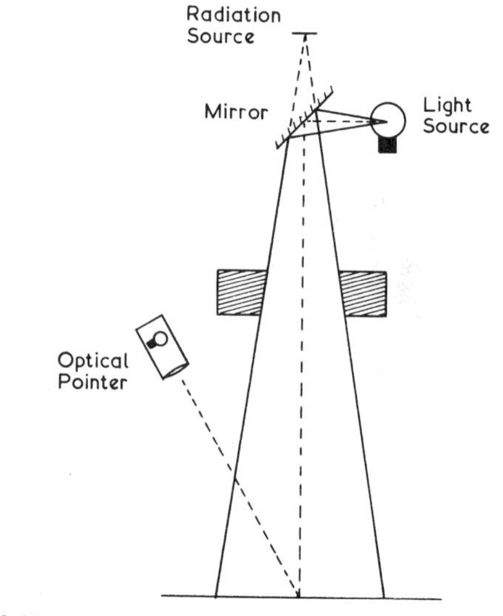

Fig. 3.23 The principle of the light beam diaphragm and optical front pointer.

be used in conjunction with an electron beam. A suitable cross-hair 'object' is incorporated to produce an 'image' on the patient identifying the principal axes of the beam and their intersection on the beam axis. A simple optical SSD pointer is also shown projecting a spot of light or an illuminated scale which coincides with the beam axis at the predetermined SSD.

As with interchangeable applicators, periodic checks are carried out to confirm the accuracy of the light beam and the adjustable diaphragm system. As before, the accuracy of the beam axis is checked by rotating the diaphragm system and the coincidence of the light beam and radiation beam is checked using a film and markers. In addition, it is necessary to ensure that each pair of blades is symmetrical about the beam axis and in particular about the axis of rotation.

Penumbra

If the source of radiation in the therapy unit was a point source, the precise direction of the radiation would be known and the size of the beam would be uniquely defined. In practice it is not a point source. The X-rays may emanate from a small area on the target. Gamma-rays, on the other hand, emanate from all points within a finite cylindrical source. This presents a problem. In Figure 3.24a the point marked A is in the beam of radiation and point B is outside the beam. In Figure 3.24b this is also true, but point C is neither inside nor outside the beam, it is said to lie in the *geometric penumbra*. It is in a region where it can receieve radiation from some parts of the source but not from the whole source. In the case of a point source, the dose gradient between points A and B is sheer, while in the case of a finite source, the dose gradient is gradual and depends principally on the size of the source. Applying the principle of similar triangles to Figure 3.24b, the geometric penumbra p is defined by:

$$p = s\left[\frac{f - c}{c}\right]$$

where s is the diameter of the source, f the source-skin-distance and c the collimated length. The geometric penumbra is independent of field size. It is easy to see from the equation above that to reduce the size of the geometric penumbra the source diameter must be reduced ($s = o$ for a point source) or the collimated length must be increased and made more nearly equal to the source-skin-distance. The collimated length is measured to the distal (most distant) edge of the diaphragm blade from the source. It should be noted that although the diagrams show the geometric penumbra as defined at the skin surface, it increases as the beam passes through the patient. If the penumbra is to be calculated at any other position, then f should be taken as the distance from the source to the point of interest.

In practice, the diaphragm blade may be very thick and unless its face is parallel to the rays from the source of radiation, the rays 'just outside the beam' will not traverse the full thickness of the collimator blade and will give rise to a further blurring of the edge of the beam which can be referred to as a *transmission penumbra*. By changing the angle of the collimator face as the field size is changed, the face can be maintained parallel to the rays, thereby presenting the full thickness of the blade to the rays to be absorbed. However, there will be some transmission penumbra effect as the source is of finite size whereas each blade can only rotate about a single point.

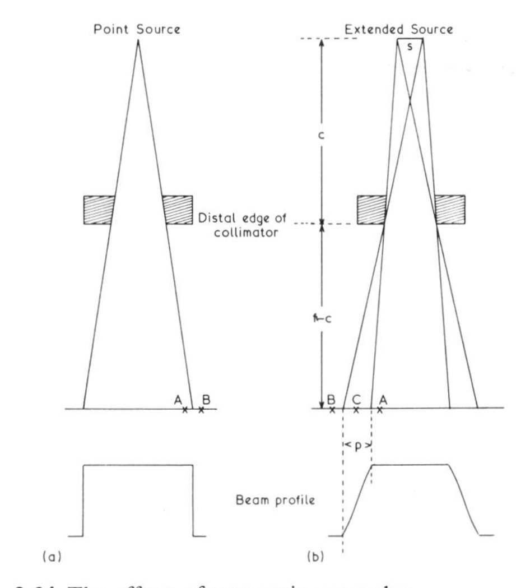

Fig. 3.24 The effects of geometric penumbra.

Penumbra Trimmers

Where the geometric penumbra is large and where the treatment technique permits, it is possible to increase the effective collimated length by the use of *penumbra trimmers*. These are heavy metal bars mounted between the diaphragm system and the patient (Fig 3.25) and positioned in such a way that the inner surfaces lie in the same

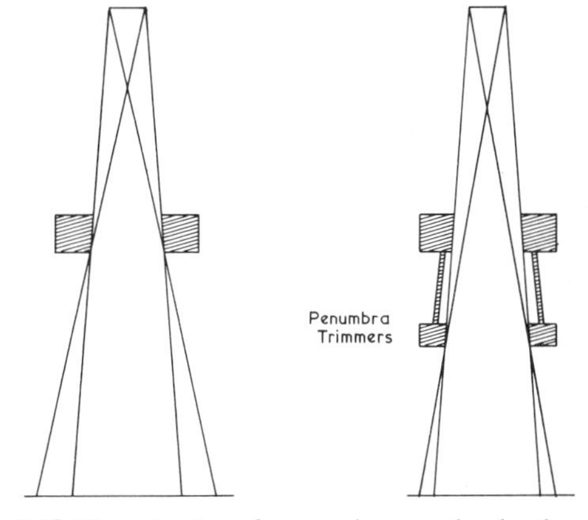

Fig. 3.25 The reduction of geometric penumbra by the use of penumbra trimmers.

plane as the faces of the diaphragm blades themselves. Penumbra trimmers are often offered as an accessory, particularly for cobalt or caesium teletherapy units.

Beam Shaping and Local Shielding Blocks

Adjustable diaphragms, in general, limit the choice of field sizes to squares and rectangles. Any modification to the shape of the field must be done by the addition of heavy metal blocks (lead, tungsten, heavy metal or one of the special alloys). Areas in the centre of the field may be protected in a similar manner. The thickness of the blocks will be determined by the degree of protection required having taken into account the fact that the dose in the shadow will be increased by the scatter from the adjacent irradiated tissues. For megavoltage radiations the thickness will be at least 5 cm lead equivalent (Table 3.4).

In practice these blocks may be suspended from the front of the unit or mounted on a 'shadow tray' similarly suspended. Alternatively, they can be supported on a table over the patient. The shadow tray or table may be a sheet of perspex or a perforated aluminium plate. The actual means adopted must be decided from the consideration of penumbra width, the electron contamination and the weight of the blocks. The sharply defined shadow produced by the light beam can be deceptive in that it gives no indication of the penumbra width. Accurate shielding requires, therefore, the blocks to be close to the patient. On the other hand, electron contamination and weight consideration require the blocks distant from the patient.

The electron contamination can be reduced by fitting each block with a material which will absorb these secondary electrons and which will not itself produce additional contaminating electrons. Brass is often used both as a filter and a means of protecting lead blocks from damage—lead is soft and easily distorted. Brass has the further advantage that it can be tapped to accept fixing screws through a perforated shadow tray so that the blocks can be fixed in position when the beam axis is not vertical.

Lead shielding blocks intended for general use are often rectangular in section, ie, the (collimating) faces are parallel to the beam axis and not to the radiation they are intended to absorb. This gives rise to the maximum transmission penumbra and, therefore, to a poor X-ray shadow. Wherever possible the blocks should be tapered to minimise the transmission penumbra effect.

Low melting point lead/bismuth alloys are now available and enable shielding blocks to be tailor-made to the patients' individual requirements and later melted down and re-used. The density of the alloy is less than that of lead, usually about 9 or 10 gcm^{-3} and, therefore, the blocks need to be proportionately thicker than those of lead. The simplicity of manufacture and the accurate shaping (in all three dimensions) far outweigh the problem of thickness. One method which may be used is as follows. A radiograph is taken using the simulator so that the SSD and field size match those to be used during treatment. The area to be shielded is marked carefully on the film (Fig 3.26a and Fig 3.27). The pattern on the film is then

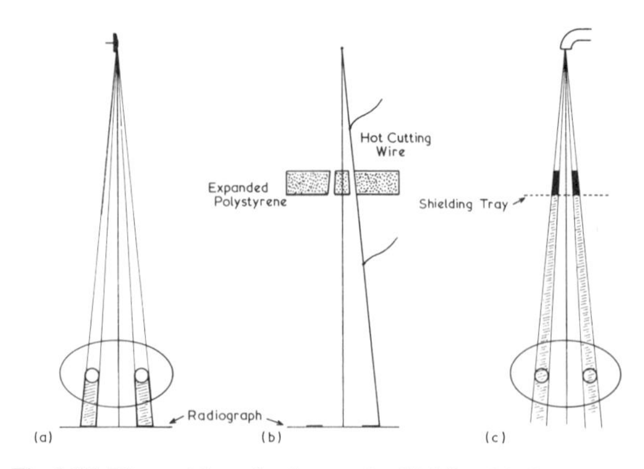

Fig 3.26 The making of tailor-made shielding blocks
(a) Patient radiographed on simulator under treatment geometry
(b) Polystyrene cut to shape using hot wire and marked-up radiograph
(c) Patient treated using shielding blocks cast in the polystyrene

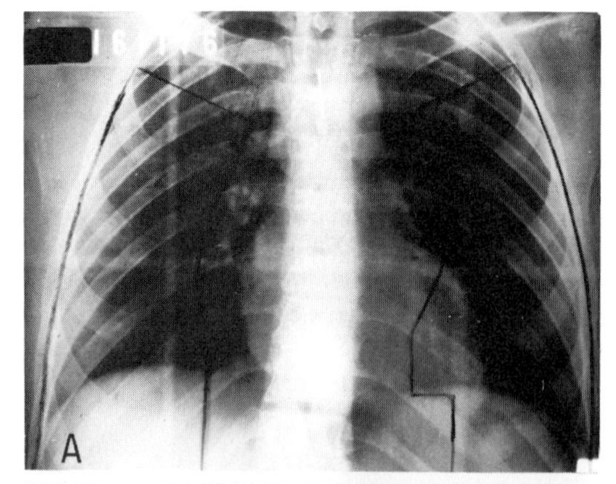

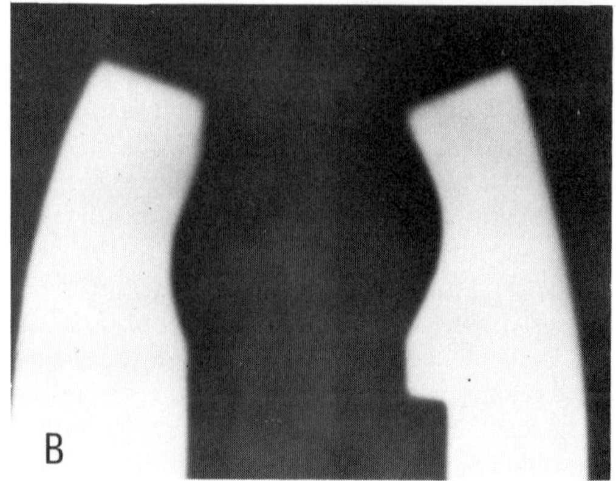

Fig. 3.27 Planning film (A) from the simulator used in the preparation of tailored shielding blocks and the check film (B) on the linear accelerator confirms their accuracy.

used in conjunction with the hot wire to cut a hole in the expanded polystyrene (Fig 3.26b) using the FFD (focal-film-distance) from the simulator and the source-shielding tray distance from the treatment unit. The polystyrene is then used directly as a cast for the low melting point alloy. The blocks then produce the precise shadows to protect the areas originally marked on the film (Fig 3.26c and Fig. 3.27). As an alternative to the alloy, lead shot may be poured into the polystyrene cast and used directly or the lead shot may be welded together using wax and then removed from the cast. Great care is needed in handling lead shot to guard against its being spilt on the floor and producing a skating rink surface.

4. The interaction of X- and gamma rays with matter

The Attenuation of X- and gamma ray beams

In this chapter the interaction between photons of electromagnetic radiation and the atoms of the material through which they pass will be described. For the purposes of this text photons of energy greater than about 5 to 10 keV are of interest, called X-rays if produced by deceleration of electrons and gamma rays if they arise from radioactive materials. Whether the photons are described as X-rays or as gamma rays is quite immaterial to a discussion of their behaviour in matter and reference to X-rays only should be taken to include also gamma rays of the same photon energy.

X-ray and gamma ray beams possess considerable power of penetration through matter, this penetration depending both on the photon energy or wavelength of the beam and on the atomic number and density of the material through which it is passing. A quantitative examination of the penetrating properties of a beam is obtained by measuring the change in the intensity of a beam when it passes through a known thickness of material. Such a measurement is made by an ionisation chamber in which the saturation ionisation current produced in a fixed volume of air is taken as proportional to the intensity of the beam at the measurement point (see p. 41).

When an X-ray beam diverges from its point of production its intensity is reduced as the distance from the source increases in a manner determined by the *inverse square law*. It will be recalled that this law indicates that for radiation spreading from a point source in a non-absorbing medium the intensity of the radiation at any point varies inversely as the square of the distance from the source. At a fixed distance from the source the intensity is reduced by the interposition of matter for two reasons. First, *scattering* of the electromagnetic radiation occurs so that some of the energy of the beam is carried by photons which deviate from the original path of the beam and thus pass outside the detecting device at the point of measurement. Second, some of the photons of the incident beam give up their energy to the material through which they are passing, again causing a reduction in the recorded intensity at the point of measurement. This is *absorption*. The beam is said to be *attenuated* by both these processes.

Measurement of attenuation

An experimental arrangement for measuring the attenuation of an X-ray beam in layers of attenuating material is shown in Figure 5.9 (see p. 49).

The radiation from an X-ray tube is collimated by a suitable diaphragms, so that a narrow beam falls on the detector ionisation chamber at a distance of about 1 metre from the source. Layers of attenuating material of known thickness are placed across the beam, midway between the target and the measuring device. Observations are made of the intensity at the detector as the thickness of the attenuating layer is increased. The ratio of the transmitted intensity compared to the intensity with no attenuator present, generally expressed as a percentage, is plotted against the thickness of the attenuator to give an *attenuation curve*.

The attenuation curve obtained in a particular experiment is dependent on a number of factors. These include the width of the beam passing through the collimators, the geometrical arrangement of the attenuators and the detector, the material of the attenuators and the spectral distribution of photon energies in the beam. The curve in Figure 4.1 is drawn for an experimental arrangement in which a narrow beam of X-rays is used and in which a beam of one photon energy—i.e., a homogeneous beam—of photon energy of approximately 100 keV is being attenuated in copper. The ordinates give the transmitted intensity expressed as a percentage of the unattenuated beam.

For homogeneous radiation, i.e., monochromatic radiation or radiation of one wavelength, the attenuation takes place according to the following *law of attenuation*.

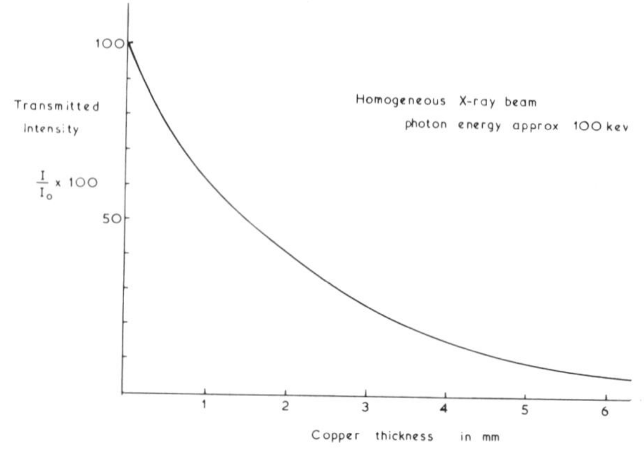

Fig. 4.1 Attenuation curve for homogeneous X-ray beam.

Equal thicknesses of the same attenuating material remove equal fractions of the radiation intensity incident on them and the fractional reduction of intensity per unit thickness of attenuator is constant (for small thicknesses of attenuator).

This law of attenuation gives rise to an exponential attenuation curve and the curve of Figure 4.1 is such a curve. It can be expressed by the mathematical equation:

$$I_t = I_0 e^{-\mu t}$$

where I_t is the intensity transmitted through attenuator thickness t

- I_0 is the intensity with no attenuator present
- t is the attenuator thickness
- e is a mathematical constant (= 2.718)
- μ is a constant for a given material and a given photon energy of the radiation

μ is called the *total linear attenuation coefficient* and is defined as the fractional reduction of intensity per unit thickness of attenuator (for small thicknesses of attenuator).

It will be noted that the mathematical form of the attenuation curve for homogeneous beams is the same as the mathematical form of the exponential decay curve of a single radioactive material (see p. 10). The exponential attenuation curve similarly has two characteristic properties. First, if plotted logarithmically, that is, if $\log I_t/I_0$ is plotted against t, a straight line is obtained whose slope is $-\mu$, where μ is the attenuation coefficient. Second, there is a constant value of thickness of attenuator which reduces the incident intensity to half its initial value. This is called the *Half Value Thickness* (HVT). It will be seen that the attenuation curve of Figure 4.1 corresponds to a HVT of 1·5 mm of copper. These two properties hold for homogeneous radiation.

The quantity μ, the total linear attenuation coefficient is a constant for a particular photon energy of the beam and for a particular attenuating material. It increases with the density, and the atomic number of the attenuator and decreases with increasing photon energy of the beam.

It can easily be shown that there is a simple relationship between the HVT of a homogeneous beam and the total linear absorption coefficient μ. This is as follows:

$$\text{HVT} = \frac{0{\cdot}693}{\mu}$$

where HVT is expressed in cm and μ is expressed in cm^{-1}.

If the constant μ is divided by the density ρ of the attenuating material a quantity (μ/ρ) is obtained which is independent of the density of the attenuator. This is known as the *total mass attenuation coefficient* and gives the fractional reduction of intensity if the beam passes through an attenuating layer of thickness 1 g per cm^2. The significance of the word 'total' in these definitions of attenuation coefficients will be shown on page 38.

The total mass attenuation coefficient changes rapidly with the photon energy (i.e. the wavelength) of the beam and the atomic number Z of the attenuator but the relation is a complicated one. This is because attenuation, that is the loss of energy from the beam, arises from a number of processes of interaction between the beam photons and the electrons in the attenuating material. A rough guide, however, is as follows. For the low photon energy, long wavelength, region of the X-ray spectrum (i.e. for diagnostic X-rays for example) μ/ρ is roughly proportional to Z^3 and to λ^3. In the high photon energy, short wavelength, region of the spectrum (i.e. for high energy radiotherapy beams) μ/ρ is roughly proportional to λ and is largely independent of Z.

In practice X-ray beams emerging from the target of an X-ray tube are, as shown in Chapter 1, composed of a continuous spectrum and have a wide range of photon energies. In this particular respect X-ray beams differ from the gamma ray beams produced for example in the gamma ray beam units described in Chapter 3, where the radiation is homogeneous. The practical implications of the heterogeneity of X-ray beams are discussed in Chapter 5. Before dealing with this it is important to understand the processes by which the beam is attenuated, and these are described in the following sections.

The attenuation processes

A beam of electromagnetic radiation loses energy in passing through an attenuating medium in two ways. Energy may be lost by the scattering of electromagnetic radiation out of the direct beam. It may also be lost if photon energy is absorbed in the medium through which the beam is passing. This second absorption process arises when energy is transferred from the photons to the electrons of the medium. These electrons, arising as fast electrons in the medium, lose their energy by producing ionisation and excitation along their tracks. Ionisation will occur if the impinging electron removes an electron completely from the orbital structure of the parent atom. Excitation will result only in the shift of an electron to an orbit of higher energy in the same atom. Both of these processes will continue until the fast electron comes to rest after a path which is very small for low photon energy radiation but may be many millimetres for high photon energy radiation. Absorption is always achieved by the intermediate production of fast electrons in the attenuating medium.

Several processes contribute to the attenuation phenomenon each with their own particular relationship between the contributions of scattering and absorption.

Classical scattering

For long wavelength radiations, that is for soft diagnostic X-radiations and in materials of high atomic numbers, such as most metals, the bound electrons in the attenuator atoms scatter radiation in all directions. This radiation is of the same wavelength as the incident radiation. This scattering phenomenon is similar to the scattering of light from dust particles. Since it can be explained by well established theories which treat X-rays as a wave propagation it is often referred to as *classical scattering*, or alternatively as *elastic, unmodified* or *coherent scattering*. It does not contribute any real absorption of energy to an attenuator since no fast electrons are liberated from the attenuator atoms. This type of scattering is relatively unimportant in medical uses of X-rays since low photon energy beams and high atomic number attenuators are not the normal conditions in such use.

The Compton effect

In practice it is found that when shorter wavelength radiation beams are used to irradiate materials of low atomic numbers (which are the more usual conditions in the medical use of X-rays), the scattered electromagnetic radiation has a longer wavelength than the original beam. In addition the greater the angle through which the radiation is scattered the longer is the wavelength of the scattered radiation. This scattered radiation is called *modified scatter* and the process is referred to as *inelastic scattering* or *Compton scattering*.

For an explanation of this effect the quantum theory must be invoked. Inelastic scattering arises when an incident photon interacts with an 'unbound' electron in the attenuating medium. A collision-like process occurs in which the incident photon loses some of its energy to the electron and the remainder of the energy appears as a photon of smaller energy. The energy of the incident photon, therefore, is shared between the scattered photon of longer wavelength and the electron. The electron, called a *recoil electron*, appears as a fast electron which loses its energy gradually by the formation of ion pairs along its track.

The term 'unbound' electron refers to those electrons in the atoms of the attenuator in which the binding energy is small compared to that of the incident photon. The binding energy of the outer electrons in light elements, for example, may be only a few electron volts (eV). This is quite small compared to the energy of a photon of even the soft X-rays used in medical work—which is generally 50 keV or more.

The Compton process is the important process of interaction for high energy photons (short wavelength radiation) passing through low atomic number materials. It is therefore the predominant process for the attenuation of deep therapy beams in soft tissues.

The following properties of the Compton process are important:

1. The wavelength of the scattered radiation increases as the angle of scattering relative to the incident beam increases.
2. The energy of the incident beam is shared between recoil electrons in the attenuator and scattered photons of long wavelength. If the photon energy of the incident beam is low the proportion of the energy appearing as scattered electromagnetic radiation is high, but for incident beams of high photon energy the majority of the energy lost from the beam appears as energy of recoil electrons.
3. As the incident photon energy increases the scattered electromagnetic radiation is concentrated more and more in the forward direction.
4. The recoil electrons are emitted mainly in the forward direction and are increasingly concentrated in the direction of the incident beam as the energy of the incident photons increases.
5. The total attenuation of a beam produced by the Compton process is independent of the atomic number of the attenuator but increases with the wavelength of the radiation, that is it decreases as the photon energy of the radiation increases.

The photoelectric process

When the photons of an incident beam react with a bound electron classical scattering results, as we have seen, if the incident photon energy is small. If, however, the incident photon has an energy greater than the binding energy of the electron a complete transfer of the photon energy to the electron occurs. The electron escapes from the atom with an energy equal to that of the incident photon minus the energy required to release the energy from its orbit. For instance if an electron is ejected from the K shell, the energy of the *photoelectron* as it is called, will be $(hn - E_k)$, where hn is the photon energy (being the product of Planck's constant and the frequency) and E_k is the binding energy of the electron in the K shell. Photoelectrons would similarly be released from other shells such as the L shell if the incident photon has sufficient energy, though the photoelectric process from the K shell is the more important since the electrons are mainly removed from shells in which they are relatively well bound.

The electrons emitted from the atoms of the attenuator in this process emerge in all directions but mainly at about right angles to the direction of the incident beam.

The photoelectric process is the predominant attenuation process for low energy photons i.e. long wavelength radiation, in attenuating materials of high atomic numbers. It is thus the main effect in the interaction of diagnostic quality X-rays in bone or metals. It is possible to assign to this attenuation process an attenuation coefficient as defined on page 35 but relating only to the attenuation by the photoelectric process (see p. 38). The mass attenuation coefficient for photoelectric attenuation is very dependent on both the photon energy of the radiation and on the atomic number of the attenuator. It is approximately proportional to the cube of the atomic number and inversely proportional to the cube of the photon energy that is proportional to the cube of the wavelength. The photoelectric process thus increases very rapidly with wavelength and with atomic number and is the important factor in producing high contrast between bone and soft tissues in diagnostic radiography, and in producing high bone absorption in superficial X-ray therapy.

Since the photoelectric process results in the emission of an electron from a position in one of the shells of the atom of the attenuator, a vacant space is left in that shell. This vacant space is filled by an electron falling into it from an outer shell or from outside the atom. This gives rise to the emission of characteristic radiation from the atom. Photoelectric absorption therefore is accompanied by the emission of electromagnetic radiation characteristic of the element of the attenuator.

The pair production process

One further interaction process occurs for very high energy photons if the photon energy is greater than 1.02 MeV. Photons of energy greater than this can interact with the electric field around the nucleus. This interaction results in the complete absorption of the energy of the photon and the simultaneous creation of two electrons, one negative and one positive. The positive electron is the positron, the particle of short life appearing in some nuclear disintegration processes and emitted from some radioactive nuclides (see p. 8).

The production of these two electrons represents the creation of matter from energy. The equivalence of matter and energy was set out by Einstein in quantitative form by the equation:

$$E = mc^2$$

where c is the velocity of light. Using this equation the energy equivalent to the mass of one electron can be calculated as 0.51 MeV. In order to create two electrons therefore a minimum energy of 1.02 MeV is necessary. Since the electrons are opposite in charge no creation of charge is involved in the pair production process. Any energy of the initiating photon in excess of 1.02 MeV appears as kinetic energy shared equally between the two electrons. This kinetic energy is dissipated in the production of ions along the tracks of the two fast electrons which are thus brought to rest after a short range in the medium.

The positive electron does not have an isolated existence when at rest, and when it comes to rest it combines with a neighbouring negative electron. The two charges neutralise each other and the mass of the two electrons is converted back into two photons of electromagnetic radiation each of 0.51 MeV energy and travelling in opposite directions to each other. This is *annihilation radiation*.

The pair production process occurs as we have seen only for radiations of photon energies greater than 1·02 MeV. Above this energy however the probability of the process occurring increases rapidly with increase of photon energy. This is the only attenuation process therefore in which attenuation *increases* with increasing photon energy. The pair production process also increases with increasing atomic number of the attenuator. In practice this process is of little importance in the low atomic number elements of soft tissues except above about 20 MeV radiation. In industrial radiography, however, when high energy beams are used to penetrate metals it can become the major attenuation process.

Summary of interaction processes

Figure 4.2 summarises the four processes which are involved in the loss of energy from a beam of photons passing through an attenuating material. Some of these processes give rise to deposition of energy in the attenuator through ionisation: they are (a) the photoelectric process (b) that part of the Compton process that gives rise to recoil electrons (c) the pair production process. The rest of the energy lost to the main beam escapes from the attenuator as scattered radiation. This includes (a) classical scattering of long waves (b) the characteristic radiation following the photoelectric process (c) the scatter component of the Compton effect and (d) the annihilation radiation arising from the pair production process.

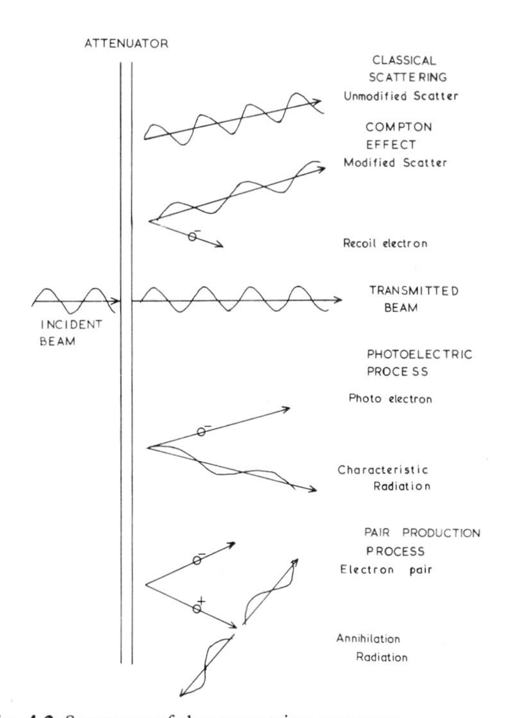

Fig. 4.2 Summary of the attenuation processes.

Partial attenuation coefficients

A better quantitative idea of the relative importance of the three main attenuation processes can be obtained by allocating to each an attenuation coefficient defined as on page 35 but referring only to one attenuation process. The total linear attenuation coefficient μ is then the sum of the attenuation coefficients for the various separate processes. We may therefore write:

$$\mu = \sigma + \tau + \pi$$

where σ is the linear attenuation coefficient for the Compton process
where τ is the linear attenuation coefficient for the photoelectric process
where π is the linear attenuation coefficient for the pair production process

Again σ the coefficient for the Compton process can be split up into $\sigma_a + \sigma_s$ where σ_a is the attenuation coefficient corresponding to energy absorbed in the production of recoil electrons and σ_s the coefficient corresponding to energy appearing as modified scatter radiation.

Corresponding mass attenuation coefficients independent of the density of the attenuating medium can be also given for these separate processes. Thus:

$$\frac{\mu}{\rho} = \frac{\sigma}{\rho} + \frac{\tau}{\rho} + \frac{\pi}{\rho}$$

where σ/ρ, τ/ρ and π/ρ represent mass attenuation coefficients for the Compton effect, the photoelectric effect and pair production respectively.

The mass attenuation coefficients for each process vary in different ways with the photon energy (E) of the incident radiation and with the atomic number Z of the attenuator. The mass attenuation coefficient for the Compton effect (σ/ρ) is almost independent of Z and is approximately proportional to the wavelength (λ), i.e., inversely proportional to the photon energy (E). For the photoelectric effect τ/ρ is approximately proportional to $1/E^3$ and Z^3. For pair production π/ρ is proportional to Z and increases rapidly with the increase of photon energy above 1·02 MeV.

The *total* mass attenuation coefficient, being the sum of the three partial attenuation coefficients, varies in a complicated way with both E and Z but for high Z and low E the rapid variation of τ/ρ (the photoelectric effect) with E and Z is the predominating influence. For higher photon energies (short λ) and low atomic numbers (Z) σ/ρ, (the Compton effect) is the important influence. This remains so until very high photon energies and high atomic numbers are involved when the pair production process becomes important.

Figure 4.3 shows the relative magnitudes of the three partial mass attenuation coefficients for soft tissues over the complete range of X-ray photon energies, together with the total coefficient. It illustrates the predominance

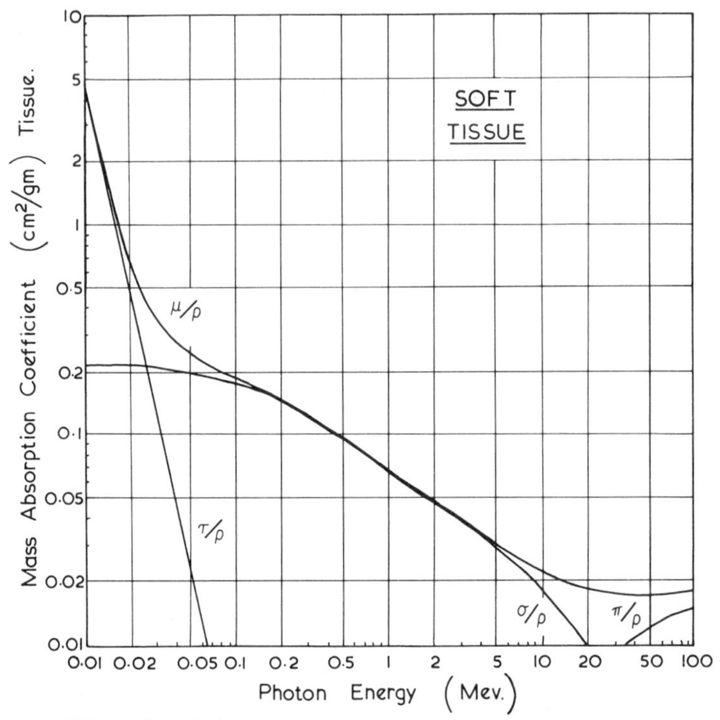

Fig. 4.3 Mass attenuation coefficients for soft tissues.

of the photoelectric effect in soft tissue only below about 30 keV photon energy and the unimportance of the pair production effect below 20 MeV photon energy.

Real Energy Absorption

When a beam of electromagnetic radiation is attenuated in a medium it will be seen that all the interaction processes involved give rise to some form of electromagnetic radiation as a secondary product. This radiation does not cause the direct deposition of energy to the attenuating medium, though as scattered radiation it has important effects both in diagnostic radiology and in radiotherapy and is very important from the point of view of radiation hazard control. Energy is only transferred to the attenuating medium through the production of fast electrons which produce ionisation in the medium. Real energy absorption arises therefore from photoelectrons, the recoil electrons of the Compton process and the pair of electrons in the pair production process.

In terms of partial attenuation coefficients therefore real energy absorption over the main range of photon energies used in radiotherapy can be represented by the coefficients: $\tau/\rho + \sigma_a/\rho$ where τ/ρ is the mass photoelectric attenuation coefficient and σ_a/ρ is the mass attenuation coefficient representing the conversion of the incident photon energy to energy of recoil electrons in the Compton process. We have seen that σ_a is an increasingly high proportion of the coefficient σ for the whole Compton attenuation process as the photon energy increases. For very high photon energy beams where the pair production process becomes important an additional contribution to real energy absorption in the medium will arise from the kinetic energy of the two electrons produced in the interaction. Calculations on the real energy absorbed using these partial attenuation coefficients are relevant to the problem of dose measurement discussed in Chapter 5.

Electron Ranges in Tissue

The ranges of the electrons responsible for the ionisation in the attenuating medium and therefore for the deposition of energy, vary with the incident photon energy. For X-rays in the diagnostic range the ranges in tissue are very small but at high quantum energies used in megavoltage therapy the electron ranges become larger and of the order of several millimetres. They thus become comparable with the size of the coarse structure of the tissues being treated. The range then has profound effects on the distribution of absorbed energy in tissues. This will be dealt with in Chapter 5.

Table 4.1 gives some examples of the electron ranges involved in the attenuation processes.

Table 4.1

Photon energy of incident radiation	Energy of electrons	Range in tissue
50 keV	photoelectron 50 keV	42 μm
	recoil electron 4–8 keV	0.5–1.7 μm
250 keV	recoil electron 50 keV	42 μm
1 MeV	recoil electron 500 keV	2 mm
8 MeV	recoil electron 5 MeV	28 mm

Measurements of radiation

Later chapters of this book will describe in detail the methods by which reliable quantitative measurements of X- and gamma rays can be obtained and used in radiotherapy. It can be seen from the description of the processes of interaction between photon beams and matter given in this chapter that two concepts of measurement are possible. It is possible, for example, to seek a quantitative estimate of the total flow of photons at a point in space or in tissue being irradiated by an X- or gamma ray beam. In practice this flow of photons is measured by ionisation produced in air or in an appropriate air cavity within an irradiated medium. The quantity so measured is referred to as *Exposure*. On the other hand energy is only transferred to a medium subject to this irradiation when secondary electrons are produced which deposit energy. This energy is the absorbed energy and a quantitative estimate of this is referred to as the *absorbed dose*. Both of these two concepts of measurement are in use and it is important to distinguish them and the measurement units involved. An illustration might help to clarify the difference. When one moves round a garden in sunlight one is bombarded from all sides by photons of light of a variety of wavelengths. Some of these are the primary photons from the sun and some are scattered photons from the environment. This photon flux is equivalent to exposure as referred to above. However some of these photons enter the eye and by producing chemical changes on the retina impart visual information. Similarly some enter the skin and produce chemical changes leading to pigmentation. This is equivalent to absorbed dose. Exposure and absorbed dose are clearly related to each other. The second is proportional to the first but the constant of proportionality will depend on a variety of conditions. In tissues being irradiated the constant of proportionality will depend on the relative magnitudes of the different processes of attenuation and therefore on the photon energy distribution in the irradiating field and on the atomic number of the medium. Figure 4.3 has illustrated that the relationship between exposure and absorbed dose can be expected to be complex.

The concept of exposure—the roentgen

Exposure is a measure of the flux or flow of X- or gamma-ray photons to which the point of interest is subjected in a given time. It is measured by observation of the amount of ionisation produced in a unit volume of air under standard conditions. The formal definition of exposure as adopted by the International Commission on Radiological Units in 1962 is as follows:

Exposure (X) is defined as $\Delta Q/\Delta m$ where ΔQ is the sum of the electrical charge on all the ions of one sign produced in air when all the electrons liberated by photons in a volume element of air whose mass is Δm are completely stopped in air. The special unit of exposure is called the *Roentgen* (R) and corresponds to 1 esu of charge per 0·001293 gm of dry air or $2{\cdot}58 \times 10^{-4}$ coulombs per kilogram of dry air.

The definition as quoted above is a strict and formal one. It replaced an earlier definition which expressed the same meaning in less precise terms. The old definition made more clear the experimental arrangements under which the measurement is to be made. For this reason the old definition is quoted as follows. 'The roentgen is that amount of X-, or gamma, radiation such that the associated corpuscular emission per 0·001293 g of air, produces in air ions carrying one electrostatic unit of charge of either sign.' The description of the standard ionisation chamber (see p. 41) will illustrate how the measurement can be made in practice.

The concept of absorbed dose—the rad

In X-ray or gamma ray therapy it is possible by methods described in Chapter 5 to estimate the exposure or the exposure rate at any point in the radiation field. The actual energy deposited by secondary electrons at that point is the quantity of real interest. This is the absorbed dose and the formal definition of this is the energy imparted to matter by the ionising radiation per unit mass of the irradiated material at the point of interest. The ICRU have defined absorbed dose formally as:

$$D = \frac{\Delta E_d}{\Delta m}$$

where ΔE_d is the energy imparted by ionising radiation to the matter in a volume element of mass Δm. The special unit of absorbed dose is the *rad* where

1 rad = 100 ergs per gram or 0·01 joule per kilogram

A new unit of absorbed dose is now being introduced which is a simple multiple of the rad.

1 Gray = 1 joule per kilogram i.e. = 100 rads.

Absorbed doses are in many cases very difficult to measure and have to be calculated. The basis of the calculation is that the exposure of one roentgen corresponds to the production of a certain amount of ionisation, i.e. the expenditure of a certain amount of energy, in a unit mass of air. In media other than air, such as tissues for example, a different amount of energy will be absorbed if the medium is subjected to the same exposure. This energy absorption is defined by the appropriate mass absorption coefficients for the particular medium and for the particular beam since these coefficients depend on Z of the medium and λ of the radiation. Conversion of exposure in roentgens to absorbed dose in rad, therefore, involves a comparison of the real mass absorption coefficient for air and the medium. This conversion is described in Chapter 5 (see p. 57).

5. The measurement of X- and gamma ray beams

Introduction

The interaction processes discussed in the last chapter showed that the effect of radiation interacting with matter is to produce ionisation. It is this ionisation which precipitates biological damage in living tissues and which forms the basis of measurement of radiation.

Radiation beams can be produced in various ways and in various forms and the only way to be sure of the beam intensity is to measure it. The most direct method is to measure the ionisation produced in air by counting the number of electrons released from the defined mass of air contained in an ionisation chamber. The concept of *exposure* was introduced in the last chapter along with its unit of measurement, the *roentgen* (R). It will have been noted that the exposure and the roentgen are only defined in terms of the effect of X- and gamma radiation in air. The concept is of limited value, but, nevertheless, valuable in the measurement of X-ray beam intensity and in the standardisation of dose measuring instruments.

A more versatile and significant quantity is based on the concept of *absorbed dose* and its measurement in *rads* (rad). It is more versatile in that it may be used to measure any form of ionising radiation (photon beams as well as particle beams) in terms of the specific energy absorbed in any medium. The energy absorbed per gramme is a more accurate measure of the biological effect than the charge liberated per gramme of air (exposure). The conversion from exposure to absorbed dose will be considered later in the chapter.

The dependence of the attenuation coefficients on photon energy and atomic number is seen in this chapter in two forms of differential absorption – (i) different tissues absorbing different fractions of energy from the same beam of photons and (ii) the same filter absorbing different fractions of photon beams of different energy.

Finally the distribution of dose absorbed in homogeneous and inhomogeneous media will be discussed in this chapter while the use of this information for radiation treatment planning will be the subject of the next chapter.

The standard free air ionisation chamber

The standard free air ionisation chamber is a purely laboratory instrument and is of no practical value in the radiotherapy department. It is, however, a fundamental instrument in that it is a practical interpretation of the definition of exposure enabling measurements to be made of the charge liberated from a known mass of air. Ionisation chambers of this type can be found at national standardising laboratories and form the basis for the standardisation of the day-to-day ionisation chambers used routinely in radiotherapy departments and elsewhere (p. 55).

The measurements are made under strictly standardised conditions. A beam of radiation is passed between two parallel plates (electrodes) under conditions such that the volume where the primary ionisation is taking place is known accurately. The ions produced are collected and measured. The apparatus is illustrated in Figure 5.1.

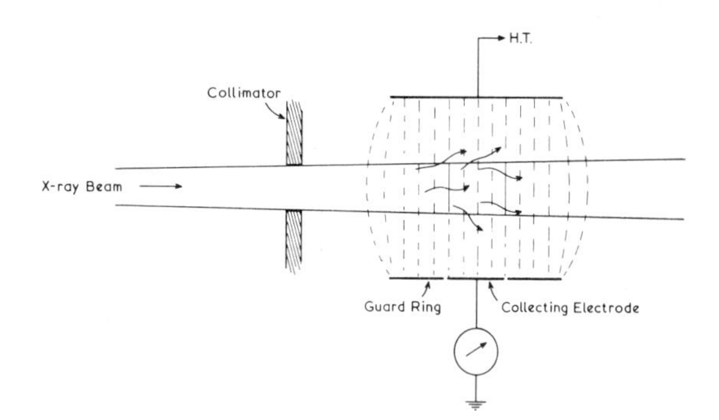

Fig. 5.1 The standard free air ionisation chamber.

A narrow and accurately defined beam of radiation passes between two parallel electrodes. Between the two electrodes a potential difference is maintained so that ions of one sign are collected on one electrode while those of the opposite sign travel to the other. The potential difference between these electrodes must be sufficient to collect all the ions liberated—and to prevent any recombination of the (negative) electrons and positive ions—that is the ion chamber must operate under *saturation conditions*. This is a very important precaution that must be observed when ionisation chambers of any form are employed. It should also be noted that if the rate at which the air is ionised is increased then to maintain the saturation condition, the potential difference will need to be increased. One electrode consists of two sections, an inner collecting electrode and insulated from it an annular electrode known as a *guard ring*. The collecting electrode and the guard ring are maintained at the same potential by an external connection. When an electric field is maintained between parallel plate electrodes, its strength is uniform over the central area but weaker at the edges. Alternatively, it can be said that the lines of electrostatic force over the central area are straight and perpendicular to the electrodes while at the edges they are not. Isolating the collecting electrode from the guard ring by a narrow insulated region guarantees the uniformity of the electric field and, in particular, the lines of force will become straight and perpendicular over the whole area of the collecting electrode. By this means a volume of air is defined. The cross-sectional area of the beam is predetermined by the accurate collimation and the length of the beam within the measuring system is determined by the size of the collecting electrode. The mass of air in the volume so defined can be calculated from the known ambient conditions (of atmospheric temperature and pressure) and the electric charge liberated is measured directly on the collecting electrode. This charge is small and very sensitive instruments are required.

In using the free air chamber, care must be taken to ensure that the distance between the plates is so large that the high-speed electrons produced in the central region of the chamber lose all their kinetic energy in the production of ions before they hit the electrodes. For X-rays generated at 200 kV, the plates will be of the order of 20 cm apart, the separation increasing with photon energy. The overall chamber size is, therefore, large and it is this size which places the rather arbitrary 3 MeV limit on the definition of the roentgen.

In addition the collimating system and the exit port of the chamber must be far enough from the collection volume so that electrons released from them lose all their kinetic energy before reaching the collection volume. This isolation of the collection volume from the surrounding components by a wall of air of a thickness which is greater than the range of the most energetic electrons, is the reason for describing the chamber as 'free air' and an appreciation of this requirement will assist the reader to understand the need for an *air equivalent wall* in more practical chambers.

It will be appreciated that photo- and Compton electrons will be released at all points within the beam and not only in the collection volume and that these electrons will follow tortuous paths of considerable length before losing all their kinetic energy by producing further ionisations—each will produce several hundred further ionisations. It is inevitable that some electrons originating outside the collection volume will enter that volume and contribute to the collection of charge, but fortunately the converse is also true, i.e., electrons originating inside the volume will stray outside and the ions these produce will be collected by the guard ring and lost; what is gained is balanced by what is lost. A state of *electronic equilibrium* is said to exist. It can, therefore, be assumed that the ions collected by the collecting electrode and therefore measured, all originate from the collection volume.

The thimble ionisation chamber

Three aspects of the free air chamber must be applied to practical chambers, these are:

1. The chamber must operate under saturation conditions
2. There must be an air (equivalent) wall
3. The air mass must be corrected for the ambient temperature and pressure.

For clinical purposes the exposure at a point can be measured using a *thimble ionisation chamber*. This term describes an ionisation chamber in which a small volume of air is enclosed in a thimble-like conducting cap with an insulated axial collecting electrode (Figure 5.2). The cap is often made of a unit density material such as graphite, nylon or bakelite made conducting by a layer of colloidal graphite on the inner surface. The central electrode may be aluminium. The insulator may be amber but perspex and polythene are often used. If a potential difference (usually a few hundred volts) is maintained between the cap and the collecting electrode, then radiation falling on the chamber will cause the air to be ionised and an ionisation current will flow between the two electrodes.

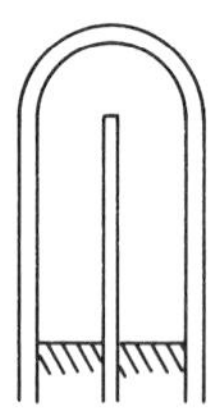

Fig. 5.2 A thimble ionisation chamber.

Such a chamber will give an accurate measure of the exposure providing (a) the walls are made of a material with the same atomic number as air and (b) that the wall has a thickness greater than the range of the electrons produced in the material. A chamber wall of unit density material 1 mm thick would be equivalent to approximately 75 cm of air (the density of dry air being 0·001293 gcm^{-3}). The actual wall thickness required is determined by the energy of the radiation to be measured. For low energy radiations a very thin-walled chamber is required as the electron range in the material is small and the radiation would be attenuated by a thicker wall before reaching the enclosed air. As the energy increases, the thickness of the wall must be increased, for example, for Cobalt-60 radiation some 4 mm would be required.

Errors are introduced into the measurement of exposure if the wall material has an atomic number which differs from that of air, especially in measurements at depth in tissue where low energy scattered radiation is present due to the pronounced photoelectric effect. At higher energies discrepancies in atomic number are less serious.

The size of the chamber depends on the sensitivity of

the measuring system and on the intensity (or exposure) of the radiation to be measured. The exposure rate in the useful beam of the therapy unit may be conveniently measured using a chamber volume of approximately 0·5 ml, while the leakage radiation through the treatment room walls will require a chamber of 500 ml.

Since the measurement is dependent on the air mass enclosed by the chamber, it is important that the chamber contains either a fixed mass of air or a fixed volume of air at the ambient atmospheric pressure. In either case the electrodes must be rigid. If the chamber is sealed so as to enclose a fixed mass of air, then fluctuations in the ambient temperature and pressure will not affect the air mass enclosed and the reading obtained will not require the air mass correction factor to be applied. If the chamber is unsealed fluctuations in the ambient conditions will affect the magnitude of the reading obtained and two correction factors will be required, namely the temperature correction and the pressure correction—together they are referred to as the air mass correction factor. The density of air at 0°C and 760 mm mercury (101.3 kPa) is 0·001293 gcm^{-3} but if the temperature rises the density falls while if the pressure rises the density also rises. The density of air and, therefore, the reading obtained in the measurement of exposure, has to be corrected by multiplying by

$$\frac{760}{P} \times \frac{(T + 273)}{273}$$

when the temperature is T°C (or $T + 273$°K) and the atmospheric pressure is P mm Hg. Some standardising laboratories use 20°C as a reference temperature, in which case the temperature factor is

$$\frac{(T + 273)}{(273 + 20)}$$

Despite the need to apply correction factors most thimble chambers are designed to be unsealed (but see p. 44).

Methods of using thimble chambers

The use of a thimble chamber under the conditions described makes possible a measurement of the exposure in roentgens at a particular point—usually the centre of the chamber. A *calibration* of the chamber is, however, required. This calibration enables the readings from the thimble chamber to be used to calculate the exposure that would be recorded at the same point using the free air chamber. Such a calibration is required because no practical chamber will have a wall which is air equivalent for more than one photon energy. The standardising laboratory will, on the other hand, compare the thimble chamber with the free air chamber over a wide range of photon energies and issue a calibration curve of the calibration factors, N, (Figure 5.3) for different photon energies or radiation qualities.

The thimble chamber may be used to measure *total exposure*, X, or to measure *exposure rate*, $\dot{X}$. From the definitions of exposure and exposure rate

$$X = \frac{\Delta Q}{\Delta m} \text{ and } \dot{X} = \frac{\Delta Q/t}{\Delta m}$$

it can be seen that to measure exposure, the total charge released, ΔQ, must be measured, while to measure exposure rate the ionisation current, $\Delta Q/t$ is required. Thus, to measure exposure, the potential difference developed across a *capacitor* is measured and for exposure rate, the pd developed across a *resistor* is measured. It is possible to determine the exposure rate by measuring the exposure and dividing by the duration of the exposure, but this assumes the exposure rate is constant—an assumption that is not always valid.

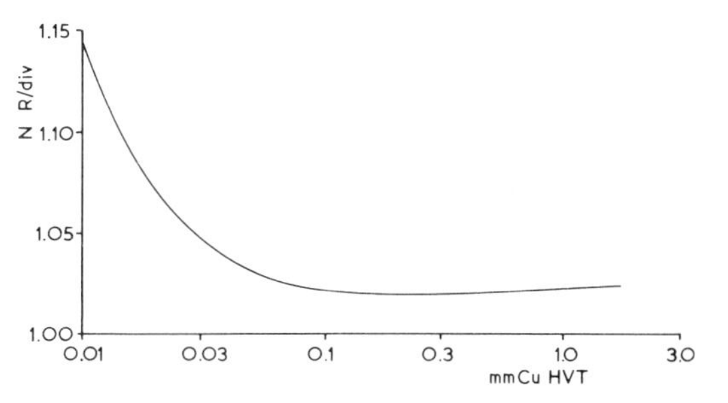

Fig. 5.3 A typical correction curve for the energy response of a thimble ionisation chamber.

The measurement of exposure

An ionisation chamber consists of two electrodes separated by a very good insulator. It may be treated therefore as an electrical capacitor and used as a means of storing electrical charge. When the chamber is irradiated, the ionisation will cause some of the stored charge to leak away and providing the chamber remains *saturated*, the charge leaked will be directly proportional to the total exposure given to the chamber and this may be measured by determining the *change* in the potential difference across the *capacitor-chamber*. This may be interpreted mathematically as follows. If the electrical capacitance of the ionisation chamber is C, then the storage of a charge Q will produce a potential difference across the capacitor of V where $V = Q/C$. If now an exposure X is given to the chamber resulting in an ionisation of ΔQ then the pd will fall by ΔV where $\Delta V = \Delta Q/C$. Thus, the exposure, X, will be given by $C\Delta V/\Delta m$, where Δm is the mass of air enclosed in the chamber.

This type of capacitor-chamber has the particular advantage that it is possible to use many chambers with only one charging/measuring instrument. The chambers will not be connected to the instrument during exposure. This system can, therefore, be used to measure the exposure at various points in the beam during a single irradiation.

On the other hand, the direct connection to the measuring device via a *screened lead* enables the exposure to be integrated during the irradiation which can then be terminated at a predetermined exposure. Such a simple direct reading exposure meter is illustrated in Figure 5.4.

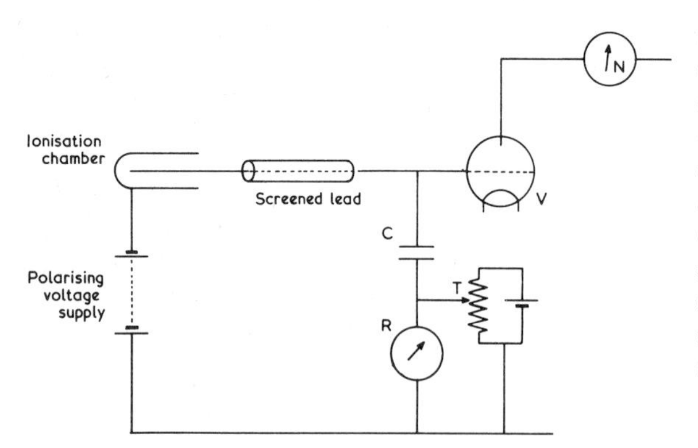

Fig. 5.4 A simplified schematic circuit for an integrating exposure meter.

The circuit shows a simplified version of that used in the popular Farmer exposure meter—for many years used as a secondary standard instrument. The chamber is connected via a screened flexible cable to the exposure meter. The polarising voltage causes the ions to flow through the collecting electrode onto the capacitor C, thereby increasing the potential difference across the capacitor. This rise in the grid voltage applied to the (triode) electrometer valve V, causes the null meter N, to deflect. This deflection is corrected by adjusting the potentiometer T (sometimes referred to as a Townsend balance) to lower the grid voltage to its initial value. The voltmeter R then measures the change in the pd across the capacitor C and can be calibrated in roentgens. The function of the electrometer valve V is to amplify the small change in grid voltage, thereby increasing the sensitivity of the instrument. As mentioned before, the ionisation in the chamber is very small.

The measurement of exposure rate

If the integrating capacitor C and the Townsend balance T in the circuit shown in Figure 5.4 are replaced by a very high resistor (say 10^{10} ohms), then the ionisation current will flow through the resistor. The potential difference developed across the resistor is proportional to the current flowing through it (Ohm's law) and, therefore, proportional to the exposure rate. The deflection of the meter N is then a measure of the exposure rate.

The parallel plate ionisation chamber

So far only a small thimble chamber has been discussed. In fact, provided two insulated electrodes enclose a volume of air, an ionisation chamber can be almost any desired shape. In practice the radiation beam monitor or dose rate meter incorporated into the X-ray head of a therapy unit is invariably a parallel plate ionisation chamber.

Two or three thin electrodes of aluminium or graphite coated perspex are used to enclose a volume of air close to the X-ray target but on the patient side of any added filters. The electrodes of the ionisation chamber may constitute part of the added filtration. The electrodes are large enough to extend over the whole of the useful beam defined by the primary collimator. If only two electrodes are used then one will be at HT (the polarising voltage) and the other at earth potential and either one used as the collecting electrode (Fig. 5.5a). Where three electrodes are used, then the centre electrode will be at HT and used as the collecting electrode, enabling the whole of the outer surface of the chamber to be connected to earth potential (Figure 5.5b).

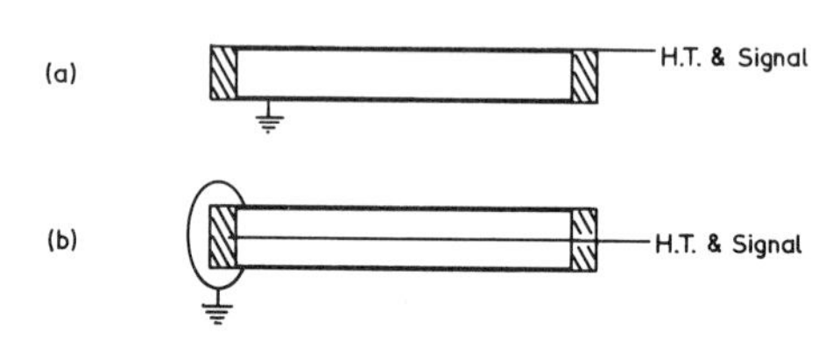

Fig. 5.5 Typical beam monitor chambers used to monitor the exposure rate from therapy units, (a) two plate, (b) three plate.

Parallel plate chambers may be used to monitor exposure or exposure rate as before, but if they are to be used to actually measure the dose given to the patient, then the chambers must be *sealed* to obviate the need to correct for temperature and pressure changes—the temperature of the air in the chamber may be higher than that of the ambient room temperature owing to its position in relation to the target.

Alternative methods of dose measurement

The primary effect of the radiations in which we are interested is the production of ionisation. This ionisation may be revealed as electrial conductivity as in air, or it may be revealed in a number of other ways such as physical or chemical effects in solids or liquids. Any of these physical or chemical effects which can be estimated quantitatively can be used as a dose-measuring system and in the past most of them have been tried. A satisfactory dose-measuring system requires, however, that the effect observed does not show too great a variation of sensitivity with wavelength. This is because of the continuous nature of the X-ray spectrum and the presence of scattered radiation in the beam under certain conditions. For high-energy gamma-rays the beam consists initially of radiation of a single wavelength or of a few discrete wavelengths but scattering again changes the spectral distribution and in measurement a reasonable wavelength independence is desirable. Some of the methods used in the past have now only a historical interest but some methods, other than that of ionisation in gases, are quite important. A brief account of these will now be given.

Biological method
A biological method of measuring dose was attempted in the early days of radiotherapy using the property of the radiations to produce a reddening of the skin—the skin erythema. This biological reaction is described in detail in Chapter 16. In using this reaction as a dose-measuring system, a unit of dose was defined as the amount of radiation which just produced an appreciable reddening of the skin. The method was extremely inaccurate because of the difficulty of determining the threshold erythema and its variability from one person or one site to another. The method, however, could not be continued as a dose-measuring system since it was soon established that different amounts of X-ray energy are necessary to produce the erythema when beams of different effective wavelength are used.

Chemical Method
Chemical methods of measuring dose were in use for some time in the so-called 'Pastille Dose'. A layer of crystalline barium platino-cyanide changes colour from yellow to brown under exposure to an X-ray beam. A unit of dose was defined as the amount of radiation necessary to produce a given change of tint in a specimen of the material held half-way between the skin and the target. As a method of dosimetry for radiotherapy, this method has long since been abandoned because of the fact that since the material contained elements of high atomic number, it did not respond equally to radiations of different wavelengths. It did not, therefore, record equal dose to tissues when a given colour change was produced by beams of different spectral distributions.

The production of chemical effects by X- and gamma-ray beams requires in general very large doses and chemical methods of measurement are, therefore, relatively insensitive. There has been a great revival of interest in chemical doesmeters recently, however, since they can be used under two conditions in which ionisation methods are somewhat unsatisfactory, both arise out of the development of nuclear energy. They are the measurement of doses of thousands or millions of rads, such as might be experienced in the neighbourhood of massive radioactive sources and the measurement of extremely high dose rates such as might be experienced in the neighbourhood of a nuclear explosion. Chemical dosimeters can be made relatively simply to work under these conditions.

The ability of ionising radiations to promote the oxidation of ferrous compounds has been known for a long time as one of the many chemical effects of radiation. For dose-measuring purposes a solution of ferrous sulphate in sulphuric acid is irradiated and the amount of ferric sulphate produced is estimated by making a quantitative estimate of the colour change. This is generally done by measuring the absorption by the ferric ions of a beam of light of suitable wavelength using a spectrophotometer.

A more complicated chemical dosemeter is a mixture of chloroform and an aqueous solution of a suitable indicator dye sealed in a glass container together with a small amount of a stabilizing agent, resorcinol. Radiation produces acids in the chloroform. Shaking the mixture separates the acids into the aqueous layer and a visible colour change is produced in the dye. This type of dosimeter is much more sensitive than the ferrous sulphate one and can be used to measure doses in the range 50 rad to 10 000 rad. These dosemeters are useful for measuring bursts of radiation of several hundred rads at high intensity.

Calorimetry
A direct method of measuring the real energy absorbed from an X-ray beam in a medium would be to measure the total amount of heat developed. There are great difficulties in making a direct measurement of absorbed energy in this way, the major one being the small temperature changes involved. A beam of X-rays giving a soft tissue dose of 500 rad, for example, causes a temperature rise of only 10^{-3}°C. In spite of the small temperature rise, however, methods of heat measurement are now sensitive enough for a new interest to be taken in this method of measuring dose. This interest arises because calorimetry provides a measurement of radiation energy in terms of fundamental energy units, that is, in units to which the rad is directly related.

The calorimeter has been used, for example, to measure the energy in the beam of gamma-rays from a cobalt-60 therapy unit. The absorber was a block of lead suspended in a vacuum inside a well-insulated box. The heat-detecting element was a thermistor—a semiconducting material whose resistance changes very rapidly with temperature. Its resistance changes were detected electrically by using it as one arm of a Wheatstone bridge. Temperature changes of about 10^{-4}°C per minute were measured.

Scintillation detectors
The property of ionising radiations to cause flashes of light or scintillations in suitable crystals has been used to produce a very sensitive type of dosemeter for gamma ray measurements. The scintillations in a small crystal which is itself transparent to ordinary light, are detected by placing the crystal near to the cathode of a photomultiplier. A photomultiplier is a photosensitive device in which the electrons emitted from the photocathode are accelerated on to a series of grids. Each grid is itself sensitised so that each electron falling on it produces some three or four more secondary electrons which themselves fall on the next grid with similar effect. The result is that after pass-

ing through eight or ten grids the electron stream arising from a single scintillation in the crystal is large and can easily be detected. The rate at which the scintillations occur can be measured by electrical counting of the charge pulses coming out of the photomultiplier. Alternatively, the total light produced in the scintillation can be measured as a current proportional to the rate at which energy is absorbed in the crystal. A robust current meter can be used for this purpose if some amplification of the output current is applied.

This device is of extreme importance as a detector of gamma-radiations from radioactive materials. For the measurement of gamma-radiations the scintillation device is one of extreme sensitivity but suffers from the disadvantage that it is wavelength dependent. An anthracene crystal of 1 cm^3 volume can be used to measure dose rates as low as10^{-3} rad h^{-1} and there is no real limit to the high dose rates that can be measured since the crystal volume can be made very small and the photomultiplier can be used under conditions of relatively small amplification. The photomultiplier, however, must be shielded against the radiation being measured.

If an organic crystal like sodium iodide (activated by the introduction of a small percentage of thallium) is used, still greater sensitivity is obtained. This increased sensitivity arises because of the photoelectric absorption by the atoms of high atomic number. It is, therefore, very wavelength dependent but as a detector this crystal is of extreme importance.

An instrument for the measurement of dose rates in body cavities during radium treatment uses a crystal mounted at the end of a thin stainless steel tube which can be inserted so that the crystal is at the point at which the dose is to be measured. The light flashes from it are conducted to the sensitive photocathode of the multiplier either by reflection along the inner polished walls of the tube or along a transparent rod of a material such as perspex inside the tube. This 'light piping' does involve some loss of light but there is generally ample sensitivity in hand in these cases to allow a probe length of perhaps 25 cm and still give good readings of the dose rate at the point required.

Scintillation counting is dealt with more fully in Chapter 9.

Photographic methods

The effect of X- and γ-rays in producing blackening on a photographic film may also be used to estimate dose, especially if the amount of blackening is measured with a densitometer. A densitometer causes a pencil beam of light to pass through an area of the film on to a photocell which measures the intensity of the light transmitted by the film in the form of an electric current, the current being proportional to the intensity. The degree of blackening is expressed as an *optical density*, which is the logarithm (to base 10) of the light intensity transmitted by an unexposed area of the film divided by that transmitted by the exposed part of the film. Thus, an optical density of 1 means that only one-tenth of the light is transmitted, while a density of 2 transmits only one hundredth. Over a limited range density is proportional to dose. The properties of photographic film exposed to ionising radiation are dealt with more fully in Chapter 7.

The photographic method suffers from several disadvantages. It is difficult to obtain reproducible and accurate results even though great care is taken to develop films under standard and controlled conditions. For instance, variations of developer strength and temperature and of processing techniques make great differences to the density. Films also show great differences in sensitivity to radiation over the wavelength range in which measurements are needed. A typical film produces a density of 1 with an exposure of about 0·24 rads of 100 kV radiation, or with an exposure of about 9·0 rads of gamma-rays, the difference arises from the photoelectric absorption in the silver halide crystals of high atomic number (Z).

The photographic method is nevertheless a very sensitive one and is very useful when small quantities of radiation are to be measured, particularly if the time involved in the exposure is long and if a high degree of accuracy is not required. These are the conditions involved in the measurement of radiation exposure for protection purposes and the use of photographic film for this purpose is described in Chapter 11.

Thermoluminescence

The transition of an electron from an outer orbit to an inner orbit and the consequent emission of a characteristic X-ray photon has already been discussed (Ch. 4). Similar transitions take place between the outer orbits with the corresponding emission of characteristic radiation. In a crystal, these outer orbits are shared by the atoms and again transitions can occur between them. If the emission is in the visible part of the spectrum the phenomenon is known as *luminescence*. For example, sodium iodide has been cited above as a crystal which produces scintillations under irradiation. In Chapter 7 it will be seen that intensifying screens enhance the photographic action of X-rays by emitting visible light of a wavelength to which X-ray film is particularly sensitive and fluoroscopy (or 'screening') is a means of making an X-ray image visible to the human eye. In each of these instances, the visible light emission is an immediate and instantaneous response to the absorption of an X-ray photon. This is *fluorescence*. Another form of luminescence of particular interest in this section is that of *thermoluminescence*.

Certain crystalline materials such as *lithium fluoride* can absorb X-ray energy and store that energy at room temperature for a very considerable time, many months in fact. On the atomic level this process can be explained briefly as follows. When X-ray energy is absorbed, secondary electrons are lifted from the outer electron orbits into another orbit which is normally empty in a non-conducting material. (In a conducting material, this orbit will contain a certain number of *free* electrons which carry the flow of charge we refer to as an electric current.) Since the electron orbits are not uniquely defined they are referred to as *bands*, the normally *filled band* and the empty *conduction band* respectively with a *forbidden zone* between them, as between any two electron shells. In a pure crystal the electrons will fall back from the conduction band into the *holes* they left in the filled band. If an impurity is deliberately added to the crystal—manganese, for example—some of these electrons will fall into the *impurity traps* which now lie in the forbidden zone. Here the energy is stored. This energy is only released when sufficient heat is applied to lift these trapped electrons into the conduction band a second time from where they may return to the filled band emitting their excess energy in the form of photons of visible light. This is thermoluminescence. The intensity of the light output is small but it is proportional to the X-ray energy previously absorbed and hence the value of the technique as a means of measuring radiation dose.

The dosimeters may be in the form of *powder*, impregnated plastic *discs*, extruded *chips* or *rods*, each containing some 50 to 100 mg of phosphor material. After irradiation, the phosphor is placed in a planchet in the 'TLD Reader' where it is heated to 300°C in an oxygen-free (nitrogen) atmosphere (Fig. 5.6). The light output is measured using a photomultiplier and amplifier feeding a digital display. The system is essentially a comparator and any measurement of dose relies on the readout of identical dosimeters exposed to known doses of the same quality radiation. Lithium fluoride has an atomic number close to that of soft tissue and therefore its response is almost independent of the photon energies encountered in the radiotherapy department. Doses from tens of millirads to hundreds of rads may be measured enabling the system to be used for measuring patient doses, doses to sensitive organs outside the treatment beam and for personnel monitoring (see Ch. 11). Crystals of lithium borate and of calcium fluoride may be used as alternatives to lithium fluoride.

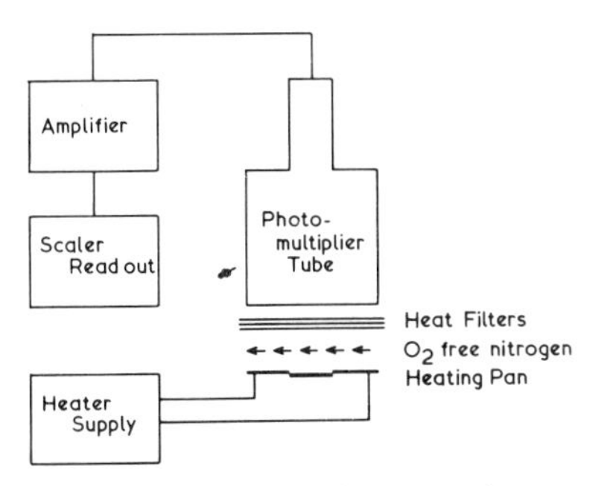

Fig. 5.6 A thermoluminescent dosimetry read-out system.

The quality of an X-ray beam

To be able to measure the exposure in air at a point outside the patient is one thing, but to be able to predict the dose at a point inside the patient is another and one of equal importance. The dose at such a point depends on many factors:

1. The ability of the primary radiation to penetrate the tissues
2. The contribution made by the radiation scattered by the surrounding tissues
3. The dose delivered to the surface (or just below the surface) of the patient.

The penetrating ability of the primary radiation is often referred to as the *quality* of the beam as it emerges from the X-ray head. It has been seen in previous chapters that the X-ray spectrum contains photons of many wavelengths and the extent to which any of these interact with matter depends on the wavelength and on the atomic number of the attenuating medium. If the photons are not absorbed they will be transmitted to subsequent layers of matter. Therefore, to know fully the penetrating ability of an X-ray beam requires a knowledge of the number of photons of each wavelength and the method by which each wavelength is absorbed and attenuated. Fortunately, for radiotherapeutic purposes such a detailed knowledge is not required and some approximation is adequate.

The beam quality of an X-ray beam generated at less than one million volts is stated in terms of either:

1. The generating voltage (kVp) and the HVT, or
2. The HVT and the homogeneity coefficient.

Above one million volts the statement of the X-ray beam energy in MV is adequate and in the case of gamma-ray beams the isotope and its mass number should be specified.

The HVT is the thickness of a stated material required to reduce the intensity of the beam to half its value. The *homogeneity coefficient* is the ratio of the HVT divided by the thickness of the same material required to further reduce the intensity of the beam from 50 per cent to 25 per cent. This is sometimes referred to as the second HVT. For a homogeneous (monochromatic) beam (eg, Cs-137) the first and second HVT will be the same and the homogeneity coefficient will be equal to one. For X-ray therapy beams it is generally between about 0·5 and 0·7.

The X-ray beam quality depends on various factors:

1. The accelerating voltage
2. The voltage waveform
3. The target material
4. The inherent filtration
5. The added filtration

Taking these very briefly in turn, the accelerating voltage (kVp) determines the minimum wavelength in the spectrum and all the longer wavelengths will be present. If the generator is a constant potential then this minimum wavelength will be present throughout the exposure, whereas if a half-wave rectified circuit is used, it is only present for one instant in every 1/50 s giving the spectrum proportionately more lower energy photons. The target material will superimpose on the 'white' spectral distribution its own characteristic line spectrum (Chapter 1). Filters can be used to remove the low energy photons and their effect will now be discussed in detail with particular reference to the HVT.

Filtration

The mass attenuation graph of Figure 5.7 shows that in the photoelectric region an absorber will attenuate low energy photons more readily than the higher energy photons. This property is exploited in the use of *beam hardening filters* applied to radiation beams accelerated at voltages of less than 1 MV.

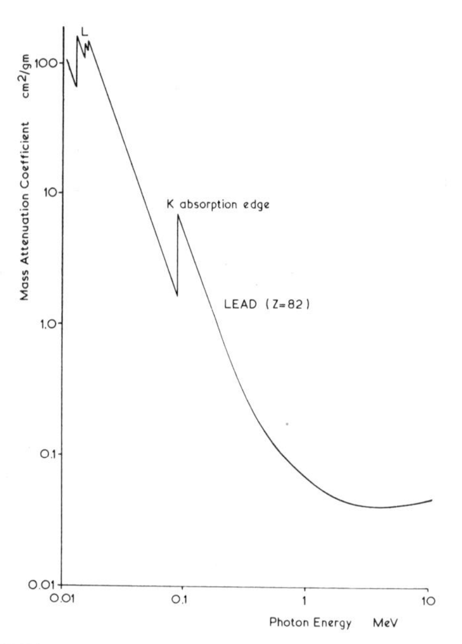

Fig. 5.7 The mass attentuation curve for lead.

In the range from about 1 to 10 MV, the absorption graph is almost flat and there is no preferential absorption. Filters, therefore, have no beneficial effect in removing the lower energy photons. Above 10 MV in the pair production dominated region, the higher energy photons are absorbed in preference to those of lower energies. Again there is no beneficial effect. Beam hardening filters are, therefore, not used at megavoltages and must not be confused with beam flattening filters (p. 24) and beam modifying filters (p. 63).

In Figure 5.8 the horizontal axes of the two graphs cover the same photon energy range. The upper graphs show the linear attenuation coefficients for aluminium, copper and tin. The lower graph shows a 250 kVp X-ray spectrum under different degrees of filtration. The unfiltered spectrum shows that the majority of the photons are of such low energy that they will not penetrate to a deep seated tumour but will be absorbed in the skin. In order to preserve as many high energy photons while removing the unwanted low energy photons, a filter which is highly attenuating below 40 keV and offering little attenuation to the photons in the 200 keV region is required. Aluminium ($Z = 13$) is of little value in this situation, but may be used as a filter for X-rays generated at energies up to approximately 150 kVp. Copper ($Z = 29$) may be used up to 250 kVp, but tin ($Z = 50$) is better for energies above 200 kVp. Copper and tin are never used in isolation. In the lower graphs of Figure 5.8 the upper curve represents the spectrum of a 250 kVp beam emanating from the tube housing, that is, filtered only by the inherent filtration. If a filter of 0·5 mm tin is added to the beam it hardens the beam very adequately, the spectral peak moving from 45 keV to about 115 keV, but there is still a lesser peak at 29 keV owing to the K absorption edge of tin. For this reason a copper filter, which is highly attenuating at 29 keV, is added. Copper, however, has a K-absorption edge at 9 keV which has to be filtered out using aluminium—the K-edge of aluminium is less than 2 keV and is negligible. It needs to be remembered that tin and copper not only have a low attenuation coefficient at 29 keV and 9 keV respectively, but that they can become sources of their own characteristic radiations with photon energies 1 or 2 keV below these edges. These are not shown in the diagram, but constitute an addition to the spectrum. It is important therefore, that when using a *compound filter*, the metal filters are fitted in the correct order—highest Z close to the target, lowest Z to the patient.

The tin-copper-aluminium filter is often referred to as a *Thoraeus filter* after the Swedish physicist who found the most efficient filter for his 200 kVp X-ray beam was 0·4 mm Sn + 0·25 mm Cu + 1·0 mm Al. By 'efficiency' in this connection we mean getting the desired hardening effect with the minimum loss of beam intensity. Thoraeus found that his combination gave the maximum output dose rate.

Beam hardening filters can, therefore, improve the relative penetration of the beam and raise the value of the homogeneity coefficient. The actual measurement of these parameters will now be considered.

The measurement of HVT

The method for measuring HVT is the same over the whole range of energies found in the radiotherapy department providing the characteristics of the radiation are observed, namely that the ionisation chamber wall is air equivalent over the range of photon energies being measured—the same chamber could not be used for both 10 kV and 10 MV. It should also be remembered that the penetrating ability of the beams will vary considerably and

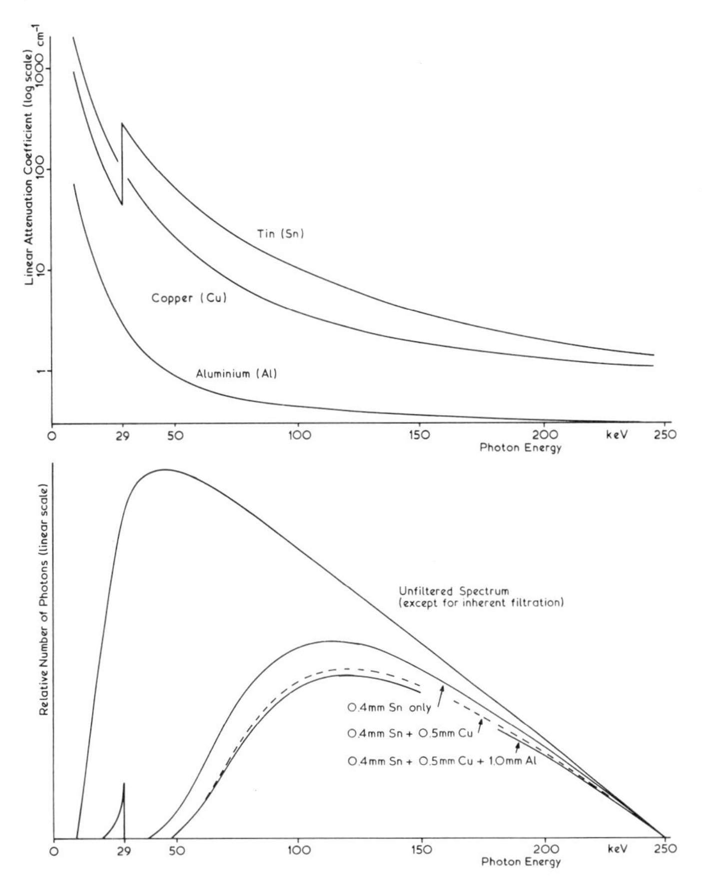

Fig. 5.8 The effect of filters, the upper curves show the linear attenuation coefficients for the filters used on the unfiltered X-ray spectrum shown below. The copper filter reduces the 29 keV peak.

the choice of absorbing material is important—those normally used are shown in Table 5.1.

The apparatus required is shown schematically in Figure 5.9. The HVT is determined by measuring the intensity of the beam transmitted through various thicknesses of material and interpolating to find the thickness that would transmit 50 per cent of the intensity measured with zero thickness of material added. Therefore, the following form the basic requirements. A means of collimating the beam, C, to a small size but large enough to cover the ionisation chamber, D, which will be referred to as the detector. The error resulting from the use of a large beam size will be explained later. If the exposure cannot be given very accurately a second monitor chamber, M, will be required on the tube side of the small beam collimator to detect the variation in the exposure incident on the filters,

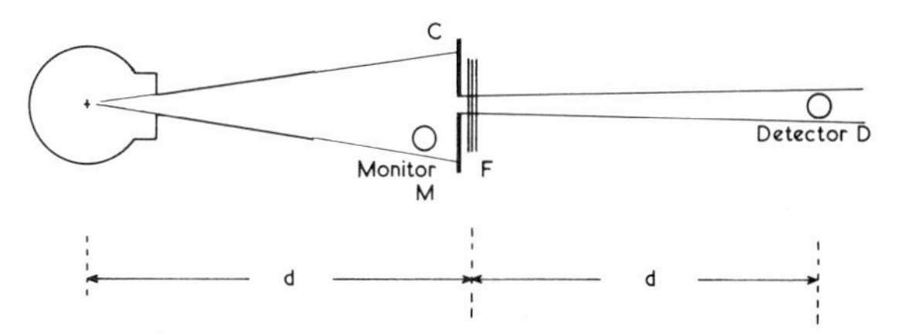

Fig. 5.9 The layout for the measurement of half value thickness.

Table 5.1 Typical values of added filtration and the specification of beam quality

	kVp	Added filter	HVT	Effective photon energy	Equivalent wavelength
Grenz rays	10	Nil	0.02 mm Al	6 keV	0.207 nm
Superficial X-rays	100	1 mm Al	2.5 mm Al	30 keV	0.041 nm
Orthovoltage X-rays	250	1 mm Cu + 1 mm Al	2.1 mm Cu	108 keV	0.011 nm
	300	0.8 mm Sn + 0.25 mm Cu + 1 mm Al	3.8 mm Cu	155 keV	0.008 nm
Cobalt 60 γ-rays	—	Nil	10.4 mm Pb	1.25 MeV	0.001 nm

F, as they are added during the experiment. The focus-to-filter distance should be approximately equal to the filter-to-detector distance and greater than the range of the electrons emitted from the filter material. With the apparatus carefully and rigidly aligned and a table in which to record the results (Table 5.2), the measurements can begin. (If the monitor, M, is not used then it is assumed that the readings that would have been obtained would all be equal, namely $M_0 = M_1 = M_2$ etc. in Table 5.2 and columns 4 and 5 would not be required.) The maximum thickness of the filters to be added will depend on the radiation and material being used and the exact purpose of the experiment but the transmission should be reduced to between 25 and 10 per cent. The calculation of percentage transmission for the final column of the table is:

$$\frac{D_n}{D_o} \times \frac{M_0}{M_n} \times 100$$

Table 5.2 Measurements taken to determine the HVT

Added filter	Total filtration added (t)	Detector Reading D	Monitor Reading M	Ratio D/M	Percentage transmission
Nil	Nil	D_0	M_0	D_0/M_0	100
t_1	t_1	D_1	M_1	D_1/M_1	
t_2	$t_1 + t_2$	D_2	M_2	D_2/M_2	
etc.	etc.				

where D_n and M_n are the readings taken on inserting a thickness t_n making the total filtration equal to t. On completion of the readings and calculations, a graph of the percentage transmission through the chosen filter material is plotted as in Figure 5.10a. If possible, the logarithm of the percentage transmission should be plotted as in Figure 5.10b. From these graphs it can be seen that 1·0 mm is required to reduce the intensity to 50 per cent and 2·6 mm to reduce it to 25 per cent. The HVT is, therefore, 1 mm and the homogeneity coefficient is (1·0/1·6) or 0·67. This completes the measurement of HVT, but this graph reveals a good deal more. First, it has been said that the homogeneous radiation from a gamma-ray source would obey the law of exponential absorption and would have a homogeneity coefficient of unity. The law of exponential absorption states, using the same nomenclature:

$$D_n = D_o e^{-\mu t}$$

which is equivalent to saying:

$$\ln (D_n/D_o) = -\mu t$$

The straight line graph (curve B) in Figure 5.10(b) may be the graph of this equation and its gradient is $-\mu$. The gradient of our curve A is also $-\mu$, but the magnitude of μ varies from one point to the next; in fact μ decreases with increasing filtration. Now consideration of the photoelectric region shows that the linear attenuation coefficient μ (or rather μ/ρ, but ρ is constant) decreases with increasing photon energy (Figure 5.8). This, therefore, is another way of explaining that the peak of the spectrum is moved to a higher photon energy by the preferential absorption of the low energy photons by the addition of filters.

Effective photon energy and equivalent wavelength

Secondly, if $t_{\frac{1}{2}}$ is the HVT, then D_n/D_o in the above equation is 0·5 by definition and the equation simplifies to:

$$t_{\frac{1}{2}} = 0{\cdot}693/\mu \text{ or } \mu = 0{\cdot}693/t_{\frac{1}{2}}$$

since $-\ln(0{\cdot}5) = 0{\cdot}693$. From the attenuation data appropriate to the filter material μ can be used to derive the

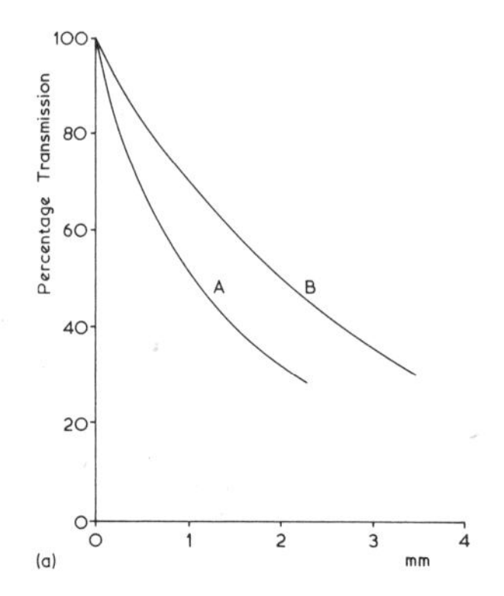

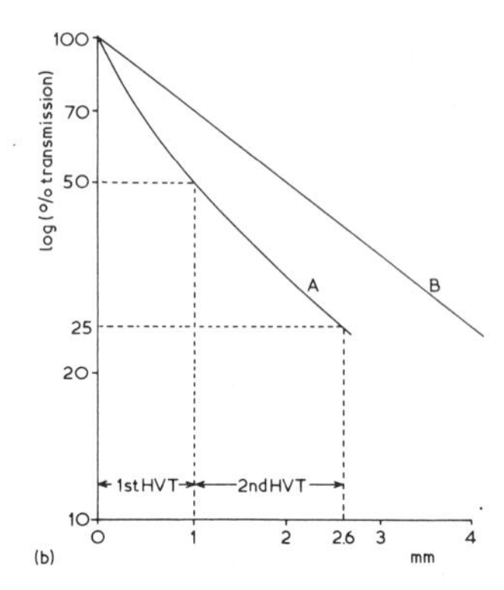

Fig. 5.10 The attenuation of a radiation beam by the addition of filters plotted (a) on a linear scale and (b) on a logarithmic scale. Curve A is for an inhomogeneous X-ray beam and curve B for a homogeneous γ-ray beam.

energy of a homogeneous beam of photons having the same HVT as the X-ray beam under examination. This energy (in keV) is known as the *effective photon energy* of the X-ray beam. This is particularly useful when attempting to compare X- and gamma ray beams.

A third alternative specification of the quality of the beam can be derived using Duane-Hunts law which relates the wavelength of the radiation, λ, to the photon energy E:

$$E = \frac{hc}{\lambda}$$

where h is Planck's constant and c the velocity of light. Thus, if E is the effective photon energy of a beam of X-rays the *equivalent wavelength*, λ, can be defined. Some typical values of effective photon energy and equivalent wavelength were given in Table 5.1.

Broad beam attenuation

Thirdly, it was stated earlier that a narrow beam should be used for the measurement of HVT. This is necessary to minimise the radiation scattered by the added filters by an excessively large beam of radiation. Such scattered radiation would increase the reading D_n which when plotted on the transmission graph would give an over-estimate of the HVT. Such *broad beam* data has its usefulness in radiation protection (see Chapter 11) but for radiotherapy treatment, narrow beam attenuation is required.

Tissue equivalent materials

Tissue equivalent phantom materials are those materials which behave in much the same way as the body tissues when irradiated. From the analysis of the absorption processes it is clear that a phantom material must have (i) an effective atomic number very close to that of the tissue it simulates because of the dependence of photoelectric absorption and pair production processes on Z, (ii) an electron density close to that of the tissue simulated because of the dependence of the Compton scattering process on the number of electrons per gramme, and (iii) because spatial measurements are to be made in the phantom material the density or specific gravity should be as close as possible to that of the tissue simulated.

Table 5.3 lists these parameters for the principal body tissues and some of the phantom materials in common use. Soft muscle tissue phantom materials are the most commonly used and water the most common of these. In many ways, water is the ideal material in that it is homogeneous and yet permits ionisation chambers (or other radiation detectors) to be moved freely within it. Mix D is a mixture of paraffin wax, polyethylene and other materials and is often used in the form of 1 cm thick slabs. Temex ® rubber is a polymerised rubber and can be obtained in homogeneous slabs or moulded onto phantom bone as an actual body phantom. Temex® has the advantage that it is flexible whereas Mix D is brittle. Lincolnshire Bolus is a mixture of sugar and magnesium carbonate made up into small spheres. While the density of the material is greater than that of tissue, the packing density of the spheres produces good tissue equivalence. It is important, therefore, to check from time to time the condition of the bolus and to separate fragmented or powdered spheres from it. Perspex is useful but only at megavoltage energies. Phantom materials for other body tissues are not in common use, but cork and chalk may be used to simulate lung and bone respectively.

Table 5.3 Tissue equivalent materials

	Effective atomic no.	Electron density	Specific gravity
Soft muscle tissue	7.1–7.4	3.36×10^{23}	0.98–1.07
Water	7.42	3.34	1.0
Mix D	7.46	3.39	0.99
Temex rubber	7.05	3.27	1.01
'Bolus'	7.33	3.32	1.0
Bone	13.8	3.0	1.85
Fat	5.92	3.5	0.91

The distribution of X-ray dose in a uniform phantom

In the earlier part of this chapter the quality of the beam was discussed in terms of its ability to penetrate through attenuating materials which have been conveniently referred to as filters. Applying our understanding of the absorption processes, the photoelectric absorption process has been exploited in the use of beam hardening filters to improve the penetrating power of the beam. It is now necessary to investigate the penetration of the beam into the tissues of the body. Owing to the complex structure of the body and the problems of measuring the radiation within the body itself, most measurements are made using a *phantom* which is carefully chosen to be *tissue equivalent*.

The calculation of dose within a patient is invariably based on the dose distribution in a uniform soft tissue equivalent phantom and the rest of this section will be devoted to this. For the purpose of measurement the phantom will be a *semi-infinite rectangular water tank* and the beam axis will be at right angles to the entry surface of the phantom. A phantom is said to be semi-infinite if the measurements being undertaken are not affected by any further increase in the size of the phantom.

Percentage depth dose

The term *percentage depth dose* expresses the dose at any point within the phantom as a percentage of the maximum dose on the central axis of the beam, i.e.

$$\%DD = \frac{\text{Dose at a point within the phantom}}{\text{Maximum dose on the central axis of the beam within the phantom}} \times 100$$

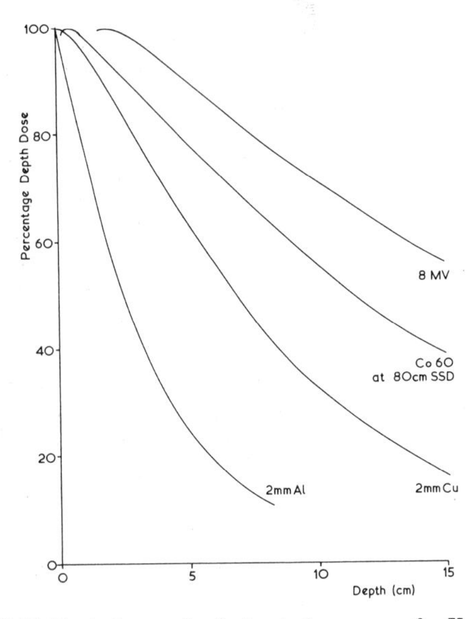

Fig. 5.11 Typical central axis depth dose curves for X- and γ-ray therapy beams.

Apart from a few exceptions (p. 64), the percentage depth dose at all points will be less than 100 per cent. The *central axis depth doses* are often plotted against depth as a means of comparing one radiation beam with another (Fig. 5.11), while a map of the dose distribution within the irradiated volume will be shown as an *isodose chart* where each line represents points of equal dose; in many respects they are analogous to height contours on a geographer's map. Typical isodose charts are shown in Figure 5.12, where the charts represent the distribution of dose in one of the principal planes of the beam, i.e., a plane parallel to the edge of the beam but containing the central axis. (In Figure 5.12 only half of each chart is shown.)

Central axis depth dose

There are many factors which affect the distribution of dose along the central axis of the beam. The dose will fall with increasing depth owing to absorption in the successive layers of tissue and the increasing distance from the target. (The inverse square law has to be obeyed.) Superimposed on this *primary* radiation (from the target) will be the *scattered* radiation which results from the Compton scattering processes taking place within the tissues.

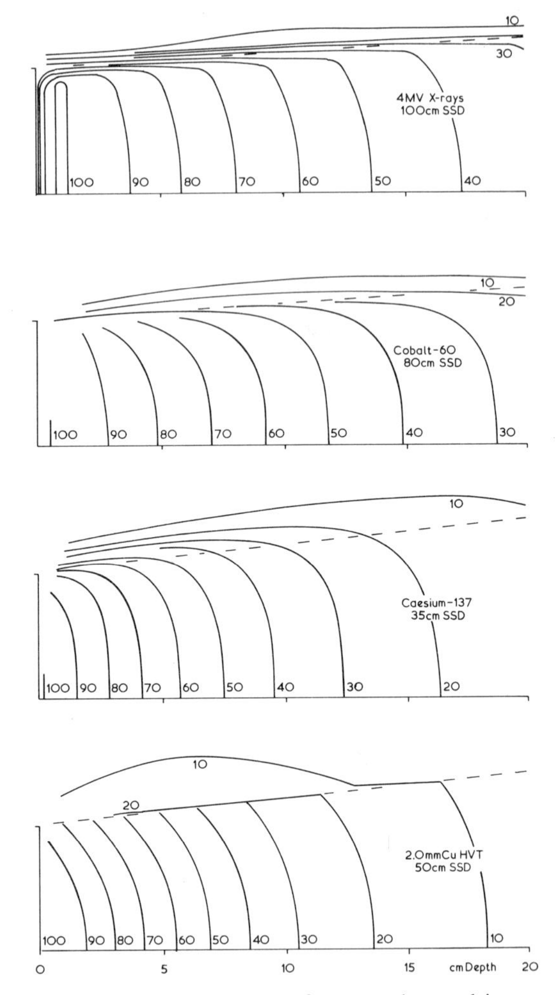

Fig. 5.12 Typical isodose charts for megavoltage, teleisotope and orthovoltage beams.

The magnitude of these effects will depend on the quality and the size and shape of the radiation beam and the nature of the tissue itself. Figure 5.13 shows the central axis depth dose curves for a 250 kV and a cobalt-60 beam with the contribution from scatter plotted separately. The important role played by scattered radiation in orthovoltage therapy is clearly visible and emphasised when compared with the relatively minor role it plays in megavoltage therapy. Two other factors are also clearly shown, first at orthovoltage energies the exposure to the surface (zero depth) is significantly affected by scattered radiation or *back scatter* and second, at megavoltage energies the surface dose is less and builds up to a maximum at some distance below the surface. *Build-up* is a physical phenomenon which explains the *skin sparing effect* of megavoltage radiations and is one of the principle advantages of these higher energy beams in radiotherapy.

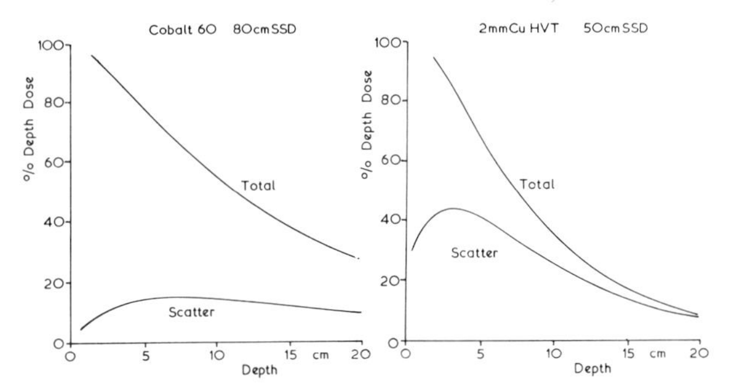

Fig. 5.13 Percentage depth dose curves for cobalt-60 and 250 kV beams showing the relative magnitude of the dose contributed by the scattered radiation.

Back scatter

The exposure at the surface of the phantom is substantially greater than the exposure at the same point if no phantom were present. The phantom material is scattering back to the surface a considerable amount of radiation and this contribution is known as back scatter. The contribution due to back scatter at the point where the beam axis enters the phantom is expressed as a percentage of the contribution due to primary radiation and called *percentage back scatter*. The percentage back scatter increases with the area of the field irradiated and with the thickness of the underlying tissues. This is to be expected, although with very large fields and very thick tissues the extremeties of the irradiated tissue are a long way from the centre of the incident field and contribute very little. The back scatter, therefore, approaches a maximum. For the same reason, the back scatter from a square or circular field will be greater than that for an elongated field of the same area.

The amount of back scatter at the surface of an irradiated phantom varies in a complicated way with radiation quality. It is largest for a beam quality of 0·8 mm Cu HVT—the deep therapy region—and falls as the quality increases, that is as the Compton scatter becomes increasingly in the forward direction as well as decreasing in magnitude. It also falls with softer quality beams where Compton interactions are swamped by photoelectric interactions by which most of the radiation is absorbed. At its maximum the percentage back scatter can reach 50 per cent and is, therefore, very important.

Equivalent square

It has been seen above that back scatter cannot be directly related to the area of the field. Elongated fields produce less back scatter as measured on the central axis than square fields of the same area. Nevertheless, the rectangular field will give rise to back scatter of the same magnitude as a square field of a *smaller* area. It is therefore possible to define for a rectangular field an *equivalent square* field which produces the same percentage back scatter. This concept is of particular value when treating non-standard field sizes. It is not only the back scatter that is the same, but the forward scatter is also the same and this gives rise to the same central axis depth dose. The primary contribution is the same for all fields (of the same quality and SSD). The scatter contribution is the only variable.

Tables are available which give the size of the equivalent square for all rectangular fields commonly encountered.

Build-up

At megavoltage energies, the scattered radiation is more in the forward direction and gives rise to less scattered radiation outside the edges of the beam. This is clear as soon as you compare the isodose charts for orthovoltage and megavoltage beams (Fig. 5.12).

Analysis of the Compton scattering process also shows that the recoil electron is ejected in the more forward direction and with increasing kinetic energy. The recoil electron range at 200 kV is extremely small but at megavoltages it is considerable (Table 4.1) and as each layer of tissue produces its quota of fast electrons, the number of electrons (and the ionisations they produce) increases with depth. This ionisation continues to increase until the electrons from the surface come to the end of their range, at which point the ionisation levels out as the electrons produced simply replace those which are lost. Decreased intensity due to absorption and attenuation (inverse square law) means the ionisations being lost slightly exceed those produced and the depth dose curve falls after reaching the peak (Fig. 5.14). The depth of the peak in centimetres is approximately a quarter of the X-ray energy in megavolts (2 cm at 8 MV, etc), Cobalt 60 being approximately equivalent to 2 MV X-rays has a peak at approximately 5 mm.

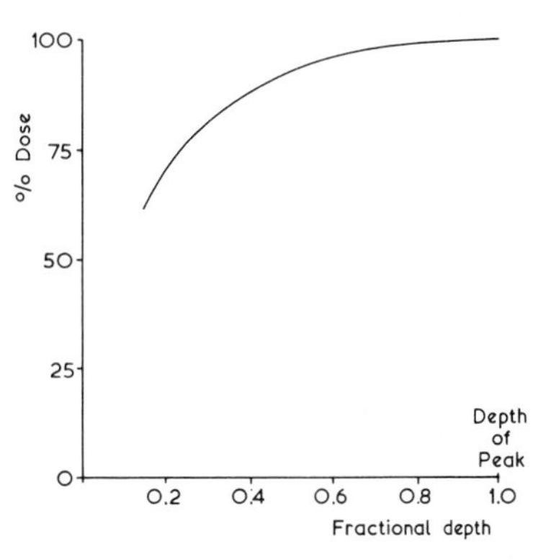

Fig. 5.14 Percentage depth dose in the build-up region of megavoltage beams.

This analysis of build-up suggests that the dose at the surface is zero. The surface dose in practice is variable and is mainly due to electrons scattered from other material in the beam—diaphragms, penumbra trimmers, beam shaping blocks, etc.—together with the intervening air itself. In practice the surface dose will be in the region of 20 per cent to 50 per cent for most teletherapy units. This problem of electron contamination cân be so pronounced that if, for example, Cobalt 60 is used at very short distances (5 −20 cm SSD, see p. 74), the skin sparing build-up is completely lost and may even be reversed, i.e., the contaminating electrons may raise the surface dose to a value higher than the dose at 5 mm depth, even after the effects of absorption and inverse square law are taken into account. In this situation *electron* filters may be used to advantage.

Definition of field size

Another characteristic of isodose charts is the width of the penumbra and here we are referring to the penumbra as measured in the water phantom and, therefore, include the geometric penumbra, the transmission penumbra and the effects of scattered radiation. It will be noticed that megavoltage X-ray beams have a narrower penumbra than cobalt-60 or caesium-137 beams owing principally to the differences in the source diameter. The question then arises, 'how do we define field size?' The design of the diaphragm system and the light beam is such that the edge of the light beam roughly corresponds to the width of the beam defined by the 50 per cent isodose line at the depth of the peak dose. In many ways this is the preferred definition, but clinically the width defined by the 80 per cent isodose is considered more useful. With such variations in the definition of field sizes and there can be up to 2 cm difference, it is important that the definition used is clearly understood so that allowances can be made.

Effect of changes in source skin distances

It has already been stated that the central axis depth dose is dependent on the inverse square law and, therefore, on the source-skin-distance. If primary radiation from the source alone was involved then the simple application of the inverse square law would be all that is necessary to derive depth dose data for different source-skin-distances. However, the field width measured at depth varies with the source-skin-distance owing to the beam divergence. Therefore, assuming the same field size at the surface, the scatter component at depth will decrease with increasing source-skin-distance. Furthermore, at megavoltage energies the 100 per cent reference point is at a depth t below the surface and this must also be taken into account. To quantify the basic assumption that the *depth dose at depth d increases with source-skin-distance f*, we define a factor

$$F = \left[\frac{f_1 + d}{f_1 + t}\right] \cdot \left[\frac{f_2 + t}{f_2 + d}\right]$$

in which t is equated to zero when the 100 per cent is at the surface. The modified depth dose D is then given by:

$$D\,(f_2, S) = D\,(f_1, S/F) \cdot \frac{B(S/F)}{B\,(S)} \cdot F^2$$

where $B(S)$ is the back scatter factor for the equivalent square field of side S. This simple formula is only approximate giving results accurate to about 2 per cent. Since F is a function of d, its application is usually limited to evaluating a few points in any clinical situation, e.g., the mid-point and exit dose for a parallel-opposed-pair.

The measurement of isodose charts

Owing to the complexities of the isodose distribution within the radiation beam, many approximation methods have been devised to minimise the number of measurements required before the isodose chart can be uniquely specified. Mathematical models have been devised which, with the help of a computer, can calculate the shape of isodose lines with the minimum of input data—and some of these models are entirely empirical. These techniques are beyond the scope of this book. The outline given below is based on a simple rectilinear system for measuring the distribution directly which will be available in some form or another in almost every department.

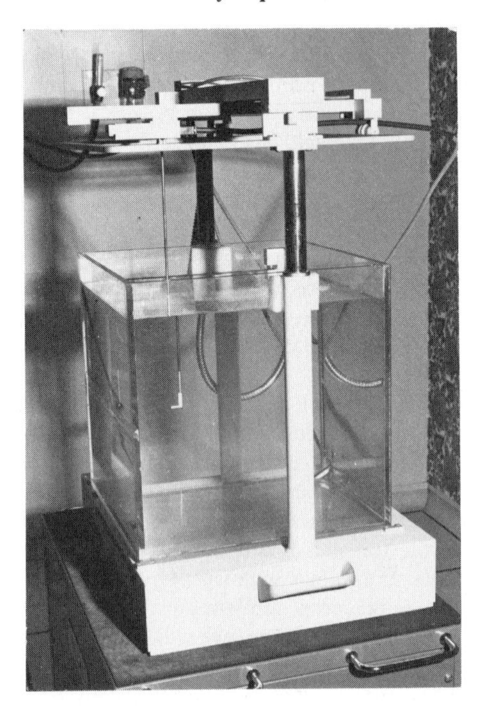

Fig. 5.15 The water phantom for use with an automatic isodose plotter. The horizontal beam to be measured enters the side of the tank on the left. The solid state detector is suspended from and moved about the horizontal plane by the XY mechanism above. It may also be raised and lowered to traverse parallel planes.

The basic requirement is a water tank, probably 30 × 30 × 30 cm^3 or larger, in which a small detector can be moved in two directions by remote control (Fig. 5.15). The plane of movement will contain the central axis of the beam. If the output of the radiation beam is likely to fluctuate (as in any X-ray unit) a monitor will be rigidly mounted in the beam but out of the plane of movement of the detector—to avoid collisions. The movement of the detector will be followed exactly by a pen—through a mechanical or electrical linkage—which will trace the position of the detector on paper. Now the pen marks the position of the detector in the phantom tank, whereas the object of the exercise is to record the magnitude of the ionisation at that position. It is at this point where the systems differ. The laborious, but very practical, method is for the operator to write alongside the pen the dose measured at each point in the beam. The automatic device electronically causes the pen (and the detector) to move along lines of equal dose, thus drawing the isodose curve directly on the paper and switching from one isodose level to another (Fig 6.20c). Whatever 'electronic black box' is used, isodose lines are charted based on the measurement of the ionisation produced at different points in the uniform tissue equivalent phantom.

Where a water tank is not available, useful information regarding the dose distribution in a megavoltage beam can be determined using film. If a slow radiographic film is sandwiched firmly between sheets of tissue equivalent material of adequate size—e.g. Mix D or Temex®rubber—and exposed in such a way that the film contains the beam axis, then isodensity lines may be plotted using a densitometer. The dose calibration of these isodensity lines must be based on ionisation measurements along the central axis. The technique may be improved by darkening the room so that unwrapped film may be used to ensure good contact with the phantom and by inclining the beam axis at a few degrees to the plane of the film (Fig. 5.16) to ensure the primary radiation traverses the phantom and not only the film. At conventional X-ray energies, the energy response of the film and the changing X-ray spectrum with depth in the phantom make the interpretation of the film too difficult to be of any practical value.

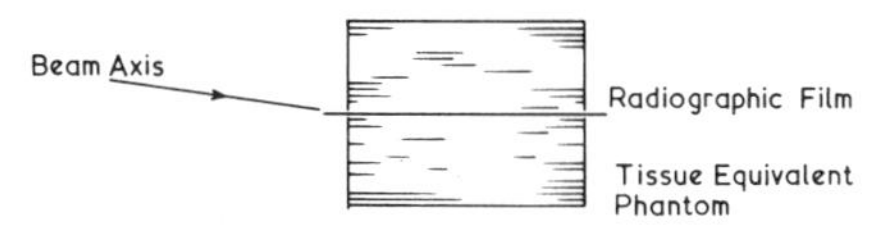

Fig. 5.16 The use of film for the measurement of megavoltage isodose charts.

Some isodose curve characteristics

Before leaving the subject of isodose charts it is worth summarising some of the characteristics of the different radiations met in the radiotherapy department.

Any isodose chart (Fig. 5.12) may be identified by examination of the following three aspects

1. Central axis depth dose distribution, particularly the depth of the maximum dose
2. Beam profile or the dose distribution across the beam
3. The penumbra region

The orthovoltage X-ray beam has a central axis dose which falls fairly rapidly from 100 per cent at the surface to 50 per cent at between 4 cm and 6 cm, the profile is curved and the penumbra is very sharp and clearly diverges from the source while outside the penumbra there is the distinct contribution due to side scatter. The caesium-137 chart has a very rounded appearance with broad penumbra, the peak dose on the central axis is 2 mm below the surface and the beam 'edge' appears to be undefinable. Cobalt-60 is readily identified by the 5 mm depth of peak, a well defined beam edge but fairly wide penumbra, the profile is less rounded except close to the beam edge, the 50 per cent depth lies between 7 and 12 cm depending on field size and SSD. Megavoltage X-ray beams are characterised by their narrow penumbra, flat profile and a depth of peak dose greater than 5 mm, the 50 per cent depth will be between 10 and 20 cm depending on energy and SSD but largely independent of field size.

The student should compare these isodose curves with those shown in Figure 6.20.

The measurement of output dose rate

There have been different recommendations laid down on how to measure the output dose rate for a therapy unit. The techniques described below are those currently laid down in the United Kingdom. The technique used is based on the use made of the therapy unit in question, namely at energies above 150 kVp, the unit is generally required to treat tissues at depth and the output is measured at depth in a standard phantom. At energies below 150 kVp, the unit is generally required to treat the skin surface and the output is measured in air (Fig. 5.17). In both cases an ionisation chamber is used for which a known quality correction factor, N, has already been determined by comparison with a secondary standard exposure meter (which in turn has been compared with the standard free air chamber, p. 41). The measurement is then suitably corrected to give the actual dose rate at the point of maximum ionisation on the central axis of the beam entering the patient.

In each case the output is measured under the operating conditions used clinically and using an average field size (10 × 8 cm^2 or 10 × 10 cm^2 are recommended). If more than one combination of kV, mA and filter are used, then the output must be measured for each combination. Regular calibrations are recommended, the frequency being at least once every four weeks.

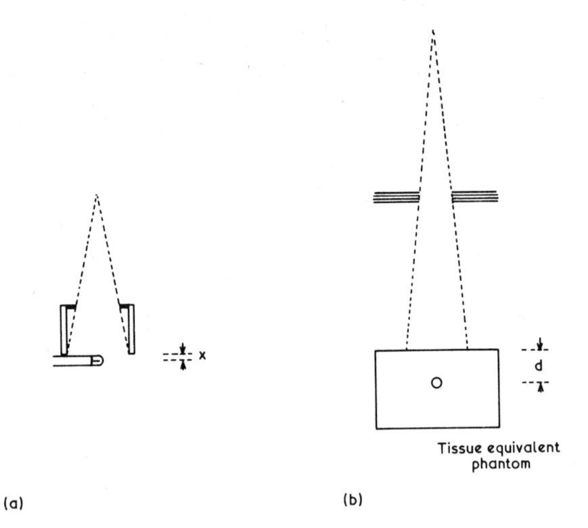

Fig. 5.17 The calibration of output dose rate, (a) in air for superficial X-rays and (b) in phantom for megavoltage X-rays.

Measurements in air (below 150 kVp)

The ionisation chamber is aligned close to the surface of the applicator and centred on the axis of the beam. The distance, x, from the centre of the chamber to the surface must be measured accurately. The chamber is exposed for a period of time, t, equivalent to a typical treatment time and the ionisation charge integrated in the exposure meter, as a reading R. The ambient air temperature (in the treatment room) and atmospheric pressure is measured. The reading R is corrected by multiplying by the air mass correction factor, A, (p. 43) and then by the quality factor, N, to give the total exposure in roentgens in air, at the centre of the chamber.

The inverse square law correction factor $(\{s + x\}/s)^2$, where s is the source-skin-distance, corrects for the stand-off of the chamber, x, and the back scatter factor, B, takes care of the lack of tissue behind the chamber.

Thus, the exposure rate at the surface of a patient in contact with the applicator surface is given by

$$\dot{X} = R \cdot A \cdot N \cdot (\{s + x\}/s)^2 \cdot B/t \text{ roentgens/min}$$

A further multiplying factor, f rad/roentgen, converts the exposure rate, $\dot{X}$, into dose rate, $\dot{D}$

$$\dot{D} = f \cdot \dot{X} \text{ rad/min}$$

It should be noted that the factors N, B and f are all energy dependent and the values used must be appropriate to the HVT of the beam.

Measurement at depth (above 150 kVp)

At megavoltages an 'in-air' measurement is not relevant as the roentgen is undefined at high energies. Instead the output is measured at depth. The depths used vary with the photon energy but for the majority of deep therapy units 5 cm is used. The other values are given in Table 5.4.

Table 5.4 Calibration depths in water for output dose rate measurements

Radiation energy	Depth (cm)
150 kVp–10 MV X-rays	5
Cobalt-60, Caesium-137	5
11 MV–25 MV X-rays	7
26 MV–50 MV X-rays	10

A phantom of adequate size, at least $20 \times 20 \times 10$ cm^3 is used such that the ionisation chamber is positioned with its centre at the appropriate depth below one of the larger faces of the phantom. The top face of the phantom is positioned at the normal treating distance with the ionisation chamber on the central axis of the beam. The chamber is exposed for a typical treatment time or for a known 'set dose'—the nominal dose set on the integrating dosimetry system of the therapy unit. As before, the instrument reading, R, is corrected using the air mass correction factor based on the atmospheric pressure and the temperature of the phantom if that is likely to be different from the ambient temperature of the room. The inverse square correction factor and the back scatter factor are replaced by the percentage depth dose factor relevant to the depth of the chamber, beam quality and the field size set. The quality factor, N, and the rad/R factor, f, are used as before but at megavoltages N is the 2 MV (or Co-60) factor and f converts from the exposure at 2 MV to absorbed dose at the operating megavoltage, thus

$$D = \frac{R \cdot A \cdot N \cdot f}{\%DD} \text{ rad or } \dot{D} = \frac{R \cdot A \cdot N \cdot f}{\%DD \cdot t} \text{ rad/min}$$

Output factors

The routine output dose rate measurements are made using an average field size. At the initial calibration of the unit, the output dose rate will be measured for all field sizes which will be encountered in the clinical use of the unit. The variation in output dose rate is principally due to the increase in scattered radiation with field size. It is customary to tabulate the output dose rate for each applicator size available on an orthovoltage unit or to tabulate output factors for (equivalent) square fields on megavoltage units. The output factor is unity for the reference field size. For other field sizes the output factor is defined as:

$$OF_s = \frac{\text{Output dose rate for field size, } S}{\text{Output dose rate for the reference field size}}$$

The conversion from roentgens to rads

In the previous two sections reference has been made to the *f*-factor for the conversion from exposure in roentgens to dose in rads. The derivation of this factor may be explained briefly as follows. From the definition of mass absorption coefficient, we have for any absorbing medium:

$$\text{Energy absorbed per unit mass} = \text{Energy in the beam} \times (\mu_a/\rho)$$

(μ_a/ρ) being a function of the atomic number of the medium and the energy of the radiation beam. If this equation is true for tissue it is also true for air at the same point in the beam, so we can say

$$\frac{\text{Energy absorbed per unit mass of tissue}}{\text{Energy absorbed per unit mass of air}} = \frac{(\mu_a/\rho)_{\text{tissue}}}{(\mu_a/\rho)_{\text{air}}}$$

Now the average energy to be absorbed in air to produce one ionisation is 5.39×10^{-18}J, the charge on one electron is 1.6×10^{-19}C and one roentgen is defined as the release of 2.58×10^{-4}C/kg. If an exposure of one roentgen releases $(2{\cdot}58 \times 10^{-4})/(1.6 \times 10^{-19})$ electrons per kg of air and each of these requires an energy absorption of 5.39×10^{-18}J, it follows that the total energy absorption in air per kg of air resulting from an exposure of one roentgen is

$$\frac{2.58 \times 10^{-4} \times 5.39 \times 10^{-18}}{1.6 \times 10^{-19}} = 0.00869 \text{ J/kg}$$

or 0.869 rad. The equation above can therefore be rewritten in the form

$$D(\text{rad}) = f.X(\text{roentgen})$$

where

$$f = 0.869 \frac{(\mu_a/\rho) \text{ tissue}}{(\mu_a/\rho) \text{ air}}$$

Although for most purposes the *f*-factor is that for soft muscle tissue (Table 5.5) it is instructive to realise that this will be different from that for other body tissues—bone and fat for instance—owing to the differences in their atomic numbers. Bone, for example, has an atomic number nearly twice that of soft muscle tissue and a lower hydrogen content. Gramme for gramme, therefore, the dose to bone is several times greater than that to tissue at low photon energies due to photoelectric absorption, while at higher energies it is less (Figure 5.18).

Table 5.5 Typical values of the *f*-factor for soft tissue

Beam quality	*f* rad/R
1 mm Al HVT	0.88
4 mm Al HVT	0.87
2 mm Cu HVT	0.94
Cobalt-60	0.95
4 MV	0.94
8 MV	0.93

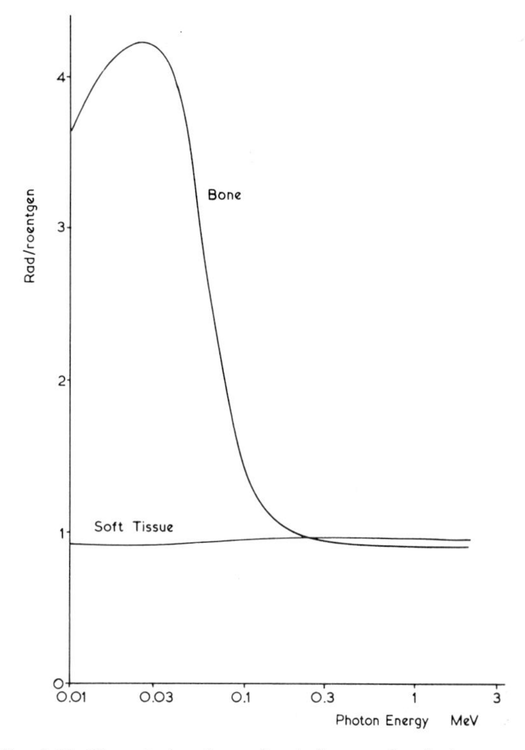

Fig. 5.18 The relative absorption in bone and soft tissue.

The effects of inhomogeneities on depth dose

The fundamental measurements on the relative distribution of dose and the absolute magnitude of the dose in a beam of radiation have all assumed that the patient is accurately represented by a rectangular water tank. This is far from true! Yet in most of our treatment planning little cognisance is given to this fact, largely because of the complexities which arise as soon as any inhomogeneity is introduced.

Where treatment planning is carried out using manual techniques, the sheer complexity of the problem makes any adjustment of the dose distribution for the presence of inhomogeneities prohibitive. Where computer techniques are available, the ability to adjust the distribution is there but often the patient data is not. Several attempts have been made using atlas's of body sections to provide data on the size and position of the principal inhomogeneities. It was not until computerised transverse axial tomography became available that any measurement of the size and position of any inhomogeneity was possible. Although the early work was confined to neuroradiology, body scanners are now available which may provide the required information—at a price! An alternative—considerably cheaper and correspondingly less informative—is the use of ultrasound for the detection and mapping of tissue interfaces. This has distinct possibilities particularly where inhomogeneity corrections are important, namely the chest. These techniques are discussed in greater detail in Chapter 12. The effect of an inhomogeneity can be described in qualitative terms.

A beam of radiation passing through healthy lung will be less attenuated than it would be through soft tissue. The result is that soft tissues beyond the lung will receive a greater dose than that predicted from the standard isodose charts. Similarly bone will absorb more radiation than soft tissue with the result that the soft tissue beyond the bone will receive less dose. Remember that although bone and tissue may absorb equal doses of megavoltage radiation gramme for gramme, bone is nearly twice as dense as soft tissue. Remember too that while healthy lung tissue has a density of about 0·3 g cm^{-3}, malignant lung tissue density is closer to 1 g cm^{-3}. This relatively simple situation is further complicated by the fact that the radiation scattered by bone and lung is different and so is the electron flux. At orthovoltages the dose to soft tissue beyond lung is higher due to increased transmission through the lung and lower due to the reduced scatter produced by healthy lung, while beyond a bone inhomogeneity the dose is reduced by the high absorption in bone and increased by the increased scatter. Similarly at megavoltages the secondary electron flux is greater in bone than in soft tissue and since the electrons have a range of several millimetres, the electrons generated in the bone close to the bone–tissue interface, will increase the dose to the tissue until electronic equilibrium is restored. Conversely lung. Soft tissue close to bone and, therefore, the soft tissue elements actually living within bone, always receive a higher dose than they would if the bone was replaced by soft tissue. So, although there may be clearly defined interfaces between the soft muscle tissues and the inhomogeneities, their effect on the dose distribution is far from clear. The student is referred to more advanced texts for further discussion on this subject.

Integral dose

Dose is the energy absorbed per unit mass of tissue in a small element of tissue. Integral dose is the summation of the dose to all elements of the irradiated tissue and represents the total energy absorbed. The unit of integral dose is the gramme-rad. In practice the mega-gramme-rad is used. In SI, the Mg-rad will be equivalent to 10 kg-Gray.

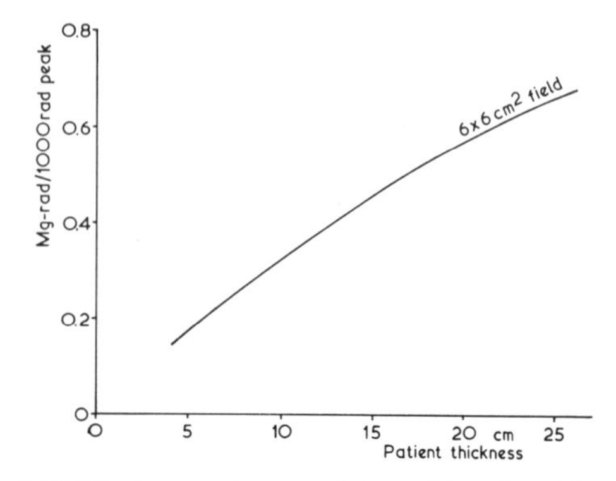

Fig. 5.19 The increase in integral dose with patient thickness for a given field size and beam energy.

The detailed evaluation of integral dose is complex in practice but it is easily appreciated that it increases with incident dose, field size and thickness irradiated. For a particular beam, the integral dose per 1000 rad peak takes the form shown in Figure 5.19. In order to estimate the integral dose for a particular treatment, the length of the central axis of each beam within the patient is measured and the integral dose per 1000 rad is read from the graph for the corresponding field size. This is multiplied by the peak dose, in krad, for the field. The results for each beam are added together to give the total integral dose. Figures

Table 5.6 Examples of the calculation of integral dose based on the dose distributions shown in Figures 6.1, 6.2 and 6.3

Figure Technique	6.1 2-field	6.2 4-field	6.3 3-field
Prescribed tumour dose (say)	5000 rad	5000 rad	5000 rad
Prescribed percentage tumour dose (say)	135%	240%	180%
Peak dose to be given to each field	3.70 krad	2.08 krad	2.78 krad
Thickness irradiated by each field and corresponding integral dose per krad peak[a], field 1	20:0.57	24:0.65	20:0.57
field 2	20:0.57	24:0.65	26:0.68
field 3		24:0.65	26:0.68
field 4		24:0.65	
Total integral dose per krad peak	1.14	2.60	1.93
∴ Integral dose	4.22 Mg rad	5.41 Mg rad	5.37 Mg rad

[a] Taken from the data given in Figure 5.19. In these examples weighting factors are not being used, if they were these figures would be modified accordingly before determining the total integral dose per krad peak.

6.1, 6.2 and 6.3 show the dose distribution resulting from the use of a 2-, 3- and 4-field technique for treating the same volume of tissue. Table 5.6 illustrates the steps involved in calculating the integral dose given to the patient when 5000 rads are given to the tumour-bearing volume. It will be noted that although the total volume of tissue irradiated increases markedly with the number of fields, the integral dose is not very different. In practice, the integral dose is more dependent on beam energy than on the number of beams used and over the range of beam energies in common use the integral dose given during the treatment of deep-seated lesions falls with increasing beam energy.

This assessment of integral dose is confined to the tissues deliberately irradiated during the course of radiotherapy. In addition there is a component of integral dose arising from the fact that the patient receives a smaller dose to a larger volume of tissue from the radiation leakage through the tube or source housing and from radiation scattered from the treatment volume. It is in consideration of this that the limits for the exposure rate from the leakage radiation are determined (Table 3.2).

6. Principles of radiation treatment planning

Introduction

The phrase *radiation treatment planning* as applied to external beam therapy is used to describe the work involved in displaying graphically a dose distribution which results when one or more radiation beams converge on the treatment volume. It does not include any assessment of the dose fractionation or of dose rate effects (Ch. 17).

For most purposes the dose distrubution is a two-dimensional distribution in the midplane of treatment which also contains the principal axes of all the treatment fields. Although estimates of dose can be made for non-coplanar fields, this is best achieved using a computer, as is the planning of three-dimensional distributions. The two-dimensional distribution has limitations, but the majority of treatment sites are cylindrical in section and the distribution in planes parallel to the midplane are not markedly different.

The criteria by which a dose distribution is judged vary from one centre to another although the following give a guide as to what is required. For the sake of clarity the words *treatment volume* are used to refer to that volume defined by the clinician as being the tumour or lesion surrounded by any margin he chooses to include in the treatment; the words *treated volume* are used to refer to that actually covered by the radiation beams and raised to a prescribed level of dose.

1. The dose throughout the treatment volume should be uniform to within ± 5 per cent of the tumour dose
2. The treated volume should as near as possible be the same as the treatment volume in position, size and shape
3. The dose to the treated volume should exceed that to any other area by at least, say 20 per cent
4. The dose to neighbouring radiosensitive sites (e.g. eyes, spinal cord, etc) should, where practicable, be kept below their tolerance dose
5. The integral dose should be kept to a minimum (p. 58).

These criteria also provide guidelines on how the treatment planning of any site should be approached. Criterion (3) suggests that unless the treatment volume includes the skin surface, then two or more fields should be used to converge on the treatment volume. Criteria (2) and (5) suggest that the radiation beam size should be kept to a minimum and the beams should enter the patient as close to the treatment volume as possible providing criterion (4) is not broken. The first and foremost criterion is dose uniformity and examples will be given to show how this can be achieved.

Later in Part 2 examples are given for different treatment sites, whilst here we concentrate on the effects of different field combinations. It should be borne in mind that the relative distribution is not altered appreciably by changes in the size of the patient section or the size of the radiation beams used, e.g. four oblique fields to the head treating the pituitary produce the same basic distribution as four oblique fields to the abdomen treating the bladder.

Simple distributions

The parallel pair

Two directly opposing fields are known as a parallel-opposed-pair and the distribution is typified in Figure 6.1. In general, the distribution is symmetrical about the beam axes. In particular, the midpoint receives the minimum dose *along* the central axes (except within the build-up region) and the maximum dose *at right angles* to the central axes. The actual variation of dose along these axes of symmetry depends on the type of radiation and on the separation of the fields. The variation in dose is reduced by reducing the separation and by increasing the energy of the radiation. Within these limits, the tissue irradiated receives a uniform dose.

It should be remembered (Fig. 5.12) that the dose to the edges of the beam, while being lower than the central axis dose where the beams are incident normally on the skin, may be increased where there is excessive curvature of the entry surfaces (p. 62), e.g. frontal and occipital fields to the head.

This technique is valuable where large volumes of tissue are to be irradiated or where the treatment is palliative.

The box technique

If a parallel pair gives uniform irradiation to the enclosed tissues, then two intersecting parallel pairs will produce an even distribution over the volume enclosed by all four fields. This is true regardless of the angle between the axes of the two pairs, i.e., the box is not necessarily rectangular. For example, two pairs at right angles produces a 'square' distribution (Fig. 6.2), while two pairs at, say 120 degrees produce a 'diamond' distribution. The disadvantage of the box technique is seen when small fields are used at large separations since the irradiated tissue outside the treated volume receives half the dose to that inside. This may violate the criterion (4) above. Although there is no reason against using 3 parallel pairs, the improvement is rarely justified.

The three field technique

By arranging fields so that they overlap *only* in the treated volume, it may be possible to bring most of the tissue outside the treated volume to less than half the tumour dose. In Figure 6.3 it will be seen that except for regions close to the surface the dose in general is less than half of that to the treated volume. It will be also noticed that the dose gradient at the edge of the treated volume is less steep than that in the box technique.

The three and four field techniques are widely used in radical treatments using quite small fields—providing the beam direction can be guaranteed to be accurate.

Atlases of dose distribution

Where the planning work has to be minimised through lack of facilities it may be helpful to refer to an atlas of these three basic treatment techniques. Each chart in the

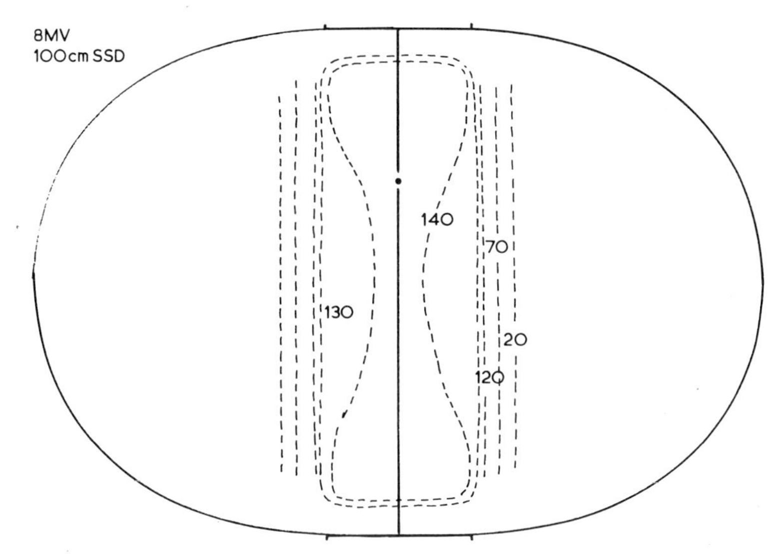

Fig. 6.1 The parallel opposed pair.

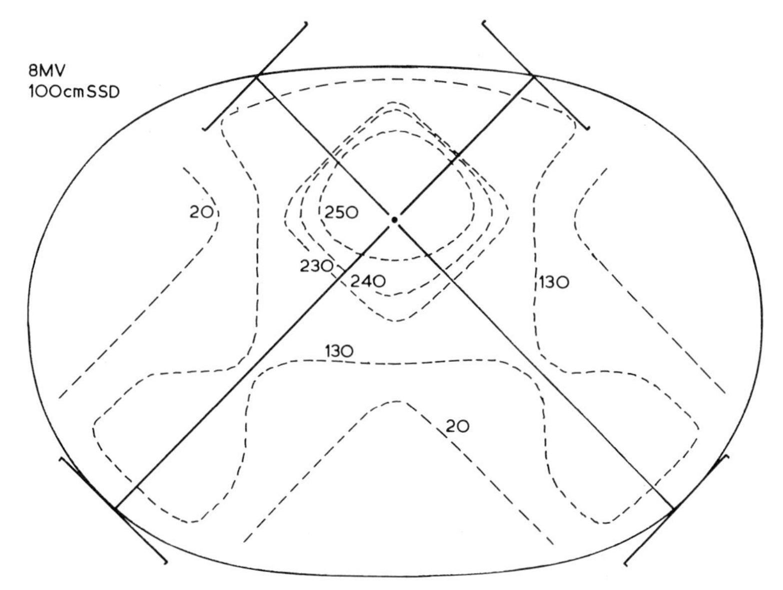

Fig. 6.2 The four field box technique.

atlas provides the dose distribution resulting from fields of a stated size, relative angle, separation and beam quality. In practice, the chart giving the 'best fit' is used and may be modified slightly as required. Atlases are not to be recommended for the routine planning of radical courses of treatment because of the limitations inherent in such distributions but may often assist in the planning of difficult cases by showing the effect on the distribution of varying one or other of the many parameters involved. The radical treatment plan demands the distribution to be tailored to the individual patient. An accurate distribution may require the use of oblique incidence corrections, compensators, wedges or weighting factors. Patient shells (moulds) or other immobilisation devices may also be required. These will now be considered.

If the build-up region is to be retained then the isodose distribution may be corrected from that of normal incidence to that for oblique incidence. The computer will invariably use a mathematical approach to the problem and use either an exponential correction factor based on the lack of attenuation through the tissue deficiency, or an inverse square law correction factor based on the change in source skin distance. For routine manual treatment planning these approaches are too time consuming and an approximation method has to be used.

A 'rule-of-thumb' is frequently used. A rule-of-thumb is an approximation to the truth which is sufficiently accurate for the purpose in hand providing it is used within its limitations. The rule here is to move the charted isodose curves through a distance related to the thickness

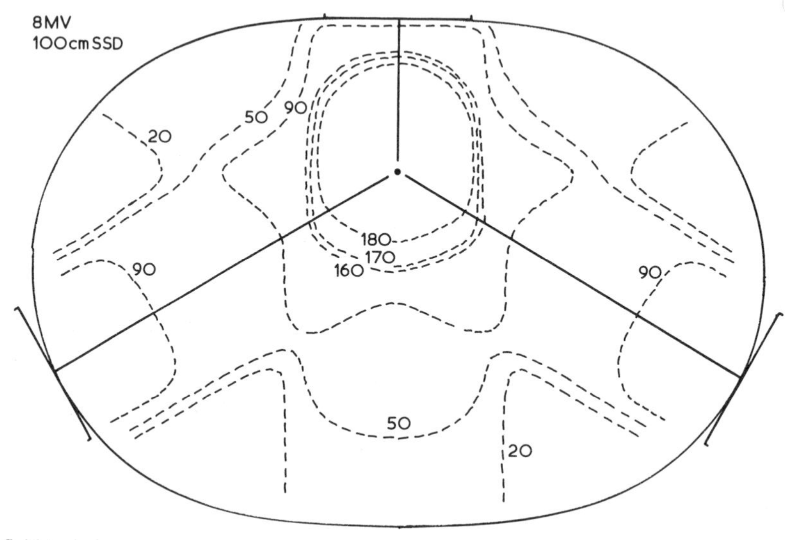

Fig. 6.3 The three field technique.

Oblique incidence and its correction

Owing to the curvature of the skin, most radiation beams applied to the patient will enter a curved surface at an angle of incidence which is not zero (normal incidence). Such oblique incidence may be quite deliberate or simply due to the fact that the patient's skin is not flat, so that in either case the dose distribution in the patient will not be that shown on the appropriate isodose chart. There are two alternative ways of dealing with this, either to correct the distribution or to correct the curvature.

At orthovoltage energies where there is no build-up or skin sparing effect the space between the applicator and the skin can be filled with 'bolus' or tissue equivalent wax, thus making up the tissue deficiency. This may be done at megavoltages also where skin sparing is not required (Fig. 6.4 and Fig. 6.16f). In either case normal incidence is restored and the appropriate isodose chart can be used without correction providing the entry point is correctly positioned.

of the missing tissue. The distance moved depends on the beam energy and the depth of the isodose line, but for most purposes, small deficiencies can be corrected by the

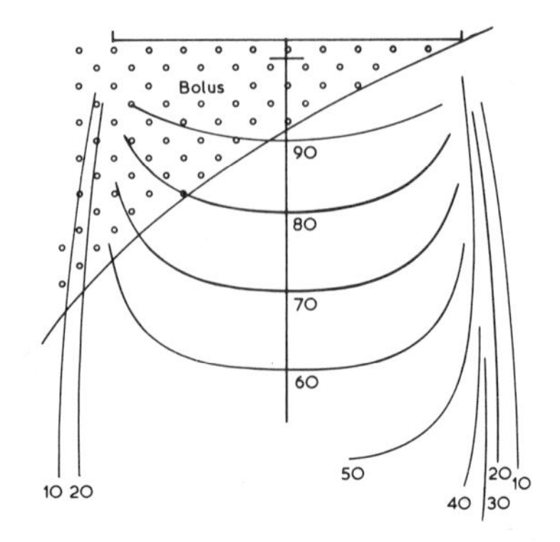

Fig. 6.4 The use of bolus to restore the normal incidence and so maintain the charted dose distribution, but note the loss of skin sparing.

'half-shift' technique. That is, the isodose lines on the chart are shifted by half the thickness of the tissue deficiency as measured along the ray from the source of radiation (Fig. 6.5). This does not apply in the build-up region where the build-up follows the expected pattern, i.e., the isodose lines remain essentially parallel to the skin surface.

Where the treating distance is measured to the skin on the central axis, it is often found that on one side there is a tissue deficiency while on the other there is excess tissue. The same correction technique may be applied—but making the reverse effect.

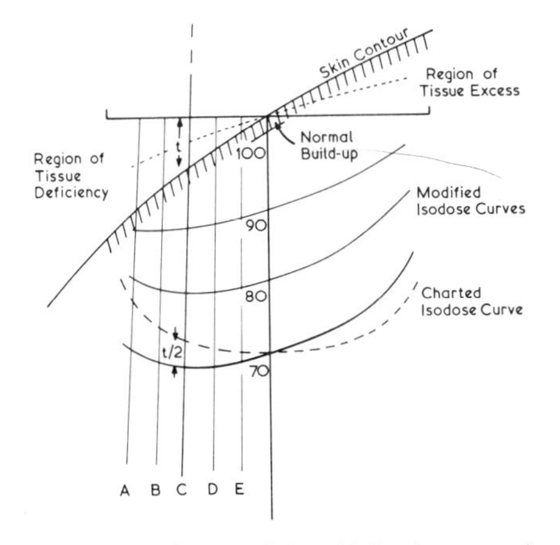

Fig. 6.5 The half-shift 'rule of thumb' for the correction of the isodose chart for skin curvature.

Tissue compensators

The charted isodose curve may be used while maintaining the skin sparing properties of the megavoltage beam by the use of *tissue compensators*. Here the lack of attenuation in the region of tissue deficiency is made good by using an attenuator in the beam sufficiently removed from the skin that the build-up region is unaffected (Fig. 6.6). This technique is particularly valuable in regions where the skin curvature is complex, e.g., in the region of the neck.

Where the curvature is sufficiently great to justify a tissue compensator, the accurate alignment of the compensator and patient during treatment demands the use of a mould to immobilise the patient. If the mould extends over the treated region, then the data required to make the compensator can be obtained from the mould—a considerably easier exercise than measuring directly from the patient. Having determined accurately the position and direction of the beam axis, the stand-off or thickness of tissue deficiency is measured at one centimetre intervals over the whole of the field area and preferably one centimetre beyond. For simplicity each measurement is rounded to the nearest one centimetre. Using this information it is possible to build up the compensator with 'bricks', each equivalent to a cubic centimetre of tissue. These bricks will be less than 1×1 cm^2 in area owing to the fact that the compensator will be mounted about half way between the source and the skin. Their height will be 1 gcm^{-2} in whatever material is chosen, e.g., aluminium. The bricks will be glued to a base plate, say 3 mm aluminium, which can be uniquely positioned in the beam to guarantee the correct alignment (Fig. 6.7). In most cases two or more compensators will be required, one for each treatment field. In use, the normal source skin distance will be increased by the compensated distance as measured along the central axis of the beam and an output correction factor applied based on the attenuation through the base plate.

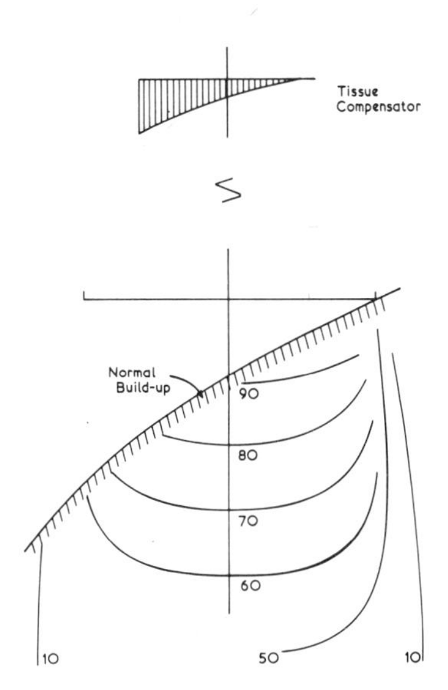

Fig. 6.6 Using a remote tissue compensator, the skin sparing effect and the charted dose distribution are both maintained. Note the source-skin-distance is increased by the compensated thickness.

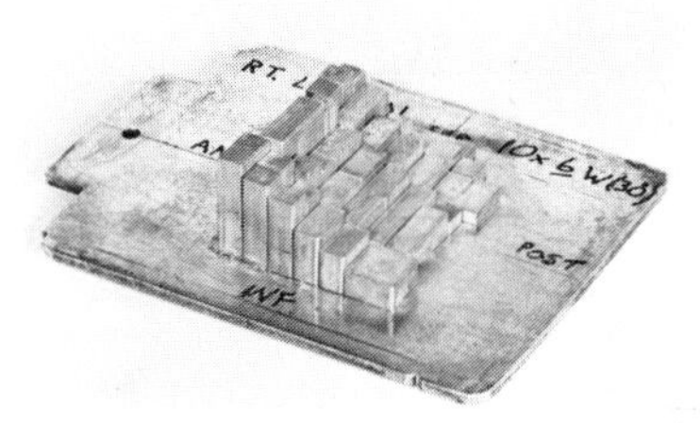

Fig. 6.7 A tissue compensator.

Wedges

Where the treatment volume is close to the surface of the patient but too thick to be adequately treated by a single field, it is often convenient to use two beams with wedges. A *wedged pair* can be used to treat regions such as the antrum, the larynx and the middle ear.

The wedge is a specially machined wedge-shaped piece of metal. It can be made of any metal although there is some preference for those of low atomic number especially where the wedge has to be positioned close to the patient.

The wedge is more often positioned in the beam immediately behind the mirror of the light beam diaphragm (see Figs. 3.14 and 3.19). The purpose of the wedge is to attenuate the beam on one side of the beam axis relative to the other, thereby tilting the isodose curves through an angle. This *wedge angle* is defined as the angle through which the 50 per cent isodose line is turned (Fig. 6.8). In the design of a wedge it is often expedient to straighten out the curvature of the (un-wedged) beam profile as well as tilting it through the required angle. The wedge profile will take the form shown in Figure 6.9.

The wedge attenuates the beam and thereby reduces the useful beam dose rate. The treatment time, therefore, has to be increased. The increase is determined by the *wedge factor* which is defined as the time required to deliver a given dose to a point on the central axis when using the wedge, divided by the time required to deliver the same dose to the same point without the wedge. It is sometimes advantageous to use a wedge appropriate to the field size in use and so minimize the wedge factor. This is sometimes referred to as *wedge efficiency* in that to use a wedge which is too large requires the use of an unnecessarily large wedge factor. For example, in Figure 6.9 the field size AA requires a much thinner wedge than the field size BB to produce the same wedge angle with the result the wedge factor could be reduced for field AA if a separate wedge was available.

The clinical use of wedges is such that wedge angles up to 45 degrees are used in fields up to about 15 cm wide and above 45 degrees in fields up to about 8 cm wide. In all cases, it is advisable to extend the thickest part of the wedge beyond the maximum field size BB to minimise the effects of overdosage should an over large field be used accidentally. The wedge will usually cover the whole range of field sizes perpendicular to the wedged direction.

The ideal dose distribution is obtained when two beams of wedge angle θ are used with a hinge angle of $(180 - 2\theta)$ degrees, where the *hinge angle* is the angle between the beam axes at their point of intersection (angle *qps* in Fig. 6.10). This only applies when the beams enter a flat surface at normal incidence. The 'wedging effect' produced by oblique incidence invariably reduces the wedge angle and increases the optimum hinge angle.

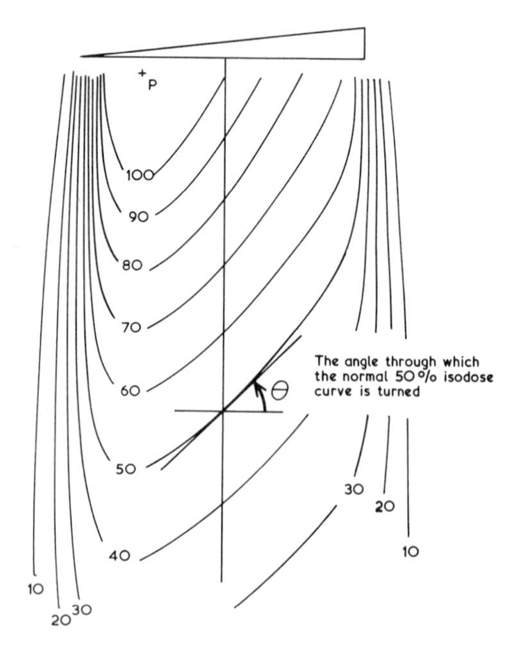

Fig. 6.8 The definition of wedge angle.

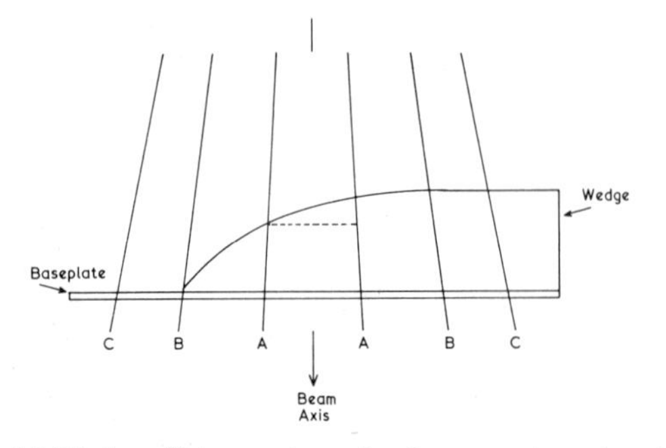

Fig. 6.9 Wedge efficiency: the wedge factor can be reduced by using a wedge no bigger than the field.

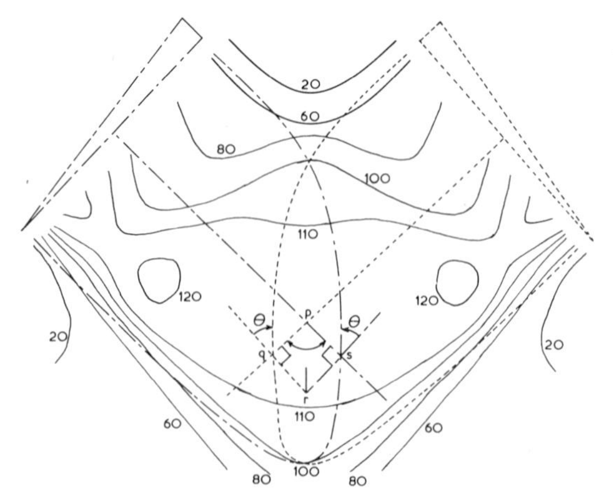

Fig. 6.10 The optimum hinge angle for a wedged pair.

Wedge fields are always used in pairs (Fig. 6.11) with their thick ends together. Useful dose distributions can be also obtained by adding a third un-wedged field between the wedged pair (Figs. 19.10 and 20.8), the hinge angle will be modified but the thick end of the wedges are still positioned together.

The *wedging effect* of oblique incidence can sometimes be compensated by the use of wedges of shallow wedge angle, say 15 degrees. For example, when treating a larynx with a parallel opposed pair it may improve the final dose distribution if wedges are used such that their thick edges are to the anterior. This is a technique of limited value but can with care be used to advantage wherever the skin curvature is predominantly in one plane.

This section would not be complete without some reference to wedges at orthovoltage energies. Wedges can be made for orthovoltage units. There are two basic problems. Firstly, the thick end of the wedge not only attenuates the radiation but also increases the filtration on that edge of the beam, thereby making it relatively more penetrating and negating its wedging effect. Secondly, the dose at depth is very dependent on the scattered radiation generated in the tissues and this again counteracts the wedging effect. Thus, the isodoses produced are wedged close to the surface but the wedge angle decreases with increasing depth. A practical problem also exists in that a different wedge has to be made for each applicator.

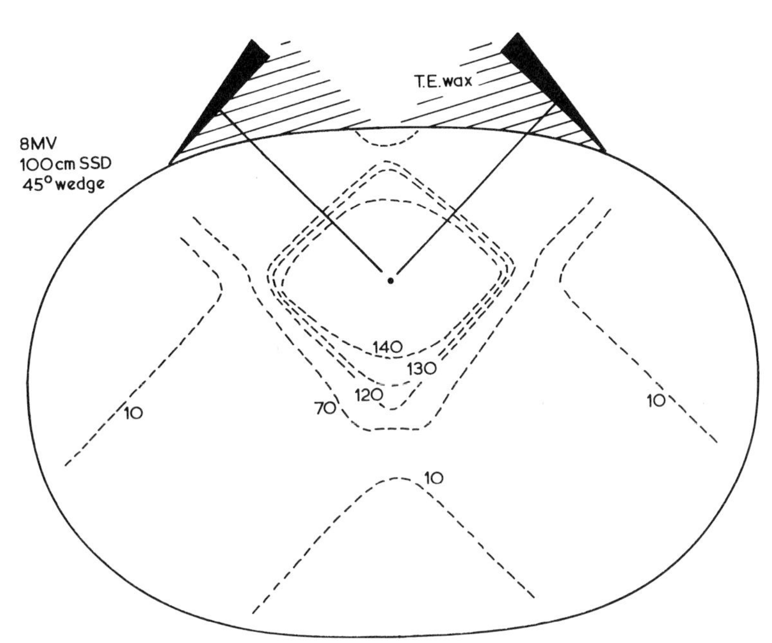

Fig. 6.11 A wedged-pair distribution.

Weighting factors

So far it has been assumed that each field in a distribution is given the same dose (100%). Some distributions may be improved by reducing the dose given to one or more fields in the multifield treatment plan. These fields are said to be weighted against the others. Weighting factors of 50 per cent (half dose) or 75 per cent (three-quarter dose) are simple to handle but any factor can, in theory, be used. The effect is to reduce the dose gradient along the central axis.

For example, if the box technique is used to treat a volume which is off-centre (as in Fig. 6.2), then the distribution may be improved by weighting those fields closer to the treated volume so that they only contribute the same actual dose to the centre of the volume as the more distant fields (Fig. 6.14). Some centres use weighting factors on each field so that the dose distribution is normalised to 100 per cent at the tumour (treated) volume.

Rotation therapy

A logical extension of the multiple fixed field distribution is to have an infinite number of fields all aimed at one centre. This is achieved by moving the source in an arc whilst irradiating the patient. In *arc therapy*, the source of radiation moves through a prescribed angle, while in *rotation therapy* the source moves through a full 360 degrees. Rotation therapy was developed in the 1920s as a means of achieving an adequate dose at depth without exceeding the tolerance dose to the skin using 200 kV radiation. This was prior to the introduction of megavoltage radiation. Although skin tolerance is no longer a problem, rotation therapy is still used in some centres, particularly where cobalt-60 provides the only megavoltage radiation. The dose distribution resulting from the rotation of the source of radiation is essentially circular or elliptical—the major axis of the ellipse being at right angles to that of the patient contour (Fig. 6.12). There are no steep dose gradients. The actual dosimetry is complicated by two facts, one that the source-axis-distance is constant, therefore, the source-skin-distance will vary and, two, that a full rotation needs to be represented by at least 18 fields (at 20° intervals)—a problem best solved using a computer. However, on the basis that the maximum dose will be close to the

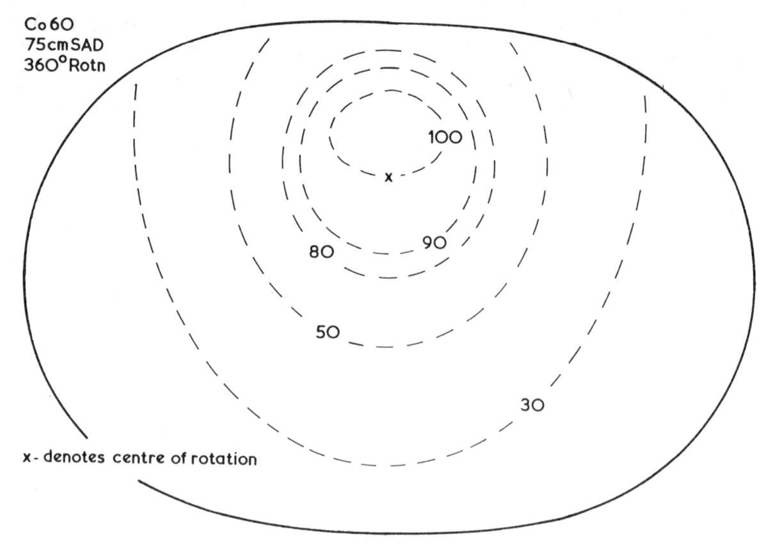

Fig. 6.12 A dose distribution resulting from rotation therapy.

centre of rotation, it is often adequate to calculate the dose at that point only. This is most conveniently done using Tissue-Air-Ratios (TAR).

Tissue-air-ratio

The tissue-air-ratio is the ratio of the exposure rate at a point in the patient to that at the same point in air when there is no patient present. The exposure rate at a point is dependent on the distance from the source (the inverse square law), the thickness of the overlying tissues and the scatter contribution which varies with field size. Since the TAR is the ratio of exposure rates at the same point in space, it is independent of distance from the source—the inverse square factor cancels out. For a given field size (and, therefore, scatter contribution), it is possible to tabulate or represent graphically the TAR against the thickness of the overlying tissue (Fig. 6.13).

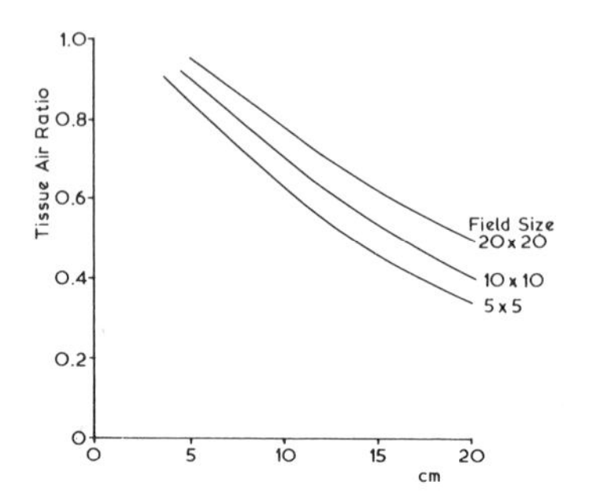

Fig. 6.13 Tissue-air-ratios for cobalt-60.

Dose calculation

By definition the exposure rate on the axis is the exposure rate in air at that point multiplied by the TAR. The centre of rotation is marked on the patient contour and eighteen radii are drawn at 20 degree intervals. The length of each radius is measured from the centre to the patient's skin and the TAR noted. The mean value of the TAR is calculated and used to determine a mean exposure rate on the axis. The mean dose rate can then be determined.

The calculation of exposure rates at points other than the centre of rotation is complex and the student is referred to more advanced texts for a detailed explanation. Atlases of rotation dose distribution are frequently normalised so that the point at the centre of rotation is 100 per cent, as in Figure 6.12.

If a full rotation is not possible, then arcing over a limited range of angles can be used, this increases the skin/tumour dose ratio and brings the maximum dose point away from the axis of rotation towards the centre of the irradiated arc.

Sieve (Grid) therapy

A sieve or grid for 200 kV X-rays consists of a sheet of lead (say 3 mm thick) in which there are small regularly-spaced apertures. Many grid patterns are possible, usually arranged so that the area blacked-out is 50–60 per cent of the total area. A chess-board pattern using 1 × 1 cm² squares occludes 50 per cent of the field. It results in a bundle of narrow 'pencil' beams with the intervening areas protected. Its use is known as *sieve or grid therapy* and differs from the usual continuous field by being discontinuous. As the beams penetrate the tissue, the discontinuity tends to be evened out, owing to scattered radiation and the geometrical divergence of each beam. The method is comparable to a multiple very-small-field technique, but in this case the small fields do not converge on a single small area.

The object of the technique is to take advantage of the untreated areas of skin and subcutaneous tissue to assist in healing of the radiation effects on the small 'open' fields. The dose is fractionated in space as well as in time. In this way much bigger doses can be given than with ordinary large fields, for each small area of radiation damage has a surrounding zone of relatively undamaged tissue and will, therefore, recover after a high dosage that would cause necrosis to a large open field. This technique is based on the unconventional assumption that it is not essential to deliver a high dose to all parts of a tumour in order to treat it satisfactorily.

Another advantage of the method is that it can be used to treat through skin that has already received high dosage and that would not tolerate further dosage by ordinary fields without necrosis.

Sieve therapy is particularly useful for palliative treatment of advanced growths, e.g., in the abdomen, where conventional therapy to high dosage would not be tolerated. Fields from front and back may be used. Using large anterior and posterior circular fields of 30 cm diameter, at 200 kV, daily doses of 500 rad may be given to the abdomen, one field per day, to a total of 10 000 rad. The maximum skin dose will be about 8500 rad and the central depth-dose will vary from 2500 to 4000 rad. Higher dosage can, of course, be delivered on smaller fields.

Computer assisted radiation treatment planning

The first application of computers in medicine was in radiation treatment planning—the planning of rotation therapy, to be precise. Since then a wide variety of computer programs have been developed taking advantage of the computer technology and applying it to radiation treatment planning in all its aspects. Since computers can only handle isodose charts in some mathematical form, various computer models have been devised which faithfully represent the same data. Today the isodose charts for all the beams available from a megavoltage unit can be adequately represented by a few hundred numbers.

The patient data required by the computer is again largely numerical, but the patient contour and other anatomy can be reduced to numerical coordinates using a rho-theta device or a graphics tablet (these are electromechanical devices for converting the position of a pen to electrical signals which the computer can recognise).

Now that quite small computers have large processor and storage capacities it is common to find a small computer dedicated to radiation treatment planning either in the radiotherapy department as a part of the simulator suite or in the medical physics department. These computers are used in the *interactive mode*, that is to say, the operator observes the dose distribution on a video display unit and adjusts the parameters—field size, angle, wedge, etc.—until he is satisfied with the result. The final distribution can be produced as a hard (permanent) copy on paper using the line printer or the line plotter as required (Fig. 6.14). This visual optimisation of the distribution is thought to be adequate when the operator has some experience of radiation treatment planning. The whole procedure may take only a few minutes per patient. In particular, several distributions can be produced from which the clinician can make his final selection.

The technique first described is considered to be one of the best at the present time, that is where an operator experienced in treatment planning observes, criticises and modifies the computed distribution until a satisfactory result is achieved. The alternative is to use a non-

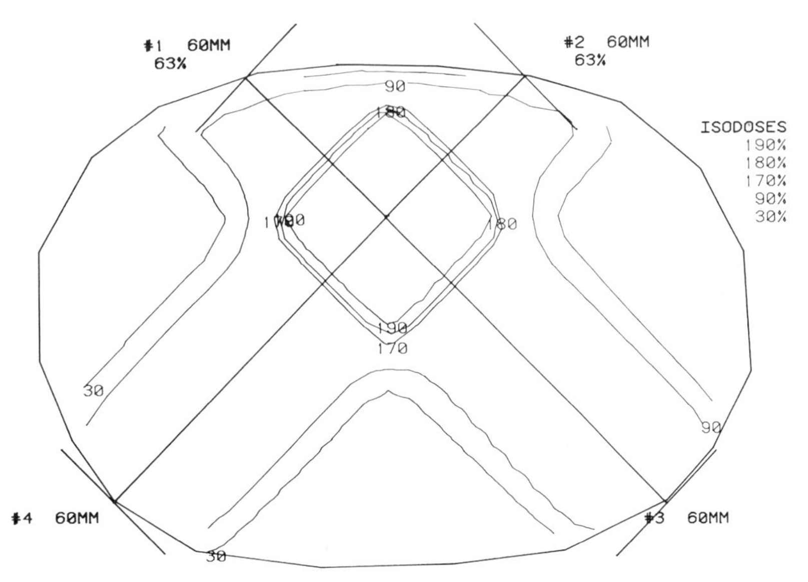

Fig. 6.14 A computed dose distribution.

interactive technique where the one experienced in treatment planning prescribes one or more possibilities and then sends the data by telephone or through the post to a (commercial) computer service where the program is run. The results are then posted back. If the results are not satisfactory, the procedure is repeated using some alternative data. Clearly the procedure is time consuming—from despatching the data to receiving the results may take several hours if not days—but it is practicable where there is no computer readily available on site.

Two potential advantages of the computer lie in the handling of inhomogeneities and in three-dimensional treatment planning. Neither of these can be fully realised at the present time owing to the problems of obtaining the necessary data from the patient—the size and shape of the inhomogeneity in the first instance (p. 141) and the three-dimensional size and shape of the treatment volume in the second.

Patient contouring devices

The value of radiation treatment planning depends on the accuracy of the position and shape of the patient on a day-to-day basis throughout the course of treatment. The couch on the simulator (Ch. 7) where the patient contour will be taken should be identical in every respect to the couch on which that patient will be treated. The obese patient will have a contour which changes from day-to-day, while the patients with less 'padding' will find the hard top couch so uncomfortable that they will find it hard to lie still for the duration of the treatment. In either case immobilisation devices (e.g., patient moulds or shells) become important especially for treatments of the head and neck. For treatments in the abdomen and thorax reliance is placed on the radiographer's ability to position the patient in the same way each day of treatment, paying particular attention to the position of the limbs and extremities. It is imperative that the patient is so positioned before the contour is taken for planning purposes. A variety of devices are available for taking patient contours. The very simple lead strip (say 10×3 mm^2 $\times$ 600 mm long) is widely used. Such a strip can be easily bent round the patient and skin marks transferred to the lead using a piece of chalk. The lead should be varnished in preference to being covered with a plastic sheath as the latter can slide along the strip (moving the marks with respect to the contour) or stretch or crack. As with other devices, it is advisable to check the principal dimensions of the contour by an independent means, e.g., calipers or a height bridge.

Adjustable templates are valuable for taking complex contours (e.g., round the ear). These are made up of a large number of parallel rods which can be adjusted lengthwise to fit any shaped contour and then clamped in position (Fig. 6.15a). The contour can then be traced on paper round the ends of the rods. Three such templates can be mounted on a frame and used to take complete contours (Fig. 6.15b).

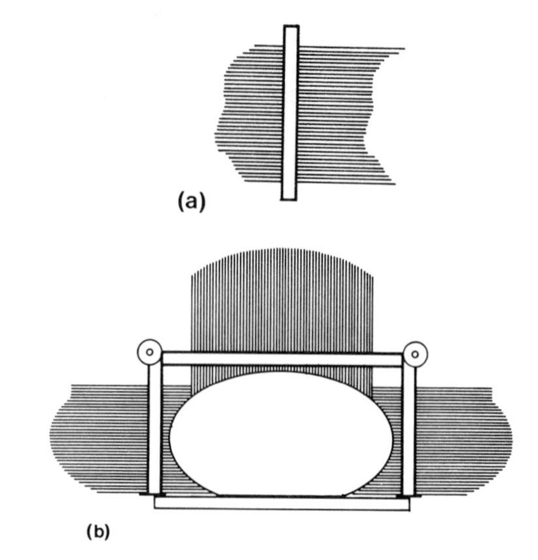

Fig. 6.15 The use of adjustable templates.
(a) Round the ear.
(b) For complete body contours.

A simple jig is shown in Fig. 6.16d for taking the contour from the outside of a patient's shell. The shell is supported on four pins which define the plane of the contour. A fifth pin is free to rotate about a vertical axis and scaled in degrees of rotation about the axis and centimetres radius from the axis. By measuring these two parameters for a variety of points round the surface of the shell, the contour can be readily plotted on polar graph paper.

Patient shells

A patient shell which fits a patient and immobilises that patient enables the planning and treatment to be carried out with greater accuracy. There is no place for a badly fitting shell. The good shell

1. Enables the localisation of the lesion to be accurately determined from surface markers attached to the shell
2. Enables an accurate position of the patient to be achieved each day
3. Provides an accurate and constant patient contour
4. Can be labelled and marked clearly and avoids the need to draw lines on the patient
5. Provides accurate beam entry and exit points
6. Provides a base on which wax build-up and local lead shielding can be built (Fig. 6.16f).

Space does not permit a detailed description of all the

procedures involved in making a shell and the student should get practical experience in a Mould Room wherever possible. The principles of making a shell using a plaster bandage impression are outlined below.

1. Preparation of the patient. When the impression is taken, the patient should assume the position to be adopted in the treatment room (Fig. 6.16a) as near as possible. The clinician should be consulted if there is any doubt. The mould of a patient's head will be more accurate if his long hair is cut short or shaved if he is likely to lose his hair during treatment. Hair should be covered with a tight swimming cap or stockinette and the face well coated with a separating medium such as petroleum jelly. The patient should be warned that the plaster will be cold at first, but will become quite warm as it becomes hard.
2. If a full head mould is to be made, the back half should be made first. The edge of the plaster bandage should be neatly finished to define the coronal plane through the centre of the ears. The impression should be allowed to harden and then a separating medium should be put on the outside extending about 3 cm from the edge.
3. Before commencing the front half, it must be decided how far the impression should extend—upwards and downwards—bearing in mind that the finished mould needs boney protuberances on which to locate, e.g., the nose, the eyebrows, the chin, the shoulders. Secondly, how is the patient going to breathe? He may breathe normally through the nose in which case the mould will be made with the mouth closed—with or without dentures. Alternatively, he may breathe through the mouth in which case a mouth insert and short breathing tube should be made to fix the position of the chin and to depress the tongue.
4. The impression can now be commenced. The plaster bandage should be shaped carefully round the boney points and round the greased edge of the back half of the impression (if present, Fig. 6.16b). The two halves should fit uniquely together after they have been removed from the patient. The impression will usually require about six thicknesses of plaster bandage suitably overlapped to give rigidity to the finished impression. Weaknesses can be strengthened after the impression has been removed from the patient—distortions cannot be corrected. Once the impression has hardened—after about five minutes—it should be eased off carefully. The patient will require assistance to remove the separating medium and splashes of plaster.
5. The two halves should now be fixed together by adding a further three layers of plaster bandage over the joint. If only one half has been made then the open ends should be closed using plaster bandage taking care not to distort the impression. The mouth or nose aperture should also be closed. After rinsing with a separating medium it is possible to fill the impression with a fairly thin mix of plaster of paris (adequate support round the impression must be provided so that it is not distorted under the weight of plaster). This should be allowed to set without applying additional heat.
6. The impression can now be peeled off the cast, leaving an exact replica of the patient. Small pimples on the cast resulting from air pockets in the impression can be pared away easily with a knife. The cast of a full head should be sawn into two halves along a diameter chosen to avoid the proposed treatment beams. The cut must be straight. Any deviations may cause the casts to crack and in any case they will result in the two vacuum-formed half shells being at too great a separation when reassembled.
7. The casts are now ready for vacuum forming (Fig. 6.16c). The flat surface of each cast is placed on the platform of the vacuum forming machine. The plastic sheet is inserted, clamped around its edges and heated until it is pliable. Compressed air is used to pre-stretch the plastic into a bubble and the cast is raised into that bubble. The vacuum is applied removing all the air and forming the plastic tightly round the cast.
8. The plastic is trimmed off the cast leaving a flange approximately 3 cm wide all round. Press-studs fitted into this flange will hold the two halves of the full mould together. If only a half mould is being made the excess plastic will be cut away completely. The breathing aperture can be cut away using a dental saw.
9. The patient shell is thus completed. Supports will be moulded (by hand) and attached to the shell to allow it to be fitted to a head rest or the couch of the treatment unit so that the patient is held secure (Fig. 6.16e).

If the treatment demands the skin sparing effect to be retained, the entry ports for the treatment beams will be pared away. The need for this should be borne in mind from the beginning as it will tend to weaken the shell making it less effective as an immobilisation device. There is no reason why the plaster bandage impression should not be used as the mould, providing it is well made it can be similarly pared away and fixed to the head rest or couch. The plastic vacuum-formed shell is more pleasant to look at and to handle, it stands up to daily use rather better and the transparency of the plastic enables the accuracy of the fit to be checked.

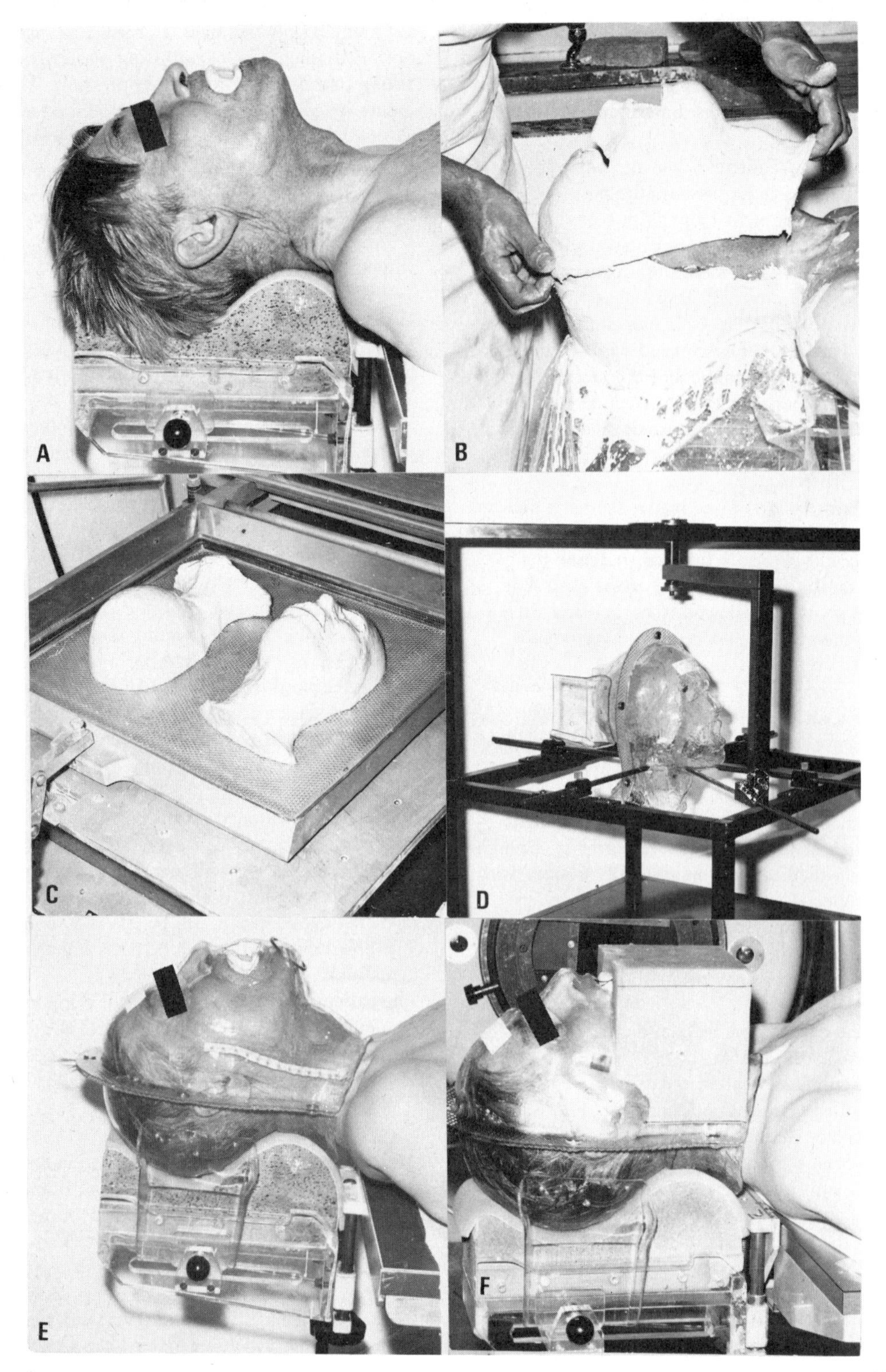

Fig. 6.16 The production and use of patient shells in treatment planning. A: the patient in the proposed treatment position with a breathing tube. B: the removal of the front half on completion of a whole head mould. C: the two halves in position for vacuum forming. D: the complete shell in the contour jig. E: the patient in position for the localisation films on the simulator. F: the shell complete with tissue equivalent wax block in use on the linear accelerator.

Alternative materials

Although plaster bandage has been used exclusively as an impression material in the discussion above, other materials are available and can be used to advantage in certain cases. Invariably they are used in conjunction with plaster bandage. Certain materials (e.g., dental compression compounds) are used dry, relying on heat to soften them during the impression stage and hardening as they cool. Other water based materials (e.g., alginates) are poured on to the patient and set into a gel which reproduces the very texture of the skin. These materials are valuable if impressions are required from moist open wounds.

Beam direction devices

The radiation treatment planning and the immobilisation of the patient is of little value unless the day-to-day treatment can be set up accurately. This requires that the treatment unit and the patient are correctly aligned for each radiation field. In simple terms *beam direction devices* are those which enable the beam axis to be directed at the lesion in the planned direction at a predetermined point and distance from the source of radiation. The simplest and most widely used devices are the front and back pointers.

Front and back pointers

Front and back pointers are mounted on the unit in such a way that they lie and move along the beam axis, the front pointer being calibrated to show the distance between its tip and the source of radiation (Fig. 6.17).

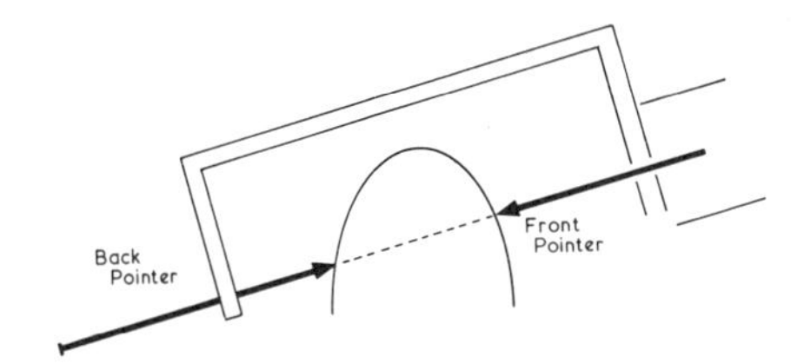

Fig. 6.17 The principle of the front and back pointer.

The accuracy of any pointer system needs to be checked periodically. In this case, this is done by rigidly mounting a reference marker such that it coincides with the tip of the pointer to be checked and rotating the beam diaphragm system about the beam axis—the tip of the pointer should not deviate from the reference marker by more than one millimetre. This check should be repeated with the pointer set at several positions along its length and at several gantry angles.

In use the front pointer is set at the SSD prescribed and the patient positioned so that the entry point (for the beam axis) on the patient coincides with the tip of the pointer. The angle of the beam is then adjusted until the back pointer coincides with the exit point, the entry and exit points being clearly marked on the patient's skin or mould. The pointers should be removed before the beam is switched on.

The front pointer may in practice be replaced by a light pointer system or by a cross hair mark on the face of an interchangeable applicator. Similarly back pointers have been replaced by other devices, too numerous to mention but which fulfil the same purpose, namely to define the direction of the beam in relation to the entry point on the patient.

Pin and arc

The pin and arc device has largely been replaced over recent years by the isocentric gantry mounting of megavoltage equipment. Not all units are so mounted and a brief description of this once popular device is not out of place. The principle is shown in Figure 6.18.

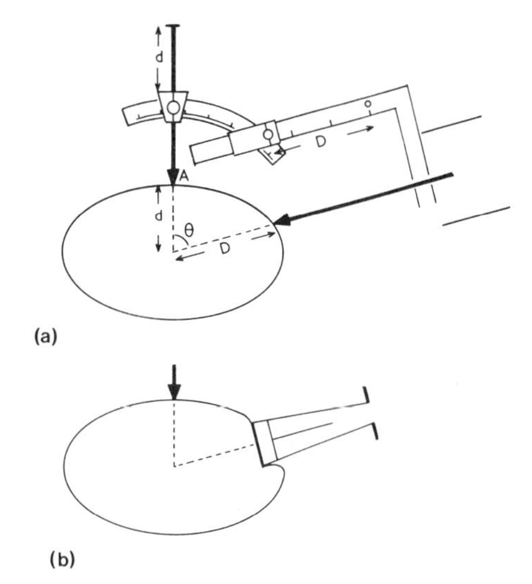

Fig. 6.18 The principle of the pin and arc technique.
(a) Using beam diaphragms and front pointer
(b) Using beam applicators and compression.

A bar is attached to the unit parallel to the beam axis. A 90 degree arc is mounted on the bar in such a way that its centre (the centre of the circle of which the arc is a part) lies on the beam axis. The pin is mounted on the arc and lies along a radius of the arc. Both the pin and the bar are graduated in centimetres so that the point of the pin lies on the beam axis at the normal skin distance when both scales read zero.

The distances d and D and the angle θ are determined from the treatment plan. The treatment head is rotated so that the beam axis lies at angle θ to the vertical. The depth d of the centre of the tumour vertically below the skin mark A is set on the pin, the distance D on the bar and the angle θ on the arc. When the pin coincides with the skin mark A, the beam axis will be accurately centred on the tumour.

The mechanical checking of this device is similar to that

for front and back pointers. With the pin set at $d = 0$, its tip should lie on the beam axis for all the available values of D and θ.

The isocentric gantry

Reference has already been made to the isocentric gantry but its description is included here for it is essentially a beam direction facility. It is a development of the pin and arc.

The isocentric gantry has three principal axes of rotation, the beam axis or axis of rotation of the diaphragm system, the horizontal axis of rotation of the gantry and the vertical axis of rotation of the couch. These three axes of rotation intersect at a point called the *isocentre* (Fig. 6.19). The position of this point in space may also be identified by two or three light pointers mounted on the walls and ceiling of the treatment room.

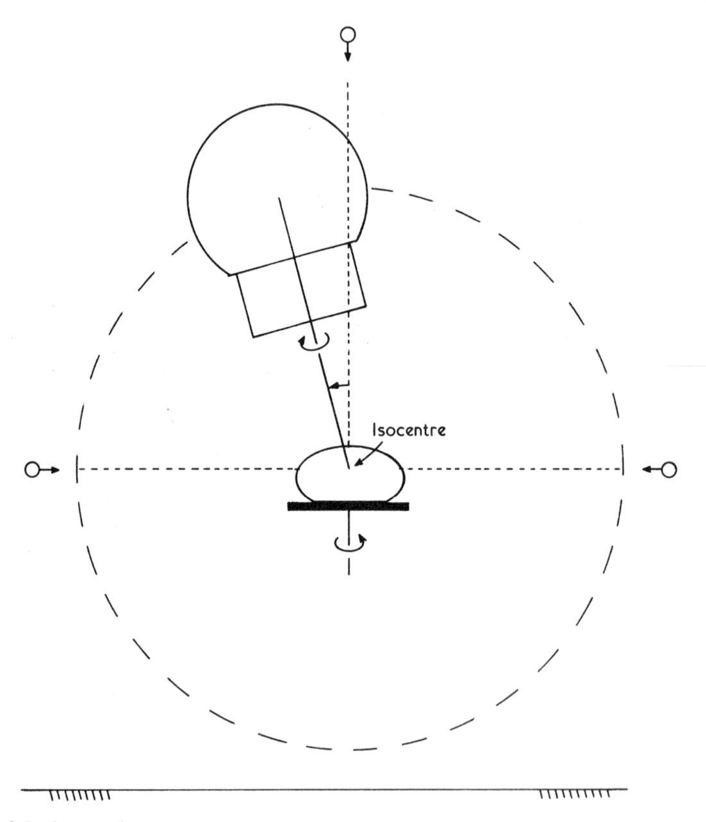

Fig. 6.19 The principle of the isocentric gantry.

The value of the isocentric gantry lies in the fact that in setting up the patient a rotation about any of these three axes can be carried out knowing that the position of the isocentre is not altered. In practice, this is a considerable help whether the patient is being treated with the isocentre on the skin or inside the lesion.

To set up to a skin mark. Using the optical or mechanical front pointer the skin mark is positioned at the isocentre using the *linear* movements, i.e., couch height, lateral and longitudinal movements. This fixes the source-skin-distance. The beam direction is then set using the *rotational* movements, i.e., gantry and couch rotations to the specified angles or until the back pointer locates the beam exit mark. The diaphragm rotation completes the set-up. Do not use the linear movements after the entry mark has been set up. It is easy to remember: linear movements set the entry point, rotation movements set the exit point.

To set up to a point inside the lesion. At the planning stage the intersection of the beam axes is located as being 'd cm vertically below' an appropriate skin mark. The front pointer is set at the source-axis-distance *less* the value of d. With the beam axis vertical, the pointer and the skin mark are brought together using the *linear* movements as before, so positioning the planned intersection of the beam axes at the isocentre. The patient couch is now rotated to align lateral skin marks with the wall-mounted light pointers or to define the plane containing the planned beam axes. The gantry and diaphragm rotations can now be set for each field in turn. Once the patient has been positioned there is no need to move the patient or the couch between the treatment of different fields.

A further advantage of mounting a Cobalt unit or linear accelerator on an isocentric gantry lies in the fact that the massive weight required to counterbalance the head on the gantry rotation can be put to good use in the form of a *primary beam stop*. This is a shield mounted on the gantry in such a way that it will attenuate the primary beam after it has passed through the patient irrespective of gantry angle. In certain situations this can be more economical than providing primary barriers in the structure of the room (Ch. 11) but at the expense of some inconvenience to the radiographer setting up the patient.

Finally, a word of warning. The rotational movements referred to above are those about the principal axes which define the isocentre. The unit may have other rotational facilities, the main ones being the rotation of the couch about an eccentric support and a limited rotation of the treatment head within its supporting arm or yoke. The latter enables the beam axis to move away from the isocentre in special circumstances—it should be returned to and locked in the isocentric position at all other times. Similarly, the eccentric rotation should be locked in its central position when it is not being used.

Treatment of superficial lesions

So far in this chapter we have dealt with the treatment of deep seated lesions using two, three or four fields. If the lesion is superficial and only involves the skin or underlying tissues to a depth of a few centimetres, then a single treatment field is all that is required.

Depth dose is a function of photon energy and inverse square law and it may be reduced by reducing either photon energy or the source-skin-distance or both. Traditionally, superficial X-rays are those generated at 50 kV to 150 kV and used at a source-skin-distance of 15 to 25 cm. Owing to the predominance of photoelectric absorption in tissues at these photon energies, the irradiation of surface lesions overlying bone has lead to very high doses being given to the bone. Today, higher photon energies may be used at even shorter source-skin-distances, e.g., Co-60 gamma rays or heavily filtered orthovoltage X-rays at about 5 or 10 cm SSD. (Radium-226 at similar distances had been used for many years prior to the change to Cobalt-60, see Ch. 3). Furthermore, electron beams up to about 10 MeV and beta rays plaques are valuable in certain circumstances.

Typical isodose curves for these types of radiation are shown in Figures 6.20 and 8.11. The students should note the three characteristics mentioned earlier (p. 55) in respect of these isodose charts. The advantages and disadvantages of each will be dealt with briefly.

Superficial X-ray beams

Superficial X-rays still treat over 90 per cent of the superficial lesions in the radiotherapy department. The advantages are the simplicity of the X-ray unit both in design and operation, the ease of collimation and field-shaping to individual requirements with only one millimetre of lead

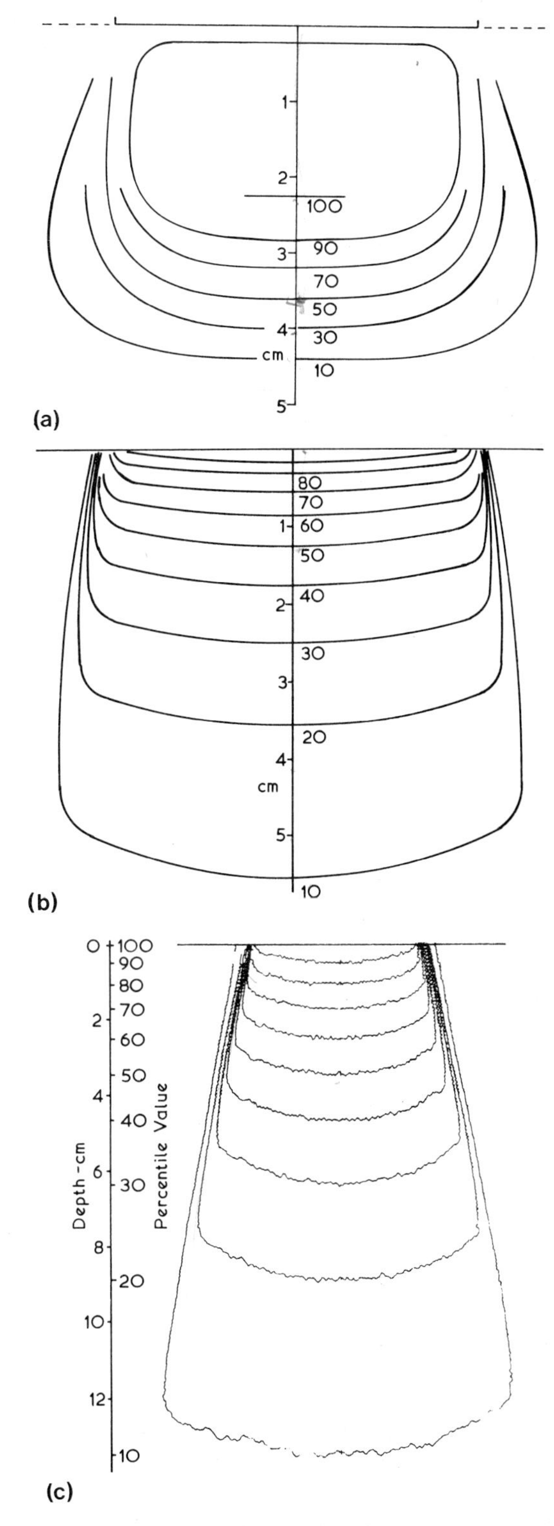

Fig. 6.20 Isodose charts for superficial treatments
(a) Electron beam: 9 MeV
(b) X-ray beam: 1 mm Al HVT, 15 cm SSD
(c) Cobalt-60: 10 cm SSD.

and a wide range of field defining applicators. The disadvantages are the high bone absorption and the minimal penetration—over the range of energies and field sizes in common use the 90 per cent depth varies from only 2 mm to 8 mm (see also Fig. 5.11).

High energy photon beams

Cobalt-60 gamma rays at 5 or 10 cm SSD can only be collimated adequately by using 5 cm of heavy alloy which means there can be no individual beam shaping for each lesion to be treated. Skin sparing is not required as the lesions are superficial although electron contamination of the gamma beam must be filtered out to prevent an excessive skin dose. The small source (say 50 Ci, 4 mm diameter) produces a narrow penumbra but the short SSD gives rise to a wide divergence of the beam. The shortness of the SSD also means that the setting up of the patient is critical. A 1 millimetre error in a 5 centimetre source-skin-distance produces a 4 per cent error in doserate. A patient shell is essential to maintain the SSD as well as to immobilise the patient.

The heavily filtered orthovoltage X-ray beam has certain advantages over the cobalt-60 gamma beam but shares the same disadvantages of the short source-skin-distance. It has the similar bone sparing properties as cobalt-60 but collimation is possible using several millimetres of lead instead of several centimetres of heavy alloy. Therefore, beam shaping is possible and electron contamination is less of a problem. The small field sizes and low penetration means there is a negligible amount of low energy scattered radiation and the heavy filtration cuts out the low energy primary photons giving the beam a high homogeneity coefficient.

Orthovoltage X-rays (say, 2 mm Cu HVL) may be used at normal SSDs for the treatment of some superficial lesions.

Electron beams

High energy beams for clinical use may be produced from linear accelerators and from betatrons over a range of energies from about 3 MeV up to about 50 MeV. For clinical purposes the nominal energy of the electron beam is defined in terms of its half value depth, i.e.

$$\text{Energy (MeV)} = 2.5 \times \text{HVD (cm)}$$

where the HVD is the depth of the 50 per cent isodose line on the central axis. The depth dose curve (Fig. 6.21) for electrons is characterised by its flat peak and the steep dose gradient which flattens in to the X-ray tail (the X-rays being generated at the same peak energy as the electrons using the electron window, ionisation chamber, the air and the tissue as X-ray targets). The gradient of the depth dose curve is less steep for the higher energies since these electrons undergo many more interactions (ionisations and excitations) and their paths are correspondingly more tortuous.

The advantages of electron beams (up to 10 MeV) for the treatments of superficial lesions are clear cut. The dose is satisfactorily uniform to a depth determined by the energy of the electrons (approximately half the HVD) and the sharp fall off of dose leads to the sparing of tissues at depth. Collimation is relatively simple over a wide range of field sizes, local beam shaping requires a few millimetres of lead depending on electron energy.

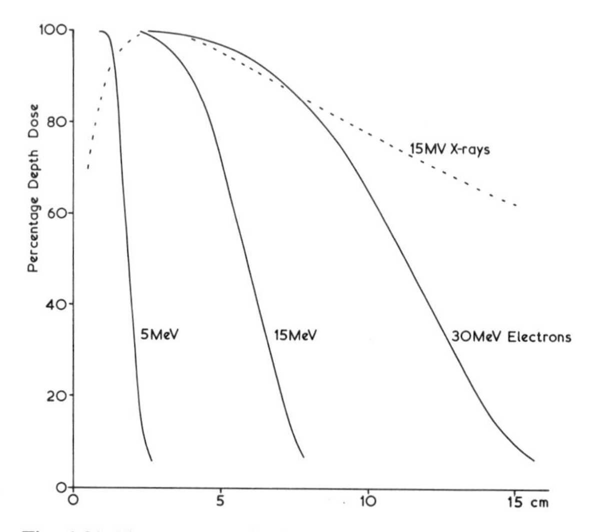

Fig. 6.21 Electron central axis depth dose curves.

The higher electron energy beams (10 to 50 MeV) may be used in single or multi-fixed field treatments or in combination with photon beams.

Extensive skin disease (e.g. mycosis fungoides) can be treated using the large low energy electron beams (say 3 MeV) covering the whole surface of the patient. In this case the patient may be treated with four very large (2×2 m^2) fields at long source-skin-distances, or with more than four smaller overlapping fields. The minimum of four fields is required to cover the anterior, posterior and two lateral aspects of the patient. At long source-skin-distances, the X-ray contamination of the electron beam may contribute a significant if not prohibitive whole body dose to the patient.

Beta particle beams

An alternative to low energy electron beams for the treatment of mycosis fungoides is the use of the beta particle emmision from strontium-90. A high activity (24Ci) strip source covering the width of the patient can be scanned over the length of the patient, delivering an acceptably uniform dose to a depth of approximately 1 millimetre falling to 50 per cent at 2 mm.

Calculation of dose

No treatment plan will be complete without a specification of the dose to be given to the tumour and how this is to be achieved. The dose prescription will be the responsibility of the clinician and will be decided in terms of the tumour dose in rads, D, to be given to a specified percentile value, TD, in a number of fractions, F. The number of fractions per week of treatment will be an important factor in determining the overall dose, but does not come into the calculations.

The clinician's choice of the percentage tumour dose, TD, will be based on the detailed dose distribution and whether he will require the prescribed dose to be the *maximum*, *minimum* or *modal* dose to the tumour. The first two of these are unique points and easily identified. The percentage modal dose is not so easily determined. It requires the treatment volume to be divided into a large number of elements and the percentage dose to each elemsent to be recorded. From this record one percentage value will be found to occur more frequently than any other and this is called the *percentage modal dose*. Whichever of the three possibilities is chosen, it expresses the tumour dose as a percentage of the peak (or applied) dose to each field, this peak dose, P, can, therefore, be determined as

$$\text{Peak (applied) Dose, } P = D \times \frac{100}{TD} \text{ rad}$$

If this is delivered in F fractions, then the peak dose per fraction is P/F rad. If the peak dose rate, $\dot{P}$, is known in rads per minute, then the exposure time, t, is calculated as

$$t = \frac{P/F}{\dot{P}} \text{ min}$$

Where weighting factors (p. 65) have been used in the derivation of the final dose distribution, then the peak dose to the weighted fields is not 100 per cent and it must be modified accordingly. If the weighting factor is W, then the peak dose will be WP, the peak dose per fraction WP/F and the exposure time $WP/F\dot{P}$.

Now the peak dose rate $\dot{P}$, may have been measured and listed for all the available beam defining applicators and the appropriate value will be used. If the actual field size is not listed and an overlarge applicator is used in conjunction with a lead cut-out (p. 29), to reduce the irradiated area to the required size and shape, the peak dose rate is determined by reducing that for the applicator by the ratio of the back-scatter factors.

$$\dot{P} = \dot{P}_{\text{app}} \times \frac{BSF_{\text{cut-out}}}{BSF_{\text{app}}}$$

Where adjustable diaphragms are used the available field sizes are too numerous to list all the corresponding peak dose rates and the peak dose rate, $\dot{P}$, is quoted for one field size only, say 10×10 cm^2 (p. 55). The peak dose rate for any other field size is determined by multiplying the quoted value of $\dot{P}$ by the *output factor* for the equivalent square of the field size. This output factor is defined as

$$= \frac{\text{peak dose rate for the equivalent square}}{\text{peak dose rate, } \dot{P}\text{, for the reference field size}}$$

This peak dose rate may be further modified owing to the use of wedges, compensators (p. 63) or the use of a non-standard source-skin-distance.

These calculations are based on dose rate as in this way they are simple to understand. On certain types of equipment (e.g., linear accelerators) reliance is placed on a dose integrating system rather than a treatment timer. The dose level at which the integrator is set to trip will be calculated on the basis of the peak dose required and the relevant correction factors listed above.

7. Diagnostic radiography and the treatment simulator

Introductory note
In previous chapters the word *exposure* has been given to a clearly defined scientific concept, namely the ionisation produced per unit mass of air ($X = \Delta Q/\Delta m$), while in every day usage it takes the meaning of *being exposed to*. In this chapter the word will be used in the latter sense unless it is accompanied by the symbol for exposure (X).

Introduction
The most precise method of checking the accuracy of a treatment plan is to expose an X-ray film during treatment and so produce a picture of the tissues included in the treatment field. This film is known as a *check film*. The film should show the entire treatment volume enclosed by the adjacent anatomy thereby confirming that the treatment field is correct in position, size and direction. In addition, it is convenient if the film can be left in position for the whole treatment fraction rather than have to disturb the patient to remove the film before completing the dose.

At orthovoltage X-ray energies, the image quality of the check film is poor but adequate providing the exposure factors and processing combine to produce the maximum contrast. The advent of megavoltage radiotherapy brought several potential improvements in treatment not least

1. Improved penetration (greater depth dose)
2. Reduced skin dose (the build-up effect)
3. More sharply defined beams (narrow penumbra)
4. Reduced bone absorption

The reduced bone absorption is an advantage to the radiotherapy but not to the check film. The only appreciable contrast on a check film taken using megavoltage beams is between the tissues and the air passages (where the density differences (in gcm^{-3}) are very great, Fig. 3.27b). The only way open to continue the use of check films is to have a machine which simulates the movements of the therapy unit but uses a source of low energy X-rays. The *treatment simulator* is, therefore, a unit incorporating a

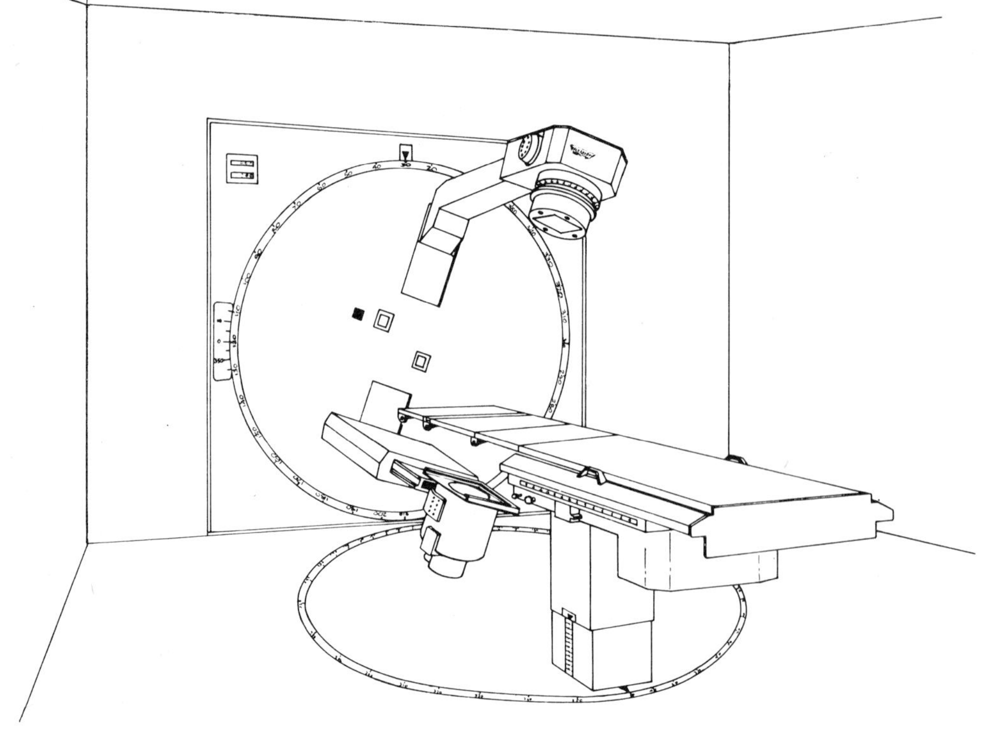

Fig. 7.1 A typical treatment simulator (courtesy of TEM Instruments Ltd).

diagnostic X-ray tube in a gantry similar to the treatment unit, particularly in regard to beam collimation and direction and the patient's couch. Early simulators were home-made conversions of a therapy unit gantry to accept a portable diagnostic X-ray tube and generator. The modern simulator (Fig. 7.1) is a sophisticated unit and often considered essential to the planning of any course of radical radiation treatment using megavoltage radiation. The unit is best situated close to the treatment units it simulates but perhaps more important, it should be close to the mould and planning room facilities. It is often the focal point in the therapy department where staff in these various disciplines converge and bring their expertise to bear on the patient's treatment. In most departments, the simulator will be in the hands of the radiographic staff.

This chapter seeks to outline some aspects of diagnostic radiology which are relevant to the use of simulators. In no way is it to be taken as a comprehensive treatise on the subject. It is only an introduction to encourage the radiographer to begin to understand the simulator.

The radiograph

The latent image

The radiographic film is a transparent plastic base coated on both sides with a gelatine layer containing silver halide crystals which will absorb X-ray energy. After exposure to X-rays, the film appears to the naked eye to be unchanged, but within the crystals of the emulsion a *latent image* has been produced. This image is made visible by the action of the developer (a reducing agent which reduces the activated silver halide to silver) and the fixer (an acid solution) which dissolves away any remaining silver halide crystals. The processed film shows a pattern of varying degrees of greyness from the clear transparent film base to the black opaque silver. The effect is illustrated in Figure 7.2. The degree of blackness or the *density* of the film depends on the exposure (X) received by the film.

The characteristic curve

The characteristic curve is a graphical representation of the way the optical density varies with the exposure (X) received by the film. At very low exposures (X) there is little density change and the density measured is mostly that of the *fog level* of the film. At very high exposures (X), any further increase produces no density change and the density measured is that of the silver deposit on the film when all the silver halide crystals have been reduced—a situation which exists when the film is *over-exposed*. Between these two extremes, there is a region of *correct exposure*—the relatively straight portion of the characteristic curve where the gradient is at a maximum (Fig. 7.3). The correctly exposed film has densities within this region of the curve, that is, within the range of approximately 0.4

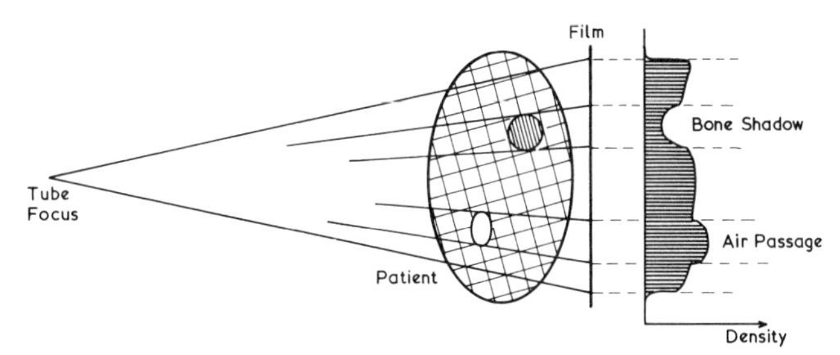

Fig. 7.2 The principle of the radiograph.

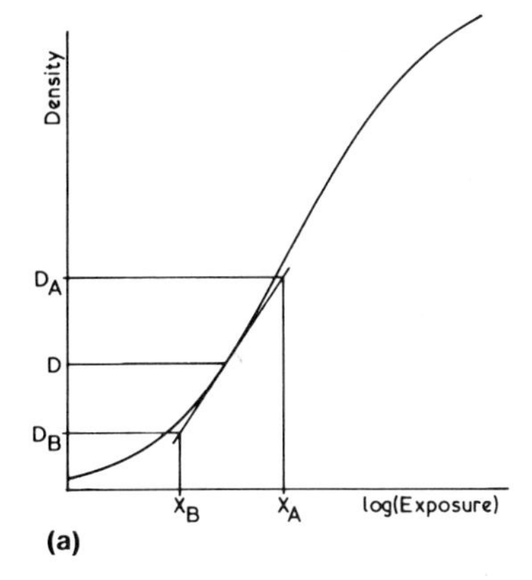

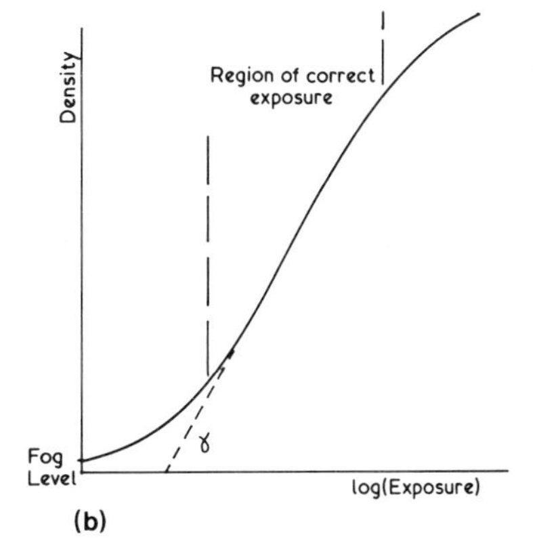

Fig. 7.3 The characteristic curve for X-ray film.
(a) Film contrast.
(b) Region of correct exposure.

to 2.0. It is important in producing a radiograph to obtain the maximum density difference between the points of interest. The difference in density at two points is known as the *contrast*.

$$\text{Contrast } (C) = D_A - D_B$$

The human eye under ideal conditions can detect density differences as small as 0.02. Although contrast between two points of the film is the difference in their densities, the *film contrast* is the slope of the characteristic curve at the point of measurement, D

$$\text{Film Contrast} = \frac{D_A - D_B}{\log X_A - \log X_B}$$

while the maximum film contrast, the slope of the straight portion of the characteristic curve, is known as the *film gamma* (γ) or contrast index. Another important parameter of the film is its *speed*. If only a small exposure is required to produce a given density, the film is said to be *fast*, whereas a *slow* film requires a longer exposure to produce the same degree of blackening.

The emulsion of an X-ray film is very thin and only absorbs a fraction of the X-ray photons incident upon it to form the latent image. In order to improve the efficiency, a pair of *intensifying screens* are used, one in front and one behind the film. These screens are coated with a layer of crystals (usually calcium tungstate) which fluoresce when irradiated with X-rays converting the X-ray energy into light—usually blue or ultraviolet—to which the emulsion is also sensitive. The action of the screens is, therefore, to supplement the direct effect of the X-rays on the film thereby reducing the X-ray exposure (X) required to produce an adequate density on the film. This *intensification factor* is a measure of the increased speed, namely

$$\text{IF} = \frac{\text{Exposure required to produce a given density when screens are not used}}{\text{Exposure required to produce the same density when using screens}}$$

The use of intensifying screens also greatly enhances the contrast on the film. Rare-earth screens (so called because they use the rare earth elements of gadolinium and lanthanum) have recently become available and have increased the intensification factor to 2 to 4 times that of 'fast' tungstate screens. A typical pair of fast screens have an intensification factor of 30. Since long exposure times have limited the use of simulators in certain examinations in the past, the rare-earth screens will undoubtedly be used in the future.

Because the developer plays an important role in the production of the visible image, it is important to use the correct type of developer under the prescribed conditions of dilution, temperature and duration. Over-development increases slightly the effective speed, the gamma and the fog level of the film. Automatic processing prevents our using the old trick of over-developing the film to compensate for an earlier error of under-exposure.

Geometric factors

Any X-ray image will be a magnified image. Just as an optical shadow is larger than life size, so the image on the X-ray film is larger than the anatomy it depicts. Since X-rays travel in straight lines from a small focal spot, the magnification factor is defined (Fig. 7.4) as:

$$m = \frac{XY}{xy} = \frac{f}{s} = \frac{f}{f-h}$$

It will be noted that the object, xy, lies parallel to the film. If this was not the case, then the magnification factors applicable to the two ends of the object would not be equal and the image would suffer *distortion*. It should

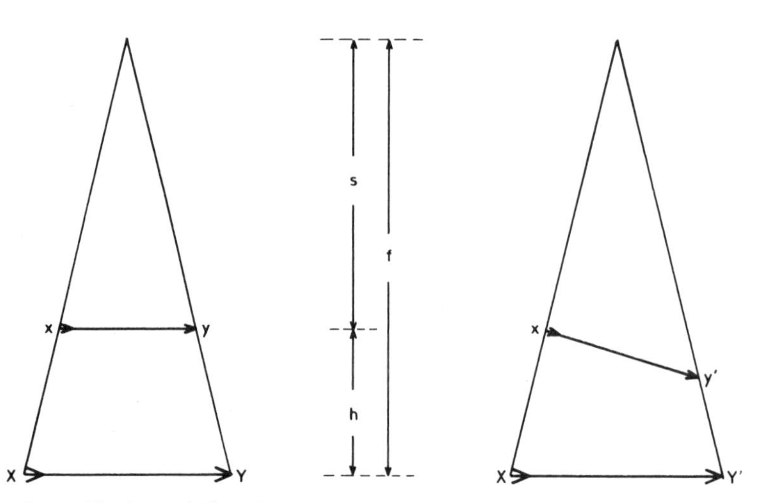

Fig. 7.4 The principles of magnification and distortion.

also be noted that f, h and s are measured parallel to the beam axis, not along the ray from the focal spot. If the focal spot is not negligibly small, a situation arises analogous to the geometric penumbra of the therapy beam. Here, there will be a *geometric unsharpness* at the edge of the image owing to a gradual fall-off of density. This is minimised by using a large focal-film-distance (FFD), f, minimising the distance between the object and the film, h, and selecting a small focal spot. On simulators, this effect is most clearly shown in the image of the wire diaphragms used to define the field size. The wires are of small diameter at a long distance from the film and cast a broad shadow on the film—particularly where a broad focus tube is used. (Note, a double image is sometimes visible on one pair of wire diaphragms but not visible on the other pair. This is due to the nature of the focal spot. The focal 'spot' usually consists of two relatively intense line sources of X-rays. When the wire of the diaphragm is parallel to the line foci, two images of the wire are produced under conditions of large magnification. This effect is not apparent in the tissues of the patient because the magnification is small and the effect reduces to one of geometric unsharpness.)

The loss of contrast

All the attention so far has been on the differential attenuation of the X-rays passing through the patient, reaching the film and causing differential blackening on the film. The contrast produced is the result of the photoelectric attenuation being dependent on the atomic number and on the density of the tissues. Inevitably, Compton scattering will be present. The scattered radiation does not reflect the attenuation pattern of the primary radiation, but always uniformly irradiates the whole film, increasing the fog level and reducing the contrast. One way of improving contrast is to lower the tube kilovoltage or to reduce the effects of scatter by:

1. Minimising the production of scatter by reducing the overall field size
2. Preventing the scatter from reaching the film by using a grid.

Just as back scatter increases with field size (p. 53), so forward scatter (which reaches the film) increases with field size—and for the same reason. It is common practice on simulators to have two sets of diaphragms—opaque diaphragms to limit the irradiation of the patient and wire diaphragms to delineate the edges of the treatment beam. It is good practice to keep the area defined by the opaque diaphragms as small as possible to minimise the scattered radiation.

A *grid* (sometimes called a scatter grid) is a series of equally spaced lead strips, usually between 20 and 40 strips per centimetre, such that X-rays travelling from the tube focus can pass unhindered between the strips while scattered radiation travelling in random directions will be absorbed in the lead (Fig. 7.5). For most purposes a *parallel grid* is adequate providing the *grid ratio* is not too large, the grid ratio being defined as:

$$\text{grid ratio} = h/D$$

where h is the height and D the width of the space between the lead strips. If the grid ratio is too large, even the primary radiation from the focus will get absorbed at the edges of large fields. To prevent this a *focussed grid* can be used in which the lead strips are mounted so as to lie along the rays from the focus rather than parallel to the central axis.

The alternative to the use of grids is the acceptance of very large magnification factors with the film removed a

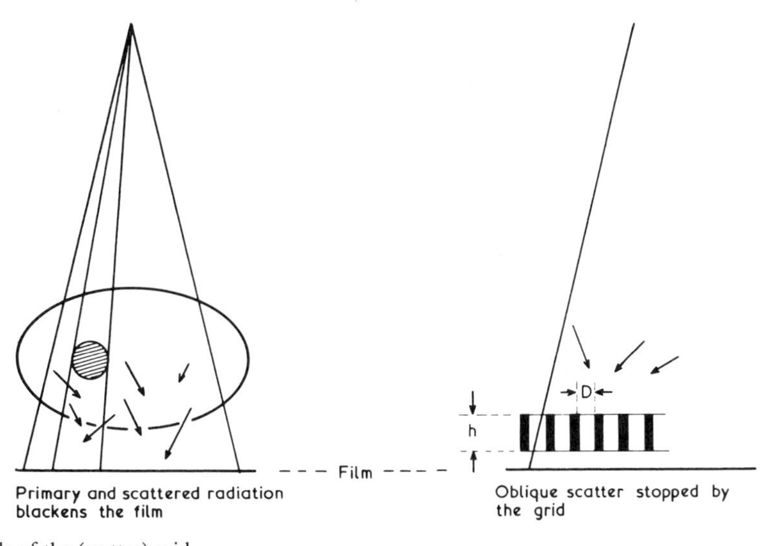

Fig. 7.5 The principle of the (scatter) grid.

long way from the patient. In this way the scattered radiation is attenuated by the inverse square law and much of it will shine past the film altogether. Unfortunately, geometric unsharpness is exaggerated by this technique.

Radiographic exposure

Several hints have already been given as to the choice of the *exposure factors* required to produce a good radiagraph—e.g., to use a low kV for maximum contrast and to adjust the total exposure to keep the range of densities within the straight portion of the characterisitic curve. There are others: to use the minimum exposure time to arrest any voluntary or involuntary movement of the patient and to use a fast film-screen combination to minimise the dose to the patient. The exposure (X) from a diagnostic X-ray unit at the surface of the patient is related to the exposure as follows:

$$X \propto \frac{(kV)^2 \times (mA) \times (s)}{d^2}$$

where d is the SSD, kV the peak voltage of the tube, mA the tube current and s the exposure time in seconds. Since the exposure is directly proportional to both the tube current and the time in seconds, it is often convenient to quote the exposure in *mAs* leaving the final choice of tube current and exposure time to the radiographer—usually she chooses the maximum mA available from the unit at the chosen kV and, therefore, the minimum exposure time—but how does she ascertain the correct mAs in the first place? In any one department, the film-screen-grid combination will be decided upon in advance and there will be little or no choice. Secondly, radiographs will have been taken many times before and satisfactory exposure factors will be available based on experience. Published 'Exposure Tables' may be used as an alternative guide, but is must be remembered that on a simulator the source-skin-distance and the FFD may be very different (up to a factor of 3) from those used in a diagnostic X-ray department and the mAs recommended may need to be increased to obtain comparable films on the simulator. It is for this reason that X-ray generators giving up to 800 mA or even 1000 mA are used on simulators. Once the mAs has proved adequate for a given examination, then patient-to-patient differences will usually be accommodated by small changes in kV.

Unlike the policy adopted in diagnostic X-ray departments, it is often said that in simulation the patient dose is unimportant in that the dose to be given in the treatment situation will be many times greater than that from the simulator. This is true of the lesion and the surrounding tissues. However, if the patient receives an extensive radiographic examination in regions where every effort will be taken to minimise the dose received during treatment (e.g., the lens of the eye), then the radiographic dose may be a significant fraction of the whole.

In all this discussion it seems to be 'the fastest, the shortest and the lowest'. This duly qualified by a satisfactory radiographic result forms a good maxim on radiographic exposure.

Image intensification

So far we have concentrated on the production of an image on film and much of what has been said is true for the formation of any radiographic image. The early simulators used only film. It was soon realised that the immediate display of a dynamic image was far superior to the image produced using a film, which, after all, needed processing. The image intensifier takes the rather faint image formed on a fluorescent screen and through electronic amplification produces a brighter image on a television screen. In no way can the intensifier improve the quality of an image, but it can make a good quality image brighter and easier to see. Diagnostically, the intensifier reduces the dose to the patient in that it requires a lower intensity image than the human eye. On a simulator, the intensifier has its value in that the image is displayed on a television screen remote from the patient and independent of the direction or movement of the X-ray beam.

In addition to the television display the same dynamic picture may be recorded on a *video-tape* recorder connected to the television system, or by a *cine camera* attached to the intensifier. Alternatively, static pictures may be recorded using a 70 or 100 mm camera attached to the intensifier. All these enable the radiotherapist to look again at the pictures at a later date.

Under normal operation the television display is dynamic. This is ideal for, for example, the barium meal examination where the radiotherapist is localising an oesophageal obstruction. However, in many examinations a static picture gives the operator time to absorb more of the detail displayed which would otherwise be lost. Equipment is now available to feed the television display with a series of static pictures (as in cine photography) in place of the dynamic one. By the use of a *video disc* to record a single television frame, or by the use of a *storage display tube*, a static picture can be produced by a flash of X-rays and displayed for any length of time. A press of a button will initiate another flash of X-rays and a new picture on the screen. This updating of the picture may be triggered automatically at some prescribed frequency if required. This *flash radiography* again helps the diagnostician to reduce the patient dose and at the same time to freeze the picture he wishes to scrutinize at length. The real value of the immediate television display over the delayed radiographic film will become clear when the use of the simulator is discussed below, but both facilities are required if the simulator is to be used to advantage. The full-sized radiograph is required in the planning of the treatment and the dynamic television picture in the initial localisation stages. The miniature pictures may be used for record keeping and follow-up purposes.

Localisation technique

It will be noticed that the emphasis on the simulator is on localisation rather than diagnosis. A lesion cannot be localised until it has been diagnosed. Although the simulator incorporates many of the features of a diagnostic X-ray unit, it is not intended to be used for the diagnosis of disease. This is best done by the radiologists in the diagnostic X-ray department.

By localisation we mean using the radiographic image in order to ascertain the position of an internal structure with respect to an external marker. In order to do this the radiographs need to be taken under prescribed conditions to show the relationship between the internal structure and the external marker. Two techniques are common—the orthogonal film technique and the tube shift technique. The former is prefered for tumour localisation as tumours are not sufficiently differentiated to be adequately discerned on the double image of a tube shift film. Two films are taken of the patient, one at 90 degrees to the other. These are usually, but not necessarily, the AP and lateral views of the patient. For convenience we shall refer to them as 'AP' and 'Lateral'. Ideally, the films are taken simultaneously, but in practice this cannot be done and the patient is asked to remain motionless between the two exposures.

Under screening conditions the marker M (Fig. 7.6) is positioned at the isocentre vertically over the centre O of the tumour-bearing volume. Then, $S_1M = S_2M = 100$ cm (say). The distance of the film from the isocentre will be noted in each case, i.e., F_2M and F_1M, they may or may not be equal. From these data, the magnification factor for the lateral film at the level of the centre O will be:

$$M_{\text{lat}} = \frac{F_2M + S_2M}{S_2M}$$

Thus, the distance of the centre O below the marker M will be given by:

$$MO = \frac{M_2O_2}{M_{\text{lat}}}$$

The magnification factor for the AP film at the level of the tumour centre O will be:

$$M_{\text{AP}} = \frac{F_1M + S_1M}{S_1M + MO}$$

If O is the centre of a tumour bearing volume the actual dimensions of that volume within the patient may be determined from the dimensions on each radiograph divided by the appropriate magnification factor. If it is not possible to measure the distances F_2M and F_1M, then the marker M should be replaced by a ring of known diameter (say 5 cm). The reason for using a ring is that on the radiographic image the maximum dimension is always a diameter. The two magnification factors are then:

$$M_{\text{lat}} = \frac{d_2}{d}$$

$$\text{and } M_{\text{AP}} = \frac{(d_1/d) \times S_1M}{(S_1M + MO)}$$

where d is the diameter of the ring and the suffices have the same meaning as before.

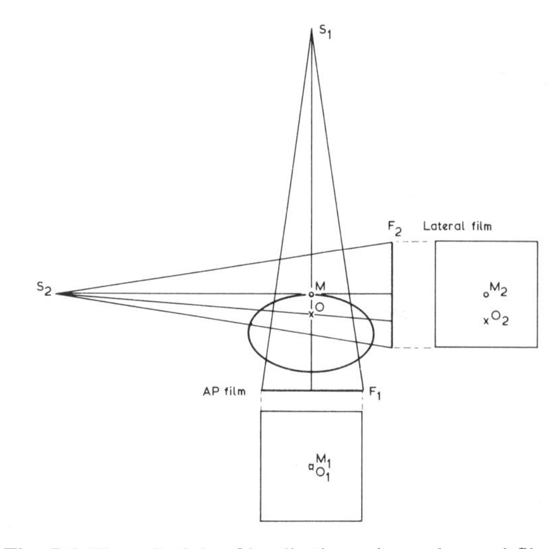

Fig. 7.6 The principle of localisation using orthogonal films.

The simulator in practice

External radiation treatment planning

The patient's visit to the simulator will usually follow the diagnosis of the disease, the proposal for radiation treatment and the preparation of a patient shell or mould. At the first visit then the patient can be placed in the position to be assumed during treatment and immobilised by the shell as appropriate. Using the screening facility that patient can be positioned so that the centre of the treatment volume lies on the plane containing the beam axis. Surface markers will be placed in position on the skin or shell and if necessary the ring to determine the magnification. Figure 6.16e shows the ring and surface markers made from ball bearings and adhesive tape.

Orthogonal films will be taken as described above and a contour taken in a plane containing the centre of the treatment volume. The size of the tissue volume to be treated will be marked on both films. Using the localisation technique these marks will be transferred to the patient contour to indicate the actual size and position of the treatment volume. Other important anatomical detail can be added. The treatment can then be planned in detail, as described in the last chapter and the isodose distribution drawn or computed.

The patient's second visit to the simulator is to verify all the details of the proposed treatment—particularly the size, position and direction of each beam. Any inaccuracies can be corrected. The position of local shielding blocks can be identified and checked (Fig. 6.16f). All this is carried out before the treatment is commenced and it is at the second (or subsequent) visit to the simulator that the patient is conveniently transferred from the care of the planning/mould/simulator staff to the radiographer who will be giving the treatment. The radiographer should see the final simulation so she knows the exact treatment set-up and the patient's position. This preparatory work may take several days to complete, but should prevent the need to make corrections during the course of treatment. If adequate records are kept, the progress of treatment can be followed up both during and after completion of the course using the films or video tape records from the simulator. Some centres may choose to reduce the field sizes part way through the course.

The localisation of sealed sources

The use of the simulator is not confined to external beam therapy. It provides an accurate means of localising any internal structures including implanted sealed sources. It will become evident in the next chapter that the dose distribution around any implant containing more than one sealed source is critically dependent on the relative positions of those sources. Since the sources are enclosed in platinum or silver they will show up clearly on a radiograph or image intensifier.

Under screening conditions, the accuracy of the implant can be examined in detail before the localisation radiographs are taken. For example, a single plane implant can be screened and the X-ray beam positioned to give the edge-on view of the implanted plane. If the calculation is to be done manually using the Paterson-Parker tables (p. 87), then a film taken in a direction perpendicular to the edge-on view will give a true projection of the plane thereby simplifying the calculation. If the calculation is to be done on a computer then two films at 45 degrees to the edge-on view will enable the two ends of each needle to be readily identified thus simplifying the data input to the computer. Where sealed source implants are done regularly, a small operating theatre adjacent to the simulator can be very useful. If the implanted sources are checked immediately on completion, any deviations from the planned insertion can usually be corrected without delay.

In the case of gynaecological insertions of sealed sources, the simulator not only provides accurate data on the relative positions of the sources but also on their position relative to the body contours, thus enabling the post operative X-ray treatment to be planned with greater accuracy.

8. The use of discrete sealed sources

Introduction

Radioactive materials are used in a wide variety of ways in medical practice. Chapter 3 has dealt with the use of isotopes for the production of gamma ray beams for radiotherapy. Chapters 9 and 10 deal with the use of unsealed radioactive materials, i.e., liquids, gases, etc. The present chapter will describe the physical aspects of the use of small quantities of radioactive material in the form of *sealed sources* where the active material is contained in a metal capsule. The use of small radioactive sources for producing gamma ray effects in limited regions of tissue is one of the oldest established practices of radiotherapy. Both radium and radon have been used for this purpose for a long time. Radium-226 and radon-222 are themselves alpha emitters but within their common decay scheme are the daughter products radium B and radium C. Radium B and C are both radioactive solids emitting gamma ray photons with energies in the range from 0.2 to 2.4 MeV. With an effective photon energy of approximately 1.0 MeV and a half life of 1600 y, radium was ideal as a small gamma ray source. The specific gamma ray emission for radium (and radon) is 8.25 R cm^2 h^{-1} m Ci^{-1} when filtered by 0.5 mmPt. More recently radium substitutes, e.g., cobalt-60, caesium-137, are used and these will be detailed later in the chapter, but for the most part what is said here about radium is equally applicable to its substitutes. In these techniques irradiation of a lesion by gamma rays is normally achieved by using several small tubes of active material distributed inside or around the tissue to be irradiated and arranged according to a precise pattern.

Radium needles and tubes

Radium needles are intended for surgical insertion into tissues. Such treatment, with the needles actually buried in tissue, is called *interstitial*. The needles may also be used for *surface* or *intracavitary* application. Radium tubes are intended for intracavitary or surface application. In these containers the radium is generally in the form of radium sulphate, an insoluble salt of the metal. To ensure uniform and complete filling of the container the radium salt is usually mixed with some inert powder or 'filler'.

Various metals are used for containers, including platinum, silver, gold, steel, monel metal (an alloy chiefly of nickel and copper), etc. Heavy filtration is essential to absorb the alpha and beta particles and also the softer gamma rays unless the source is to be used for beta ray therapy. Platinum is the densest metal conveniently available, i.e., it gives the maximum degree of filtration for a given thickness of metal wall. Platinum has an atomic number of 78 and a density of 21·5 g cm^{-3}, hence its common use, especially for needles, in which thinness is a great advantage for insertion into tissues. Moreover, it has a very high melting-point and is not destroyed if accidentally thrown into the incinerator along with dressings from the ward; it is also non-corrosive and so can easily be cleaned.

Radium needles

A radium needle is a hollow container, usually made of platinum alloy with a sharp trocar point at one end to pierce the tissue and an eyelet at the other end to carry a thread. The radium salt can be inserted directly into the container, but it is normally enclosed in one or more thin-walled subsidiary containers of platinum alloy (0·2 mm thick) called *cells* and the needles are then said to be *cell loaded*. Cell loading is desirable to minimise the spillage of the radium salt or the leakage of the enclosed radon gas in the event of the needle being damaged. The point and the eye are screwed tightly on to the body of the needle and the joint sealed with gold solder. Figure 8.1 shows the details of construction and Figure 8.2 a typical isodose chart. A platinum alloy with 10 per cent or 20 per cent

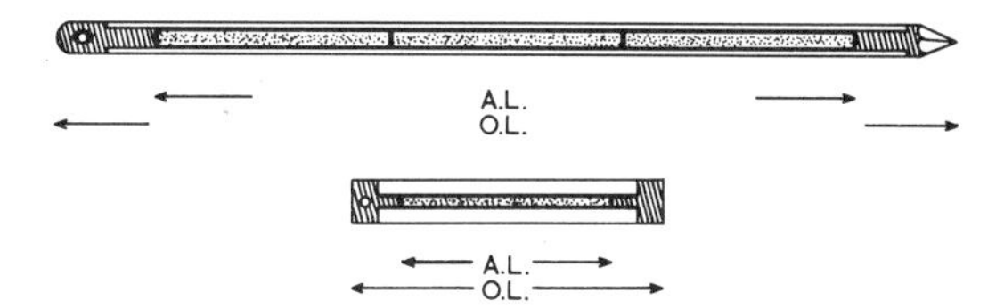

Fig. 8.1 The construction of radium needles and tubes.

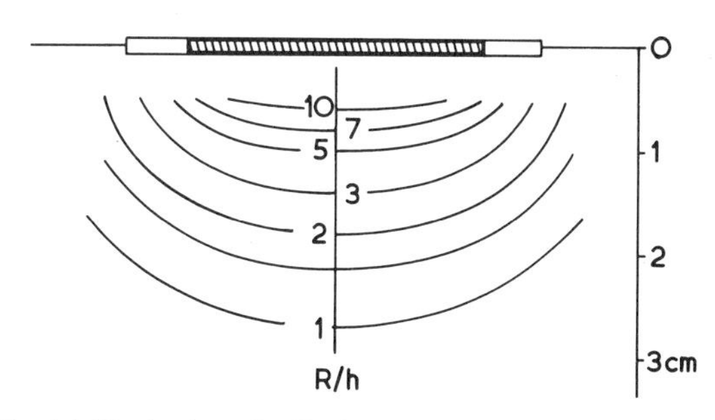

Fig. 8.2 The isodose distribution round a typical 1 mg (half strength) radium needle.

iridium is used in this construction to give increased hardness to the metal.

The length from end to end is called the *overall length*, in contrast to the length of the actual radium-containing portion, which is called the *active length*. The thickness of platinum wall is at least 0·5 mm to provide adequate filtration and mechanical strength to the needle (Table 8.1).

The actual distribution of the radium inside the containers may be checked by placing them on pieces of wrapped X-ray film for about two minutes. The gamma-rays affect the emulsion in the same way as X-rays and the developed film will give a picture of the distribution of the radium. If a flash of soft X-rays is given to the film at the same time, the outline of the container is also shown.

Table 8.1 Typical radium needles (0.67 mg/cm)

Nominal Radium content	External length (mm)	Active length (mm)	External diameter (mm)	No of cells
1 mg	25	15	1.85	1
2 mg	42	30	1.85	2
3 mg	58	45	1.85	3

Radium tubes

Radium tubes are essentially similar to needles but have no sharp points as they are not intended for insertion into tissue (see Figs 8.1 and 8.3) but for use in applicators for insertion into body cavities or in surface applicators. They are generally thicker for the same reason, often with 1 mm or more of platinum wall thickness. Platinum may be replaced by a greater thickness of a cheaper metal e.g., silver, since thinness is not a prime consideration. Their radium content is generally higher than in the case of needles, e.g., 5 or 10 mg or more (Table 8.2).

Each tube or needle has an identification number or mark engraved on it. This is necessary not only for identification purposes but also for the routine tests that are carried out periodically (especially the 'leakage test' p. 94).

Radon seeds

The production and properties of radon and the emission of gamma-rays from the decay products which accumulate as the radon decays has been described in Chapter 2. Radon seeds have been used for a long time as substitutes for radium sources under appropriate conditions. They are made from fine capillary tubing of pure gold of 0·5 mm wall thickness. The amount of radon gas in each seed can be adjusted so that, allowing for radioactive decay during delivery, the activity is a specified value at the time of use. Typically they have a total length of 5·0 mm, an active length of 4·0 mm and a maximum diameter of 1·35 mm. Thus, the seeds are small enough to be inserted into tissues and left permanently buried forming harmless 'foreign bodies' after their radioactivity has decayed to a

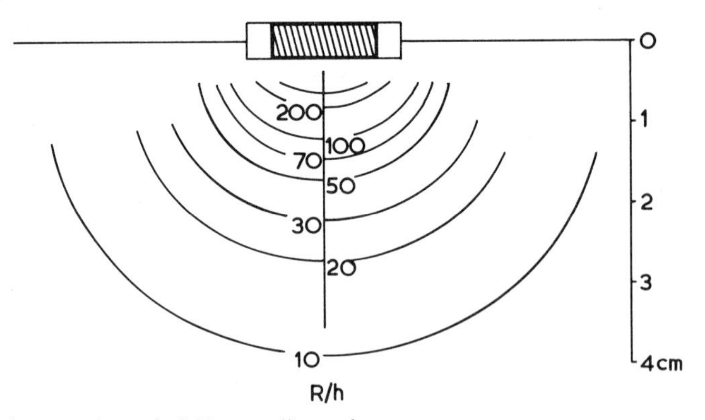

Fig. 8.3 The isodose distribution round a typical 20 mg radium tube.

Table 8.2 Typical radium tubes (for cervix use)

Nominal Radium content	External length (mm)	Active length (mm)	External diameter (mm)
10 mg	20	13.5	2.65
15 mg	20	13.5	2.65
20 mg	20	13.5	2.65
25 mg	20	13.5	2.65

negligible value. The half life of radon is 3·8d. Grains of gold-198 are now often preferred to radon seeds (p. 90).

Radiation dosage for radium distributions

It is a simple matter to deliver a known radiation exposure from an isolated point source of radium to a particular place on the skin or in tissue. Accurate physical measurements on standard radium sources have established that a point-source of 1 mg of radium, screened by 0·5 mm of platinum, will deliver an exposure of 8·25 roentgens in 1 hour in air to a point 1 cm distant from the source. (Where the screening exceeds 0·5 mm Pt, the calculated exposure rate should be reduced by 2 per cent for every additional 0·1 mm Pt.) The application of the inverse square law to this situation gives the equation for the exposure rate from a point source of s mg of radium at a distance d cm from that source, namely:

$$\text{Exposure rate, } X = \frac{8{\cdot}25 \times s(\text{mg})}{d(\text{cm})^2}\ R/h$$

This equation may be used with reasonable accuracy for point sources or for points at a distance greater than three times the maximum dimension of the source. Multiplying the above equation by the exposure time in hours, t, we have:

$$\text{Exposure, } X = \frac{8{\cdot}25 \times s \times t}{d^2}\ \text{roentgens}$$

$$\text{and for radium, } f = 0{\cdot}957\ \text{rad/R}$$

$$\therefore \text{Dose, } D = \frac{8{\cdot}25 \times 0{\cdot}957 \times s \times t}{d^2}\ \text{rad}$$

Examination of this equation shows that the total dose, D, is the milligram-hours ($s \times t$) multiplied by a factor which is dependent only on the isotope (radium) and the geometry of sources (in this case a simple inverse square factor). This concept of the milligram-hour forms the basis of the Paterson-Parker system for the calculation of dose from an array of sealed sources which will be described later.

If the dose distribution from a point source is given by this equation, then the isodose lines are circular and the isodose surfaces spherical with the source as centre. (Since the absorption of radium gamma-rays in tissue is only about 3 per cent/cm and almost entirely compensated by multiple scatter, these effects are normally neglected in calculations). Figure 8.4 represents a line of five equally spaced point sources, each emitting gamma-rays in all directions and producing a complex pattern of isodose lines, but each line representing the cumulative effect of the inverse square law applied to each source. From this one figure several important conclusions may be drawn.

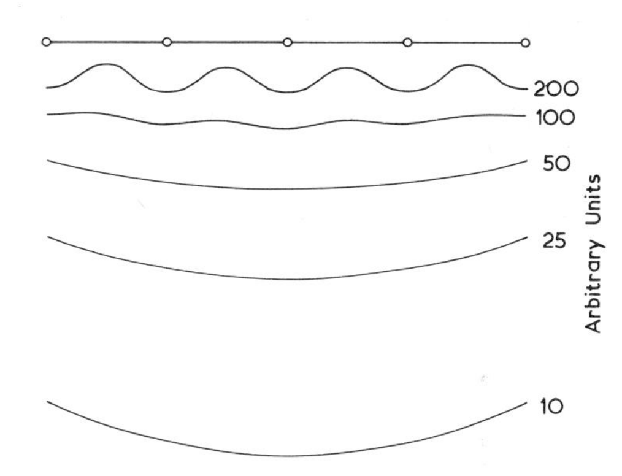

Fig. 8.4 The complex isodose distribution from a line of five point sources.

1. Despite the complex nature of the isodose lines shown, each one will be circular when viewed from one end with the line as axis
2. At points close to any one source the isodose curve will be essentially circular, the influence of the neighbouring sources being negligible
3. The dose distribution along a line close to and parallel to the line of sources will be very uneven, high close to each source and low between any two sources.
4. Each low point could be raised by inserting a further source half way between each existing source, making the distribution more uniform.
5. As the distance between the line of interest and the line of sources is increased, the uniformity of dose improves until the isodose line becomes the arc of a circle with the middle source as centre.
6. Keeping the distance between the lines the same and bringing the sources closer together has the same effect as in (5).

The student will appreciate that the extension of (4) is to make a continuous line source and the extension of (5) and (6) is the five sources become effectively a single point source. The corollary of (1) is that the distribution shown is essentially that of a section through five parallel line sources viewed end-on. The three valuable consequences drawn from the above analysis are:

(a) The uniformity of dose achieved is dependent on both the distance from the plane containing the sources and the spacing of those sources
(b) A line source may be thought of as a line of closely spaced point sources (some computer programs for evaluating the dose distributions from interstitial implants treat each source as a line of five or more point sources)
(c) The uniformity of dose is further improved by increasing the activity of the two end sources

The Paterson-Parker system

The Paterson-Parker system sets out rules relating to the distribution of sources in various geometric configurations to give a uniform dose to the prescribed plane or volume and data for the calculation of this dose based on the activity used. Typical arrays are shown in Figure 8.5. The Paterson-Parker system outlined below is the system widely adopted in the United Kingdom. There are other systems in use (e.g., the Quimby system) and the student should seek to master the one he will have to practise and ignore the others to avoid becoming confused.

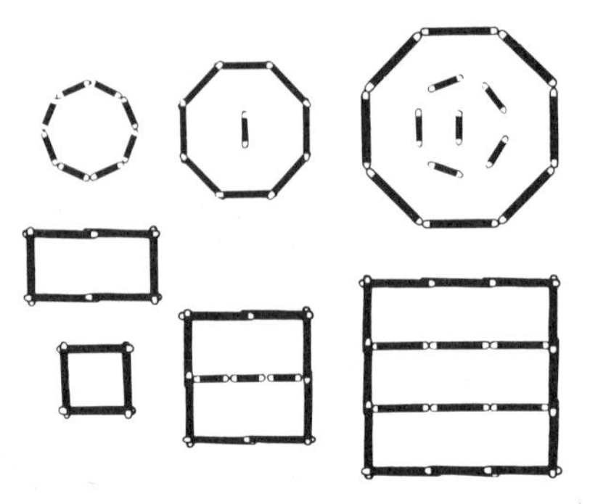

Fig. 8.5 Source arrays following the Paterson–Parker rules.

Uniformity of dose

Where sources are implanted into the tissues a very high dose is inevitably delivered to the immediately adjacent tissue. It is, therefore, important to understand what is meant by *uniform* dose. Where the sources are actually implanted into the tissue, the dose quoted for a technically perfect implant is 10 per cent above the minimum dose encountered in the specified volume of tissue. For surface applicators where only a surface dose is quoted the non-uniformity of dose for the applicator which obeys the rules will not exceed ± 10 per cent. (This is consistent with the first definition in that the quoted dose is again 10 per cent above the minimum but adds that the maximum dose to the tissues will not exceed the quoted dose by more than 10 per cent—the very high dose region close to each source being within the material of the applicator.) The inverse square law shows that errors of 0·5 mm in a radium-surface-distance of 5 mm can also produce dose discrepancies of up to 10 per cent, thus demonstrating the need for accuracy in the construction of the applicator or in the implantation of the needles, whilst at the same time demonstrating the inevitable variation of the dose over any clinically defined surface.

The actual distribution of the sources used is closely related to the *treatment distance* or radium-surface-distance (RSD) which is given the symbol, h. It is this distance, h, which determines the spacing of the sources, for example (i) the spacing of the sources in Figure 8.5 should not exceed $2h$ and (ii) one continuous line source may be made up of several smaller sources provided the spacing between their active ends does not exceed h.

The Paterson-Parker rules are divided into five groups, each for a different type of treatment, these treatments being either by surface applicators, planar implants, volume applicators or implants or linear sources. The principles are the same for each group. The rules for surface applicators and planar implants are very similar and will be outlined in some detail. The rules for linear sources will be given. Space does not permit more than a comment on those relating to volume applicators and implants.

Surface applicators

A plane surface applicator is used for treating very superficial lesions with a distribution of radium sources arrayed over an area the same size as the lesion but at some distance, h, from the surface of the lesion. The applicator is best made using a procedure very similar to that outlined in Chapter 6 for patient moulds, except that once the plastic has been formed, the area to be covered by the radium will be thickened by adding further layers of plastic until its overall thickness is the RSD less the radius of the sources to be used. In addition, to guard against the loss of a source, a lid will be made to totally enclose the sources after they have been fixed in place (Fig. 8.6).

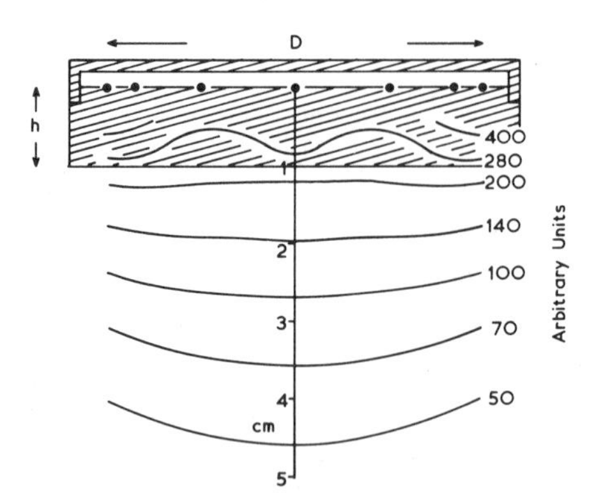

Fig. 8.6 The isodose distribution in a plane containing the axis of a circular array of 12 sources (95%) in the periphery and a centre spot containing 5% of the total activity.

Planar implants

Unlike the surface applicator, the planar implant is designed to treat a volume of tissue, usually 1 cm thick with 0·5 cm either side the plane containing the radium. Alternatively, the radium may be divided into two planes 1 cm apart to treat the tissue enclosed. If the separation of the planes is greater than 1 cm then the dose at the mid-plane will be low and the radium required will need to be increased as given in Table 8.3.

The rules for surface applicators and planar implants

In general, the array of sources will be circular or rectangular with elliptical areas being approximated to either according to their eccentricity. Implants are more conveniently made with needles arranged as a rectangle and surface applicators with small discrete 'point' sources arranged in a circle.

Table 8.3 The requirements of planar implants at separations greater than 1 cm and areas less than 25 cm^2

Separation	Increase mgh by	Mid plane dose
2.0 cm	40%	−20%
2.5 cm	50%	−30%

Sources used end-to-end to make up a line source should be sufficient in number to ensure that the gap between the active ends does not exceed the RSD, h. A minimum of six sources is required to define a circle.

If the surface applicator is applied to a curved surface—the rules may be applied up to hemispherical or hemicylindrical surfaces—the radium area is defined by the perpendicular to the edge of the lesion, i.e., for a surface which is concave the radium area is smaller than the lesion, while for a convex surface it is larger than the lesion. The radium will be distributed over that area according to the appropriate rules, but the activity will be calculated on the basis of the *smaller* area in each case.

1. *Circles.* Normally surface applicators are of such a diameter, D, that the ratio D/h lies between 3 and 6 and they require the radium to be distributed so that 5 per cent lies in the centre and 95 per cent round the periphery. If the ratio D/h is less than 3, all the radium is placed round the periphery. The ideal distribution is said to be obtained when $D/h = 2{\cdot}83$. If D/h is 6 or larger, then the centre spot is reduced to 3 per cent and an 'inner circle' of half the diameter is added, in which case the remaining 97 per cent of the radium is distributed as in Table 8.4. If the area is elliptical then D is the mean diameter.

Table 8.4 The distribution of radium for circular applicators when D/h lies between 6 and 10

D/h	6	7.5	10
Centre spot	3%	3%	3%
Inner circle	17%	22%	27%
Periphery	80%	75%	70%

2. *Rectangles*. For square distributions, radium is distributed uniformly around the periphery and then the area divided into strips of width not greater than $2h$. If one extra line is required, a half strength needle is used, for more than one extra line then two-third strength needles are required. In the case of rectangles the added lines are parallel to the longer sides and their activity is relative to that of the longer sides. Where the rectangle has sides in the ratio of 2, 3 or 4, the milligram-hours should be increased by 5, 9 and 12 per cent respectively.

In certain implant sites it is not possible to get the radium all round the periphery, i.e., it is not possible to 'cross' the ends of the needles. To offset the shortage of radium resulting from *uncrossed ends*, special needles are available. *Indian club* needles have additional activity at the pointed end for use where it is only possible to cross the one (eyelet) end; as is often the case in the implant of the tongue. *Dumbbell needles* have extra loading in both ends and crossing needles are not required. If neither indian club nor dumbbell needles are available and the ends remain uncrossed, 10 per cent should be deducted from the implanted area for each uncrossed end before determining the milligram-hours.

3. *The calculation*. The calculation in practice is twofold. Stage 1 is a theoretical calculation of the ideal and Stage 2 is a practical calculation based on the available sources. In both, a clear understanding of the milligram-hour concept is all that is required. Paterson and Parker have produced tables and graphs (e.g. Fig. 8.7) which give the milligram-hours of radium filtered by 0·5 mm Pt required to deliver 1000 rads to a particular geometrical situation. From this point on all we have to do is to find two numbers (the milligrams and the hours) which, when multiplied together, give the figure derived from the graph.

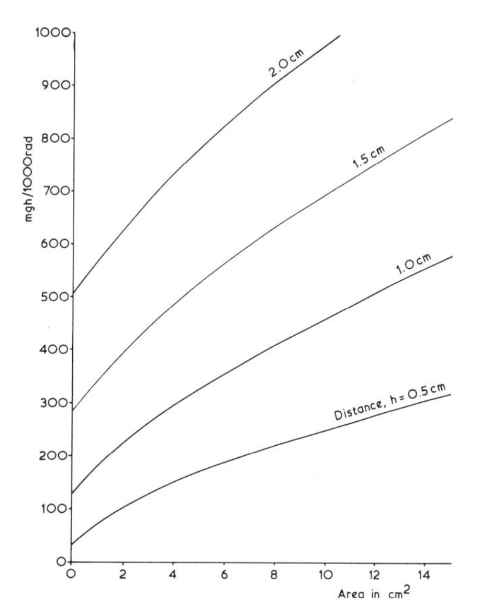

Fig. 8.7 The mgh/1000 rads for radium filtered by 0·5 mm Pt for planar surface applicators of different areas.

For example:

A needle implant is required to cover an area of $3 \times 4 \cdot 5$ cm^2 and deliver 6000 rad to a plane at 0·5 cm distance in 5 days.

Stage 1 To calculate the required activity for the prescribed treatment.

Area of implant $= 3 \times 4 \cdot 5\ cm^2 = 13 \cdot 5\ cm^2$

To deliver 1000 rad at 0·5 cm distance requires 300 mgh (Fig. 8.7)
To deliver 6000 rad at 0·5 cm distance requires 300×6 mgh
To deliver 6000 rad in 5 days ($\times$ 24 hours) requires

$$\frac{300 \times 6}{5 \times 24} = 15 \text{ mg}$$

Stage 2 To calculate the actual treatment time using the practical sources. Let us assume the radium stock comprises the needles listed in Table 8.1. We shall add extra lines parallel to the longer side to divide the 3 cm width into strips not greater than $2 \times 0 \cdot 5$ cm. We shall, therefore, require two extra lines each containing two-thirds the activity of the longer side.

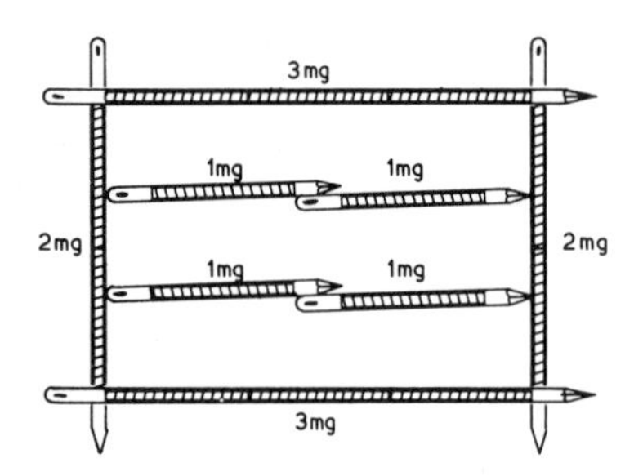

Fig. 8.8 A typical needle implant.

∴ We must use the above layout of needles (Fig. 8.8)

Periphery	2×3 mg + 2×2 mg	= 10 mg
Extra lines	$2 \times 2 \times 1$ mg	= 4 mg
	Total activity used	= 14 mg

The dose of 6000 rad at 0·5 cm distance requires 300×6 mgh

$$\therefore \text{ Treatment time} = \frac{300 \times 6}{14} = 128 \cdot 5 \text{ h}$$

i.e., 5 days $8\frac{1}{2}$ hours

The dose rate per mg is the reciprocal of the mgh multiplied by 1000 rad. In the above example, 300 mgh were required for 1000 rad, the dose rate, therefore, is 1000/300 rad/h per mg, or 46·7 rad/h per 14 mg.

Because of the inherent non-uniformity of dose in the implant situation, a relaxation of the above rules is permitted, namely to implant a fraction of the total radium round the periphery and to distribute the remainder as evenly as possible over the enclosed area. The peripheral fraction for areas less than 25 cm^2 is two thirds of the total, one half for areas between 25 cm^2 and 100 cm^2 and one third for areas greater than 100 cm^2. Using this approximation in the above calculation would have lead to the same 10 mg round the periphery and 5 mg over the enclosed area.

The rules for linear sources
Line sources are made up of several collinear tubes mounted in some form of applicator (e.g., perspex) or outer tube and can be used for treating the inner surface of certain body cavities, e.g., the uterus, the nose. There are just two rules for this simple applicator. First, the active length is defined as the distance between the active ends of the whole composite source. (This is not the sum of the active lengths of the individual sources.) Secondly, the inactive space between the active ends of adjoining sources should not exceed the distance h, which in this case is the radius of the applicator (Fig. 8.9). The calculation follows the same pattern as the example above but using the line source data.

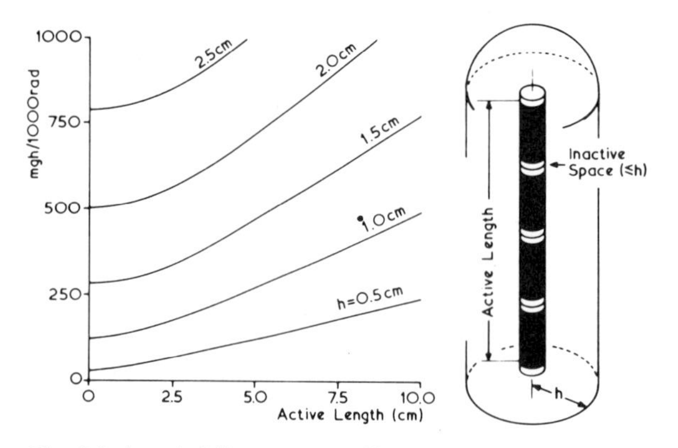

Fig. 8.9 A typical line source applicator.

Rules for a cylindrical applicator have been evolved using either a series of coaxial rings or a series of line sources. These may be used to treat parallel surfaces of smaller diameter than the applicator, e.g., the penis, the finger. Rules for the implantation of solid tissue volumes of a variety of shapes have also been derived but such techniques are rarely used where modern external beam therapy facilities exist.

The treatment of the cervix
Perhaps the most common use of sealed sources is in the treatment of cancer of the cervix and the techniques are fully described in Chapter 23. As before, the description is in terms of radium, although caesium is the radionuclide of choice in practice. The radium tubes (Table 8.2) are incorporated into applicators made of rubber or perspex to

fit the uterine canal and the vaginal vault. A variety of loadings and sizes are made available to enable the insertion to be tailored to the individual patient's requirements. Despite the possible variations, the basic shape of the dose distribution remains essentially pear-shaped (Fig. 23.6) and the duration of the insertion is based on the dose rate to two defined points, namely point A and point B (Fig. 23.5). These dose rates may be tabulated for a range of *standard insertions* rather than calculated for each individual patient. Where individual calculations of dose rate are required, then it is necessary to determine the dose rate each radium tube contributes to each point. In most cases the inverse square law approximation may be applied for point B as its distance from each tube is large with respect to the active length. For point A it is necessary to determine the precise position of the point in relation to each tube and its dose rate contribution using isodose charts (e.g., Fig. 8.3) or a mathematical approach such as the *Sievert integral* or *Young and Batho tables* (see more advanced texts). For a more complete and accurate estimation of the dose distribution, the calculation is most conveniently done on a computer using the accurate localisation films from the treatment simulator to provide the patient input data.

Radium substitutes

Two basic hazards exist with the use of radium which make the use of alternative nuclides preferable. These are (i) the gaseous daughter product radon-222 which builds up a gas pressure in any sealed radium container and threatens to leak out should the container become damaged in any way and (ii) the radium spectrum has eleven photon energies up to 2·4 MeV and necessitates very thick protective barriers. Every radium substitute however, decays at a faster rate than radium and necessitates the use of decay corrections in the calculation of dose. The nature of this correction depends on the rate of decay and the treatment time.

The properties of common radium substitutes are listed in Table 8.5, together with the types of sources usually available. The gamma ray photon energies in each case are comparable with the effective energy of the radium spectrum and as with radium, the absorption in tissue is usually ignored. In order to make the best use of the data already outlined for radium, it is common practice to define the activity of each substitute in terms of the equivalent activity of radium.

The mg Ra equivalent

If a source of a radium substitute gives the same exposure rate at a distance of 25 cm from the midpoint of the source as a 1 mg radium source (filtered by 0·5 mm Pt), then it is said to have an activity of 1 mg radium equivalent. In general terms, using the simple inverse square formula for exposure rate (p. 85) and cancelling d^2 from both sides of the equation, we have

$$8{\cdot}25 \times s(\text{mg}) = (s_1 \times k_1)_{\text{substitute}}$$

where s_1 is the activity in mCi and k_1 the specific gamma ray emission of the radium substitute. Therefore, the mg Ra equivalence is given by

$$s = \frac{(s_1 \times k_1)_{\text{substitute}}}{8{\cdot}25} \text{ mg}$$

Once this has been calculated and provided the decay during the period of the proposed treatment is negligible, all the data referred to above for radium may be used. The only radium data which have to be discarded when substitutes are used are those relating to protective barriers. The old barriers are now more effective, new barriers may be considerably thinner and less heavy.

The effect of radioactive decay

The rate of decay for the radium substitutes listed in Table 8.5 varies from approximately 1 per cent per half year for caesium-137 to approximately 1 per cent per hour for gold-198. Since treatments involving these types of sources usually take from 3 to 7 days, the decay during that period is negligible in the case of caesium and very significant in the case of gold. The dividing line is roughly at a half life of one year. If the half life is longer than one

Table 8.5 Radioactive isotopes as substitutes for radium

Isotope	Half life	Photon energy	Specific gamma ray emission	Sources available
Caesium-137	30 y	0.66 MeV	3.3	Needles and tubes as for radium. Beads for afterloading techniques
Cobalt-60	5.26 y	1.17, 1.33 MeV	13.0	Needles and tubes as for radium. Also beads and pellets for high dose rate and afterloading techniques
Gold-198	2.7 d	0.41 MeV	2.35	Grains approx 2.5 mm × 0.8 mm diameter (including a coating of 0.15 mm Pt) in magazines
Iridium-192	74 d	0.3–0.6 MeV	4.8	As for Au-198 and as wire 0.4 mm diameter (including a coating of 0.1 mm Pt)
Radium-226	1600 y	0.2–2.4 MeV	8.25[a]	Needles and tubes.

[a] In equilibrium with its daughter products and filtered by 0.5 mm Pt

year, the activity should be calculated at the beginning of the treatment and the further decay during the treatment ignored. If the half life is less than one year, then it is necessary to calculate the activity at the beginning of treatment *and also* to take into account the effect of decay during the treatment. This is done using another concept—that of the millicurie destroyed.

The millicurie destroyed

Where very small sources of short lived radium substitutes are used (e.g., gold grains), it is customary to implant them permanently into the tissues. The grains are considered to decay completely and then reside in the tissues as 'foreign bodies'. In this way a 1 mCi grain will be capable of delivering a certain maximum dose in infinite time after which we can say the 1 mCi has been destroyed. Now, we know 1 mCi of gold-198 has a mg Ra equivalence of

$$\frac{1 \times 2{\cdot}35}{8{\cdot}25} = 0{\cdot}28 \text{ mg}$$

but because the activity of gold-198 decays with a half life of 2·7 days, it has a mean life ($1/\lambda$) of only 2·7/0·693 days or 93·5 hours. In other words, a 1 mCi gold grain permanently implanted will deliver the same dose as a 0·28 mg Ra source implanted for 93·5 h, or

$$1 \text{ mCi destroyed of Au-198} \equiv 26{\cdot}6 \text{ mgh of radium.}$$

If now we consider a surface applicator applied to a patient when its total activity is 20 mCi of Au-198 and worn continuously for (say) 8 days and then removed, the activity on removal will be 2·5 mCi (8 days is approximately 3 half lives), then the number of millicuries destroyed during the treatment will be 20 minus 2·5, or 17·5 mCi and the mgh equivalent of radium will be $17{\cdot}5 \times 26{\cdot}6 = 465{\cdot}5$ mgh. By this means then, the milligram-hours equivalent of a short lived radium substitute can be determined and used in exactly the same way as for radium in the Paterson-Parker system.

The use of gold grains

In Table 8.5 the types of sources available have been listed and it will be noted that whereas cobalt and caesium follow the traditional forms of needles and tubes, gold is in the form of grains and iridium in the form of wire. Gold grains are generally supplied in magazines of 14 grains. The magazine is loaded into an implantation gun, such as the one developed by the Royal Marsden Hospital and illustrated in Figure 24.5. The gun is supplied with a variety of straight and curved hollow needles through which the grains can be deposited into the tissues, one at a time, by pulling the trigger. Normally the needle will be inserted to the deepest point in the tissue to be irradiated and the grains released at intervals during the withdrawal of the needle as required. The needles are graduated in centimetres to assist in depositing the grains at the appropriate intervals.

Afterloading techniques

The implantation of gold grains and radium or radium substitutes has for a long time been achieved only by accepting that the staff will be exposed to gamma radiation during the operation. Techniques have developed where this exposure can be drastically reduced or even considered to be negligible and they fall into two groups. The *low dose rate afterloading techniques* use activities comparable with those described above and deliver the treatment dose over a period of a few days, whereas the *high dose rate afterloading techniques* use higher activities and deliver the dose in a matter of minutes. In each case the sources are loaded only after their final positions have been determined and checked.

The low dose rate afterloading technique using iridium wire has been developed in Paris and Oxford. It involves the implantation of nylon tubes into the superficial tissues, the tubes being secured at both ends with nylon balls. The positions of these tubes can then be checked radiographically by inserting inactive 'fuse' wire into the tubes. The iridium wire is cut to the required length and inserted into a second nylon tube of a smaller diameter than the first. This tube is then threaded through the implanted tubes and secured by crimping lead discs outside each nylon ball as shown in Figure 8.10. After the required treatment time, one lead disc is cut off and the other is used to withdraw the whole assembly. Where the implant is not superficial, rigid steel guide needles can be

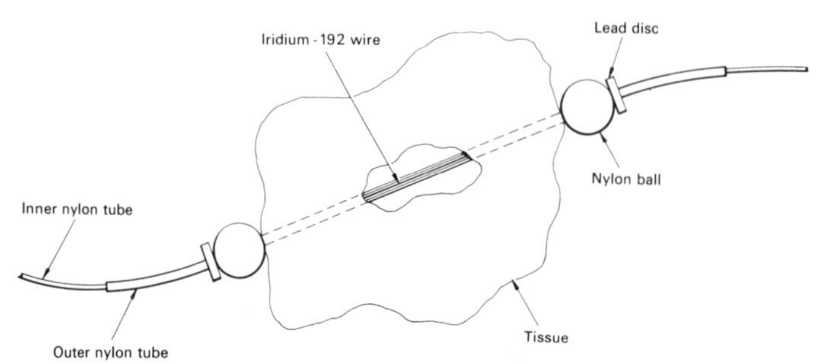

Fig. 8.10 A typical assembly of the Pierquin/Paine after-loading technique. (By courtesy of the Radiochemical Centre, Amersham.)

inserted and their position checked prior to inserting lengths of iridium wire or 'hairpins'. This latter technique is used to treat the bladder for example.

Where intracavitary insertions or surface applicators are used, rigid or flexible catheters can be used. After their position has been confirmed, they may be loaded easily and quickly with caesium sources made up from small cylindrical beads of caesium-137 separated by spherical spacers in a flexible outer sheath. This alternative low dose rate technique particularly lends itself to gynaecological insertions.

Perhaps the best known afterloading techniques are those using rigid catheters and very high activity sources of cobalt-60. These high dose rate techniques require special apparatus (Figure 23.9) installed in a treatment room similar to a megavoltage X-ray therapy room. The technique is similar to that described above except that the sources are transferred from a protected safe to the catheters through flexible tubes by remote control from outside the room. The advantage of this technique lies in the fact that the actual treatment only takes a few minutes and several patients can be treated each day using the same (or different) sources, while the radiation hazard to the staff is reduced to the absolute minimum. There is no exposure to the nursing staff on the wards.

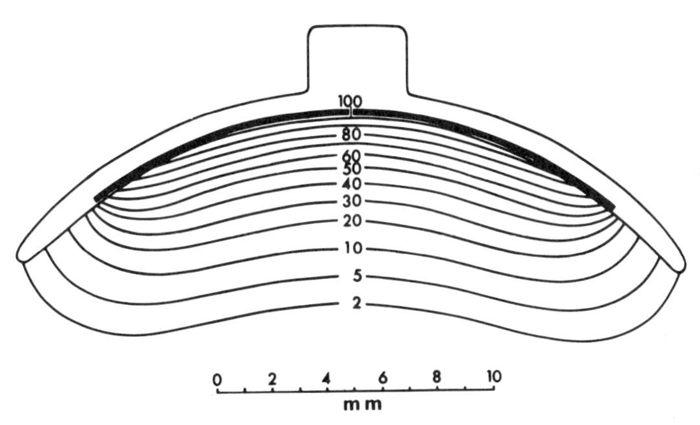

Fig. 8.11 The isodose distribution for an 18 mm diameter Sr-90 ophthalmic applicator.

Beta ray applicators

Radium in equilibrium with its decay products produces a complex spectrum of gamma rays and alpha and beta particles. The reason that radium needles and tubes have a minimum filtration of 0·5 mm Pt, is to absorb the alpha and beta particles and to leave the gamma rays. If, however, the filtration is reduced to say 0·1 mm monel metal, the alpha particles are removed and the beta particles may be used to deliver a very large dose to the tissues close to the source in comparison to that delivered by the gamma rays. Such beta ray applicators were used formerly in the form of plaques to treat very superficial lesions, for example, on the cornea. As before, there was always the hazard that radon gas would leak from the applicator—the 0·1 mm monel metal filter is easily damaged—and the high energy gamma rays were a nuisance.

Strontium-90 (half life, 28 y) is now the isotope of choice for ophthalmic applicators. The strontium-90 compound is incorporated in a rolled silver foil bonded into the silver applicator and formed so that it presents an active concave surface of 15 mm radius to the eye, with a surface filtration of 0·1 mm silver. Like radium, strontium is only useful when in equilibrium with its daughter product, namely yttrium-90. The 0·1 mm Ag filter is designed to absorb the low energy beta particles from the strontium decay (0·54 MeV max) while transmitting the higher energy beta particles from the yttrium-90 (2·27 MeV max). The back of the applicator is finished with a much thicker layer of silver to reduce the dose rate to a safe level. The surface dose rate is typically 100 rad/min and a typical isodose distribution is shown in Figure 8.11. The measurement of the dose is complex and involves the comparison of the dose rate of the ophthalmic applicator and that of a plane strontium-90 plaque which has been measured using a special ionisation chamber known as an *extrapolation* chamber. This enables the dose rate at a surface to be extrapolated from a series of readings of the ionisation at different distances from the surface.

Radiation protection and sealed sources

Handling

The use of small sealed sources presents a hazard to the staff who handle them. The principle hazard is that of their beta or gamma radiation. There is not the additional hazard of contamination as exists with unsealed sources. The main difficulty arises from the fact that gamma rays have considerable penetrating power and the sources must always be handled using forceps—for threading needles, implanting sources into patients and then cleaning and preparing them for the next patient. In some larger departments some of these operations will be done using slave manipulators (Fig. 8.12) or by remote control in afterloading units (Fig. 23.9).

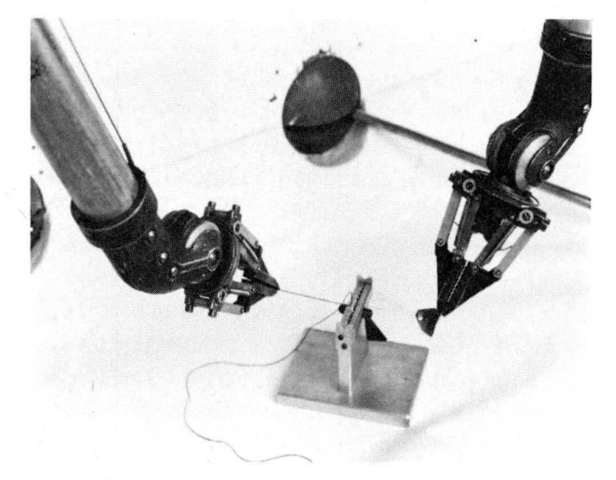

Fig. 8.12 Slave manipulators may be used to thread radium needles, note the needle threader being held in the jaws on the right.

In general, there are three methods of reducing the radiation hazard. The first is to reduce the time spent in manipulation, the second is to increase the distance between the source and the operator and the third is to use protective barriers or screens. To some extent these three measures work against each other. For example, the time spent manipulating the source may be increased by either increasing the distance or by the use of protective screens. A compromise has invariably to be reached.

A simple lead bench is shown in Figure 8.13, which consists of two pieces of lead approximately 35 × 35 × 5 cm^3 firmly secured together. Its surfaces may be finished with a plastic laminate. The lead bench can rest on a laboratory bench so that the operator can sit or stand behind it and only his head and arms are exposed to the sources being manipulated on the horizontal section. Such a simple bench is adequate for manipulating small activities. The alternative is to use a specially constructed bench where the protective material is built up all round and incorporates a protected observation window. Such a system is better where high activities are being handled or where the manipulation may take a long time. A selection of long-handled forceps should always be available.

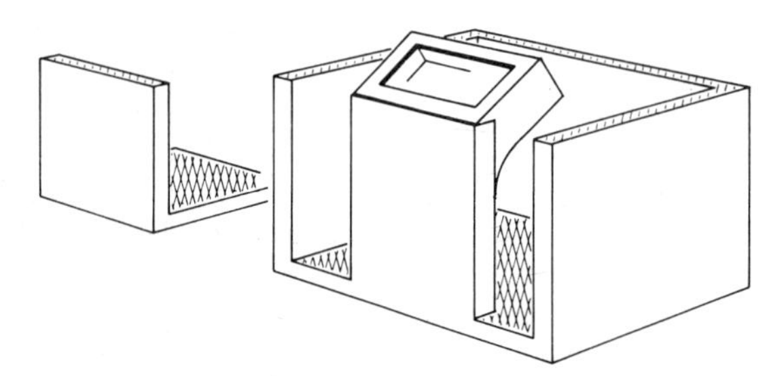

Fig. 8.13 Lead protected benches suitable for the small radium laboratory.

Care must be taken in the detailed manipulation of sealed sources to avoid excessive irradiation to the fingers. Under no circumstances should the operator pick up the source in his fingers. Where the manipulation requires skill, the procedure should be practised using inactive 'dummy' sources made to the same dimensions. The threading of needles prior to their use in theatre is a good example of this, in that considerable exposure could be accumulated if the task were not practised following the correct procedure. A simple device (Fig. 8.14) can be made to reduce the hazard involved and to hold the needles securely during the procedure. The lead block has a series of holes bored vertically to hold the needles so that only the eye of the needles protrudes through the top of the block. The horizontal pin is inserted into the hole appropriate to the length of the needle. Attached to the bottom of the block is a rubber sheet with a series of notches in the edge to hold the threads and to prevent them getting tangled. The use of a needle threader is to be recommended (one is shown in the tongs of the slave manipulator in Figure 8.12). Once threaded, the thread is knotted 5 to 10 mm from the needle's eye so that the clinician can still get hold of the needle with his needle forceps without damaging the thread and to enable the thread to be easily cut away prior to the cleaning of the needle after use.

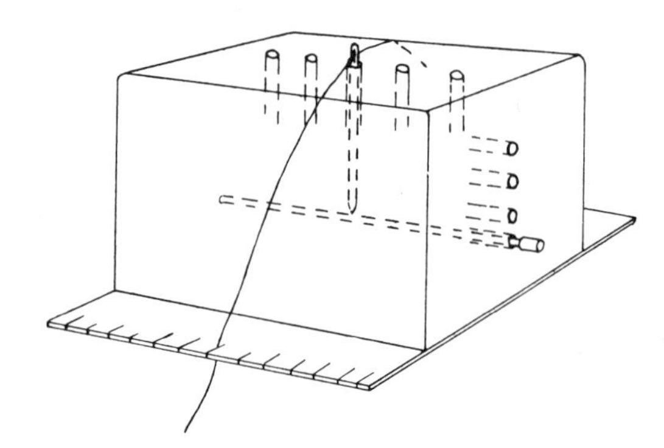

Fig. 8.14 Lead block for holding radium needles for threading.

Nursing

It has already been indicated that a patient undergoing treatment with sealed sources will have them *in situ* for three to seven days. Suitable ward accommodation is therefore essential. A single-bedded room is preferred although a general ward may be used providing the spacing of the beds is adequate—at least 2·5 m between centres is suggested. A radiation symbol should be attached to the patient's bed. Mobile lead bed shields should be placed by the bed so that the staff are protected whilst carrying out the essential nursing procedures, non-essential procedures being postponed until after the removal of the sources. Staff should not remain unnecessarily in the vicinity of the patient and the Radiological Safety Officer may restrict the time spent by the staff nursing the patient. Routine procedures should be shared by using a rota system in order to minimise the exposure to individual members of staff. The Radiological Safety Officer may also restrict the patients' visitors.

The basic principles of speed and distance and the use of protective barriers must be followed during the actual insertion of the sources in theatre and during their removal.

Sterilisation
Sterilisation of sealed sources may be achieved by a variety of means—ethylene oxide gas sterilisation, immersion in cold sterilising liquid, boiling in water, etc. In any case, care must be taken to reduce the risk of damage to the source or corrosion of the surface. Where heat is used (as in boiling) care must be taken to ensure that in the event of failure (e.g. boiling dry) the temperature cannot rise above 180°C. Radium sources present a greater risk than the radium substitutes. Sources should be examined for any defects before any sterilising procedure is undertaken and any source thought to be defective should be taken out of use until it has been thoroughly examined and tested.

Cleaning
Cleaning can contribute significantly to the exposure of the fingers unless some automatic devices are used and a rigid procedure followed. For example, the immersion of the sources in hydrogen peroxide solution or normal saline solution immediately after their removal from the patient will ensure that any blood adhering to the sources does not harden and become caked to the surface. The use of a low power ultrasonic cleaning bath—suitably surrounded with lead—reduces the handling of the sources to a minimum. The replacement of perspex applicators before they become roughened or cracked will reduce the time required to clean them. Abrasive substances should never be used for cleaning.

Storage
Normally all the above procedures will be carried out in a room specially designed for the purpose. In this same room will be the central safe for the storage of these sources. Such a safe will be so designed that sources may be divided into individually protected drawers, each drawer being fitted to accommodate specific sources in an orderly fashion (Fig. 8.15). It should be possible to check the number and type of source in each drawer at a glance. The store should be ventilated to the open air by a fan operated both before and during the transfer of sources to and from the safe. Every transfer must be accompanied by an entry in the record book and the signature of a responsible individual. The safe should be kept locked whenever it is not being used. Both the room and the safe should be marked with the black trefoil to indicate the presence (or potential presence) of ionising radiations.

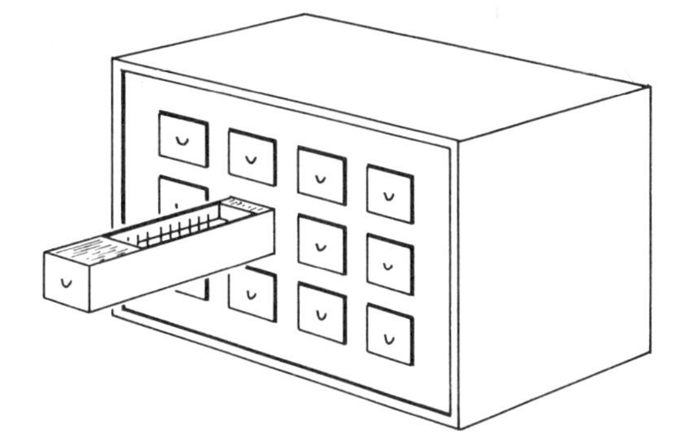

Fig. 8.15 A sealed source safe.

Movement
Movement of sources both within the hospital premises and outside must be fully documented so that in the event of fire or the suspected loss of a source the exact location of each source can be readily identified. Movements outside the hospital complex should be kept to a minimum and in any case conform to the current regulations or codes of practice relating to the mode of transport to be used. Regular movements within the hospital should follow specified routes which should be the shortest or the easiest. The design of new departments should be drawn up with this in mind, for example, the main store and laboratory could be adjacent to the theatre to be used and the sources transferred via a pass-through safe in the dividing wall. Other movements of the sources should be in long-handled containers specially designed for the purpose, that is providing adequate protection, appropriate labelling and minimising the risk of loss or damage. Patients with sealed sources in situ should not be allowed to leave their ward without permission.

Records
It is normal practice to appoint a custodian to take the overall responsibility for the safety of sealed sources in the hospital. It will be his responsibility to ensure adequate records are kept. These should include:

1. A register of stock of all sealed sources having a half life greater than a few days. This register will contain the detailed specification of each individual source, its identification marks, its serial number and the reports of leakage tests and repairs.
2. A record of all sealed sources issued from and returned to the main store and the signature of the responsible person making the transfer. This record may be accompanied by some visual display system to indicate what is in the store and available for issue.

3. A record of the administration of permanent implants.

Leakage (wipe) tests

All sealed sources should be checked for free activity at least annually and whenever there is any suspicion of damage to the source. Any source is capable of being damaged, but needles are the most vulnerable as they may be bent by mishandling, by being dropped or by being pushed against a bone during insertion into the patient. They may be easily checked for straightness by rolling them on a hard flat surface—a piece of plate glass is ideal. Even a minor crack in a radium source can give rise to a leak of radon gas and if this is allowed to build up in the closed atmosphere of the radium laboratory or of the store itself, it can be particularly dangerous. Adequate ventilation is very important.

The leakage test is readily carried out by wiping the source—using forceps—with a highly absorbent material moistened with a mild cleansing agent. The swab is then checked using a geiger or scintillation counter. In the case of radium, sealing the source in a test tube along with a swab of cotton wool overnight and then checking the swab is usually adequate. Alternatively, the radium source may be checked with a monitor which is sensitive to alpha particles but insensitive to beta and gamma radiations. When leakage testing beta plaques particular care must be taken to avoid damage to the thin protective coating on the active surface.

If the test shows a free activity greater than 0·05 μCi the source must be regarded as leaking and steps taken to get the source examined and repaired accordingly. Any source suspected to be leaking must be sealed in an air-tight container pending its repair.

Suspected loss

Every effort must be taken to prevent the loss of sealed sources and although the records keep a check on the movement of sources, it is still possible for sources to get mislaid in transit or lost from a patient. In the event of a source being suspected lost, a routine such as the following should be adhered to closely. Notices outlining the procedure should be posted in all areas where sealed sources are routinely used.

1. All avenues by which the source may have been removed from its last known vicinity must be closed immediately. There should be no flushing of toilets or sluices, no removal of dirty dressings, laundry or rubbish of any sort, no sweeping of floors, no movement of patients, staff or perhaps even visitors. No further material should be placed in the hospital incinerator.
2. The persons responsible for radiation safety in the department or ward and in the hospital as a whole must be notified as soon as possible.
3. A search for the source must be initiated as soon as possible, bearing in mind that other sources known to be in the vicinity will also be detected by the monitor used in the search. If at any stage the search suggests the lost source is damaged, then the further precautionary measures relating to unsealed sources (p. 127) should be observed.
4. If the initial search is unsuccessful, consideration must be given to calling in assistance from outside.

The use of permanently installed gamma alarms at strategic points along the corridors and at doorways to detect the passage of any source is to be recommended. Such installations need to be checked regularly.

Code of practice

The above principles of safe operation are based on the recommendations laid down in the Code of Practice. Detailed extracts from the Code are quoted in Chapter 11 (p. 126) but the student is referred to the Code itself for a fuller account of the requirements.

9. The use of unsealed radioactive sources

Introduction

It has been shown in Chapter 2 that radioactive isotopes of practically all elements can be produced. Many of these are used in medical practice. They may be used therapeutically and in this case the biological effects of the emitted ionising radiations produce therapeutic results in the tissues in which the isotope is deposited. They may be used for diagnostic tests in which case the emitted radiations serve as indicators for the presence of the active material in some particular body tissue or fluid. The diagnostic use of radioactive isotopes is by far the most frequent of the clinical techniques involving active materials. Closely allied to these diagnostic applications is the use of isotopes for investigation of physiological phenomena in man or of drugs administered to patients.

The extensive development of clinical isotope techniques depends on some special characteristics of radioactive materials. Though a large number of radioisotopes are available only a limited number have the chemical and physical properties which make them suitable for internal administration to patients or for use in other clinical tests. These properties are discussed below. First however the extreme *sensitivity* with which radioactive materials can be detected should be noted. Geiger and scintillation detectors respond to single beta particles or gamma ray photons arriving at the sensitive zone of the detector. In general only a small fraction of the particles or photons emitted from a radioactive source reach the sensitive region of a detector placed nearby, but even so the detectors can record the presence of a radioactive source containing a relatively small number of atoms. A very small mass of the chemical element can thus be detected and this is particularly so if the element is in the form of a carrier free radioactive isotope (see p. 13). An example is the use of carrier free radioactive iodine for investigations of thyroid function (see p. 110). It is quite easy to detect one microcurie of ^{131}I in the thyroid gland using a scintillation detector. If this radio-iodine is carrier free this corresponds to a mass of chemical iodine of only 8×10^{-12}g. This amount of the element is well below the amount detected by normal chemical analytical procedures and also well below the level which can produce physiological effects on the thyroid gland. It is very uncommon for the administration of a radioactive material to a patient to cause any interference with the physiological effects being investigated and it never happens if carrier free isotopes are used.

Required physical characteristics of radioactive materials

Type of radiation emitted

When radioisotopes are to be used for therapeutic purposes in which it is desirable to restrict the absorbed dose to an organ of limited extent it is clearly advantageous if a large proportion of the energy emitted during disintegration is in the form of relatively short range beta particles. If detection of the isotope from outside the body is also required the isotope must also emit gamma rays. The therapeutic use of ^{131}I for thyroid disorders (see p. 269) illustrates the use of an isotope with satisfactory beta and gamma emission characteristics for this purpose.

For diagnostic purposes an *in-vivo test* almost invariably involves external detection of the isotope using a scintillation detector. In this case the favourable physical characteristic of the isotope include a gamma ray photon energy which is high enough to avoid serious attenuation in the tissues but low enough to allow the detector to be shielded and collimated without using excessive thicknesses of lead (see p. 102). It is also advantageous if the isotope used is one which gives no beta emission, or beta emission of very low energy, so that a low absorbed dose is given to the patient. The isotope technetium 99-m (see p. 9) with a gamma ray photon of 140 keV is a favourable one for many such investigations.

For the assay of radioactivity in fluids or tissue samples, such as is involved in some *in-vivo* tests and in all *in-vitro* tests, a wide range of gamma photon energies can be used since they are detected effectively by the scintillation counter (see p. 100). For some isotopes however which give off only very low energy beta particles special detection methods are needed and liquid scintillation counting is most commonly used (see p. 105).

Half life

Radioisotopes of very long half life are in general to be avoided for *in-vivo* tests in order to reduce the possibility of producing long term harmful biological effects. For therapy it is desirable that the biological effect should be relatively rapid and half lives of not more than a week or two are desirable in this case. In diagnostic tests it is advantageous to use isotopes with half lives about the same as the time over which the test is to be continued. The radiation dose to the patient is proportional to the half life of the isotope (see p. 10) and the half life therefore should be as short as possible but technical problems are greatly increased if the half life is too short. Half lives from about 2 hours to 1 month are commonly used in diagnostic tests and the use of isotopes produced from generator columns (see p. 11) is very common.

Biological half life

The removal of a radioactive material from a site in the body is determined not only by the physical decay of the radioisotope but also by the metabolic processes to which the isotope as a chemical material is subjected. The metabolic activity may result in the elimination of the material at a fast rate or if the element is absorbed in some particular tissue it may disappear from the body only slowly. A *biological half life* can be defined which is the time during which the amount of an element in an organ or in the body falls to half its value by metabolic processes. The biological half life of a given element is the same for all isotopes of that element no matter what the radioactive physical half life might be but it does depend on the chemical state of the element in the body.

In some circumstances the biological half life of an administered isotope is very important, an example being the biological half life of iodine in the thyroid gland (see p. 99). In problems of dosimetry and radiation hazard the real or effective rate at which radioactivity disappears from a particular site is a critical factor. The *effective half life* is a combination of the physical and biological half lives and is given by the formula

$$\frac{1}{T_{\mathrm{eff}}} = \frac{1}{T_{\mathrm{biol}}} + \frac{1}{T_{\mathrm{phys}}}$$

where T_{eff}, T_{biol} and T_{phys} are respectively the effective, biological and physical half lives. Occasionally it is possible to use for human administration isotopes of long physical half life because in the chemical form used the biological half life is very short.

Radiopharmaceuticals

Radioactive elements

Radiopharmaceuticals are pharmaceutical materials prepared, with a radioactive label, for administration to patients. In some procedures it is the metabolic behaviour of an element that is of interest and in this case a radioactive form of the element is required. The two most important elements studied in this way are iodine and iron. *Iodine* is an element incorporated into the hormones produced by the thyroid gland and a study of the metabolic behaviour of iodine in the body gives valuable diagnostic evidence of thyroid function (see p. 286). Radioactive iodine introduced into the body therefore enables iodine metabolism to be studied easily. The concentration of iodine in the thyroid gland also means that therapeutic doses of ionising radiation can be given to the gland from deposited radioactive beta emitters. There are a number of possible iodine isotopes which can be used in these procedures. These include ^{131}I ($T_{\frac{1}{2}}$ 8 d $\beta\gamma$ emitter), ^{132}I ($T_{\frac{1}{2}}$ 2.3 h $\beta\gamma$ emitter), ^{125}I ($T_{\frac{1}{2}}$ 60 d γ emitter) and an isotope ^{123}I ($T_{\frac{1}{2}}$ 13 h γ emitter) now becoming available. In every case however the radioactive iodine is supplied and administered as a solution of sodium iodide, though in the body the radioactive iodine ion is the entity whose behaviour is of importance. To some extent the pertechnetate ion (TcO_4^-) behaves in the thyroid in a similar way to the iodine ion and the use of $^{99\mathrm{m}}$Tc for study of thyroid function is often an advantage since it gives low absorbed dose to the gland.

A study of the metabolism of the element iron is of great importance in the clinical diagnosis of blood disorders. This element, again in its ionic form, is built into the haemoglobin of red cells following its absorption from the blood stream to which it may be introduced directly or through the processing of food in the alimentary canal. The most useful isotope is ^{59}Fe ($T_{\frac{1}{2}}$ 45 d $\beta\gamma$ emitter) and it is administered in the form of a simple iron salt, often ferric citrate, though the precise compound used will depend on the nature of the particular test.

Labelled molecules

A second way in which radioactive atoms may be used is as a label for molecules or molecular aggregates. In this case only the behaviour of the molecule is of importance and the radioactive property of the attached atom is of significance because of the ease of detection by physical techniques. The choice of the radioactive label in this case depends on the favourable physical properties of the radioactive atom and on its suitable chemical properties from the point of view of attachment to the molecule of interest. Iodine atoms are easily attached by relatively simple chemical manipulation to protein molecules, and the use of radioactive iodine in one or other of its forms to label protein molecules, such as for example, human serum albumin, is very important both in the study of protein metabolism and in the use of the large protein molecule to outline the vascular system from which these molecules do not normally escape.

The labelling of large colloidal particles, which may be aggregates of protein molecules or colloids of sulphur, by the use of the isotope technetium-99m ($T_{\frac{1}{2}}$ 6 h γ emitter) is also an important part of the preparation of radiopharmaceuticals for isotope visualisation procedures (see p. 97). A most important use of a radioactive label for large molecules is the labelling of red blood cells with the isotope chromium-51 ($T_{\frac{1}{2}}$ 27.8 d γ emitter). Atoms of chromium under appropriate incubation procedures become attached to the haemoglobin molecule of the red cells and if the labelled red cells are then re-injected into the patient from whom the cells have been withdrawn a study of the behaviour of the labelled cells can be made (see p. 110). In this type of labelling procedure it is important to know if the added radioactive atom interferes with the physiological behaviour of the compound under study. Sometimes a changed physiological behaviour is of no great importance and an example of this is in the use of radioactive iodinated serum albumin (RISA) to define the plasma volume. However if the metabolism of the protein itself were being studied any change in the physiological behaviour of the molecule must be avoided and labelling procedures must be arranged with this in mind.

Synthesised radioactive compounds

A third type of radiopharmaceutical is one in which a radioactive atom is a basic component of the molecule and a large number of medically and biologically important compounds are now synthesised commercially in which a non-radioactive component atom has been replaced by a radioactive isotope of the same element. Organic compounds labelled with carbon-14 ($T_{\frac{1}{2}}$ 5760 y β emitter) or hydrogen-3 (tritium) ($T_{\frac{1}{2}}$ 12.3 y β emitter) are available in a tremendous variety. Two examples of clinical interest are vitamin B12 and water. Vitamin B12, of great importance in the study of anaemias, is a compound (cyanocobalamin) in which an atom of cobalt is present. The compound is synthesised on a large scale by biological processes in which radioactive cobalt is built into the molecule. The most commonly used isotopes are ^{57}Co ($T_{\frac{1}{2}}$ 270 d γ emitter) and ^{58}Co ($T_{\frac{1}{2}}$ 71 d $\beta\gamma$ emitter). Labelled water is available in which one hydrogen atom of the normal molecule is replaced with tritium, the isotope ^{3}H. This compound, *tritiated water*, is used to study electrolyte and water absorption problems in patients. It will be noted that here the isotope, ^{3}H, with relatively long half life is used. ($T_{\frac{1}{2}}$ 12.3 y) This is tolerable in this case because the beta ray emission energy is extremely small and the biological half life of this compound is short, approximately 12 d.

Sterility of radiopharmaceuticals

The preparation of radiopharmaceuticals and the form in which they are made available for use in clinical departments is influenced by the method by which it is intended to introduce the product into the patient. For some tests oral administration is the preferred method and indeed sometimes the only suitable one. In this case clean stable palatable solutions are ideal. For many tests however intravenous injections must be used and for some cases injection into the spinal fluid, (intra-thecal) is necessary. In these cases great care must be taken to produce a radiopharmaceutical in a sterile solution of the right pH value and free from pyrogens* which are foreign proteins arising from previous bacteriological activity. The details of the necessary procedures are beyond the scope of this text but these requirements do influence the laboratory facilities necessary in hospital departments where radiopharmaceuticals are prepared and reference to this problem will be made later (see p. 106).

The localisation of isotopes in the body

Consideration of the various metabolic processes through which radioactive compounds may pass will show that many of the compounds will at some stage be located in some particular organ or tissue, at least for a time. The localisation of radioiodine in the thyroid gland or of radioactive colloid particles of appropriate size in the lungs are good examples. This aspect of the behaviour of radioactive materials in the body is important in three ways.

Use for organ visualisation

The localisation makes possible the visualisation of organs if the distribution of the isotope in the organ can be plotted or pictorially demonstrated (see p. 102). Furthermore if the distribution of the administered isotope in the organ concerned is abnormal, deductions can be made about the function of the organ. As will be seen later abnormal isotope distributions may be an increase in isotope concentration above the normal revealing lesions (hot spots) as for example using technetium-99m in the brain or strontium-87m ($T_{\frac{1}{2}}$ 2.8 h γ emitter) in bones. Lesions may also be revealed by cold spots, i.e., reduced activity in the visualised patterns as in the use of ^{99m}Tc labelled colloidal aggregates for lung scanning or liver scanning.

Radiation dose and critical organs

The localisation of the isotope material even temporarily in some organ or tissue means that the absorbed dose of ionising radiation which a patient receives following the administration of a radioactive material is rarely uniform. In any isotope test it is always necessary to ensure that the whole body and each organ in the body receives an acceptably low dose. Some knowledge of the metabolic processes involved is essential in calculating the doses received by the patient in these procedures. It is possible that, because of retention of the isotope in one particular organ, that organ may receive a dose approaching the reasonable limit for that organ before the limit for the whole body is reached (see p. 121). The organ which is most vulnerable in this sense is called the *critical organ* and calculations of absorbed dose involved must take account of the possible concentration of the isotope in the critical organ.

Therapeutic possibilities

If a suitable radioactive compound is concentrated in one particular organ it seems feasible to give a very high radiation dose to the organ which could be effective for therapeutic purposes. Unfortunately this possibility is only feasible if the ratio of the concentration of the material in the tumour to that of the body in general is very high. For example if the administration of a dose of 5000 rads to a particular malignant tumour is the aim while keeping the general whole body dose to say 5 or 10 rads a *concentration ratio* of 500 or 1000 to 1 is required. This is hardly ever possible with available radiopharmaceuticals. The concentration ratios normally observed are at the most 10 or 100 to 1. These are suitable for diagnostic purposes since it is easy to detect and measure such concentration ratios. They are however impracticable for therapy.

The successful use of radioactive iodine for therapy of diseases of the thyroid gland is described later (see p. 269). The use of a radioactive isotope of iodine for this purpose depends on the fact that the concentration ratio for iodine in the thyroid is at least 1000 to 1. The search for radioactively labelled compounds which will be very highly concentrated in tumour tissue has been continued for a long time but so far with no great success.

Therapy with radioactive isotopes is however done to a smaller extent where a uniform distribution over the whole body is the aim. The most common example of this technique is the use of radioactive phosphorus, ^{32}P, for the treatment of polycythemia vera (see p. 251). ^{32}P with a half life of 14.3 d is a pure beta emitter whose beta particles have an energy of 1.71 MeV and a maximum range in tissue of 7 mm. In the treatment of polycythemia a radiation of the blood forming cells of the bone marrow is achieved by an intravenous injection of ^{32}P in the form of a sodium phosphate solution. One other form of therapy by unsealed radioactive isotopes is the use of beta emitters to suppress fluid effusions from surfaces in the abdomen (peritoneal effusions) or the thorax (pleural effusions), (see p. 222). Beta emitters in the form of colloid suspensions are used in these techniques and the beta emitter is deposited on the surfaces which are responsible for the effusion. The isotopes used are phosphorus-32, gold-198 ($T_{\frac{1}{2}}$ 2.7 d β 0.96 MeV), and yttrium-90 ($T_{\frac{1}{2}}$ 2.5 d β 2.27 MeV) and these are prepared as colloid suspensions for introduction into the body cavity under treatment. This form of therapy plays, however, only a minor role amongst other procedures of radiotherapy.

Dosimetry of internally administered isotopes

When a radioactive isotope is absorbed in a local site in the body it disappears from that site roughly exponentially with a half life determined by both its physical decay and the metabolic processes which may remove it from the site. The energy of the beta particles which it emits is absorbed almost entirely in the immediate vicinity of the active material. The gamma rays emitted by the active material, however, will not be absorbed appreciably in the immediate vicinity unless they are of very low photon energy. In general a substantial proportion will be absorbed in the body elsewhere, though if the radiations are of reasonably high photon energy most of the photons will escape altogether from the body. The photons that are absorbed will produce secondary electrons by the processes described in Chapter 4.

It is relatively simple to calculate the total absorbed dose given to any particular tissue by the beta emission of the active material that decays in it. If the initial concentration of active material in terms of microcuries per g of tissue and the effective half life is known, then simple arithmetic gives the total number of radioactive atoms which disintegrate per g. A knowledge of the physical properties of the disintegrating atoms gives both the number of beta particles per disintegration (generally one) and the average energy of the emitted particles. The total energy deposited by the beta particles per g of tissue can thus be deduced. This leads directly to a value of the absorbed dose in rad. A formula for this is as follows

$$\text{Dose in rads} = 73.8\, C\bar{E}_{\beta}T,$$

Where

C = concentration in μCi per gram.
$\bar{E}_{\beta}$ = average beta energy in MeV.
T = effective half life in days.

The calculation of the absorbed dose in a given volume of tissue arising from the gamma emission of isotope contained in it is rather more complicated. The physical information needed for each isotope is the specific gamma ray emission (see p. 10), i.e., the exposure in roentgens per hour at 1 cm distance from 1 millicurie of activity, and the absorption coefficient of the radiation in the tissue. Starting with these factors, by an integration process, the absorbed dose over a particular volume can be deduced for some simple shapes. The dose depends very much on the shape and size of the tissue volume and is not uniform over the whole volume, being greatest in the centre. In many cases the gamma ray contribution to the dose is small compared to the beta ray dose. This is true in the case of iodine isotopes absorbed in the thyroid. In some cases, however, where a large volume of tissue is involved—for example, when using sodium-24, which spreads uniformly through the body—the gamma ray dose is predominant. A formula similar to the one given for beta ray dose, however, can be deduced for simple geometrical shapes, the dose again depending on the concentration of the isotope and its effective half life.

This formula is as follows:

$$\text{Absorbed dose from gamma rays} = 33.1 \times 10^{-3}\, gkCT \text{ rad}$$

where C is the concentration in μCi per gm
k is the specific gamma ray emission—the 'k factor'
T is the effective half life in days

g is the 'geometric factor', which takes into account the size and shape of the tissue volume. The calculation of this factor is normally very complicated but has values between about 50 and 150 in various parts of the body.

The estimation of the total radiation dose to an organ or to the whole body therefore is never very simple since the isotope distribution may not be uniform and the assumption of a truly exponential decay of the radioactive content of a given tissue may not be valid. In a simple example however an absorbed dose to a thyroid gland can be estimated if its size is known, and if the total radioactive iodine content is measured and its combined physical decay and biological elimination. These factors are often estimated in thyroid therapeutic procedures (see p. 269). For example, if 8 mCi of ^{131}I is given to a patient and 10 per cent of this is retained in the gland of mass 50 g, and if further, the effective half life of the iodine in the gland is 5.4 days then a total absorbed dose of approximately 9400 rad will be administered. In this figure about 10 per cent of the total dose arises from gamma rays. The physical constants for ^{131}I needed in this calculation are $\bar{E}_\beta$ (average beta energy per disintegration) = 0.187 MeV and k (the specific gamma ray emission) = 2.23 R per hour. The dose to the rest of the body, arising in part from the small circulating activity of beta and gamma emitting ^{131}I and in part from the gamma emission from the thyroid gland, can be estimated by similar methods though the data available for such a calculation will be less precise.

Dose estimations of this kind are necessary for the control of radiotherapy by unsealed radioisotopes and also to ensure that doses administered in diagnostic procedures are below the accepted tolerable limits.

The assay of radioactive materials

Radioactive materials which emit gamma rays may be detected and measured by the ionisation techniques or other methods described in Chapter 5, if the gamma ray intensity is high enough. Occasionally, too, even beta emitting materials if sufficiently active can be detected and measured by the 'bremsstrahlung'* type of X-radiation, produced by the stopping of the beta particles in the containing medium. In general, however, beta particle emitters are detectable by ionisation chamber devices only if they can penetrate the thin window of the chamber and give a sufficiently high radiation intensity.

Unsealed radioactive materials are, however, generally used in amounts which are too small to make simple ionisation chamber methods very satisfactory and much more sensitive devices are available. The most common of these devices are the *Geiger counter* and the *scintillation counter*. These are very sensitive systems in which individual ionising particles or individual photons from the radioactive emitter can be detected. Each beta particle or photon gives rise to a pulse of electrical charge as the output of the detector. The recording system usually records the total number of individual pulses produced in the detector. Hence the use of the word 'counter' in this connection. A detector of this type therefore measures pulses, and counts them. The number of pulses recorded per unit time is a measure of the rate at which particles or photons arrive at the detector and this is proportional to the disintegration rate of a radioactive material situated in the vicinity of the detector.

The Geiger counter

The *Geiger counter* (sometimes referred to as the Geiger-Müller counter after its co-inventors) is essentially a simple ionisation chamber in which the ionisation produced by a single charged particle takes place in gas at very low pressure and under a very high electric field. In this case the ionisation current produced is not only saturated (see p. 41) but also undergoes an enormous amplification. This amplification arises by a process known as *ionisation by collision*. The ions originally produced by the fast charged particle are accelerated in the high electric field and because of the low gas pressure the path between each successive collision with the un-ionised molecules surrounding them is long enough for the ions to acquire sufficient energy to ionise these molecules. A cascade process is thus produced which gives rise to an amplification of the initial ionisation which can under suitable circumstances be as high as one hundred million times. A charge pulse is thus available generated from the arrival of one ionising particle which is quite adequate to be recorded on robust equipment.

The appropriate conditions for this cascade process are normally produced by making the counter chamber of a cylindrical conductor along the axis of which is stretched a fine wire acting as the positive electrode. This electrode system is often placed in a glass container filled with argon or helium at a gas pressure of about 100 mm of mercury. The outer electrode may however itself be the container for the gas. By applying a high potential between the electrodes of about 1000–2000 V the necessary high electric field is produced round the central wire so that in this region any production of ions initiates the cascade process and gives rise to a very high charge pulse on the electrode system. Any ionising radiation therefore which produces ions in the sensitive region triggers off the counter. Such radiation may be, for example, a beta particle which can penetrate the walls of the counter, or a fast secondary electron arising from the absorption of a gamma or X-ray photon in the gas of the tube, or more likely, in the walls of the tube. Geiger counters are made in many forms, as seen in Figure 9.1. Some have robust thick walls and are used for photon detection. Others have windows which are very thin but designed to withstand the pressure difference between the inside and outside. These are used as beta ray detectors. Radioactive solutions, particularly of beta emitting isotopes, may be assayed in Geiger counters in which liquid is contained in a jacket around the counter tube itself.

The large charged pulse initiated by ionisation in the sensitive volume of the counter is only a momentary one but some special arrangement has to be adopted to terminate the pulse since otherwise the conditions for recording the arrival of the next ionising particle would not be present. A pulse is '*quenched*' either by an external electronic circuit or often by introducing a small amount of alcohol vapour into the argon gas when the tube becomes a 'self-quenched tube'. Even so, the charge pulse from a single event does take time to be collected and during this collection the tube is not sensitive to the arrival of a further ionising particle. The Geiger counter therefore will miss some particles if it is exposed to a radiation field of too high an intensity. Counting rates of the order of 250 per sec for example will miss about 5 per cent of the received ionising particles for this reason. The *resolving time correction* rises very rapidly as the counting rate increases.

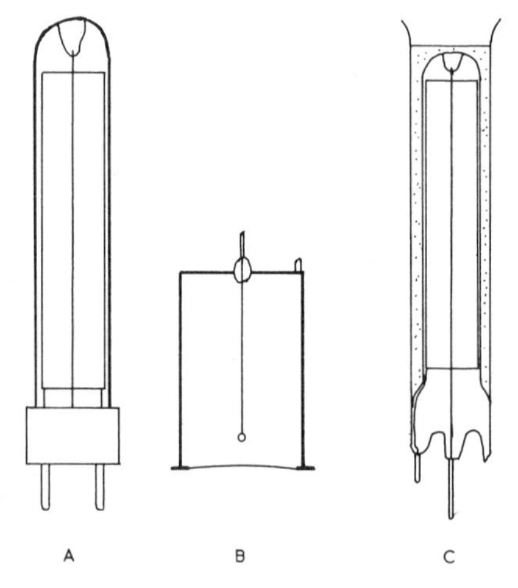

Fig. 9.1 Diagrams of Geiger Counters.
A. Cylindrical counter with lead cathode for gamma ray detection.
B. Thin window counter for beta ray detection.
C. Counter with outer annular container for active solutions.

Three properties of the Geiger counter are important for its use as a sensitive detector for radiations from radioactive material or X-rays. (a) the pulse size is dependent only on the size and shape of the detector tube and not at all on the number of ions initiating the discharge. It cannot therefore be used to distinguish, say, an alpha particle from a beta particle, or between photons of different energy. It is in fact a 'trigger' device. (b) The walls of the Geiger detectors are inevitably made of materials of high atomic number. The response to X- or gamma rays which depends mainly on the interactions of photons in the walls, will therefore be very dependent on the wavelength of the radiation. Indeed in some Geiger tubes a cathode of very high atomic number, lead for example, is introduced deliberately to enhance the sensitivity for long wavelength gamma-rays (Fig. 9.1). The Geiger counter therefore is unsuitable for accurate measurement of X- or gamma-rays but very suitable for detecting and estimating low intensities of these radiations. (c) When the voltage applied to a Geiger tube is too low the cascade process does not work and if it is too high a continuous discharge is set up which damages the electrodes. Over an intermediate range however the counting rate is nearly independent of the voltage for a constant radiation intensity. This voltage range is called a *plateau*. Under good working conditions therefore the counter is relatively insensitive to fluctuations in the voltage supply. It thus becomes a simple and fairly rugged instrument of high sensitivity easily made in a portable form.

To make a Geiger tube function additional electronic equipment is needed. This will include a variable high tension source to produce the necessary high voltage with reasonable stability. It will also include a pulse recording device. This may be a device which displays electronically the number of pulses received in a pre-set time. The number per unit time is then easily deduced. This device is called a *scaler*, since the number is normally recorded on a scale of 10 display. Alternatively, the pulses may be stored in a condenser with a high resistance leak across it. The output voltage is then proportional to the rate at which pulses are being received and a meter which records this voltage gives a measure of the pulse rate which is proportional to the intensity of the radiation being received. This device is called a *rate meter*.

The scintillation counter

The *scintillation detector* makes use of the light flashes produced in a suitable crystal when it absorbs photons, the effect described earlier in Chapter 5. Because of its inherent high sensitivity, even greater for photons than a Geiger counter, and because of other desirable characteristics it has now become the standard instrument for the assay of gamma emitting materials. The crystal material most commonly used is a single crystal of sodium iodide activated by a small addition of the element thallium and sealed hermetically in a container which is light-tight except on one face which is in close contact with the photocathode of a *photomultiplier*. This is a device in which electrons emitted from a photo cathode produce successive multiplication of charges on a series of sensitive grids so that a large charge output finally results from the last electrode proportional to the emission from the photocathode. The light pulses generated in the crystal produce emission of photo electrons from the cathode of the multiplier and hence a large charge pulse from the multiplier anode. This pulse can be recorded and successive pulses counted by scaler devices similar to those used in a Geiger counter system. To some extent therefore the scintillation counter system is similar to the Gieger counter system. Charge pulses due to absorption of photons in the detector are counted by a scaler after an amplification of some 10^8

times in the photomultiplier which is activated by a high voltage of the order of 700 – 1500V. Figure 9.2 shows a simple arrangement of a crystal and photomultiplier used as a sensitive gamma ray detector.

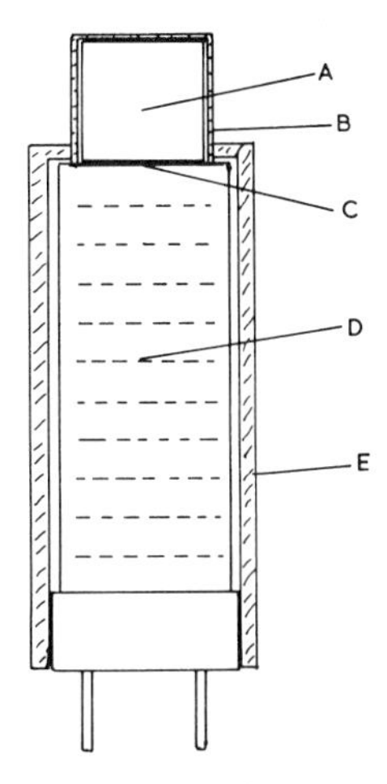

Fig. 9.2 Diagram of Scintillation Counter.
A. Sodium Iodide crystal
B. Aluminium can.
C. Photo cathode of multiplier.
D. Multiplier grids of multiplier.
E. Lead shield.

The scintillation counter differs from the Geiger counter however in one very important and significant aspect. This is in the fact that the magnitude of the charge obtained from the anode of the multiplier is directly proportional to the ionisation energy deposited in the crystal by the absorbed photon. This *proportionality* arises because the emission from the photocathode of the multiplier is proportional to the total light collected from the crystal which is itself proportional to the ionisation energy deposited in the crystal. The magnitude of the output pulse from the multiplier depends on the multiplier gain which is controlled by the voltage applied to the multiplier, but proportionality is maintained during the amplification.

This proportionality property of the scintillation counter means that the counter can be used to analyse the spectrum of the energy pulses deposited in the crystal when photons are absorbed and hence the energy spectrum of the photon field itself can be explored. This is achieved by using suitable electronic circuits between the multiplier output and the scaler. These circuits make it possible to select for counting only those pulses which have a predetermined size. Pulses may therefore be selected corresponding to particular values of the amount of energy transferred to the crystal in the absorption process. This is done in two ways. First, one may include in the count only pulses greater than a predetermined size and this method is referred to as *discrimination*. On the other hand one may include in the count only pulses of one size or of a narrow range about a predetermined size. This is referred to as *pulse analysis*. These two ways of handling the pulses allow the possibility of making the scintillation counter respond only to photons from one isotope, ignoring photons of different energy since pulses which are initiated by photoelectric absorption in the crystal of a line spectrum of gamma rays from a particular isotope will all have the same size representative of the photon energy of the radiation concerned. Similarly it is possible to reject pulses which arise from general background radiation so that pulse selection in this way makes possible a much improved ratio of source to background count. By varying the pulse magnitude selected for counting, the energy spectrum of a gamma ray beam may be explored and the assay of two or more gamma emitting isotopes in one sample becomes possible.

A further characteristic of scintillation counting is important. In a crystal any photon absorption results in a detectable charge pulse. A large mass of absorbing material may therefore be used to detect the photon beam as long as the crystal is a single crystal optically transparent so that light pulses produced anywhere inside it can reach the photocathode of the multiplier. This makes the scintillation counter of very great sensitivity compared to the Geiger counter in which gamma ray photon detection is possible only because of secondary electrons which arise either on the inner surface of the cathode or from the gas content of the tube. Extremely sensitive detectors therefore can be produced by large crystals. Single crystals have to be manufactured by an expensive and difficult process but as is shown later single crystals of a large size are produced for a variety of instruments in which the primary aim is to detect gamma ray photons with high efficiency.

Detection and measurement of radioactivity in the body

The techniques of diagnostic procedures and the control of therapy with radioactive agents both involve accurate measurements of radioactivity in either the patient or in specimens from the patient or in both. *In-vivo* measurements are now almost always made with scintillation detectors. In the simplest form a detector of this type is used to measure the activity in some particular organ of the body. In the thyroid gland for instance the total radioiodine content is required for thyroid function tests, (see p. 110) and a simple detector system can be used. Much more complex detecting systems are required for studies of the distribution of activity in various organs of the body.

Simple collimator systems

For measurements such as that of the radioactive content of a single small organ the sodium iodide crystal is mounted on the cathode face of a photomultiplier and surrounded by a lead shield which reduces the natural background radiation to reasonable level and at the same time forms a '*collimator*' so that the crystal may receive photons from a limited portion of the body only. Cylindrical crystals having a diameter of 2 to 5 cm with thickness of about 2 cm are used in these detectors. As an example Figure 9.3 shows a diagram of an internationally recognised standard collimator system for measuring the uptake of ^{131}I in the thyroid gland. The collimator will allow radiations from an area of about 13.5 cm diameter to reach the crystal at a working distance of 25 cm from the crystal and the lateral shielding round the crystal reduces both the natural background and the background due to circulating activity in the tissues outside the thyroid gland.

obtained of the time variation of the activity in the organ. This is particularly suitable when the activity is varying rapidly with time as in the case of tests of renal function (see p. 112).

Scintiscanners

A second very important use of a scintillation detector is to observe the distribution of an isotope in an organ, or in a part, or whole of the body. Such information can be obtained by moving by hand a suitable simple collimated detector and recording and subsequently plotting the resulting detector count rate at various points of the field of interest. Use of such simple distribution tests is often of great value. However, for organ scanning very great use is now made of automatic exploring devices in which a scintillation detector follows a scan pattern over the area to be examined and a dot recorder attached to it follows an identical pattern over a sheet of paper and prints on to the paper a dot for each recorded scintillation in the detector head. The recorder may also be a photographic film on which the activity distribution is displayed by a series of dots, produced by a light source. Such a device is often referred to as a *scintiscanner* or *linear scanner*. The scanning pattern adopted is generally one similar to that employed in building up a picture in a television display.

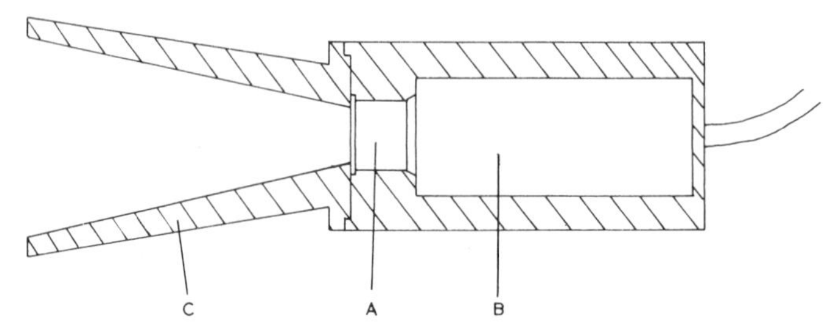

Fig. 9.3 Standard detector system for thyroid uptake measurements.
A. Crystal detector.
B. Photomultiplier.
C. Collimator and shielding in lead.

Variations of this basic design of detector mounting are made to suit the size of the organ in which the activity is to be measured, and to suit the isotope involved. The shielding thickness of a collimator for use with ^{99m}Tc for example with a 140 keV gamma ray emission can be much thinner than one designed for use with ^{131}I which has its main gamma ray energy of 360 keV.

Accurate measurement of the radioactive content of an organ is difficult and is only attempted in routine procedures for the thyroid gland. Many precautions must be taken such as a correction for attenuation in the tissues and the elimination of scattered radiation in the measurement field of the collimator. In practice for thyroid uptake measurements a calibration procedure is adopted to check the sensitivity and the accuracy of the counting system. This is normally done by using a simple model of the neck in the form of a cylinder of perspex filled with water and containing, in an appropriate hole, a small bottle holding a known quantity of the isotope to be measured, generally iodine-131.

Such simple collimator systems may be connected to a scaler-timer unit so that the count rate is simply deduced and this gives a figure proportional to the activity in the organ. Alternatively the system may be connected to a ratemeter and a *chart recorder* so that a direct display can be

The most important part of a scintiscanner is the detector head. This is often composed of a large crystal 3″ or 5″ in diameter mounted with its photomultiplier in a heavy lead shielding and fitted with a multi-hole focussing collimator. This collimator, shown diagramatically in Figure 9.4 is designed to use as efficiently as possible the large absorbing volume of the crystal but to receive at any one time photons from only a small volume of the organ being scanned. In this way variations of activity distribution in an organ can be detected with a resolution often of the order of 1 or 2 cm.

A second important part of the scintiscanner is the display system by which the activity distribution in the body is presented for clinical interpretation. This display may be dots on paper, or black spots on a photographic film as mentioned above. In many cases the displayed pattern is printed out by a multicoloured typewriter ribbon so that a coloured picture is presented in which different colours represent different count rates. This is aimed at enhancing the ability of the observer to detect abnormalities in the activity distribution in the displayed pattern. A considerable amount of electronic processing of the signals between the detector and the display unit is done including selection of pulses to suit the isotope being used. The processing also helps to enhance the detection of abnormalities. Applications of scintiscanning procedures will be described briefly in chapter 10. They are very widespread and specialist text books should be consulted for examples of the available types of display and details of the focussing detectors.

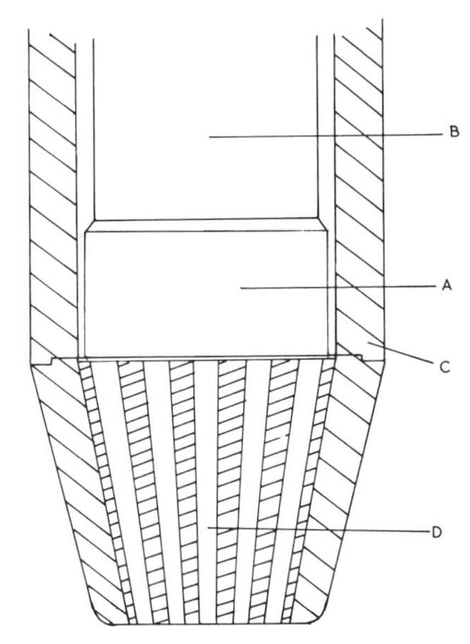

Fig. 9.4 Multi-hole focussing collimator.
A. Detector crystal.
B. Photo multiplier
C. Lead shielding
D. Multihole collimator.

Gamma cameras

It would be advantageous to collect radiations from all the field of interest at one time instead of by the sequential point by point scanning described in the previous paragraph. A device for doing this is the *gamma camera*. In this device a very large diameter crystal, 12″ diameter and 0.5″ thick receives the gamma ray photons from the patient through a grid of about 1000 or more holes drilled parallel to each other in a 2″ thick lead plate. If this multihole collimator is placed near the patient a distribution of scintillations is produced in the crystal which corresponds to the distribution of isotope in the field covered by the crystal and collimator. The scintillations in the crystal are detected by an array of 19 photomultipliers. The output from each multiplier corresponding to a particular single flash in the crystal depends on the position of the flash relative to the multiplier. By summing the outputs from particular groups of multipliers, signals can be obtained which indicate the position of each flash. These signals are passed to a cathode ray tube together with a signal summed from all the photomultipliers which is used to produce a bright spot on the cathode ray tube screen. A distribution of spots on this screen is thus built up corresponding to the activity distribution in the field of view. This pattern is often recorded on a Polaroid film photograph, though other ways of display are also available.

Developments in gamma cameras are being made. They include an increase in the field of view by using crystals of large diameter and the use of a larger number of photomultipliers. This improves the uniformity of response over the field of view and also improves the ability of the camera to resolve small details of the activity distribution.

Figure 9.5 shows a diagram of the detector head of a gamma camera. The crystal, collimator and multiplier array are mounted inside a substantial lead shield of considerable weight, which is itself mounted on a stand so that the face of the collimator can be adjusted in any suitable position relative to the patient under study.

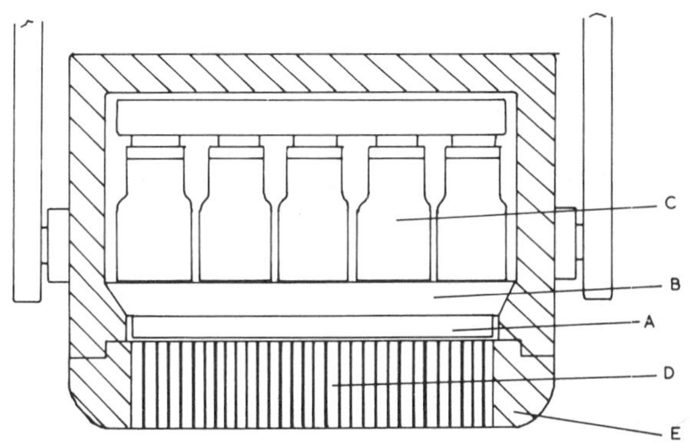

Fig. 9.5 Gamma-Camera Head.
A. Crystal detector
B. Light guide
C. Photomultipliers
D. Parallel hole collimator
E. Lead shield.

A discussion of the relative advantages of the linear scintiscanner and the gamma camera is beyond the scope of this text, and the developments particularly in gamma camera systems will influence the relative importance of the two methods. Gamma cameras however find a great use in activity distributions in which rapid variations with time are involved. In these rapid dynamic studies the scintiscanner is of little use but important observations can be made with the camera even if the pattern of activity is changing over a time of only a few seconds.

The information from a gamma camera can be processed by digital computer techniques and much information regarding the dynamic function of an organ can be obtained in this way. It is also possible to process the information from a scanner or a camera to improve the detection of distribution abnormalities in an organ. Specialist treatises on this work should be consulted for further details.

Whole body counting

One further use of scintillation counting is the measurement of the total radioactive content of the body. This is of value in the estimation of hazards following the accidental intake of radioactive material into the body, and in some clinical studies particularly of absorption problems. In this latter case the fate of an administered isotopic label in the body can be followed without the necessity of measurements of faecal or urinary excretion of the active material.

It is possible to get an estimate of the total radioactive content of the body by relatively simple detecting equipment if enough activity is present to give readings much larger than the natural background. Most whole body counters, however, are very sensitive devices in which the patient and the detector are enclosed in a thick attenuating shield inside which a low background count is obtained. In this case very high sensitivities can be achieved and for the common isotopes activities of the order of 10^{-8} curies can be detected. This means that a radioactive content of the body well below the maximum permissible body burden can be measured (see p. 122), absorption tests with small amounts of activity are possible for children, and in some equipment the natural potassium-40 content of the body (and therefore the total body potassium) can be estimated with accuracy.

A typical arrangement of a whole body counter is illustrated in Figure 9.6 Here an array of four large crystals each 5″ diameter is arranged around the patient inside a low background enclosure.

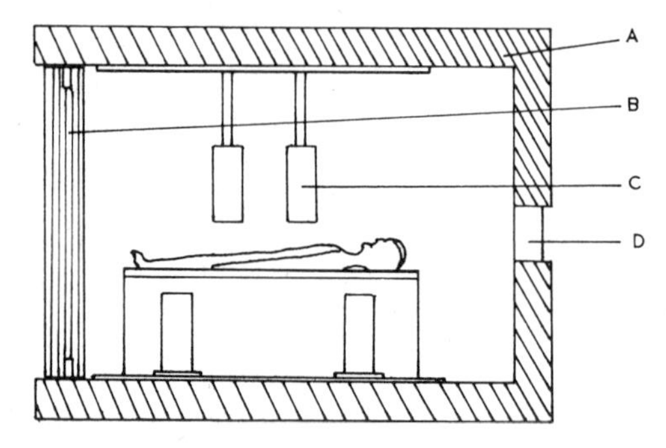

Fig. 9.6 Whole Body Counter
A. Steel Enclosure
B. Thick Door
C. Crystal-photomultiplier detectors
D. Window.

Many variations of detector and shielding arrangements of whole body counters are in use. In particular in one arrangement the patient, on a couch, is passed slowly between two highly shielded and sensitive detectors in an arrangement called a *shadow shield counter*. This device is cheaper than more elaborate systems but serves some valuable clinical purposes.

Whole body counting is a very specialised use of scintillation counting and involves complicated procedures for background assessment, analysis of multiple isotope content of the body and calibration of the counter sensitivity for patients of different weights. Details are beyond the scope of the normal studies of the radiographer.

Measurement of activity in liquid and solid samples

The measurement of the radioactive content of fluid samples, such as blood or urine, is necessary in many procedures involving the use of unsealed sources both for therapy and diagnosis. Both Geiger counters and scintillation counters are used for this purpose. A form of Geiger counter specially for the assay of activity in 10 ml liquid samples has been shown in Figure 9.1 C. This counter can be used for both beta and gamma emitting samples. It is most useful for the assay of liquids containing pure beta emitters where the beta energy is high such as phosphorus-32. The activity is assessed by comparing the counting rate with that of a standard solution of the same isotope. For counting the Geiger tube is assembled in an appropriate lead shield in which the background counting rate is reduced to a low value so that small amounts of activity in the fluid under test can be measured.

The activity of urine samples is often assayed in bulk if it is due to a gamma emitting isotope such as ^{131}I. For this purpose the whole day's excretion may be collected in a suitable container and the activity estimated by placing the bottle inside a ring of Geiger counters enclosed in a suitable shield.

Scintillation counters are more sensitive for gamma emitting isotopes and are almost always the preferred instrument for assay of such isotopes. The liquids may be placed in a suitable container in a standard position on the top of the detector crystal or inside a small hole drilled into the crystal. The crystal and sample must be mounted inside a substantial lead shield which also encloses the photomultiplier tube. Tissue samples of small volume may also be assayed in such an arrangement either in a solid form or dissolved in a suitable solvent.

Figure 9.7 illustrates three detector systems for the assay of activity in liquid samples. These are the simple Geiger counter for beta active liquids, the Geiger counter array for bulk gamma emitting liquids and the well type scintillation counter for small gamma active samples.

Some isotopes used in clinical work are pure beta emitters in which the beta ray energy is so small that the beta rays can only penetrate the walls of the counting system with very low efficiency or in some cases do not penetrate even the thinnest walls. The two isotopes of this type of greatest importance in biological and clinical investigations are hydrogen-3 (tritium) and carbon-14. These isotopes are assayed by dissolving the sample in a fluid which itself produces light flashes under the influence of ionising radiations. Such liquid scintillators, often consisting of sensitive organic phosphors dissolved in a base such as toluene, are placed in a suitable transparent container and the scintillations are detected by one or more photomultipliers placed close to the container. *Liquid scintillation counting* is a highly developed technique and often uses automatic sample changing devices and appropriate electronics to process the charge pulses from the multipliers so that assays can be made of more than one isotope in any one sample under test.

must be ordered regularly so that the delivery fits in with a carefully designed programme of diagnostic procedures. A third group of isotopes consists of those in which the half life is so short that they are generally only available by production in the hospital from relatively long lived generator columns (see p. 11). The most common example is ^{99m}Tc ($T_{\frac{1}{2}}$ 6 h) produced from a molybdenum column (^{99}Mo $T_{\frac{1}{2}}$ 67 h). Frequent renewal of the generator column is necessary in this case.

A second consideration regarding supplies of radioactive materials is that they must be transported and stored under conditions such that the leakage radiation from the container presents no hazard either to any worker involved or to any member of the general public. A further aspect of this is the necessity to ensure that when such supplies reach a hospital the leakage radiation does not

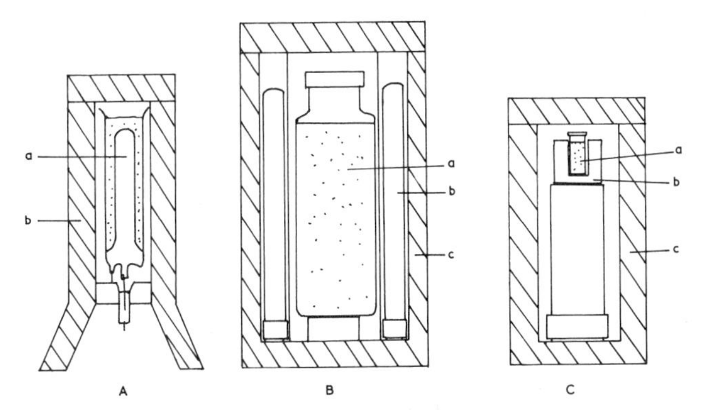

Fig. 9.7 Equipment for assay of radioactivity in liquids
A. Geiger Counter for 10 ml samples
(a) counter with annular container for samples
(b) outer lead shielding
B. Geiger array for bulk samples
(a) container for radioactive liquids
(b) ring of Geiger counters
(c) outer lead shielding
C. Well scintillation counter
(a) plastic container for 4 ml sample
(b) crystal detector and multiplier
(c) outer lead shielding.

Arrangements for handling isotopes in hospital practice

Supply of unsealed sources

The use of radioactive isotopes in liquid form for either diagnosis or therapy involves a number of procedures which are different from those in other hospital departments, and which arise from the radioactivity of the materials used. Arrangements must be established for the regular receipt of supplies from a firm manufacturing radiopharmaceuticals. In the U.K. this is frequently the Radiochemical Centre at Amersham, Bucks. For radioactive materials of rather long physical half life, such as ^{51}Cr ($T_{\frac{1}{2}}$ 27.8 d), no great problem is encountered in dealing with the day to day diagnostic programme from supplies stored in the hospital dispensary. Other isotopes with a short half life, such as sodium-24 ($T_{\frac{1}{2}}$ 15.0 h) however

interfere with any sensitive radiation measurement equipment which may be being used. The transport of a pure beta emitter such as ^{32}P is generally an easy matter but for isotopes with a penetrating gamma radiation such as ^{24}Na the protection during transit is difficult and involves large and heavily shielded containers. There are national and international regulations governing the transport of radioactive materials which take care of the leakage of ionising radiation from the intact container and also prescribe packing methods so that in the event of an accident, leakage of the contents, and therefore the spread of active contamination, is most unlikely. On arrival at the hospital a suitable place for the reception of the containers must be arranged and the sources themselves must be placed when unloaded in a suitable shielded store. Small cupboards with 3″ or 4″ of lead wall are convenient for many of these sources.

Laboratory arrangements

The laboratory arrangements for handling radioisotopes in a hospital are designed to prevent hazards arising both from the external irradiation of the workers and from the spread of radioactive contamination. The contamination may give rise to an internal hazard to workers as well as interfere with sensitive tests with active materials. Detailed consideration of the problem of radiation hazard control is dealt with in Chapter 11, but the present chapter will conclude with a brief account of the laboratory arrangements which are necessary to handle unsealed radioactive materials efficiently and safely in a hospital.

Unsealed radioactive sources are ideally handled only in working areas specially allocated and equipped for the work. The several clearly distinguishable steps in the whole programme of handling are best kept in their own laboratory or clinical space. For example, dispensing and storage of the basic materials should be done in a special room where arrangements will be needed for handling amounts of activity of many millicuries. A storage space for excreta and specimens is necessary though in general arrangements for the effective and prompt disposal of waste form an important part of the general procedures (see p. 107).

Sample measurements on patients are normally made in an appropriate clinical area away from the laboratory, with a special place allocated for administration of the radioactive material to the patient. It will be recognised however that measurements on samples deal with levels of activity generally in the microcurie or even milli-microcurie range and for this reason it is necessary to set up assay procedures in a room away from the storage or clinical rooms.

The previous paragraphs suggest some necessary physical separation between the rooms allocated to different aspects of the whole work so as to avoid unnecessary hazard both from external radiation and from the possible spread of radioactive contamination. However, the whole clinical unit involved in the use of unsealed radioactive materials ideally should be arranged so that the unnecessary transport about the hospital of radioactive materials or of patients to whom these materials have been administered is avoided. It is possible to arrange a unit in which these separate but closely allied procedures can be undertaken and in many hospitals departments of 'nuclear medicine' are so organised.

Laboratory techniques

The techniques adopted in clinical isotope laboratories are concerned with the safe handling of the radioactive materials. Shielding from gamma radiation is in general a problem very similar to that encountered in dealing with small sealed radioactive sources (see Ch. 8). The gamma ray exposure rates normally met with are however smaller than when working with sealed sources so that quite simple shielding arrangements are often adequate. The use of simple lead bench shields (see p. 92) for example will keep the gamma ray exposure down to appropriate levels in most cases.

On the other hand three problems are of importance which do not occur appreciably in dealing with sealed sources. These are (a) the very high dose rates arising from the escape of beta rays from the unshielded surfaces of radioactive solutions, (b) the risk of entry into the body of radioactive material from the laboratory environment and (c) the hazard and nuisance value of a spill of radioactive material or of contamination of working spaces.

Beta radiation from the surface of active solutions is very easily absorbed by 2 or 3 millimetres of a transparent shield such as glass or perspex or by the glass of spectacle lenses. The necessity to handle vials and other containers indirectly and to avoid direct contact between the hands and radioactive solutions is obvious. Both the beta and gamma ray hazard is thus reduced. The use of surgical-type thin rubber gloves is essential in most of these working procedures.

The risk of taking into the body radioactive materials involves a technique which is similar to that used in many laboratories working with toxic but non-radioactive products. No pipetting is ever done by mouth and some indirect method, such as the use of a compression bulb or a syringe must always be adopted. Eating and smoking are strictly forbidden in these laboratories. In general laboratory coats or protective gowns are worn as well as the gloves mentioned above and these are reserved for this particular kind of work.

Some radioactive materials may be handled in conditions under which active products may escape into the atmosphere as volatile products from the surface of solution, as dust from dry preparations or as gaseous products. To prevent ingestion from any of these products work with radioactive materials must be conducted in a fume cupboard and to give adequate containment of escaping products an efficient fume cupboard is necessary. This requires an exhaust draught system which produces an air flow inwards across the whole of the open working window at a rate of at least 0.5 metres per second. A glove box is sometimes a possible alternative method of containing radioactive material liable to present an inhalation hazard.

It has been mentioned earlier (see p. 97) that the preparation in a hospital laboratory of radiopharmaceutical products must be done under rigorous sterile conditions if the materials are to be administered to a patient by injection. A great deal of attention is now being given to the arrangements of a working space in hospital isotope laboratories in which sterile products can be produced. For this purpose the immediate working area must be bathed in a pure sterile air supply at a pressure higher than that immediately outside the working area so as to avoid leakage inwards of non-sterile air. Design of fume cupboards which provide adequate containment of the radioactivity, and yet provide sterile working conditions is not easy but these are now available and are being installed in many hospitals.

The third problem of minimising the risk of a spill of radioactive materials and of facilitating a cleaning operation if it does occur is dealt with partly by suitable techniques and partly by laboratory arrangements. Double-walled containers, the outer one being non-breakable, are often used and work with small vials is often done in stainless steel trays so that the spread of any spilled liquid is prevented. Bench tops and floors are always covered with a continuous non absorbing surface such as sheet linoleum or P.V.C. and disposable covers on benches may be used. Walls and ceilings are covered with good quality high gloss paint which can be cleaned easily. The design of laboratory furnishings to prevent trapping or absorption of spilt active materials must be done with great care but much experience in this type of laboratory design is now available.

Chapter 11, which deals with Codes of Practice relating to the avoidance of radiation hazards gives some further details quoted from the current U.K. Code of Practice with regard to the techniques of dealing with unsealed radioactive materials. There is an extensive literature on this subject and the student is advised to read further about it (see Bibliography).

Ward procedures

In wards where unsealed radioactive materials are used in patients the general techniques adopted as described in Chapter 8 for sealed sources are again relevant. In addition the necessity to avoid problems arising from spillage of, or contamination by, radioactive materials must be considered.

For patients undergoing therapy with isotopes it is best that they should be in specially designed wards with one or two beds only, and that there should be available a toilet and bathroom for the exclusive use of these patients. It may also be advisable for separate crockery and cutlery to be available for these patients. Nursing procedure will include care to avoid contamination so that protective clothing and surgical gloves may have to be used in some of the work. In wards housing patients undergoing diagnostic procedures the emphasis is on avoidance of the spread of contamination or ease of cleaning if it occurs. In these, as in therapy wards, therefore, floors, walls and ceilings must be covered with continuous non-absorbent surfaces.

Waste disposal

The disposal of radioactive waste from a hospital unit must be thoroughly organised and carefully controlled, not only to avoid hazard to staff and patients in the hospital but also to avoid any appearance of hazard to the general public outside the hospital.

In most cases the disposal of radioactive excreta from patients is best dealt with by letting it go into the sewage disposal system because in this way hazard to staff involved in collection and storage is avoided. It is possible to do this because a very high dilution of the active material and therefore sufficiently low concentration levels are normally rapidly achieved in the routine effluent of a hospital. It is important however to ensure that there is adequate inactive effluent flow and that the toilets reserved for radioactive waste are connected directly to the main drainage system of the hospital. Similar conditions must also be applied in disposing of any radioactive fluid waste from the laboratories. Some storage facilities are however always required for samples, sometimes excreta, needed for measurement purposes or for the highly concentrated residues of stock active solutions. Such storage facilities must be adequately shielded and designed to prevent risk of escape of material.

Solid waste from hospitals, such as paper tissues, glass-ware and so on are often of very low activity and may be collected separately in a specially labelled container in the laboratory or ward. At a somewhat higher level of activity some local authorities have arranged a burial site at suitably controlled refuse tips again with imposition of strict activity limits. For activities above these limits storage of solid wastes becomes necessary and this is true for example for spent generator columns. Fortunately the decay rate of hospital radioactive waste is usually rapid so that long storage times are not often required. For difficult problems of disposal of very active or slowly decaying or very hazardous products national disposal schemes are available and the Radiological Protection Advisor must be consulted about these schemes (see p. 120). The escape of radioactive gases from fume cupboards or from the chimney stacks of incinerators where some lightly contaminated refuse may be burnt is also strictly controlled and appropriate authorities must be consulted. In the United Kingdom the Department of the Environment and the National Radiation Protection Board have authority in this field.

Decontamination procedures

In spite of every care and in spite of attention to technique accidental spillage of radioactive materials does occur and a clinical unit which handles unsealed radioactive sources will have organised procedures and equipment for dealing with emergency spills.

The responsibility for dealing with radiation hazards is discussed in Chapter 11 and that chapter will also contain some quotations from the U.K. Code of Practice for dealing with hazards of the kind we are now considering. This section will describe briefly some of the techniques which should be familiar to the isotope worker who has to keep his working environment clean and free from the risk of a contamination hazard.

It is important that workers in radioactive laboratories should take great care about routine washing of hands before leaving a laboratory to ensure that contamination is not carried out of the working area. Normal soap and water are the first weapons and the cleaning of finger nails is to be included here.

Routine monitoring of laboratory areas, benches and floors in particular, and of the hands is necessary and a laboratory should have adequate monitoring equipment for this purpose (see p. 130).

For the removal of contamination from benches and floors soap and water are also the first things to try. On the skin suitable detergents may remove more firmly fixed contaminants and an even more rigorous cleaning can be accomplished by use of potassium permanganate solution followed by decolourisation with a 5 per cent sodium bisulphate solution. The techniques of removing contamination from broken skin or from heavily contaminated laboratory utensils or furnishings are complicated and easily misused and reference must be made to the local Radiological Safety Officer as the expert (see p. 120).

It is important however to recognise that decontamination procedures must be carefully arranged to limit the spread of contamination. On the skin for example care must be taken to avoid spread to the eyes or elsewhere, during any decontamination work. In the laboratory care must be taken to keep the contamination localised to as small an area as possible.

A second important matter is that the workers in a laboratory must be familiar with the appropriate procedures to take in case of emergency. There should certainly be readily available a kit of some kind containing appropriate tools and equipment to deal with a radioactive spill. It could include, for example, overshoes and protective clothing, appropriate decontamination materials, tools for handling contaminated articles and portable monitoring equipment. The procedures for limiting access to the contaminated areas should also be known to some extent in advance so that the spread of contamination can be avoided. A further point is that arrangements might have to be made for checking by whole body counting or otherwise, whether any workers have received significant quantities of ingested radioactive material in their body, though this is a longer term measurement than the monitoring of the laboratory facilities to ensure that the cleaning procedures have been effective.

Fortunately most spills in hospitals do not involve very dramatic emergency procedures since the activities involved are generally small and the materials of relatively short life. But sometimes consideration must be given to the priorities in which emergency procedures must be taken. After dealing with personal injury decontamination of staff and patients comes first, the restriction of spread of contamination second and then the cleaning of the contaminated area. Using suitable protective clothing cleaning can generally be accomplished satisfactorily without risk to the operators provided that the laboratory or ward was adequately planned and arranged in the first place. Fortunately in hospital work it is very rare for contamination to be so serious that a laboratory or ward has to be closed to allow the contamination to disappear by natural radioactive decay. Such a procedure which might mean a delay of perhaps 10 half lives of the active material would involve a very serious interruption of hospital facilities.

10. Diagnostic procedures with unsealed sources

Introduction

Diagnostic techniques employing many different radiopharmaceuticals are now used routinely over the whole field of clinical medicine. A full description of these techniques is beyond the scope of this text, and for detailed accounts reference should be made to one or other of several excellent text books on the topic. (see Bibliography). Since, however, radioisotope techniques do offer some help in the management of malignant disease and in the preparation of patients for radiotherapy they do come within the experience of the therapy radiographer. In this chapter therefore a brief description is given of a few diagnostic procedures which illustrate the application of the techniques described in Chapter 9. The five sections of this chapter describe techniques in each of the following groups (a) dilution techniques (b) metabolic studies (c) dynamic function tests (d) isotope imaging and (e) *in-vitro* tests. Such a classification cannot be a rigid one. When a radiopharmaceutical is administered to a patient the subsequent tests may be in more than one of these groups.

Dilution techniques

If a known amount of a radioactive material is introduced into a vessel containing an unknown volume of liquid and thoroughly mixed it is easy to deduce the unknown volume by measuring the concentration of activity in a sample taken from the mixture. This simple principle is used to investigate many aspects of body composition by the introduction of one or other of a number of radiopharmaceuticals into the body and measuring the radioactivity per unit volume of a suitable fluid sample removed from the body at a subsequent time.

An important example of the dilution technique is the measurement of the *volume of the circulating red cells* in the blood. For this purpose the patient's own red cells are used and labelled with chromium-51 ($T_{\frac{1}{2}}$ 27·8 d). A blood sample of about 10 ml is withdrawn from the patient into a bottle containing acid-citrate dextrose (ACD) solution, mixed, and the red cells separated by centrifugation. They are then labelled by mixing the suspension of red cells with about 20 μCi of ^{51}Cr as a sterile isotonic sodium chromate solution at room temperature for about 10 to 20 minutes. Washing twice with saline removes the free ^{51}Cr and the labelled cells are resuspended in about 12 ml of saline. An accurately known volume (say 10 ml), of this suspension is then injected into the patient and a smaller accurately known volume (say 1 ml) is put into a standard 100 ml flask filled with a dilute unlabelled sodium chromate solution. Fifteen minutes after injection a venous blood sample is withdrawn from the patient and placed in a tube with a dry anticoagulent. The radioactive assay consists in comparing the activity of equal volumes of the whole blood and the diluted standard solution. In this technique special attention must be paid to the maintenance of sterility in the withdrawing of the blood, its labelling and the subsequent injection of the labelled material into the patient. The red cell volume is calculated from the formula:

$$\text{RCV} = \frac{A_S}{A_B} \cdot D \cdot V \cdot \frac{H}{100}$$

where A_S and A_B are the count rates of the standard and blood sample, D is the dilution factor used in preparing the standard solution, V is the volume of labelled cells injected and H is the percentage haematocrit* of the venous blood sample corrected for trapped plasma. This must be obtained by a separate measurement of the haematocrit by routine haematological methods.

The dilution technique illustrated by the red cell volume measurement is of wide application, though in each type of investigation modifications to the technique will be necessary. It is possible for example to measure the *total exchangeable sodium* content of the body by a similar method. In this case however it cannot be assumed that an injected dose of sodium-24 given as sterile isotonic solution of sodium chloride mixes immediately with the exchangeable sodium in the body. Instead of sampling the mixture after 15 minutes, as in the blood volume studies where mixing is prompt, a period of at least 24 hours is required for the labelled sodium to reach equilibrium with the sodium of the body which takes part in normal metabolic changes (exchangeable sodium). During this equilibration period some sodium is excreted in the urine and an estimate of this loss of sodium must be obtained by measurement of the total excreted activity. The effective 'space' through which the sodium diffuses in the time to equilibrium is easily deduced from a knowledge of the total activity remaining in the body at this time and the concentration of activity in any suitable body fluid at the same time. This may be a urine or plasma sample. If the concentration of total inactive chemical sodium in this sample is estimated by routine biochemical methods such as flame photometry the total exchangeable sodium in the body is easily deduced.

The investigation of body electrolyte composition by isotope dilution techniques of this kind is a routine procedure and some other similar tests are listed in Table 10.1.

An extension of the ^{51}Cr labelling study with red blood cells should be mentioned here. If blood samples are taken at intervals over several days subsequent to the injection, information can be obtained about the rate of destruction of the red cells in the circulation. Some correction must be applied for the loss of the radioactive label from the circulating red cells which occurs to some extent even without the destruction of the cell itself. If this correction is made a comparison of the time in days over which the circulating activity concentration reduces to half its initial value in normal and abnormal cases is a valuable diagnostic investigation in haematological diseases. Further, since ^{51}Cr is a gamma emitter the accumulation of ^{51}Cr in the liver or the spleen can be detected by a collimated scintillation counter (see p. 101) placed on the body surface over the organ in question. Such *surface counting* carried out daily over 2 or 3 weeks is of diagnostic value in cases of haemolytic* anaemia.

Metabolic studies

The introduction of any labelled compound into the body followed by the investigation of its distribution about the body or its excretion and the way in which these vary with time provides valuable information about the metabolism of the compound and many investigations of metabolism can be achieved by isotope techniques which would not be possible in other ways. In these investigations care must be taken to ensure that the label is firmly attached to the compound under study and does not escape from it in ways which would give false information about the metabolic behaviour of the compound.

A very common and important metabolic study is that of the element iodine which is involved in the function of the thyroid gland. Knowledge of iodine metabolism gives diagnostic information about thyroid diseases (see p. 268). In this case the radioactive material used is one of the radioactive isotopes of iodine. The problem of the different behaviour of the radioactive label and the chemical entity under study does not therefore arise in this case.

The normal daily intake of iodine in the diet is about 100 micrograms. When the thyroid gland is functioning normally about half of this iodine is abstracted from the circulating blood by the thyroid gland and about half is abstracted by the kidneys and excreted in the urine. The thyroid gland synthesises the thyroid hormones of which the elemental iodine is one of the atomic constituents and the hormones are transported throughout the body in the blood circulation to control the bodily functions influenced by these hormones.

The administration orally, or by injection, of radioiodine in the form of a solution of sodium iodide with the radioiodide in carrier free form (see p. 13) gives such a small amount of chemical iodine that it has no influence on the physiological processes involved but since it mixes readily with the normal chemical iodide in the blood observation of the radioactivity in the thyroid or in the blood makes possible very simple and reliable thyroid diagnostic studies of thyroid function.

Thyroid function may be studied by observing the uptake of the radioiodide in the gland by a collimated scintillation counter suitably calibrated to give a reasonably accurate estimate of the amount of activity present in the gland (see p. 102). The isotope most commonly used is iodine-131 ($T_{\frac{1}{2}}$ 8 d) and the curves of Figure 10.1 show the variation of uptake in the thyroid gland over three days following the administration of a small tracer dose of approximately 10 μCi of ^{131}I. The ordinates represent the percentage of the administered dose taken up by the gland, and the three curves A, B and C represent typical results obtained; A in a patient with normal thyroid function, B in a case of hyper-activity of the gland, or thyrotoxicosis and C in a case of hypo-activity of the gland leading to myxoedema.*

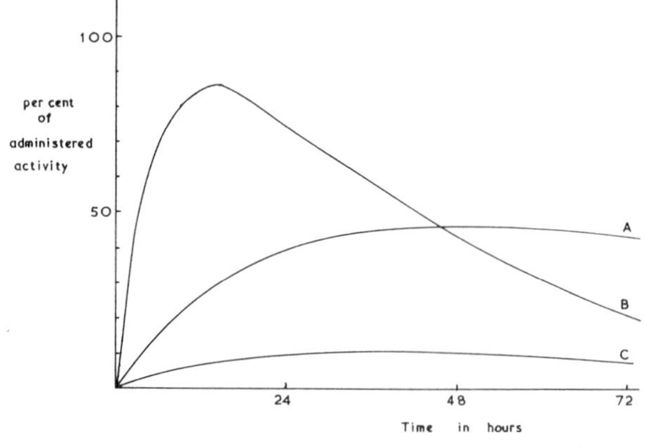

Fig. 10.1 Thyroid gland uptake of radioiodine in diagnostic tests

A. Normal thyroid function
B. Hyperactive thyroid gland—thyrotoxicosis
C. Hypoactive thyroid gland—myxoedema.

It should be noted that during the time of the test, sometimes up to 72 hours, there will have been a physical decay of the ^{131}I and in the curves of Figure 10.1 and the similar Figure 10.2 the percentage of the administered dose has in each case been corrected for this decay.

Diagnostic information may also be obtained by sampling of blood at intervals after the administration of the test dose. In the normal case the blood is cleared of its radioiodine by both the thyroid gland and the kidneys and the activity in the blood stream decreases rapidly falling to half its initial value in a few hours. If the thyroid gland is underactive as in myxoedema the clearance is due only to kidney performance and is very much slower. The blood activity in the case of the hyperactive gland is of much greater interest. The iodide ions are cleared in this case very rapidly into the thyroid but the iodine then becomes incorporated into the thyroid hormone compounds—thyroxine, tri-iodothyronine etc. and these then appear in the circulating blood so that the activity increases. The activity however in this case is no longer in the form of the simple iodide ion but in the form of the thyroid hormone compound and is now bound to protein molecules in the circulating plasma. By precipitating the proteins by standard biochemical techniques it is possible to determine whether the activity in the blood is or is not protein bound and this in itself gives useful diagnostic information. Figure 10.2 shows the variation with time of the plasma radioactivity in these tracer tests for the same three typical clinical conditions.

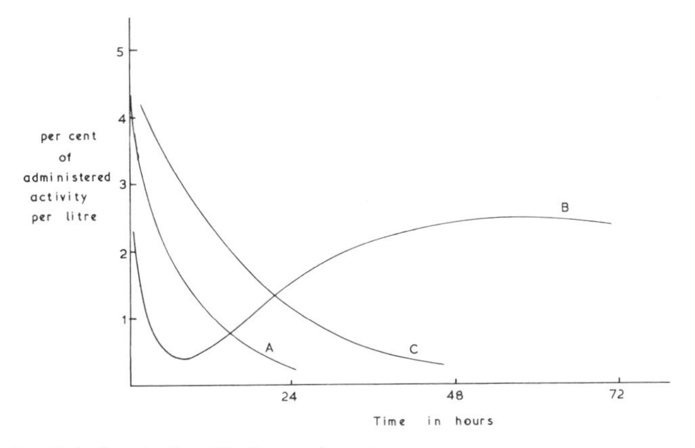

Fig. 10.2 Plasma clearance of radioiodine in thyroid diagnostic tests

A. Normal thyroid function B. Thyrotoxicosis C. Myxoedema.

It will be clear that in the clinical conditions which are the subject of investigation and treatment many variations in these typical curves will occur since thyroid malfunction may appear in a variety of forms. In particular malignancies of the thyroid gland may occur and malignant thyroid tissue may function as normal thyroid tissue or may be non-functioning in the thyroid sense. The use of the information presented by ^{131}I measurements in diagnostic techniques is discussed later dealing with the clinical handling of treatment of thyroid diseases (pp. 208, 268).

Before leaving this illustration of the use of radioiodine in diagnostic metabolic studies reference must be made to the importance of *in-vitro* studies of the blood plasma. The thyroid status of a patient is revealed by investigation of the level of non-radioactive hormone circulating in the plasma. This can be estimated by chemical methods which though elaborate and difficult do provide very important diagnostic information. The level of circulating non-radioactive hormone can however also be estimated by the competitive binding techniques, described later under *in-vitro techniques*, (see p. 113). The importance of these tests is that no radioactivity is administered to the patient but it is used only in examination of the blood sample. The significance of the *in-vitro* tests in the overall clinical management of thyroid disease is outlined in Chapter 29.

Dynamic studies

Dynamic studies are isotope techniques in which the distribution of a radioactively labelled material in the body is observed when the distribution is varying rapidly with time. Observations of this kind give information about the physiological behaviour of the organ in which the radiopharmaceutical is being concentrated, or through which it is passing.

Mention has already been made (see p. 103) of the advantages of the gamma camera for investigations where the distribution of a radioactive material in the body is varying very rapidly during the course of a few minutes or even seconds. The flow of a radioactive bolus through the brain following injection into the internal carotid artery or through the heart following intravenous injection into an arm are examples.

A much more normal type of dynamic study is one in which a collimated scintillation counter placed over an organ records changes of the radioactivity in the field of view of the detector over several minutes. The output of the counter is received by a ratemeter feeding into a chart recorder. A common example of this technique is the use of radioisotopes for assessing renal function. The compound most commonly used for *in-vivo* studies of this function is the compound ortho-iodohippuric acid (hippuran) which contains one atom of iodine per molecule and which can be labelled by synthesising with one of the iodine radioisotopes, generally ^{131}I. This compound is rapidly removed from the circulation by the kidney and actively concentrated in the kidney tissues and then also rapidly secreted into the urine. A scintillation counter placed over the kidney will therefore, following an injection of the active compound, first indicate activity arriving in the renal and other local circulation, then detect the tubular concentration of the material and later reveal a fall in activity in the kidney as the material is excreted. This is the basis of *isotope renography*. Two detectors are used one over each kidney and the variation of count rate with time is plotted over some 20 minutes, the curves being presented on a recording ratemeter.

Figure 10.3 shows the typical three phasic curves for a case in which the right kidney is functioning normally and the left abnormally. The curve (A) for the right kidney shows first the initial vascular phase lasting some 15–30 seconds, then a second phase corresponding to glomerular filtration and tubular secretion during which the total activity in the kidney rises to a peak in about 2–3 minutes. The last phase shows the activity being excreted roughly exponentially. In curve (B) for the left kidney the first two phases of the curve appear normal but the excretion is abnormal and the curve is typical of an obstruction of the evacuation mechanism in the kidney.

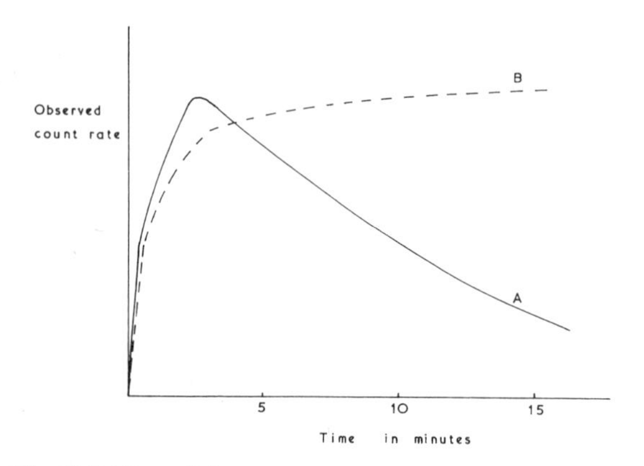

Fig. 10.3 The radioisotope renogram
A. Normal kidney function Right Kidney
B. Left Kidney showing urinary tract obstruction.

The isotope renogram is most commonly used for a qualitative comparison of the function of the two kidneys and much information can be obtained from a study of the detailed shapes of the two curves. Quantitative information is more difficult but can be obtained by more elaborate techniques including the use of a third external detector over the heart which makes it possible to subtract the blood activity from each individual kidney activity curve. Care in positioning the counters, is necessary. These counters are simple single crystals fitted with appropriate collimators to cover the kidneys, and positioning is often done by auxiliary X-ray or isotope techniques.

Many other investigations of dynamic function can be similarly undertaken with a crystal of 1″ to 2″ diameter suitably collimated and feeding into a recording ratemeter. In each case qualitative information is easy but quantitative results involve considerable difficulty associated with the problem of the variation of the distribution of counter sensitivity in the tissues under examination and of the attenuation of the gamma rays in these tissues.

Isotope imaging

Apparatus for isotope imaging has been described in Chapter 9. The technique of isotope visualisation is now very extensively used and in this chapter only two of the more common uses will be described to illustrate diagnostic procedures of value to radiotherapy.

One valuable technique is the demonstration of space occupying lesions in the brain. The radiopharmaceuticals used for this purpose are ones which are carried by the vascular circulation but are materials which do not penetrate from the blood into normal brain tissue. If the permeability of the brain tissue is disturbed by a tumour or by disease or by accidental trauma, an isotope circulating in the blood will be abnormally concentrated in these regions and the lesion will be revealed by the scanning or gamma camera examination of the brain. Many isotopes have been used for this visualisation since no selective absorption in brain tissue is involved, but only a change in permeability from the blood to brain tissue. The isotope most commonly used is technetium-99m which has a convenient half life and gamma photon energy. This generator produced isotope is introduced as the pertechnetate ion by injection of the sterile effluent from the generator column. Preparations of the alternative generator produced isotope indium-113m are also sometimes used.

Figure 10.4 shows two pictures produced by a conventional rectilinear scanner. These are lateral views of the brain obtained by scanning following an injection of ^{99m}Tc using 10 mCi of the isotope. The picture (A) is a lateral view of a normal brain showing the normal vascularisation in the skull. The second picture (B) reveals a large area of abnormal uptake in the frontal lobe, representing a gross lesion in this region of the brain.

The lesion in this particular lateral view is very obvious but the interpretation of brain scans, as with other types of isotope visualisation results, is often not easy. The lesion may be ill defined, and further, the normal pattern of isotope distribution against which the abnormal is to be compared may itself have wide variations. Finally unless the differential uptake in the lesion is big the statistical variations of the radioactive emission may make differentiation of the tumour from the background quite difficult.

A second illustration of isotope imaging procedures is the use of the technique for the visualisation of bone tumours. Increased concentration of a radioactive material relative to its surroundings will occur if the labelled material can be built selectively into newly developing tissue. There are no very suitable isotopes of calcium which can be used for imaging techniques. Of the two available calcium radioisotopes ^{45}Ca ($T_{\frac{1}{2}}$ 165 d) is a pure beta emitter and ^{47}Ca has a gamma ray emission of 1·3 MeV photon and a half life of 4·7 d. The high photon energy of the gamma ray makes collimation difficult and scans can only be done with a high radiation dose to the patient. Strontium however is metabolically similar to calcium and is built readily into newly growing bone. Strontium-87m with a 2·8 h half life and a gamma emission of 388 keV has therefore had a considerable use as a substitute for calcium in bone investigations. The comparison of an isotope scan using radio-strontium and an X-ray radiograph of a bone lesion is of great interest. The strontium scan will yield a positive result when new bone is being deposited rapidly. In such cases the actual amount of bone deposited may be insufficient to produce an absorption adequate to show as an abnormality on a radiograph. On the other hand a bony mass readily visible in an X-radiograph may give no response in an isotope study unless at the time of the scan new bone is actually being deposited.

The isotope ^{87m}Sr has however now been largely replaced by one or other of a variety of phosphate compounds which also give good selective absorption in bone tumours. These compounds can be labelled with the isotope ^{99m}Tc with its very favourable radiation and half life characteristics and bone scanning now is done almost entirely with ^{99m}Tc labelled pyrophosphate or diphosphonate.

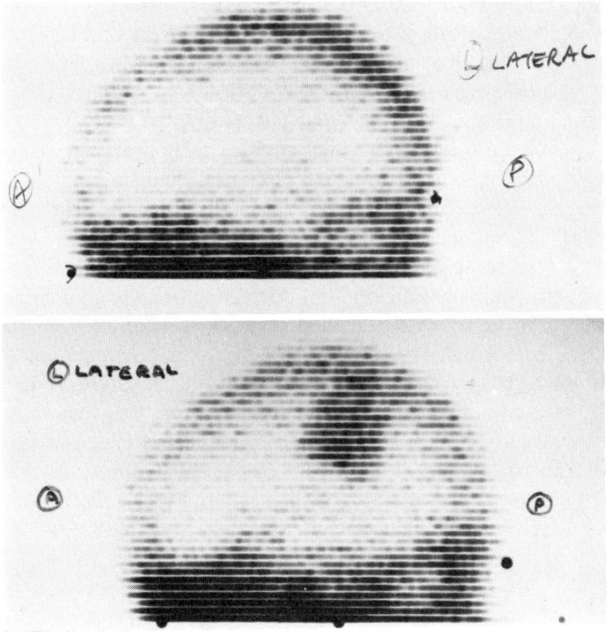

Fig. 10.4 Rectilinear Scan of Brain, lateral view. (A) Normal, (B) Abnormal uptake in posterior parietal region. Injection: 10 mCi^{99m}Tc. (Courtesy of Sheffield National Centre for Radiotherapy.)

Figure 10.5 illustrates a picture produced by a gamma camera technique using technetium-99m labelled methylene diphosphonate. The picture illustrates the uptake of the labelled material in the complete skeleton of the patient. The field of view of the gamma camera is quite small and to cover the whole of the body a technique known as a scanning gamma camera is used to produce the complete picture. In this technique the patient is moved along under the camera and the information from the camera is assembled to produce a complete picture after one passage of the patient under the instrument.

In-vitro isotope techniques

The sensitivity of detection of radioactive materials (see p. 95) has great advantages when applied to biochemical procedures. Only one example can be given here but over the whole range of clinical diagnosis there is increasing development of isotope techniques in which radioactive materials are used to assay important compounds in a sample of blood taken from the patient. These are *in-vitro* tests which involve no radiation dose to the patient.

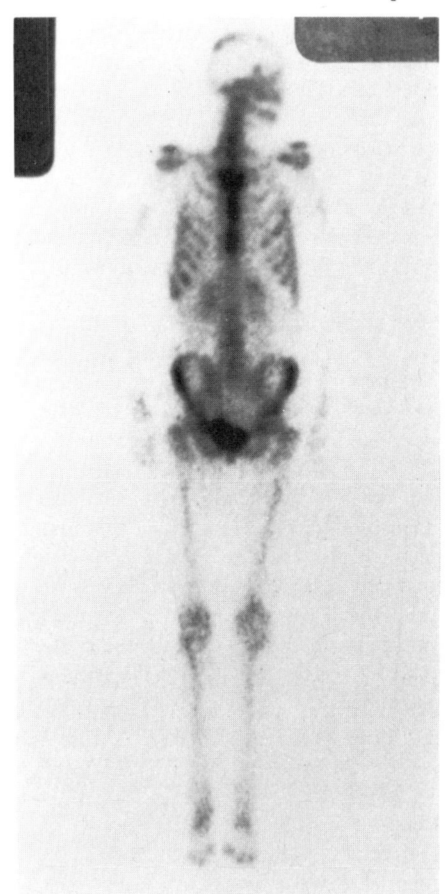

Fig. 10.5 Skeletal Uptake of ^{99m}Tc methylene diphosphonate. Injection: ^{99m}Tc MDP. Scanning Gamma Camera Technique (Courtesy of Dr. M. V. Merrick, Western General Hospital, Edinburgh.)

Diagnostic information relating to thyroid function can be obtained by an estimate of the level of the thyroid hormone tri-iodothyronine circulating in the blood. This hormone (T_3) circulates in the blood plasma bound to specific protein molecules. The hormone T_3 normally fills only a proportion of the available protein binding sites, and the availability of unoccupied binding sites therefore is inversely related to the amount of circulating hormone and indicates the thyroid status of the patient. If sufficient labelled T_3 hormone is added to the plasma the unoccupied binding sites become filled with the radioactive hormone and the excess labelled material is left free. A method therefore of measuring the relative amounts of free and bound radioactivity following the addition of the labelled material to a sample of plasma will serve as an indication of the amount of circulating hormone already present in the blood.

The principle of *competitive binding* is of very wide application. In the case of the estimation of triiodothyronine T_3 the separation of free and bound fractions is very easy, and has been achieved in a variety of ways. The red cells in the blood can be used to mop up the free activity in the plasma and the partition of the activity between the red cells and the plasma proteins becomes a diagnostic index, the activity in the red cells being increased in thyrotoxicosis and decreased in hypothyroidism compared to normal. More simple techniques of separating free and bound activities are however available using ion exchange resins and these represent the standard method of assay *in vitro* of this particular thyroid hormone.

Methods for measurement *in vitro* of the thyroid hormone thyroxine (T_4) are also available, and together these *in vitro* techniques provide such a valuable index of thyroid function that the use of *in vivo* tests on many patients with suspected thyroid problems is now no longer necessary.

A great range of investigational biochemistry can be undertaken using *in vitro* tests along this pattern. Two requirements must be met. First a preparation of the compound to be assayed must be available with a suitable radioactive label. Second a convenient technique must be possible to separate bound and unbound portions of the active material when it is added to the plasma. Some of these techniques are elaborate and involve detailed biochemistry using immunological methods. The future of this kind of procedure is however very great indeed.

Autoradiography

A very important method of examination of the radioactive content of tissues is the method of *autoradiography*. If a piece of the tissue is sectioned and a slice placed in contact with a photographic film, then after suitable exposure the radiations from the active material in the tissue act on the photographic emulsion and on development blackening of the film is produced, indicating the distribution of activity in the tissue slice.

This device is a very useful one to find the distribution of radioactivity in many materials. For example, a leaf from a plant which has been fed with radioactive phosphorus will readily produce an autoradiograph on being placed in close contact with a sensitive photographic film. The distribution of active material throughout the leaf may thus be determined from the distribution of blackening on the film following development.

For medical work autoradiography has been developed to a very high degree of precision. The aim of these developments has been to observe the distribution of an isotope amongst the cells of a biological material. Several methods have been adopted. One well-known method is to prepare thin sections of the tissue and mount them on glass slides by standard histological techniques. A thin layer of photographic emulsion is then deposited on top of this section by floating the emulsion on the top of a bath of water and bringing the slide in contact with it from underneath. This is done in the dark. On drying, the film remains in permanent and very intimate contact with the histological section. After suitable time the section is developed in the normal way and microscopic examination then reveals the presence of radioactivity in the section by silver grains produced in the super-imposed layer of gelatin.

Non-radioactive diagnostic procedures

This chapter has been concerned with diagnostic procedures using unsealed radioactive sources and illustrations have been given both of the wide range of applicability of such tests and of their sensitivity and often of their simplicity. It is clear that isotope procedures represent only one method by which physical techniques have been adapted to serve the aims of clinical medicine, and that many other techniques are available which will also provide important diagnostic information.

In some cases alternatives to isotope techniques may provide the same kind of information and the clinician is then faced with a choice of method. The choice will depend on the availability of the procedures, the trauma to which the patient is subjected by the investigation, and on the reliability of the information finally obtained from the investigation. In this choice the methods which give smaller radiation dose to the patient, or even no dose at all, will normally be preferred. An illustration of such a choice is the variety of available methods for investigating the female abdomen in cases of pregnancy. X-radiography would not be used as a first choice because of the relatively high radiation dose to the foetus and this method would only be adopted if the required clinical information could be obtained only by this means. Isotope studies of placenta localisation give smaller radiation doses and of these a rather coarse scanning examination by a hand held

Table 10.1

Isotope	Symbol	Half-life	Main emissions Energies in MeV beta	gamma	Physical form	Clinical uses
Bromine-82	^{82}Br	36 h	0.44	0.55–1.48	Solution of ammonium bromide	Intracavitary use in bladder
Caesium-137	^{137}Cs	30 y	0.51	0.66	Metal: sealed in stainless steel container	Gamma ray beam units interstitial and intracavitary use
Carbon-14	^{14}C	5760 y	0.16	—	In many compounds	Label for organic compounds
Chromium-51	^{51}Cr	27.8 d	—	0.32	Solution of sodium chromate	Labelling of blood cells
Cobalt-57	^{57}Co	270 d	—	0.122	Solution or capsules of cyanocobalamin	Measurement of vitamin B_{12}
Cobalt-58	^{58}Co	71 d	(β^+) 0.49	0.81	Solution or capsules of cyanocobalamin	Measurement of vitamin B_{12}
Cobalt-60	^{60}Co	5.3 y	0.31	1.17, 1:33	Metal: sealed in stainless steel as wire or rod	Gamma ray beam units Substitute for radium in tubes needles and as wire
Gold-198	^{198}Au	2.7 d	0.96	0.41, 0.68	Colloidal suspension of metallic gold	Intracavitary use for effusions
					Metal 'seeds' or grains	For interstitial implant
Hydrogen-3 (Tritium)	^{3}H	12.3 y	0.018	—	Combined in tritiated water	Measurement of total body water
Indium-113m	^{113m}In	1.7 h	—	0.39	Effluent from generator	Isotope visualisation procedures
Iodine-123	^{123}I	13.3 h	—	0.16	Sodium iodide solution	Thyroid diagnostic procedures
Iodine-125	^{125}I	60 d	—	0.035	Sodium iodide solution	Iodination of proteins
Iodine-131	^{131}I	8 d	0.61	0.36, 0.64	Sodium iodide solution	Diagnostic and therapeutic use for thyroid diseases
Iodine-132	^{132}I	2.3 h	0.8–2.1	0.52–1.14	Effluent from generator	Thyroid diagnostic procedures
Iridium-192	^{192}Ir	74 d	2.24	0.3–0.6	Platinum-covered wire	Interstitial use
Iron-59	^{59}Fe	45 d	0.27, 0.46	1.10–1.29	Solution of ferric chloride or citrate	Haematological disorders
Phosphorus-32	^{32}P	14.4 d	1.71	—	Solution of sodium phosphate	Treatment of polycythemia
					Colloid of chromic phosphate	Intracavitary treatment of effusions
Potassium-42	^{42}K	12.5 h	2.0, 3.6	1.52	Solution of potassium chloride	Tests of potassium metabolism
Sodium-24	^{24}Na	15.0 h	1.39	1.37, 2.75	Solution of sodium chloride	Tests of sodium metabolism
Strontium-87m	^{87m}Sr	2.8 h	—	0.39	Effluent from generator	Visualisation of lesions in bone
Strontium 90	^{90}Sr	28 y	0.54	—	Incorporated in metallic foil	Beta-ray therapy application
Technetium-99m	^{99m}Tc	6.0 h	—	0.14	Effluent from generator	Isotope visualisation tests substitute for iodine in thyroid tests
Yttrium-90	^{90}Y	64 h	2.27	—	Colloid	For intracavitary suppression of effusion
					Metal rods or pellets	Pituitary irradiation

counter requires a smaller isotope dose than a linear scan using a multihole collimated detector. If *ultrasonic* examination techniques are available however a B scope scan (see p. 138) will give more detailed information of this kind with no radiation hazard to the foetus or patient. The development of the *computer assisted transverse axial tomography* (see p. 139) provides an immense amount of additional information for cross sections of the head and trunk. The influence of such diagnostic tools on the role of isotope techniques, particularly imaging methods will be very great. An account of these two alternative diagnostic procedures is given in Chapter 12.

Developments of new techniques of diagnosis in the future will influence considerably the importance of those already established including radioisotope methods. Techniques of this kind should never be considered as rival methods. They may provide additional information on which a diagnosis can be based, or they may provide alternative and easier methods for obtaining the same information. A clinician needs to know what is available to him and what are the advantages or disadvantages of each technique in the overall management of his patients.

Isotopes used in current medical work

Table 10.1 gives a list of the more important isotopes used in medical work and includes details of their main properties and some notes on their more important uses.

The Table includes the photon energies of the main gamma rays emitted by the nuclides expressed in million electron volt (MeV). One electron volt is the energy received by an electron in being accelerated through a potential of 1 volt. Gamma ray energies are normally expressed in MeV, the unit equal to 10^6 eV. Table 10.1 also quotes the maximum energy of the beta rays emitted in the continuous spectrum from each nuclide. A beta particle of 1 MeV energy has a range in water or tissue of about 4 mm so that the Table will give some indication of which nuclides give rise to weakly or strongly penetrating radiation of both beta and gamma types. The half life of the nuclide is also quoted.

11. Control of radiation hazards

Introduction

The biological effects of ionising radiation on both the single cells of which all living material is composed and on a complete organism such as the human body are described in some detail in Chapters 14 and 16 of this book. These effects have been known and studied in great detail for a long time. Arising from these studies levels of radiation dose which can be accepted in a normal working environment of a radiological department and in the environment to which the general public is subjected have been specified by authoritative international expert committees. This chapter will contain an account of the acceptable dose levels recognised internationally and the steps which must be taken to achieve these dose levels in practice. In order to appreciate however the problems involved in establishing adequate control of radiation hazard and also to understand the apparently complicated specifications of a permissible radiation dose it is necessary first to summarise the available knowledge of the effects produced in human beings by very low doses of ionising radiations. Radiation hazards are, for a variety of reasons, a matter of great concern to the general public. Radiographers are professionally in an important position to help to ensure that the public, particularly patients in hospital as well as their own colleagues in hospital staffs, have an appreciation of the hazards, of the steps taken to achieve control of the hazards and of the relative dangers of using ionising radiation compared to other activities of normal living.

Biological effects of radiation

It is known that a wide variety of bodily injuries can be received by human beings subjected to excessive, though small, doses of ionising radiation. The following list of such injuries illustrates the need to consider a wide variety of possible working conditions and techniques when safety rules are drawn up:

1. Injuries to the skin arising from exposure of the epidermis to radiation of quite low penetrating power, such as beta rays or to soft X- or gamma rays.
2. Changes in the blood-forming organs and in the number and nature of the circulating blood cells. These changes will arise from exposure of the bone marrow to penetrating radiation received externally, or from radiation produced by internally distributed radioactive materials.
3. Cancer formation in both bones and soft tissues produced by exposure of the whole body to radiation again arising either from external or internal sources.
4. Cataract formation due to irradiation of the eye lens by radiation of low penetrating power.
5. Genetic effects arising from the production of mutations by irradiation of the reproductive system.

Accounts of these radiation effects will be given later in Chapter 16. To specify levels of radiation exposure so as to avoid these injurious effects detailed quantitative information is needed relating the incidence of the effects to the radiation dose which produces them. The information available is not complete but it is sufficient to enable *maximum permissible doses* to be specified and these acceptable levels of exposure are incorporated in various Protection Codes or lists of recommendations published in many countries. The nature of the information available and the way in which it conditions the form of the safety recommendations are described briefly in the following paragraphs.

Evidence on damage by low doses

There are five main sources of information on the long term effects of relatively low doses of ionising radiation or of effects of higher doses of radiation which can be used to estimate the effects of low doses.

Early history of radiation damage

The radiation damage received by many early workers with X-rays and radium has been studied in detail. It was not realised in the early years of the century when the medical use of radiation was growing rapidly that there could be a very long time interval, often of very many years, between the radiation exposure and the observable damage. It was also not sufficiently realised that some biological effects could be cumulative, that is long periods in which the exposure rate was relatively low, could be as damaging as shorter periods at high exposure rates. The damage suffered by the early radiologists was predominantly of two types. First, damage could occur to the tissues of the hands resulting from exposing them to the direct beam of X-rays, or handling radium sources with the fingers. This damage was often very serious giving rise to malignant changes, sometimes fatal. Second, damage could arise to the haemopoietic system—that is the site of blood cell formation in the bone marrow. This arose generally from exposure of the whole body to scattered radiation over long periods, again giving rise to fatal conditions. Though these two effects were recognised adequately after some time the magnitude of the dose to which the workers had been subjected years before could never be accurately established. The extrapolation of this information to enable an accurate dose to be deduced which might be considered safe was therefore not easy.

Evidence from radiation accidents

A great deal of evidence about radiation damage has been obtained from the study of accidental exposure arising sometimes during the use of X-rays or radioactive materials in industrial or research processes and sometimes in medical practice. In spite of extreme care accidental exposures of this kind do arise and detailed investigations of such events can lead to information of two kinds. First a consideration of the circumstances which led to the accident may suggest improved techniques for safe control of radiation facilities and these may become incorporated into a *Code of Practice*. For example some of the provisions of the Code of Practice for the U.K. National Health Service, quoted later in the chapter, arise from an attempt to avoid repetition of accidental events in earlier practice in the hospitals. Secondly, if it is possible to observe carefully the clinical damage produced by the radiation exposure and to obtain also a reasonable estimate of the radiation dose involved the information can be valuable in establishing further data on the dose-effect relationship under a variety of conditions.

The long term effects of atomic bombs

The use of the atomic bombs on the Japanese cities of Hiroshima and Nagasaki at the end of World War II in August 1945 exposed large populations to a sudden acute radiation dose. The immediate extensive and fatal effects of these explosions were due in part to the intense heat and the explosive violence of the event and only in part to the high dose of radiation. However, large numbers of survivors in both cities have been studied continuously since 1945. The incidence of disease, particularly malignant disease, amongst these survivors and the incidence of possible effects on those irradiated *in utero*, or the offspring of irradiated parents, have all been matched carefully against a similar population believed to be unirradiated. The radiation dose which the victims received at the time of the explosion has been estimated by a number of ways including investigations of the position of the subject with respect to the explosion centre. The effects of a single dose of gamma rays and neutrons therefore have been followed over a period of more than 30 years, and the results have contributed valuable data on which safety levels can be assigned (see p. 173).

Results of animal experiments

Radiobiology has developed very greatly over the last 30 years and much information has been obtained by experimental irradiation of various animal species. In this case both the irradiation effect and the administered dose can be very precisely observed. The difficulty in this kind of information is the doubtful relevance of animal data to human beings, especially in investigations of the genetic effects of radiation. The fact of radiation damage to the chromosome material is well established, but a quantitative estimate of the magnitude of the effects is by the nature of things impossible to observe in humans. Experimental data must be obtained from species in which the generation life is short enough to enable adequate observation of effects. Most of the quantitative evidence comes from experimental work on the fruit fly and on mice.

The extrapolation of experimental animal data to the human species accounts for the doubt, for example, of the exact magnitude of the radiation dose which would double the naturally occurring mutation rate in humans though this is estimated to be about 50 rads. There is also some uncertainty about the exact relationship between effect and dose for very low dose rates. Conservative extrapolations are invariably used when deducing appropriate safety standards for human beings so that an error is one on the safe side.

Evidence from radiotherapy

Finally, a very great deal of careful observation and long experience in radiotherapy gives a firm correlation of radiation effects against administered dose for a wide variety of conditions. However this evidence always relates to dose levels between perhaps 100 rads and several thousand rads. The difficulty of extrapolation to a few rads is great, especially since the exact relationship between effect and dose at low dose levels is still not certain.

Radiation effects at low doses

There are two facts about radiation effects at low dose levels which make precision very difficult. The first of these is that many of the human effects of radiation at low dose levels cannot be distinguished from the same troubles which undoubtedly arise spontaneously in human populations without the influence of ionising radiation. There is, for example, a natural incidence of malignant disease and the incidence rate varies with the type and site of cancer and with the environment to which the population is subjected. There are environmental causes of the incidence of malignant disease other than radiation, though these causes are complex and only understood in a very incomplete way. The second fact which needs to be noted is that there is a *natural radiation* level in the environment to which all populations are subjected and which it is impossible to avoid. It arises from a number of causes. One cause is cosmic radiation arising from extra terrestrial sources. Another is natural radioactivity of rocks, and therefore of building materials and of water sources and so on. Finally there is natural radioactivity in the body arising from the radioactive isotope of potassium—potassium-40, which is a long life beta emitter. The natural background of the environment varies with height above sea level which affects the cosmic ray component and with the geographical location which affects the component arising from rocks and building materials. An average figure for this natural background is 100 milli-rad per year for a normal sea level environment of which cosmic rays account for approximately 30 per cent and local gamma radiation about 50 per cent.

There is a possibility that some part of the natural incidence of malignant disease can be ascribed to natural ionising radiation but the extent to which this occurs is not known with precision and radiation is certainly only one of many environmental factors which lead to malignancies. The addition of man-made ionising radiation to the environmental radiation will give rise to an increase in both the incidence of malignant disease and of genetic abnormalities. It is the aim of radiation protection to restrict this increase to the smallest possible amount while at the same time obtaining the maximum benefit from the use of radiation.

It is helpful to recognise three distinct groups of radiation effects on human populations at low dose levels. These have been described as (i) somatic certainty effects (ii) somatic stochastic effects (iii) genetic stochastic effects. *Somatic effects* are those which arise in the irradiated person at any time during that person's lifetime. *Genetic effects* are those which arise because of disturbance to the human genetic material—the gene mutations—and which produce effects on individuals only of future generations. The effect may occur in the first generation of descendants if the mutation is 'dominant'. More frequently the effect will appear in individuals of subsequent generations if mating of two people with the same 'recessive' gene mutation occurs. *Certainty* effects are those which can be ascribed without doubt to exposure to ionising radiation. *Stochastic* effects are those which cannot clearly be ascribed to a dose of ionising radiation to an individual and can only be detected by observing an increase of incidence of a known abnormality in a large group of individuals for whom an increase of radiation exposure has taken place.

Somatic certainty effects, of which examples are fibrosis of lungs due to excessive radiation dose on lung tissue, or skin erythemas, are assumed to be effects for which some *recovery processes* are possible. They therefore do not show a linear relation with dose since there is a threshold value of dose below which an effect is not detected. The effect-dose relationship of this type of radiation damage is shown in Figure 11.1A, showing a so called *sigmoid* curve. These somatic certainty effects not only have a threshold value but also vary with the rate at which the dose is given. On the other hand the stochastic effects show a *linear* effect-dose relationship as shown in Figure 11.1B. These effects have no threshold and some effect is always produced even if the dose is reduced to a very low level; there is no recovery process and the effect is dependent only on the accumulated dose and is independent of dose rate. Examples of somatic stochastic effects are the increased probability of leukaemias or malignancies observed in large population groups who have been subject to ionising radiation doses above background level. The increase of naturally occurring hereditary diseases due to exposure of a population to radiation is a genetic stochastic effect (see also p. 173).

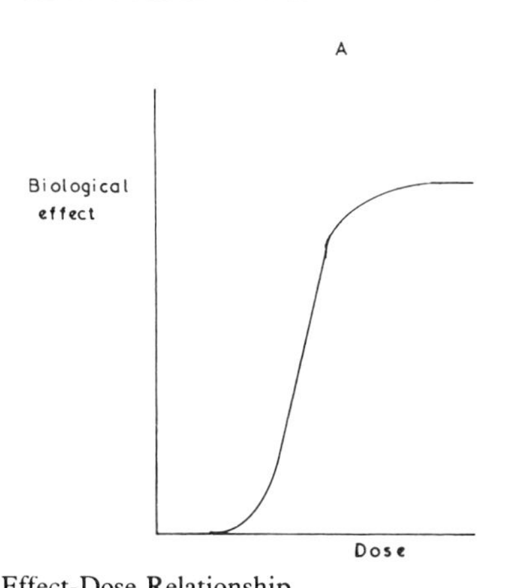

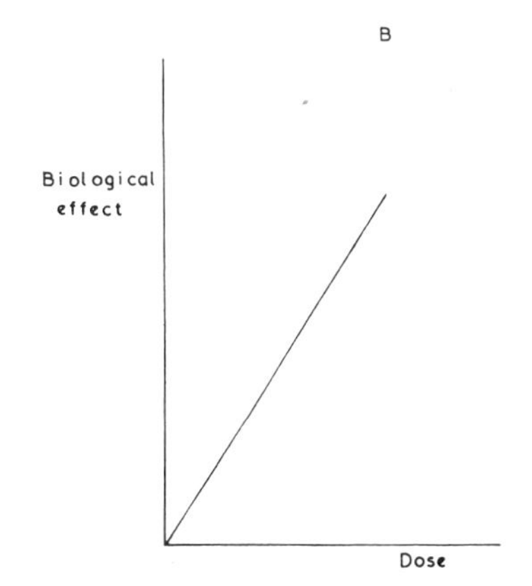

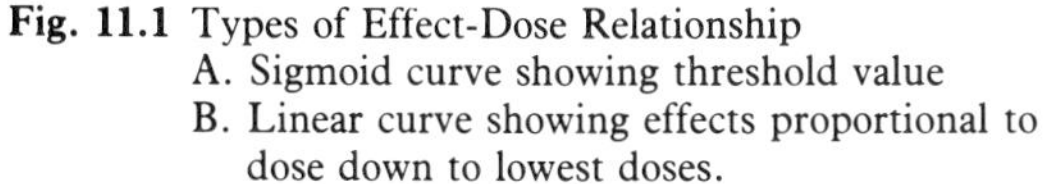
Fig. 11.1 Types of Effect-Dose Relationship
A. Sigmoid curve showing threshold value
B. Linear curve showing effects proportional to dose down to lowest doses.

Specification of tolerable dose levels

In a light of the discussion of the previous paragraph the specification of a 'safe' dose of radiation will be seen to be a complicated matter. Radiation in medical practice confers very great advantages in health care and it is inconceivable that medical work could be carried on without taking advantage of X-rays and radioactive materials. On the other hand risks do arise from their use and it is the purpose of the hazard control regulations to balance appropriately the advantages against the risks to which both the staff and the general population are subjected.

It is possible to specify radiation levels which will reduce to negligible values the somatic certainty effects on the staff engaged in radiation work. At the same time the risks of stochastic effects will be so reduced that they become commensurate with the risks of normal life. Since stochastic effects are related linearly with dose, however, it would clearly be advantageous to reduce the dose to the staff to the lowest possible value. This does involve consideration of the problems of cost of protection and of the need to perform radiation work in hospitals with great efficiency.

The general public who are exposed to man-made radiation is composed of a large group at least 1000 times greater in number than radiation workers and for this reason the reduction of the risk of stochastic effects involves a dose specification considerably lower than that specified for radiation workers. The maximum permissible dose for the general public will be seen later to be one-tenth of that for the limited numbers of people who are engaged in radiation work.

One further point might be mentioned regarding the specification of regulations for control of radiation hazards. Maximum permissible dose specifications are made for those occupationally exposed to radiation and for the general public who might be subject to exposure arising from the use of radiations in medical work and in other ways. Doses which patients receive in the course of medical care either for treatment or diagnosis are not included in this specification since it must be assumed that the radiation exposure is always justified by the clinical needs. However the dose to patients in the course of medical procedures does involve an increased probability of both somatic and genetic stochastic effects. For this reason medical procedures should be conducted with the minimum radiation dose possible consistent with obtaining accurately the required diagnostic information or producing the required therapeutic effect. There is therefore an obligation on the part of radiological workers to make sure that patients have the minimum possible radiation exposure and current codes of practice include recommendations to achieve this. Doses to patients undergoing clinical procedures however are not included in maximum permissible doses prescribed either for radiation workers or the general public.

Codes of practice

The responsibility for keeping under review all evidence concerning the effects of ionising radiations at low dose levels and of making recommendations of acceptable exposure levels for radiation workers and the general public is accepted by an international group of experts known as the *International Commission on Radiological Protection* (ICRP). This authoritative body has been working since 1950, though it was a development of an International X-ray and Radium Protection Commission established in 1928 by the Second International Congress of Radiology.

The ICRP has published a series of reports dealing with all aspects of radiation hazard control and some of these are listed in the bibliography (see p. 291). The reports of the ICRP are accepted internationally as the basis on which detailed Codes of Practice or Regulations can be set up for use in particular countries or under particular circumstances. In the United Kingdom a number of such Codes are available dealing, for example, with radiation hazards in industry or research establishments. The Code of Practice relevant to hazard control in radiation departments in the National Health Service of Great Britain and Northern Ireland is entitled, '*Code of Practice for the Protection of Persons against Ionising Radiations arising from Medical and Dental Use*'. This Code was published in 1972 and is a revision of an earlier similar Code published in 1957 and revised in 1964. To illustrate ways in which hazards in hospital work can be adequately controlled extensive quotations from this Code are included in this chapter by permission of the Controller of H.M. Stationery Office.

Responsibility for safety in radiological departments

The U.K. Code of Practice sets out an organisational procedure designed to establish a chain of responsibility to ensure that appropriate recommendations for reduction of hazard are implemented. The ultimate responsibility for safety in a radiological department rests with the *Controlling Authority*, that is with the Area Health Authority or its equivalent who administer the department. The head of a department carries the day to day responsibility for the work and safety of his department under the Controlling Authority, but the Code lists a number of officers who share in different ways the responsibility of advising the head of the department and the Controlling Authority. These are named as (a) the *Radiological Protection Adviser*, who is a qualified physicist responsible for advice on all technical aspects of the protection problem, (b) the *Supervisory Medical Officer* who is responsible for any preliminary or subsequent medical examinations, including blood counts, which may be necessary to ensure the health of workers in the department and (c) *Radiological Safety Officers* who are normally permanent members of a department, often senior and experienced radiographers, who are given the day-to-day responsibility of seeing that protection recommendations, and any local rules supplementing the provisions of the Code, are implemented.

The Code also makes obligatory the setting up of a committee to serve one or more Controlling Authorities, called a *Radiological Protection Committee*. This committee reviews local procedures, receives reports from the various officers and advises the Controlling Authority on ultimate steps to be taken.

It is worth noting before outlining some details of the Code that it is not a legal document. It makes recommendations. Some of these are obligatory in the sense that they are essential requirements for good safety precautions and must be implemented. Others are discretionary, that is they are desirable requirements but their importance and the possibility of putting them into practice will depend on local circumstances. The whole Code, however, establishes a standard of good practice in radiological departments which is expected to be the normal practice.

permissible doses. Designated persons are subject to special procedures, such as individual monitoring of exposure and special medical procedures.

The Code also specifies dose limits for members of the public who may, because of their visits to hospitals or the neighbourhood of hospitals, receive some exposure to radiations.

Practical recommendations are also included in the Code of Practice to reduce to as low a level as possible radiation doses given to patients during diagnostic or therapeutic procedures. This is rarely a relevant consideration in therapeutic practice but is very important in diagnostic departments where good practice aims at obtaining relevant clinical information with the minimum radiation dose to the patient.

Table 11.1 below sets out the maximum permissible doses for occupationally exposed persons based on the recommendations of the International Commission on Radiological Protection.

Table 11.1 Maximum permissible doses for occupationally exposed persons

Part of body	Annual	Quarterly dose	Planned special exposures	
			Single event	Lifetime
Gonads, red bone-marrow and whole body	5 rems[a]	3 rems (1·3 rems to the abdomens of women of reproductive capacity)	10 rems	25 rems
Bone, thyroid and skin of whole body	30 rems	15 rems	60 rems	150 rems
Hands, forearms, feet and ankles	75 rems	40 rems	150 rems	375 rems
Any single organ (excluding gonads, red bone-marrow, bone, thyroid and skin of whole body)	15 rems	8 rems	30 rems	75 rems

[a] If the person was occupationally exposed before 18 years of age, subsequent exposure of the gonads, red bone-marrow and whole body should be controlled so that the dose accumulated up to 30 years of age does not exceed 60 rems.

The maximum permissible dose

The Code of Practice specifies maximum permissible dose levels for workers in radiological departments who, in the course of their work, may be exposed to radiation. For the purposes of the Code these occupationally exposed persons are divided into two groups; (a) those whose work involves exposure to ionising radiations to such an extent that the resulting annual doses might exceed three-tenths of the annual maximum permissible doses. These are known as '*designated persons*', and (b) other persons whose exposure is such that the resulting annual doses are most unlikely to exceed three-tenths of the annual maximum

A number of comments should be made about the Table above. First the maximum permissible doses are quoted in *rems*. The rem is the unit of *dose equivalent*. This is the quantity of any ionising radiation such that the energy imparted to a biological system per gramme has the same biological effectiveness as 1 rad of 200–250 kV X-rays. The biological effectiveness of a given absorbed dose in rads of one type of radation is not necessarily the same as that of an equal absorbed dose of another type of radiation. The variation in biological effectiveness arises from the difference in the density of ionisation along the tracks of the ionising particles (see p. 183 and Figure 17.8). For protection purposes, X-rays, beta rays and gamma rays where the effective ionisation agent is a fast electron, the biological effectiveness is the same and the rem can be taken as equal to the rad. For radiation in which the primary ionising agent consists of heavy charged particles such as neutrons or alpha particles, the biological effectiveness is greater and factors are used called *quality factors* which convert the rad dose into rems—the unit giving the same biological effect as the rad. A quality factor of 10 is used for neutrons and alpha particles and 20 for heavy recoil nuclei arising in some ionising processes. These factors give only a rough guide of relative biological effectiveness but for protection purposes high precision in dosimetry is not required.

Table 11.1 will be seen to take account of the fact that radiation to limited parts of the body only is less hazardous than radiation received by the whole body. The extremities, for example, can receive much larger doses and for hands, forearms, feet and ankles a maximum permissible dose of 75 rems per year is specified.

Table 11.1 also illustrates that doses received in occupational exposure may be accumulated at somewhat irregular rates and the dose in any one quarter specified in the Table may be higher than the appropriate fraction of the annual dose provided that the total annual dose is not exceeded.

Planned special exposures may be needed to meet situations arising infrequently during normal practice. These are permitted as long as the total dose accumulated by the worker does not exceed a value given by the formula $5(N-18)$ rems where N is the worker's age in years. This formula was formerly used for the maximum permissible whole body exposure accumulated over the working life and it will be seen to represent a permissible accumulated dose increasing by 5 rems per year.

One additional specification of maximum permissible dose concerns the abdominal exposure of women of reproductive capacity. In the case of pregnancy the dose to the foetus of a woman occupationally exposed must not exceed 1 rem during the term of her pregnancy which remains after it has been diagnosed.

In addition to the maximum permissible doses for occupationally exposed persons specified in Table 11.1 dose limits are also specified for members of the public who may become exposed to radiation incidental to their presence near hospitals for other than radiological procedures. These dose limits are in general one-tenth of the rem doses given in Table 11.1.

Maximum permissible doses for internal radiation

It will be recognised that radiation hazards may arise not only from external irradiation by X- and gamma-rays or beta-rays but also from accidental accumulation of radioactive isotopes in the body by ingestion or inhalation of radioactive material. These dangers may arise for example from damage to sealed containers holding gamma emitting isotopes such as radium or cobalt. They arise more frequently from the use of unsealed radioactive materials for therapeutic, diagnostic or experimental purposes. These unsealed radioactive materials may be in the form of solutions, or gases or powders and may enter the body by a variety of routes.

It is possible by making appropriate assumptions about the way in which the various isotopes are metabolised in the body to deduce the concentration of isotopes in air or water which can be accepted in the working environment in an isotope laboratory or in the environment to which the general public is exposed. These concentrations are such that if a person were exposed during their normal work to that level they would receive doses specified in the tables of maximum permissible doses of Table 11.1. These concentrations are referred to as *maximum permissible concentrations* in air and water. Figures can also be deduced for the total body content of the various isotopes which would produce the maximum permissible doses of Table 11.1 and these are referred to as the *maximum permissible body burden*. These figures, as well as those for maximum permissible concentrations in air and water, are given in an appendix in the U.K. Code.

Text books on radiation protection will give further details of the specification of the tolerance dose limits for both external and internal radiation for workers exposed continuously during their working life as well as members of the general public. For the purposes of this text it is to be noted that the dose limits prescribed form the basis of recommendations for the design of radiological departments and also the basis of recommendations for working conditions and techniques used in these departments. These topics will be outlined in the following pages.

General principles for achieving appropriate safety standards

A radiotherapy service involves the use of a wide range of radiation sources applied under a wide variety of conditions. The achievement of proper safety standards in the department therefore means the adoption of a multiplicity of techniques both in the structural plan of the department and in the working methods. Some general principles can be noted. The radiation dose to the workers and the general public is reduced by (a) interposing attenuating screens between the radiation source and the person to be protected, (b) increasing the distance between the source and the protected person, and (c) reducing the time of exposure as much as possible. The cost of thick protective screens and large rooms is high and a compromise must be reached so that adequate protection is provided at a cost which is not prohibitive. The attempt to reduce drastically the time spent on some procedures involving the handling of radiation sources must not be pressed too vigorously if it results in inaccuracy in treatment and risk of accident.

Radiation protection problems arise in the hospital service in four different types of conditions. These are the use of radiation beams for diagnosis and therapy and the therapeutic use of sealed and unsealed radioactive sources. Protection in X-ray diagnostic departments will not be dealt with in this text book though many of the general principles applied in beam therapy departments apply in diagnostic work also. The main problems in protection in the use of radiation beams for therapy are the provision of adequate shielding round the radiation source and the treatment room, the adoption of appropriate working techniques and the provision of adequate safety checks and devices to avoid accidental wrong dose to the patient. In practice the problems involved in the use of radiation generated at less than about 300 kV are different from and easier than those involved in the use of radiations in the megavoltage range. This is not only because protective shields can be very much thinner, but also because radiation shielding materials in which high atomic number materials such as lead are used are very much more efficient. Efficiency here means a much greater attenuation for a given mass of shielding material (see p. 36).

A different set of conditions arises in the use of small sealed gamma emitting radiation sources such as radium or caesium tubes. In this case distances from the radiation sources to the exposed parts of the body are in general small and protective techniques involve not only shielding of the body from gamma rays or secondary beta rays but also rapid working procedures without direct contact between the hands of the operator and the source.

In hospitals the use of unsealed sources of radioactive materials is almost always in the form of solutions. The protective and shielding techniques associated with small sealed sources are involved but the major protection problem is the avoidance of risk of entry of the radioactive material into the body. Techniques to prevent spread of contamination, or escape of radioactive material outside its site of usefulness are therefore very important.

Details of the protection techniques in these different types of work will be described in the following paragraphs. Some details will also be included about methods of monitoring the working environment and exposed personnel to ensure that the maximum permissible doses are not exceeded.

Protection in beam therapy departments

Design of equipment

Adequate protection in a beam therapy unit begins with the apparatus itself. Chapter 3 has given details of the construction of both X- and gamma ray beam units in which adequate shielding is incorporated. In some cobalt units and linear accelerators when mounted on an isocentric gantry (see p. 72) a substantial lead shield is fitted to the mount to absorb the exit beam emerging from the patient. The presence of such a 'beam stop' on these units modifies considerably the protection requirements for the walls of the room in which it is housed.

Room design

In the design of treatment rooms structural shielding must be provided to ensure that the dose to the operating staff is less than the maximum permissible dose, and also that any members of the general public either in adjacent rooms or passing along near by public corridors do not get doses exceeding the prescribed dose limits. Many factors must be taken into account in deciding on an appropriate structure for the walls, floors and ceilings of a department.

In the first place it will be recognised that the barrier thickness required to attenuate the primary beam to the required low intensity will be much thicker and therefore more expensive than that required to reduce the scattered radiation to the appropriate level. It is therefore always useful to distinguish between *primary barriers* to be placed in the path of the direct beam and *secondary barriers* which do not receive the direct primary beam but only shield against scattered radiation. If the movements of the set can be restricted without undue limitation on treatment techniques then all the six sides of any treatment room need not be protected by a primary barrier. In many sets using an isocentric mounting such as linear accelerators or gamma ray beam units (see p. 72) the primary beam is limited to one plane and some of the walls and parts of the floors and ceiling could in this case possibly be made only of secondary barrier standard. When isocentric mountings for high energy therapy units incorporate a primary beam stop in the path of the exit beam from the patient the need for primary barrier protection on the walls of the treatment room will be eliminated, though at the cost of some inconvenience in setting up patients on the set. It should be noted however that leakage radiation from the housing of a set may sometimes be appreciable and a 'secondary' protective barrier may have to attenuate the sum of the scattered radiation and the leakage radiation to the required level. The scattered radiation has low intensity and is of softer quality than the primary beam, but the leakage radiation, though of low intensity, is of the same quality as the primary beam. Access to the room and observation of patients both have to be provided for and must give the required protection. It is clearly useful therefore if doors and windows can be arranged in the walls of the room which do not receive the primary beam, and behave as secondary barriers.

During therapy treatment the operator normally works in a control cubicle when the radiation beam is on. For radiotherapy using radiations generated at more than 50 kV the control cubicle is always outside the treatment room. If the observation of the patient from the control cubicle is by an observation window (see below) then it will be advantageous for this window to be part of a secondary barrier rather than a primary barrier. Below 50 kV protective screens between the operator at the control panel and the patient under treatment can usually be arranged to give adequate protection.

Having decided on the general room lay-out and the acceptable restrictions to the set movement it is then possible to calculate the required thickness of the attenuating barrier from available attenuation curves for a large variety of possible construction materials. The attenuation curves that are applicable for these calculations are broad beam attenuation curves (see p. 51). In making these calculations the possibility is taken into account that the beam itself will be pointing in a particular direction for only a fraction of the total treatment time (usage factor), and that the space to be shielded may only be occupied for a fraction of the total treatment time (occupational factor). Consideration of these factors is always made however with some care and estimates of the thicknesses of protective shields are made on the safe side to allow for unknown future changes in treatment policy.

The material of the protective barriers is a matter of importance. The thickness of a protective barrier is often quoted in terms of its '*lead equivalent*'. The lead equivalent of a protective barrier is the thickness of lead that will give the same attenuation of the radiation beam under consideration as the particular protective barrier. The lead equivalent of an ordinary concrete barrier depends very greatly on the radiation quality. At low photon energies where the photoelectric effect predominates lead is a very efficient attenuator per unit mass. In the megavoltage range lead and concrete have almost the same efficiency per unit mass since attenuation is here almost entirely by the Compton process. The lead equivalent of a concrete shield therefore increases with photon energy. At 100 kV for example 5 cm of concrete has a lead equivalent of 0·6 mm Pb but at 2 MV the lead equivalent of the same 5 cm of concrete is 6·0 mm Pb.

For radiations in the orthovoltage range or of softer quality protective barriers of adequate thickness can sometimes be made quite easily of building materials of ordinary kinds such as concrete and of ordinary constructional thicknesses. If these barriers form part of the construction of a new building all that is required is to ensure that there are no leakages through joints or inferior concrete. An adequate protective barrier however can often be made by using lead sheeting of a few millimetres thickness, over a structural support that does not itself have adequate protection. The lead may be used as sheet or it may be sandwiched between layers of plywood which makes it easier to handle. The thicknesses of protective barriers required in the lower voltage range are illustrated in Table 11.2.

For protective barriers in the megavoltage range however the thicknesses of barrier required become very great and lead loses its relative efficiency. A barrier of a given weight per unit area is enormously expensive if lead is used instead of concrete. At this radiation quality therefore concrete is used. The thicknesses of concrete required however are often so great that it becomes economically important to consider the possibility of reducing the thickness of the barrier in order to save overall space in the building. This can be achieved by incorporating in the concrete heavy material such as barytes or steel shot. The effect of this is to increase the overall density of the concrete thereby reducing, in proportion, the linear dimensions of the barrier.

Table 11.2 below gives some illustrations of representative protective barriers required in a radiotherapy department with rooms of average size using X-ray machines with outputs typical of modern practice. The concrete thicknesses quoted are for normal density concrete ($\rho = 2{\cdot}3$ g/ml). For loaded concrete which might have a density $\rho = 3{\cdot}5$ g/ml for loading with barytes or $\rho = 5$ g/ml for loading with steel shot thicknesses would be reduced in inverse proportions to the density. The lead thicknesses required for megavoltage radiation are clearly quite uneconomic and impracticable.

Doors

The doors to a radiotherapy room must be of the same protective value as the wall of which they form a part. For superficial therapy lead lined doors are easily made with lead sheeting of the required thickness attached to the wooden door material. The same is true of doors which form a secondary barrier in a deep therapy room working at 250 kV. In this range doors shielded to attenuate the primary beam become very heavy. The possible use of a motor driven door is not practicable because of difficulties

Table 11.2 Representative protective barrier thicknesses

Superficial therapy	100 kV		
Primary barrier		2.0 mm lead	13.0 cm concrete
Secondary barrier		0.5 mm lead	5.0 cm concrete
Deep X-ray therapy	250 kV		
Primary barrier		8.0 mm lead	32.5 cm concrete
Secondary barrier		2.0 mm lead	11.2 cm concrete
Megavoltage therapy	2 MeV		
Primary barrier		(165 mm lead)	108 cm concrete
Secondary barrier		(87 mm lead)	51 cm concrete

that could arise in the event of a breakdown of the electric power supply so that doors should always be sited where they need only act as a secondary barrier. For megavoltage therapy shielded doors are quite impossible even for secondary barriers. An arrangement is generally adopted so that the entrance to the room is situated in a position where the radiation intensity is sufficiently low for shielding to be unnecessary. This is achieved by placing the door at the end of a *maze*. In this position the radiation it receives will have been scattered at least twice and the leakage radiation from the source head will have been adequately attenuated. Figure 11.2 (1,2) shows typical plan drawings of the layout for two therapy rooms—one for 250 kV and one for megavoltage radiations. The second drawing illustrates the arrangement of the maze and the entrance door.

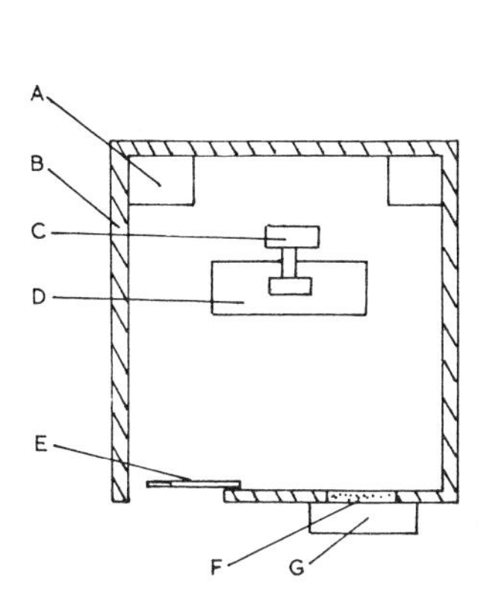

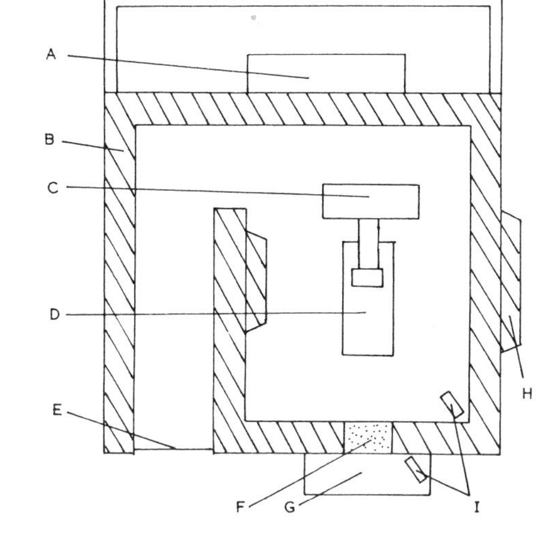

Fig. 11.2 Typical Plans for Radiotherapy Rooms.

1. 250 kV Room
   A. Generator
   B. Protective Wall
   C. Treatment Apparatus
   D. Couch
   E. Sliding Door
   F. Observation Window
   G. Control Desk

2. Megavoltage Room
   A. Auxiliary Apparatus or Generator
   B. Protective Wall
   C. Treatment Apparatus
   D. Couch
   E. Entrance Door to maze
   F. Observation Window
   G. Control Desk
   H. Extra primary beam shield
   I. Television Viewing

It is obligatory in radiotherapy treatment rooms to have on the door both a sign indicating the presence of a radiation hazard, and also an electrical interlock so that the set will automatically be put to the 'beam off' condition if the door is opened. This reduces the risk of accidental exposure arising from anyone inadvertently entering the room during exposure. The main purpose of the unshielded door at the end of a maze in a megavoltage room is to carry the interlock and warning sign. In some cases a simple 'gate' or on optical interlock may replace a more usual door in this situation.

Observation of patients

One further problem in the design of a radiotherapy room is the need for communication between radiographer and patient during treatment. A two-way speech system is easy to instal. Visual communication is generally achieved by installing windows for observation. These must have a protective value equal to that of the wall in which they are placed. They will be situated close to the control desk and preferably in a wall which is a secondary rather than a primary barrier. For superficial and orthovoltage therapy lead glass windows are satisfactory. These are made of glass incorporating lead salts. At 200 kV for example a sheet of lead glass only 16 mm thick can be made to have a lead equivalent of 4 mm Pb without excessive reduction in optical transparency. For megavoltage installations the problem is much more difficult. Blocks of lead glass of much greater thickness have been used but it is possible to produce windows of plate glass with density comparable to that of ordinary concrete or barium loaded concrete so that a window may be inserted into a concrete wall with the same thickness and attenuation as the wall itself. Windows not only give a direct view of the patient but also, if sited suitably, can give the patient a view of the radiographer at the control desk and this is often considered important.

Alternative methods of viewing in therapy rooms have been adopted including the use of a periscope system. This needs care in installation in order to avoid reduction of the protective value of the wall in which the periscope is installed. Increasing use is now made of closed circuit television systems often with a double camera arrangement giving a general view of the room as well as a detailed view of the treated area, so that movement of the patient can be detected.

The U.K. code of practice

Reference has already been made to the recommendation of the U.K. Code of Practice concerning the organisation of a chain of responsibility for hazard control in hospitals and also to its specification of the maximum permissible doses based on the recommendations of the ICRP. The 1972 edition of the U.K. Code of Practice also contains recommendations dealing with arrangements in radiological establishments and with working techniques. These recommendations cover the whole field of the use of ionising radiations in medical work. The sections of the Code most relevant to the readers of this text include those dealing with the therapeutic use of radiation beams, the therapeutic use of small sealed radioactive sources and the therapeutic and diagnostic uses of unsealed radioactive substances. The recommendations of the Code are illustrated by the following extracts relating to these three fields of hospital work. Students must, however, be familiar with the fuller details of the Code and any local additions to the national code which may relate to their own department. These additions are referred to as *local rules*. Before starting work they must have read and understood those recommendations which are relevant to their particular work and must sign a statement indicating that they have done so.

Abstracts of U.K. code concerning therapeutic uses of radiation beams

1. Treatment rooms must be prominently marked with a symbol to indicate ionising radiations.
2. Means must be provided for observing the patient and should be provided for communication with the patient from the control panel during treatment.
3. A visible signal must be provided outside the treatment rooms to indicate that the radiation beam is in operation. Wherever possible, a similar visible signal should be provided inside the treatment room.
4. Interlocks must be provided so that when any door or barrier (e.g. light beam or bar) to the treatment room is opened or interrupted the radiation beam is immediately switched off and cannot be re-energized until the interlock is reset.
5. Except for radiation generated at not more than 50 kV the operator should always be outside the room during any period whilst the beam is switched on. Only when it is unavoidable should a person, other than the patient, be in the treatment room during a treatment (e.g. when treating babies or mentally unstable persons and no alternative procedure is practicable). This person should not be the regular operator of the equipment, and must be provided with adequate protection (e.g. by shielding or distance).
6. Any person who carries out any modification to or maintenance of any apparatus which might alter the output or quality of the radiation, or the protection of the tube, must immediately attach to the apparatus an appropriate visible notification of such modification or maintenance and must, on each occasion, inform in writing the Radiation Safety Officer and radiographer-in-charge; the latter must enter the details in a record book kept for the purpose.
7. Each accessible filter must be marked with its thickness and material or in the case of a wedge filter with appropriate details which permit easy recognition of the filter in use. For low energy (50 kV or less) X-ray therapy machines an interlocked system should be employed to control the maximum kilovoltage which may be used with a particular filter thickness.
8. Unless it is possible rapidly to bring the X-ray output to the prescribed value, the housing must be fitted with a shutter, operated from the control panel, and the position (open or closed) must be indicated at the control panel.
9. The equipment must be provided with an automatic timer or an integrating dosemeter to terminate the treatment after a pre-set time or dose. For equipment generating beams of energy greater than 2 MeV and dose rates of 100 rad min^{-1} or more in air at the treatment distance special dosimetry systems must be installed. (These are specified in detail in an appendix and include a double dosimetry system to avoid accidental over exposure due to any mechanical or electrical failure.)
10. The use of low voltage equipment which requires to be held in the hand by the operator should be discouraged.
11. A physicist must be responsible for ensuring that all X-ray apparatus used for therapeutic purposes is calibrated at intervals of not more than 4 weeks.
12. Dosimeters used for calibration and check of output must be maintained in good condition and appropriate tests made to ensure that their sensitivity remains constant. They should be checked at intervals of not more than one year against a recognised secondary standard meter.
13. To avoid accidental over-exposure of patients there must be a rigid procedure in operating apparatus and each member of the staff must be aware of this procedure and must clearly understand the extent of his own responsibility.

Abstracts of U.K. code concerning therapeutic use of small sealed radioactive sources

1. A separate room must be provided for the 'make up' and cleaning of sources and applicators and this room should only be occupied during such work. The door of the room must be marked with a symbol indicating ionising

radiations. The room must be adequately ventilated. The placing of objects in the mouth, eating, smoking, drinking or the application of cosmetics whilst within the room must be prohibited.

2. A permanent record must be kept of the issue, distribution and return of all sources.
3. In order to ensure the minimum irradiation of personnel engaged in the preparation or application of sources appropriate handling tools or implant instruments must be used at all times. These tools should be constructed so as to provide maximum handling distance compatible with effective manipulation. All operators must have adequate training in these manipulative procedures. Sources, or loaded source containers must not under any circumstances be picked up directly by hand. In order to ensure that doses in excess of the quarterly maximum permissible levels are not received by members of the staff a rota system of duties must, where necessary, be instituted.
4. When multiple needles or capsules of the same appearance but of different strengths are used, they must be identified with different coloured threads, beads or other means when in clinical use.
5. The number and position of removable sealed sources in or on the patient must be checked regularly. Dressings and excreta from patients receiving treatment with sealed sources of radioactive substances must not be disposed of unless monitoring has shown that they are not radioactive or until all the radioactive sources have been accounted for.
6. All sealed radioactive sources in which the radioactive substance is encapsulated or bonded in inactive material must be tested for leakage and surface contamination initially and at least once a year thereafter and whenever damage is suspected. Records of all leakage tests must be entered in the sealed sources register.
7. Whenever there are reasonable grounds for believing that radioactive substances are leaking, or are liable to leak, from a sealed source, that source must at once be placed in an airtight container pending repair by a competent person.
8. An emergency procedure to be adopted in the case of loss or breakage of a radioactive source must be specified and notices indicating the action to be taken must be displayed in each room where such sources are handled or employed.
9. Suitable shields or baffles must be provided where required to ensure adequate protection when manipulating beta-ray sources. In order to prevent the head from being placed too near the source and to protect the eyes and face from beta-rays a transparent plate of adequate thickness should be mounted or worn between the source and the face of the operator.
10. The local rules for the protection of persons in proximity to patients undergoing treatment with sealed radioactive sources should be based on the following requirements:

    (a) Beds in which there are patients undergoing treatment with radioactive sources must carry a notice using the standard symbol indicating the fact. The notice should give details of the number and nature of source, total activity of the radioactive substances, the time and date of application and removal and relevant nursing instructions.
    (b) The Radiological Safety Officer of the relevant department must measure or otherwise estimate the maximum gamma-ray exposure rate at a distance of 1 metre from each patient undergoing treatment. If this exposure rate exceeds 5 mRh^{-1} the notice must in addition indicate that an external radiation hazard exists and the Radiological Safety Officer must give instructions regarding the daily time allowable for nursing procedures and visitors.
    (c) Patients with sources in or upon their bodies must not be permitted to leave the ward or treatment rooms without the approval of the appropriate medical officer.
    (d) Nursing staff and other persons must not remain unnecessarily in the vicinity of patients undergoing treatment with gamma-ray sources. Where possible nursing procedures should be postponed until after the sources have been removed.
    (e) It may be necessary to reallocate duties for staff who are confirmed to be pregnant and also to exclude visitors who may be pregnant.
    (f) Where possible such treatment should be carried out in wards having only 1 or 2 beds, care being taken to ensure that adjoining rooms are adequately protected. Where a general ward is used the beds of patients under treatment must be distributed as widely as possible throughout the ward and there should be not less than 2·5 metres between bed centres.

Abstracts of U.K. Code concerning therapeutic and diagnostic uses of unsealed radioactive substances.

1. It is important that unsealed radioactive substances even in very small amounts should be manipulated only in working areas recognised and equipped for this work.
2. Work with unsealed radioactive substances in hospitals should be governed by clearly understood rules of procedure. Each institution must draw up its own detailed local rules which must be supplemented by careful training of staff at all levels.
3. Adequate records must be kept including a record of all administrations of radioactive substances.
4. In laboratories and working areas:

    (a) The floor, walls and benches should be finished with smooth, continuous and non-absorbent surfaces which can be cleaned easily.
    (b) The floor and benches must be strong enough to support the weight of any necessary shielding materials.
    (c) A wash-hand-basin fitted with foot, knee or elbow operated taps should be provided.
    (d) Working procedures should be designed to minimise the spread of contamination from the working area. All dispensing of radioactive substances should be done in a fume cupboard or

glove box especially when particulates or aerosols are involved.

(e) Eating, drinking, smoking and the application of cosmetics must be forbidden in laboratories using unsealed radionuclides. The operation by mouth of pipettes and wash bottles must be forbidden.

(f) Laboratory coats (or protective gowns) and surgical or similar gloves must be worn for all procedures involving dispensing of radioactive substances and their administration to patients. For work with higher activities overshoes should be worn. Regular systematic monitoring of the hands and gloves should be carried out.

5. *Ward Design and Procedures*

(a) Patients undergoing treatment with therapeutic activities of unsealed radioactive substances should not be placed in general wards in which there are patients whose treatment does not involve such therapy. Instead they should be placed in specially designed wards which preferably should have only one or two beds; where there are two or more beds these should be arranged so that there is not less than 2·5 metres between centres.

(b) Patients undergoing treatment should be provided with a toilet and bathroom in the same suite for their exclusive use.

(c) Floors, walls and ceilings should be covered with smooth, continuous and non-absorbent surfaces which can be easily cleaned. Floor coverings should be used in sheet rather than tile form in order to reduce the number of joints.

(d) Local rules must be prepared by the Radiological Protection Adviser in consultation with the Radiological Safety Committee and the Chief Nursing Officer. Nursing procedures which are not urgent should be postponed as long as possible to take full advantage of the reduction of activity by decay and excretion.

(e) Beds in which there are patients undergoing treatment with radioactive substances must have a notice using the standard symbol indicating the fact. The notice should give details of the nature and activity of the radioactive substances, the time and date of administration and any relevant nursing instructions.

6. *Decontamination Procedures*

(a) Persons working with radioactive substances should wash their hands thoroughly with mild soap and water before leaving working areas. After washing the hands should be checked with a radiation monitoring instrument.

(b) If washing the contaminated skin with soap and water fails to remove the contamination to the required level an appropriate detergent should be tried. If this fails treatment with a saturated solution of potassium permanganate followed by decolourisation with 5 per cent sodium bisulphite may be used (potassium permanganate should not be applied to contaminated hair as there is a risk of causing temporary change of hair colour). Chemical treatment should not be applied too vigorously as the skin may become porous. Even when the contamination has not been reduced to the required level none of these procedures should be carried on to the stage of injuring the skin.

(c) When high level contamination of parts of the body other than the hands is suspected the Radiological Safety Officer of the relevant department and the Head of the Department should be notified at once.

(d) If the skin is broken or a wound is sustained in conditions where there is a risk of radioactive contamination the injury should be irrigated immediately with tap water.

7. *Emergency Procedures*

(a) Experience has shown that most incidents involving spills of radioactive substances in hospitals do not warrant any drastic emergency action but require only simple remedial action by local staff. Nevertheless a more serious accident could possibly occur and some preparation to anticipate the event is necessary. In such a case pre-arranged procedures must be instituted as soon as possible and the Radiological Protection Adviser must be informed.

(b) In any incident the first concern must be the protection of any persons involved (whether patients or staff) and the treatment of any serious injury. The second concern is to confine the contamination as far as possible to the area originally affected. Decontamination of personnel must also take priority over any plan for decontamination of working areas although arrangements must be made to restrict the spread of contamination.

(c) Local rules must be drawn up to specify (i) the persons to be notified of any incident involving the dispersal of radioactive substances (ii) the instructions to staff (including nurses) on any immediate action to be taken, (iii) the location of equipment for dealing with incidents.

(d) The best course of action in an incident depends very much on local circumstances. Until an appropriate plan has been worked out by the Radiological Society Officer of the relevant department for the particular incident only the minimum action should be taken, for example:

(i) Persons in the immediate vicinity should be warned of the incident.

(ii) Radioactive substances on the skin should be thoroughly flushed away with tap water.

(iii) Protective, and if possible, other outer clothing which is contaminated with radioactive substances should be removed and left in the affected area.

(iv) If it is necessary to evacuate all non-essential persons an attempt should be made to ensure that contamination particularly on shoes or clothing is not carried to other unaffected areas. If contaminated persons are evacuated, they should be monitored and measures to reduce surface contamination should be taken as soon as possible.

(v) Persons entering the affected area to carry out emergency procedures should wear protective clothing and equipment.

(vi) Entry to the affected area must be restricted until all the appropriate action has been taken to clear the contamination from the area and radiation surveys have satisfied the Radiological Safety Officer of the relevant department that the area may be reoccupied.

(vii) In the event of incidents involving therapeutic activities of radioactive substances the need for tests to determine the activity which has entered the body should be considered and an estimate should be made of the dose from internal and external radiation received by persons involved in the incident.

Radiation Monitoring

In order to ensure that radiation levels in a radiotherapy department are appropriately low and that maximum permissible dose levels for staff are not exceeded two kinds of measurements are required. The first of these is a measurement of the exposure rate at relevant points in the department or of total exposure at those points accumulated during a given time, say one week or one month. This is *environmental monitoring*. Its aim is to check that the design of the protective shielding in the department and the working methods can provide an adequately safe environment both for those working in the department and for members of the general public or workers in other departments who may receive radiation exposure through their presence in or near the department.

The second type of monitoring is *personnel* or *individual* monitoring. This is undertaken for members of staff who are occupationally exposed to radiation and it is aimed to check that the persons concerned receive less than the prescribed maximum permissible doses (see p. 121). This measurement will differ from the estimate of exposure that could be made from the environmental measurements since it will be affected by the distribution of duties at different places during the working day.

Environmental monitoring

The working environment in any radiation department must be checked regularly using appropriate instruments of which a brief description is given later (see p. 130). It is particularly important that this is done for any new department, or for a department which has in any way been modified so that its protection may have been changed. This latter may arise because of a rebuilding operation, or a rearrangement of working patterns or a modification or change in the radiation equipment. An environmental survey is also often necessary if personnel monitoring reveals that excessive doses are being received, since in this case appropriate modifications to the shielding in the department might be required. Environmental monitoring should check on levels of external radiation in the department and on the levels of radioactive contamination in working areas if unsealed sources are being used.

Personnel monitoring

Personnel monitoring is carried out by small integrating dosemeters which can be worn on the clothing of the radiation workers. A number of different types of dosemeter are possible including photographic film badges (see p. 131, thermoluminescent capsules (see p. 133) and quartz fibre electrometers (see p. 133). These must be worn continuously during working hours but kept away from possible radiation exposure outside working hours. They are normally worn at the waist or chest level. The exact position is not very important since personnel monitors of this kind are intended to record whole body radiation arising in the main from scattered radiation and it can be shown that under normal working conditions the scattered radiation received by workers does not differ widely at different parts of the trunk. If protective clothing is being worn the monitor should be placed under the protective apron since the whole body radiation to the trunk is the quantity that is to be assessed in this technique. Additional monitors are sometimes worn at special sites on the body if checks other than the measurement of whole body radiation are required. For example a film badge suitably placed outside the protective apron could give an indication of radiation dose to the eyes.

Of the three types of monitoring instruments mentioned above the quartz fibre electrometer is the one with which immediate readings can be obtained during the working procedure. It is usual therefore for these to be worn during particular operations attended by high exposure risk such as, for example, the loading of new radiation sources into telecurie therapy units. The film badge, or the thermoluminescent capsule is suitable for measuring the accumulated dose over long periods of many days or weeks if suitable precautions are taken to allow for the background radiation exposure during the wearing period. It is normal for these personnel monitors to be changed after four weeks, though more frequent change for staff under high risk is sometimes made.

Personnel monitoring can be adopted for all staff who work in or about a radiotherapy or a radiological department of any kind. The monitors should be available to be worn whenever staff may require assurance that the environment is adequately protected. However, personnel monitoring is obligatory for those workers named as 'designated persons' (see p. 121). These are persons whose work involves exposure to such an extent that the resulting annual doses might exceed three-tenths of the annual maximum permissible doses shown in Table 11.1. In practice all radiographers and medical staff of a radiotherapy department are invariably included in the list of designated persons and therefore are monitored continuously, even though the records of most departments show a level of exposure well below the three-tenths maximum permissible dose limit.

Personnel monitoring for internal radiation hazards aimed at estimating the radiation dose received by staff due to accidental intake of unsealed radioactive substances is more difficult than that for external radiation. The body content of gamma emitting isotopes can be measured in a whole body counter (see p. 104). Information can also be deduced, especially for beta emitting materials, by measuring the excretion of radioactive materials particularly in the urine. A common example of biological monitoring is the important checking of the accumulation of small quantities of iodine-131 in the thyroid gland in workers involved in handling this isotope. This can be done with great sensitivity with a scintillation crystal close to the neck and estimates of a gland content of radio-iodine well below the permissible body burden can be made easily.

Monitoring records

An important part of the arrangements for exposure monitoring is the keeping of *records*. Monitoring results of the environment in radiation beam departments and in departments using either sealed or unsealed radioactive sources must be kept so that reference can always be made to possible risks or changes in the protective arrangements.

For personnel monitoring a separate *radiation record* has to be kept for each designated person and the doses recorded must be added up each quarter and each year to ensure that they are within the limits set out in Table 11.1. These records are available therefore to show the complete radiation exposure history of a designated person. A particular type of record is used to pass on information to a new employer if a designated person moves from one post to another. This is the *transfer record*. On moving to another post a radiation worker would take with him such a record which gives details of his previous exposure to ionising radiation. Alternatively he should arrange for the transfer record to be sent to his new employer. Radiation records and transfer records are normally kept for a long period after the last entry since they may be of value in the study of the long term effects of low radiation exposures to individuals.

The radiation dose record is of great assistance to a Supervisory Medical Officer in the event of any person receiving a dose in excess of the maximum permissible doses of Table 11.1. The Supervisory Medical Officer must decide whether to carry out a special medical examination or to arrange remedial treatment and he must also consider if any rearrangement of the duties of the exposed person is required. In this case if the high exposure recorded, together with the total previous exposure, exceeds for the whole body a dose greater than that calculated from the formula $D = 5(N-18)$ rems, where N is the person's age in years, the Supervisory Medical Officer might advocate that the person concerned should cease radiation work for a time or indefinitely or be transferred to work which involves less risk than that normally allowed to 'designated persons'.

It should be noted that an apparent high dose recorded on a single film exposed for say, a two week or a four week monitoring period, though indicating a dose accumulation rate higher than 100 millirems per week may still indicate a dose rate per week less than that deduced from the quarterly permissible dose given in Table 11.1. Action in this case may only consist of enquiry concerning the reason for the high exposure recorded on the film and an attempt to eliminate the cause. The radiation dose record gives an immediate clue as to whether this is an isolated event or a continuing problem. In the latter case more vigorous remedial action would be required which may mean some improvement in the shielding arrangements or some modification to the working methods of the person involved.

Radiation monitoring equipment

Measurements of exposure rates around departments can be made by portable ionisation chamber monitors. Ionisation chamber instruments have been described in Chapter 5. For environmental monitoring exposure rates of the order of 1 mR per hour up to 100 mR per hour have to be measured. For this purpose an instrument with a large volume ionisation chamber is required. Portable instruments with ionisation chamber volumes of at least 500 ml with battery driven amplifiers are available that are fairly robust and can be used for measuring stray radiation around X-ray sets or radioactive sources, or leakage radiation through protective barriers.

Instruments involving Geiger counters or scintillation detectors are more sensitive than ionisation chambers and use detectors of smaller size. They do however suffer from the disadvantage that for measurement of X- or gamma radiation their sensitivity is more dependent on the radiation quality than ionisation chamber instruments. It is therefore more difficult to rely on an accurate reading of the exposure rate. These instruments however are ideal for checking contamination in an isotope laboratory. Portable instruments are available in which a number of different detector probes may be used. Some of these have thin windows suitable for the detection of alpha or beta radiation and these can be used for measurement of radiation intensity and for detection of areas of contamination on benches, floors or clothing.

Geiger counter instruments are also valuable for two other types of monitoring needs. They are available as mains or battery driven alarm monitors for installation at fixed points in laboratories or wards so that a visual or audible alarm is given if the radiation level reaches a predetermined value. In this role they can indicate an unsatisfactory radiation level in an isotope laboratory due to an accumulation of small active sources in the working space. They can also be installed to give an alarm if a gamma emitting source such as a radium or caesium needle is passing a fixed point such as the entrance to a ward and in this way can help to prevent loss of sealed sources in waste bins or in the clothing of patients. A second type of Geiger counter monitor, battery driven, is available as a personnel dose monitor to be worn in the pocket of a worker to indicate when the radiation level to which the worker is exposed rises above a predetermined value.

The film badge personnel monitor

The use of photographic film for measurement of radiation has been outlined in Chapter 5. It is easy to obtain a rough extimate of dose during normal duties by wearing a small dental X-ray film in a suitable container and measuring the blackening on the film produced by development under standard conditions. A just appreciable blackening on a film exposed to 200 keV radiation for example would correspond to a dose of about 10 millirads. For a given radiation quality the film density (see p. 77) is roughly proportional to the exposure if the exposure is small, as illustrated in Figure 11.3. This figure shows the density–exposure relationship for a typical X-ray film exposed to radium gamma rays.

Because of the variation in radiation sensitivity of films with radiation quality much smaller absorbed doses would needed to produce the same film densities for 100 keV radiation. Instead of the 800 millirads required for example to produce a density of 1 with gamma rays only about 25 millirads would be required at 100 keV.

In addition to the variability of density produced by developing procedures and the variability of film sensitivities the quality dependence of the film sensitivity makes it useless for accurate measurements of radiation exposure. The difficulty arises because any radiation field is a mixture of primary and secondary radiation and the overall quality varies from place to place in the medium exposed to the radiation. In spite of this disadvantage the photographic film is extremely useful for personnel monitoring where great accuracy is not required. In this work the photographic film has the advantage of cheapness, relative ease of processing, the production of a permanent record of the exposure, and the possibility of the film recording evidence about the radiation exposure which can help to elucidate the causes of unusual or excessive exposure. The sensitivity variation with radiation quality can be used as a method of checking the radiation quality to which the person has been exposed.

In practice photographic films used for personnel monitoring are almost invariably used in a holder or cassette fitted with a variety of metal filters designed to enable a reasonable estimate of dose to be made in spite of the difficulties due to the variation in radiation quality to which it may be exposed. The films used are specially prepared radiation monitoring films designed to match the filter system of the cassette.

The possibility of correcting for the variable sensitivity with quality is illustrated in Figure 11.4. Curve A of Figure 11.4 shows, for a typical X-ray film the variation of sensitivity with radiation quality. The sensitivity is given relative to that of the film exposed to radium

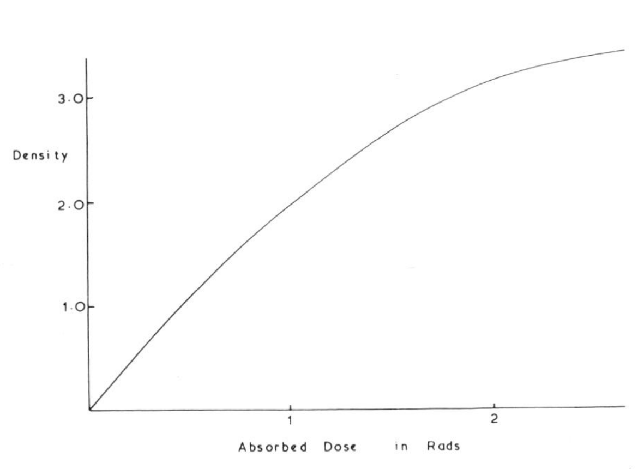

Fig. 11.3 Relation between density and absorbed dose for a typical monitoring film exposed to radium gamma rays.

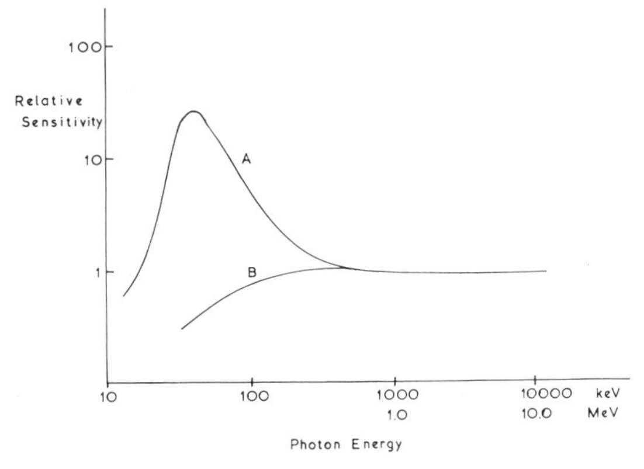

Fig. 11.4 Effect of filter on radiation sensitivity of a photographic film
A. Unfiltered
B. Film exposed through 0.5 mm Pb.

gamma rays and the curve illustrates that for X-rays of low photon energy the dose required to produce a given density on the film may be 25 times smaller than that required from radium gamma rays. Curve B illustrates the effect of exposing the film through a 0·5 mm thick layer of lead. In the soft quality range the radiation is highly attenuated and very little arrives at the film so that only small blackening is produced. For gamma ray qualities however the effect on the blackening is almost negligible since the attenuation in this thickness of lead is very small. If, therefore, one half of a film were covered in a cassette with 0·5 mm thick lead foil and the other half left unfiltered, the blackening produced under the lead covered end would be almost wholly due to hard X or gamma rays and that under the unfiltered end would be almost wholly due to soft X-rays. A comparison of the blackening at the two ends could be used not only to give an estimate of the effective quality of the radiation to which the film has been exposed but also to give an estimate of the exposure independent of radiation quality. The possibility also is suggested by Figure 11.4 of using a filter to make the sensitivity almost constant over a wide range of radiation quality.

The U.K. film holder

The film badge used for personnel radiation monitoring in the U.K. is an example of a cassette in which several filters are incorporated so that a reasonably accurate estimate (within ± 20%) can be made of the dose at the surface of the body for radiation exposures involving mixtures both of the type and of the energy of the radiation. It also gives some information about the radiation quality to which it has been exposed and the relative contribution of each type of radiation.

The badge is illustrated in Figure 11.5. The holder is a moulded polypropylene case into which is fitted a Kodak Radiation Monitoring film. The blackening under the different filter areas is a function of the quality of radiation to which the badge has been exposed. The badge has an open window (4) in the centre through which the identification number stamped on the wrapping of the film can be seen. When stamping this number on the wrapping the pressure effect on the film enables the number to be read from the film after processing.

There are two plastic filters, a thin one (5) of 50 mg per cm^2 and a thick one (6) of 300 mg per cm^2. The thin plastic filter attenuates beta rays and a comparison of the blackening under this filter with that under the open window enables beta ray doses to be estimated. The filter (3) composed of 0·028″ tin plus 0·012″ lead is chosen to give a response which is nearly independent of radiation quality from 75 keV to 2 MeV. Filter (1) made of an alloy of aluminium and copper (Dural) 0·040″ thick is chosen so that the ratio of the blackening under the dural and under the thick plastic filter gives an estimate of the dose for photon energies from 15 keV to 85 keV. Finally the filter (2) of 0·028″ Cadmium plus 0·012″ lead is used to estimate doses arising from exposure to neutrons since neutrons produce gamma radiation when they interact with the cadmium. The filter (7) is a small strip of lead to reduce errors due to leakage of radiation around the main filters and the filter (8) is a small piece of indium of use in monitoring accidental exposure in nuclear reactor plants.

The films to be assessed are developed with a set of suitable calibration films previously exposed to known doses of radium gamma rays together with unexposed films to estimate background and fog effects. A number of empirical methods are adopted by various laboratories to deduce the exposure from the densities measured under the various filters. For handling a large number of films a computer programme can be used to assess the dose on the basis of the recorded densities on the film.

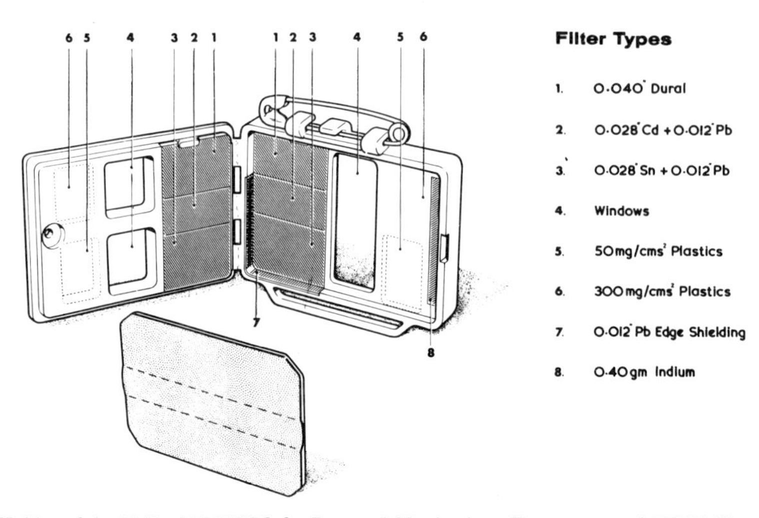

Fig. 11.5 The Film Holder of the U.K. AERE/RPS for Personal Monitoring. (By courtesy of AERE Harwell).

One valuable property of the Kodak R.M. film used in these cassettes is worth noting. The cellulose-acetate base of the film is coated on both sides with the sensitive photographic emulsion, one side being coated with a thick sensitive layer and the other with a thin, less sensitive layer. For normal use the density through both layers is measured and doses can be estimated which are only a few per cent of the recommended monthly maximum permissible dose for X-rays. For high exposures the sensitive emulsion becomes too black for measurement but it can then be stripped off and measurements made on the thin insensitive emulsion. Very high exposures up to 100 rems can thus be estimated if necessary.

Thermoluminescent dosimeters

The measurement of exposure by the thermoluminescent effect described in Chapter 5 (see p. 46) can also be used for the monitoring of personnel doses. Lithium fluoride capsules are prepared in cassettes which can be worn on the trunk. They may contain the active material in powder form or in small rods of extruded LiF, or small discs in which the active material is incorporated in a suitable plastic. The material is sufficiently sensitive so that small discs can be used inside surgical type gloves to record doses on the fingers during procedures with small sealed sources, or with unsealed radioactive materials. The advantages of this method for personnel dose monitoring are (i) freedom from sensitivity variation with photon energy because of the low atomic number of lithium fluoride, (ii) the stability of the detector over long periods, (iii) a very large dose range covered by a single dosimeter and (iv) a re-usable dosimeter. Measurements of the thermoluminescent capsules in the form of plastic discs loaded with lithium fluoride can be adapted for automatic processing so that it could be advantageous for a large personnel hazard monitoring organisation. The National Radiological Protection Board in the U.K. are planning such an automated dosimeter system coupled with a computerised record keeping service which will include printing of dose records and Transfer Records.

Quartz fibre electrometers

The pocket type quartz fibre electrometer is an instrument in which a direct and immediate reading of the exposure can be obtained. Its advantage therefore over the earlier methods of personnel hazard monitoring is that the worker himself can note the exposure to which he has been subjected during a particular operation instead of having to wait for laboratory processing of a cassette.

In this instrument, illustrated in Figure 11.6, a metallised quartz fibre is mounted in a small ionisation chamber incorporated in an instrument which is about the size and shape of a fountain pen and which can be clipped inside a pocket or some other convenient place on the clothing. At one end is fitted a microscope incorporating a scale so that the movement of the fibre image across the scale can be observed. The fibre is in turn illuminated by daylight shining through the other end. One end cap is easily removable so that the fibre system can be charged up to about 200 volts by placing it in the socket of a suitable charging device. On exposure to radiation the charge leaks away from the fibre system and the image of the fibre moves over the scale, which can be calibrated directly to read exposure in roentgens. Full scale readings may be 100 mR or 1 R for example. These instruments are very convenient for use when unusual exposures are possible or when special exposures are planned during some special but non-routine operations.

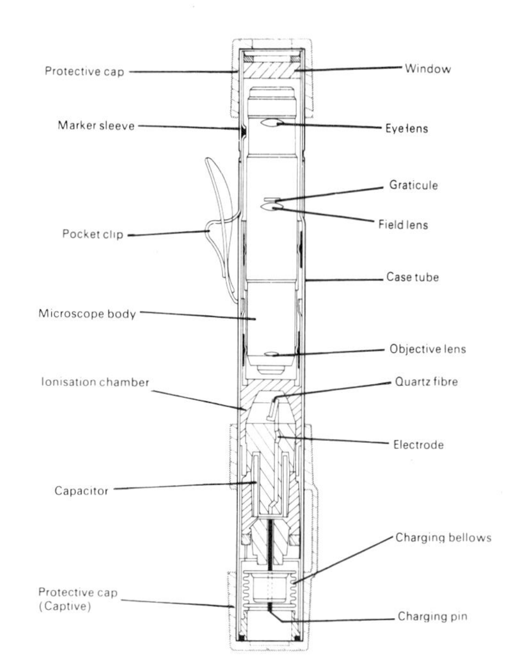

Fig. 11.6 The quartz fibre electrometer. (By courtesy of R. A. Stephen and Co. Ltd.)

Records

Reference has been made in a number of places earlier to the necessity for keeping appropriate records in order that radiation protection procedures can be adequately controlled and in order to reduce the risk of accident through a change of working conditions which are not known to the operators. Discipline in this matter is essential and good record keeping is important. The details of records which are required by the U.K. Code of Practice can best be understood by consulting the Code itself. The records required can be grouped into a few categories. First there are those dealing with the assurance of safety of personnel, such as medical records, radiation dose and transfer records and records of any unusual exposure or contamination. Secondly, there are records dealing with checks on the safety status of a department, such as the monitoring of working areas and notices about modifications to protective shields or equipment. Lastly a number of records are required dealing with the care and checking of tools, equipment and instruments whose malfunction could lead to a radiation hazard. These include notes about calibration of dosimeters both for recording therapy doses and for protection measurements, the records of movements and use of radioactive sources and the disposal of radioactive waste.

Properly organised record keeping is not an onerous unnecessary chore but a disciplined activity contributing to a high level of radiation safety in the radiotherapy department.

12. Recent developments in the technology of radiotherapy

Introduction

Some recent developments in the physics of radiotherapy have already been mentioned in previous chapters. Although radiotherapy goes back over 75 years and its physics has become well established, the last 25 years have seen advances in linear accelerators and simulators, gamma cameras and short-lived isotopes and, of course, computers. All these have developed out of existing techniques and immediately found a place in the diagnosis or treatment of malignant disease.

Other recent developments have not evolved in the same way. *Neutron therapy* is a different kind of radiation which may revolutionise the treatment of certain lesions. The first use of neutrons was not successful, but recent trials have been more successful and there are good radiobiological reasons for further investigations. The apparatus is expensive but technically possible and being developed in a few centres. *Computerised axial tomography (CAT)* was launched a few years ago as a break-through in the study of nervous diseases in that it provided detailed radiological information of the contents of the skull. The technology has now developed so that any transverse section of the body can be displayed in great detail, thus providing data for radiation treatment planning which has never been available before. It is unlikely that CAT equipment will be available in many radiotherapy departments, but where the facility exists in the hospital, it should be made available to the radiotherapists. *Echography* or *ultrasonography* has developed essentially as a means of diagnosis without the hazards associated with X-rays. The ability of ultrasound to measure chest wall thickness, to differentiate between cyst and solid growth and to localise shape and depth of certain organs has applications in the radiotherapy department. These three developments are briefly outlined in this chapter.

Neutron Therapy

The results of radiobiology experiments and the experience of practical radiotherapy have shown that cells starved of an oxygen supply (anoxic cells) are very resistant to conventional forms of ionising radiation when compared with similar cells with a plentiful supply of oxygen. Tumours whose cells are starved of oxygen, therefore, are much more resistant than those containing well-oxygenated cells (hence the use of hyperbaric oxygen for therapy, Ch. 17). However, the ionisation produced by neutrons consists of heavy densely ionising particles (p. 8) for which there is no oxygen enhancement effect. Neutron irradiation, therefore, reduces the relative resistance of anoxic cells compared to those well supplied with oxygen and improves the treatment of tumours where anoxic cells predominate. Clinical sources of neutrons include implanted californium-252 needles and tubes and the teletherapy beams from cyclotrons and D-T generators. In this section, we shall outline some of the properties of neutrons and the neutron sources being investigated at the present time.

The properties of neutrons

The importance of the neutron in the nucleus of an atom has been seen in the context of radioactive isotopes and their decay in Chapter 2. The neutron has the mass of a proton but no electronic charge. It is this lack of charge which makes the neutron both suitable for therapy and difficult to detect because, being neutral, it does not interact with the electrostatic field between the nucleus and the orbiting electrons of an atom. Any nuclear reaction in which an excess of neutrons is produced may be regarded as a neutron source. This may be by the deliberate bombardment of a light element with alpha particles, for example

$$^{9}_{4}\mathrm{Be} \; + \; ^{4}_{2}\mathrm{He} \; \rightarrow \; - \, ^{12}_{6}\mathrm{C} \; + \; ^{1}_{0}\mathrm{n}$$

or with deuterons, as in the D-T generator, for example

$$^{2}_{1}\mathrm{H} \; + \; ^{3}_{1}\mathrm{H} \; \rightarrow \; ^{4}_{2}\mathrm{He} \; + \; ^{1}_{0}\mathrm{n}$$

or the spontaneous fission of certain transuranic elements such as californium-252 (p. 8).

The simplest neutron interaction is the collision with a nucleus, which in turn is set in motion. This positively charged heavy ion loses its energy by producing intense ionisation, while the neutron continues on its path with reduced energy. The lighter the nucleus, the greater the energy transferred from the neutron to the nucleus. Hydrogen, being the lightest element, is therefore, the most efficient absorber of neutrons. Hydrogen compounds, especially paraffin wax, are invariably used as protective barriers against neutrons. Steel is also useful. The human body contains considerable quantities (10 per cent) of hydrogen and, therefore, readily absorbs neutrons.

Other light elements readily capture neutrons and produce alpha particles. These alpha particles in turn produce intense ionisation which is readily detectable. For example, boron

$$^{10}_{5}B + ^{1}_{0}n \rightarrow ^{7}_{3}Li + ^{4}_{2}He$$

This emission of an alpha particle forms the basis of several neutron detectors, boron trifluoride gas-filled ionisation chambers, boric oxide/zinc sulphide scintillation crystals and boron/silver bromide emulsions for photographic plates. The ionisation chamber is generally used for dose rate measurements in therapy. Many developments in the design of ionisation chambers now enable chambers to be made tissue equivalent in terms of the hydrogen content by the careful selection of wall material and enclosed gas.

Radioactive isotopes may be produced in any material which is irradiated with neutrons, but fortunately, the quantities and half-lives of the isotopes produced in the radiotherapy department are relatively small. The additional dose given to the patient as a result of neutron activation within the patient is insignificant. On the other hand and despite special care in the choice of materials in the manufacture of neutron therapy beam installations, the treatment couch and the treatment room walls, etc., may constitute a radiation hazard to staff unless strict controls are observed. For example, the use of the neutron beam will be restricted so as to limit the production of radionuclides, thereby keeping the background radiation down to a safe level. The beam applicators become particularly active close to the source and a means is usually provided whereby they can be changed by remote control and stored in a protected enclosure.

Californium-252

Californium-252 sources are being used in a few centres to evaluate their use as an alternative to radium for low dose rate interstitial and intracavitary treatments. Californium is a transuranic element which decays by both alpha emission and spontaneous fission, producing both gamma rays and neutrons, with an effective half life of 2.65 years. The fission spectrum neutrons have an energy of about 2 MeV and the complex gamma ray spectrum falls between that of caesium-137 and that of cobalt-60. The active content of the sources is some one thousand times smaller than for radium (i.e. the ^{252}Cf sources contain a few micrograms of the element). The sources are otherwise very similar to those described in Chapter 8.

The mixed radiation—neutrons and gamma rays—gives rise to problems in dosimetry and in general the measurement of dose is assessed separately. The actual ratio of neutron dose to gamma ray dose varies with the design of the source and the distance from the source. The radiobiological advantage of using neutrons is the increased radioresistance of normal cells relative to the anoxic turnover cells. This is measured in terms of the *oxygen enhancement ratio* (OER)) which is defined as the ratio of doses in the absence and in the presence of oxygen required to produce equal biological effects (p.182). For radium gamma rays the OER lies between 2.0 and 2.5, while for ^{252}Cf the OER value varies between 1.3 and 1.9 depending on dose rate. At low dose rates (say 10 rad/h) the OER is 1.3 and characteristic of a pure neutron radiation, while at higher dose rates, the OER increases to a value characteristic of gamma-radiation. One concludes, therefore, that the value of ^{252}Cf is only realised in the low dose rate situation and that there is little value in using ^{252}Cf sources in high dose rate after-loading techniques. The radiobiological effectiveness (RBE)—the ratio of the dose of a standard reference radiation (radium gamma-rays) to the dose of the test radiation required to produce the same biological effect—for californium-252 neutrons is approximately 7 at these low dose rates and falls with increasing dose rate. This suggests that the biological effect produced by 6000 rad of radium gamma radiation in seven days would be achieved by approximately 900 rad of ^{252}Cf neutrons over the same period of time.

So far as radiation protection is concerned, after-loading techniques are recommended for the insertion of the sources, even for low dose rate treatments. Massive bedside shields are required to protect the nursing staff as they have to include both lead to attentuate the gamma rays and hydrogenous material to absorb the neutrons.

The possibility of using californium-252 as a source for neutron teletherapy has been examined and discarded on the grounds of cost and size of the protection shield required.

Cyclotron beams

Much of the recent development work in the field of neutron therapy has been done on the Medical Research Council Cyclotron at Hammersmith, London. Although the cyclotron is not considered to be the ideal source for neutron beam therapy because of its size and the poor neutron flux (giving a low dose rate), it has been used in clinical trials on selected patients. The student is referred to other texts for details of these. For our present purposes, it is worth noting that in the cyclotron deuterons are accelerated to 15 MeV and used to bombard a beryllium target. The mean energy of the resulting neutron beam is 6 MeV. The beam is collimated using borated wood applicators—the range of field sizes available is, therefore, limited. At a source skin distance of 120 cm, the dose rate is approximately 50 rad/min and the percentage depth dose curve lies between those of 250 kV X-rays and Cobalt-60 gamma rays (Fig. 12.1). With the neutron beam there is a gamma component which contributes a small percentage to the total dose delivered.

The D-T Generator

The D-T generator appears to be the more practicable

source of neutrons for beam therapy. It is essentially a sealed tube—about the size of an orthovoltage X-ray tube—in which deuterons (deuterium nuclei) are accelerated to about 200 kV to bombard tritium nuclei. Hence the abbreviation D-T. Neutrons are produced with an energy of 14 MeV. One of the problems with this system is that the tritium target has only a limited life and so far efforts have not produced what is considered to be a clinically acceptable tube. By clinically acceptable we mean a tube which will run for several hundred hours producing an acceptable dose rate with an emission of 10^{12} neutrons per second.

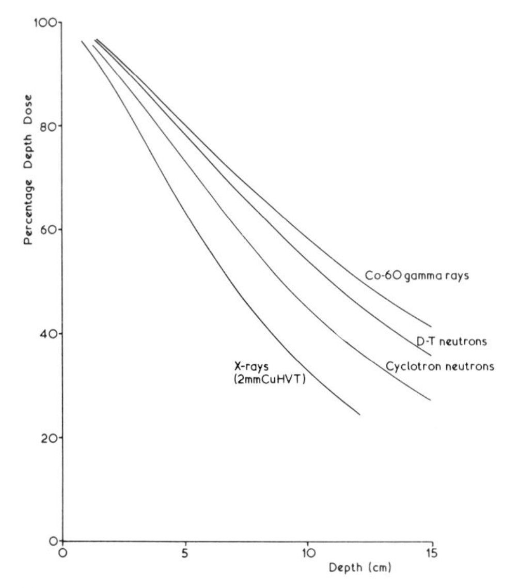

Fig. 12.1 Neutron depth doses compared with cobalt-60 gamma rays and 250 kV X-rays.

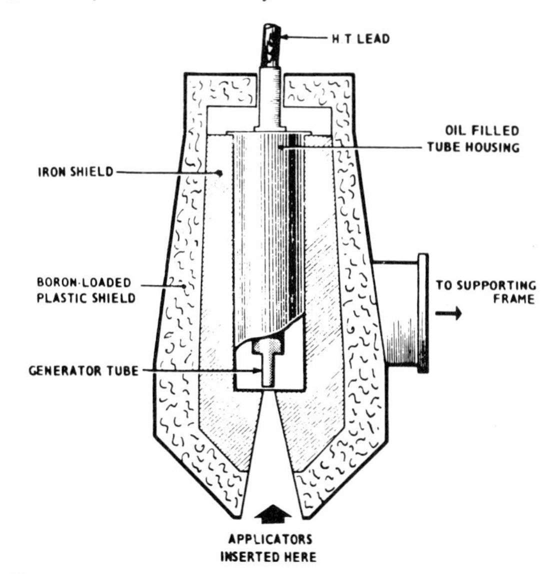

Fig. 12.2 A D-T Generator. (Meredith and Massey *Fundamental Physics of Radiology*. Bristol: Wright).

The tube has to be housed in protective shielding in which both the neutrons will be slowed down and the gamma rays will be attenuated. This is illustrated in Figure 12.2. The shield is essentially iron surrounded in boron loaded polythene. The collimating applicators are of a similar construction but using steel instead of iron. The collimators are some 50 cm long.

The D-T tube produces neutrons of a higher energy than the cyclotron and their penetration is correspondingly deeper. The depth dose is still less than that for Cobalt-60 gamma rays however.

Medical Ultrasonics

Medical ultrasonics has been chiefly developed as a diagnostic tool for obstetrics and gynaecology because there are no known hazards in its use, whereas radiology is known to be particularly hazardous in this field of medicine. The repercussions of this development are such that ultrasound is being increasingly used in the radiotherapy department as a means of locating and identifying internal structures particularly those poorly differentiated by X-rays. Before outlining these uses of ultrasound, it is useful to explain briefly the properties of these sound waves.

Properties of sound waves

X-rays and gamma rays, in common with other electromagnetic waves, are said to be transverse because their oscillation is at right angles to their direction of propagation. Sound waves are different. The vibrations in a sound wave are in the same direction as its propagation, it is a *pressure wave*. Unlike electromagnetic waves, sound waves cannot travel through a vacuum as they need the medium to carry the vibrations. Longitudinal or pressure waves, like transverse waves, can be defined in terms of frequency and wavelength, the velocity of propagation being the product of wavelength and frequency.

The human ear responds to frequencies of sound in the range from about 16 Hz to 16 kHz (the hertz, Hz, is the unit of frequency, 1 Hz = 1 cycle per second) and in air sound travels at a velocity of approximately 340 ms^{-1}. Ultrasound is a higher frequency sound. In diagnostic ultrasound, the range of frequencies used is from 500 kHz to 20 MHz, although much of the work described in this chapter has been done using frequencies of approximately 2MHz. Since the propagation of a sound wave requires the medium through which it passes to vibrate, the velocity of the wave is dependent on that medium. In the soft tissues of the body the velocity varies from 1450 ms^{-1} (in fat) to 1585 ms^{-1} (in liver) with an overall average of 1540 ms^{-1} which is close to the value for water (1500 ms^{-1}). In bone it is 4080 ms^{-1}. The amplitude of the oscillation or the displacement of the tissues propagating the sound is measured in nanometres (10^{-9}m) and is therefore very small.

Production and detection of sound waves

Waves of this frequency can be produced in certain crystals (such as quartz) which display the property known as *piezo-electricity*. This is the property of a crystal which contracts and expands in response to an applied voltage pulse and in which a voltage can be generated in response to an applied pressure. These crystals are called *transducers* for they convert electrical energy into mechanical energy and vice versa (Fig. 12.3). A pulse of ultrasound is produced by applying a pulse of about 600 V to the crystal, the crystal then vibrates at its resonant frequency emitting sound waves of that same frequency. The different frequencies are obtained using crystals of different thickness.

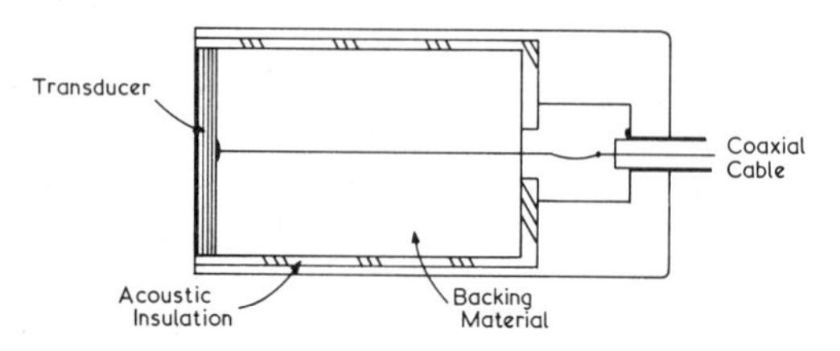

Fig. 12.3 An Ultrasound Transducer.

Sound can be propagated through a perfectly elastic medium without any loss in energy, but in tissue energy is taken from the beam to set the particles of tissue into motion and ultimately lost to the tissue in the form of heat. The intensity of the beam, measured in milliwatts per square metre (mWm^{-2}) is reduced exponentially with depth of penetration in the tissue—greater penetration being achieved with lower frequencies. Therefore, pelvic examinations require a lower frequency (2 MHz) whereas superficial structures are examined using a higher frequency (say 10 MHz) and in general terms the highest frequency compatible with the depth produces the greatest detail.

If the same transducer can be used to both transmit the pulse of ultrasound and to detect its echo, then there must be some reflection of the sound waves back along the incident beam. Reflections occur at the interface between two dissimilar tissues, the greater the dissimilarity the larger the echo. A bone-soft tissue interface reflects a large echo, while interfaces between different soft tissues reflect very small echoes. Fortunately, providing about 1 per cent of the incident beam energy is reflected, the echo can be detected. An air-tissue interface reflects nearly all the incident energy and it is for this reason that the transducer must make good contact with the patient's skin. This contact can only be ensured through some *coupling medium* such as olive oil. It also means that ultrasound examinations cannot be carried out where pockets of air are likely to exist within the patient.

A-scanning and B-scanning

The simplest ultrasonic examination is carried out using an *A-scope* and is known as an *A-scan*. The transducer is positioned in contact with the skin using an appropriate coupling medium. The pulse of ultrasound is transmitted and the echo is received back some time interval later. From the velocity and the time interval the depth of the interface providing the echo can be determined. The A-scan is displayed on an oscilloscope screen and consists of a series of pulses. On the left of the screen is a vertical pulse corresponding to the transmission of the ultrasound signal, while the echo pulses follow some time later, the time delay being measured as a horizontal deflection (Fig. 12.4). The A-scan provides information as a one-dimensional picture, i.e., the localisation of tissue interfaces along the line of the beam.

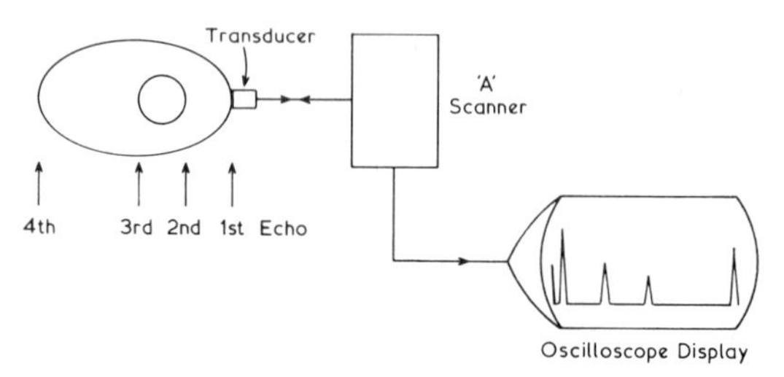

Fig. 12.4 An A-scan Display.

A *B-scan* is a two dimensional picture in which the position of the echo is displayed on the oscilloscope screen. On the more recent scanners a *grey-scale* facility is available whereby the intensity of the echo signal is represented by the brightness of the display adding a third dimension to the display. The B-scope is, therefore, more complex in that apart from the transducer, the position and direction of the ultrasound beam has to be recorded and displayed. Complex it may be, but the advantage of being able to map out the echoes through a plane section within the patient is a great advantage over the A-scope system. Most of the applications outlined below use the B-scope, with or without the grey scale facility.

Applications of ultrasound in radiotherapy

Chest wall thickness. One of the early applications of ultrasound was to measure the thickness of the chest wall, as a means of assessing the electron beam energy required to treat the tissues without causing radiation damage to the underlying lung. The tissue-air interface provided a large echo signal. In the first instance, spot measurements were made using the A-scope, but because there are quite large variations in the wall thickness and the electron beam would be incident obliquely over a large part of the wall, the B-scan has been found to be more valuable. Such examinations are reported to take about 15 minutes and cause no distress to the patient.

Pelvic examinations. The early ultrasonic work was done in obstetrics and gynaecology because of the importance of being able to examine the foetus without the hazards of ionising radiations. Even to the radiotherapist, the major role of ultrasound is the examination of the pelvis and upper abdomen. Ultrasound can differentiate the different tissues more readily than conventional radiology and the clinician experienced in ultrasonic techniques can identify the liver, gall bladder, spleen, kidneys and the aorta. In none of the literature is it suggested that ultrasound will replace radiology but there are clear reasons for the two techniques to complement each other.

One particular examination which is being used quite extensively is the use of ultrasound to determine the extent of extrinsic bladder tumours. The scan not only gives the extent of the tumour, but the position of the rectum and the degree of bladder emptying.

Radiation treatment planning. The ability to identify and differentiate between the organs of the abdomen and to locate their position is of great interest in radiation treatment planning, not least that this information can be presented in the transverse plane in which the treatment is planned. The B-scan on the oscilloscope screen is photographed and then enlarged to life size using a suitable projector. It can be positioned within a contour obtained by one of the techniques outlined in Chapter 6, or, better still, in the contour obtained using ultrasound. To obtain the contour requires operating the B-scope under slightly different conditions than those used for the internal anatomy, but the adjustment is trivial and does not prolong the examination. Taking this, as it were, a stage further, the B-scope can be used to input data to a computer where the ultrasound scan can be processed and/or simply displayed on the visual display unit. The clinician can then interpret and identify the relevant data for the treatment planning program and see the isodose distribution displayed together with the B-scan.

Two other applications are worthy of note. Ultrasound can differentiate between solid and cystic growths. It is now suggested that using the grey scale facility of a B-scanner, very small growths of only a few millimetres diameter can be detected. If such a technique can be developed whereby tumours can be detected in such early stages of development, we shall see radical changes both in the techniques and the statistics of radiotherapy. Secondly, using serial B-scans and the computer the volume of organs and tumours can be determined. Repeated volume estimates can demonstrate the rate of growth of a tumour, but more important the rate of regression during the course of treatment.

Summary

Ultrasound has, therefore, many possibilities not least in the field of radiotherapy. It is a technique which appears to present no hazard to the patient and in practice it is acceptable to the patient. There are advantages to be gained by being able to input the data to a computer either to assist the radiotherapist in his interpretation of the scan or to assist the physicist in the planning of the treatment.

Computerised transverse axial tomography

This technique provides radiological information about the contents of the skull or of the trunk in a manner quite different from normal radiological procedures. The information is presented as an image of a transverse section of the skull or trunk. The information from which the image is derived is obtained by measuring the transmitted intensity of a narrow X-ray beam passing through the transverse sections of the body in a large number of different directions and then building up a picture of the section by use of modern imaging techniques based on computer analysis. At present the transverse axial scanning of the brain is well established. Scanning of the trunk poses greater technical problems and is of more recent development. The principles of the method will be described briefly therefore, with particular reference to the transverse scanning of the brain.

Suppose a narrow beam of X-rays passes through the head and is detected by a sensitive crystal detector at D_1 on the emergent side as shown in Figure 12.5 (Beam A_1–D_1).

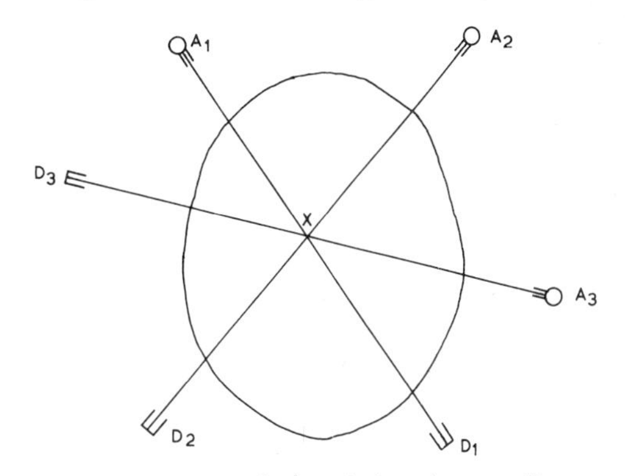

Fig. 12.5 Measurement of transmission of narrow X-ray beam.

The transmitted intensity can, in principle, be interpreted as the incident intensity attenuated by a whole series of elements of tissue along the path of the beam, each element of tissue having its own characteristic total attenuation coefficient. If now the same beam is directed through the section in a different direction A_2–D_2, the measured transmission at D_2 will be the result of the attenuation in the tissue elements along the line A_2–D_2. The tissue element X at the point of intersection of the two beams will contribute to the attenuation of both beams and the intensity readings at both D_1 and D_2 will contain information about the attenuation of the element at X. It seems possible therefore in principle, to extract from the observed transmitted readings at D_1 and D_2 some information about the attenuation coefficient for the element X. In practice, this information is completely confused by the attenuation arising from all the other tissue elements along the lines A_1–D_1 and A_2–D_2. However, if the X-ray beam is sent through the section of the skull in a large number of other directions such as A_3–D_3 etc., each passing through the element X, it is possible by using the analytical facilities of a modern digital computer to solve the problem of calculating the attenuation coefficient of the single element at the point of intersection. In practice too, by using other sets of beams intersecting at different points the attenuation coefficients of other tissue elements can be calculated and finally a complete image of the section can be constructed in which the signals at each element of the section represent the attenuation coefficient of the corresponding element in the brain. Since these coefficients are determined by the density and chemical composition of the tissues, a sectional representation of the brain can be presented in which these very small differences in the tissues are distinguished.

The practical way in which this total sectional picture is achieved is shown in Figure 12.6. The X-ray tube A produces a narrow beam at approximately 1 cm diameter which is detected by the collimated crystal system D. The transmitter-detector system is a rigid structure and is scanned across the head keeping the beam parallel to itself. In one passage the detector D records the transmitted intensity in the form of 160 transmission readings. The whole transmitter-detector system is then rotated through a 1 degree angle and the scan repeated, then through further 1 degree intervals through 180°, the single scan being repeated for each angular position. During this process the detector, therefore, records some 160 × 180 (i.e. 28800) transmission readings.

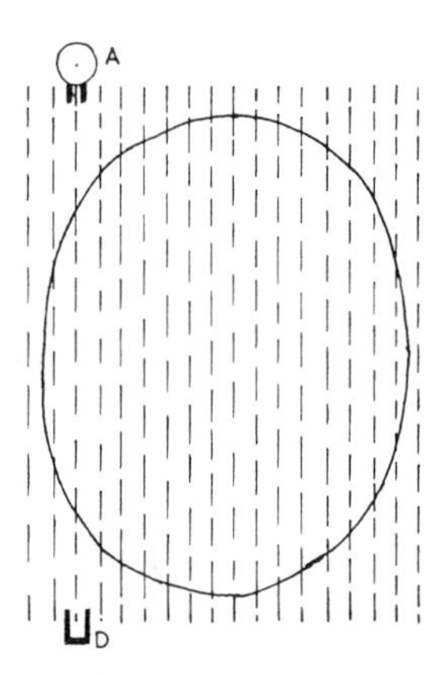

Fig. 12.6 Single scan by narrow X-ray beam.

Figure 12.7 shows a simplified illustration of the scanning process that has contributed to this enormous number of readings.

The readings contain the information about the attenuation coefficients of the tissue elements in the transverse section, each of which has been crossed by 180 beams from different directions around the periphery. The readings are fed into a computer built into the equipment which is programmed to solve the large number of simultaneous equations which represent the transmitted information readings. The output of the computer is the attenuation coefficients of the tissue elements in the section. This is presented in the form of a 180 × 180 picture matrix, each element being 1·5 mm square. The output information can be printed out in tabular form on paper. Alternatively, and more commonly, the distribution of attenuation coefficients is represented by varying blackness of tone in the corresponding element of a picture built up on a long persistance cathode ray tube. A tone range in the picture can be arranged to represent the attenuation coefficients varying over the whole range from air on the one hand to bone on the other and accuracies of the order of 0.5 per cent in the attenuation coefficients can be achieved. This tone range is analogous to the ultrasound grey scale. A presentation of this kind is a quite new quantitative achievement in radiological technique.

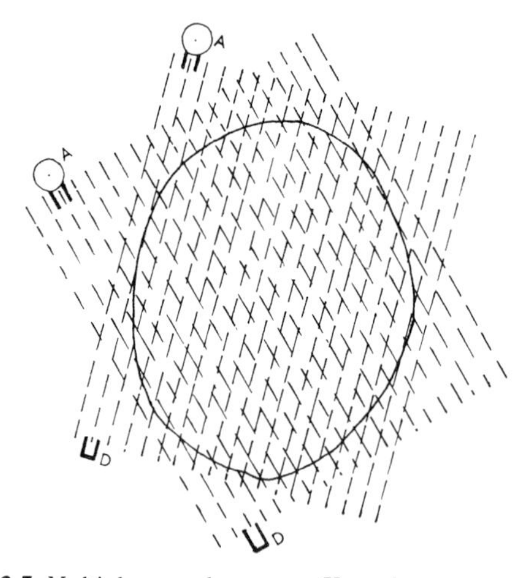

Fig. 12.7 Multiple scans by narrow X-ray beam.

To achieve a good accuracy in the computation of attenuation coefficients for the elements in the section, care must be taken to avoid movement of the patient during the scan and to reduce the total range of variations of transmission during the scan. These difficulties are solved by surrounding the head during the scan by a conical rubber cap filled with water. The scan itself takes in total about 4 minutes for the 180 linear scanning motions. The available instruments have two detector systems receiving two X-ray beams mounted on the same scanning block. This enables two sections of the brain to be produced simultaneously and the thickness of the sections can be varied from 0·8 cm to 1·3 cm long. The final display, therefore, gives an attenuation coefficient averaged over a tissue element 1·5 mm square and between 0·8 cm and 1·3 cm long. The problems of using this technique of computerised axial tomography for the trunk are considerably greater than in the head. This is because of the difficulties caused by movement of the tissues during scanning and by the inaccuracies of the computer processing produced when there is a rapid change of attenuation coefficient at the edge of the section. In the trunk it is impossible to eliminate movements due to respiration for example for more than a few seconds and a surrounding water bath cannot be used as in the head.

Rapid development of this technique is now taking place and the technical problems mentioned in the previous paragraph associated with rapid change of attenuation coefficients at the edge of the section, and with tissue movement during scanning are being solved in newer instruments. In these instruments for example an X-ray beam in the form of a fan is used with a variety of multiple detectors and the total scanning time can be reduced to a few seconds only. In this new equipment too bolus bags can be used instead of the water bag described above and this also improves the facility of obtaining accurate sections of the trunk.

It is clear that the sectional type of display achieved in this technique coupled with an ability to differentiate tissue variations very accurately and with a quite good space resolution, makes this new tool of tremendous value. With a spatial resolution of about 1·5 mm in the section it is possible to distinguish easily in the brain section, for example, necrosis, edema and tissue variations corresponding to pathological changes. The value of such a tool is enormous in the study of diseases of the brain and diagnostic information of great value is already achieved. The value in the handling of malignant disease will undoubtedly be very high partly because of the ability to differentiate between tissue types and partly because of the possibility of localisation of pathological tissues which the technique presents.

At the moment the instruments are very expensive and the demands on them from a wide range of clinical services are very high. The potential of this new tool and of the new computer based technique which it represents can, however, hardly be overestimated. It is likely to produce a complete change in radiological techniques in the next generation. The effect on isotope scanning techniques described in Chapter 10 is also likely to be very great. Brain scanning by isotopes is likely to be largely displaced by the newer techniques where both are available since it provides better spatial resolution and better differentiation of tissue types. The radiation dose given to the patient in computerised transverse axial tomography is not excessive—the beams are narrow and the total exposure for a complete skull scan is about that due to a single skull X–ray.

In summary, this new technique though of restricted availability at the moment represents a diagnostic potentiality of great importance and one which will have applications in the radiotherapy field of considerable value. For example, the display of the tissues in a transverse section provides detailed information about the anatomy in the treatment plane which has not previously been available and, in particular, it outlines inhomogeneities (e.g. lungs) and critical organs (e.g. eyes). Furthermore, the computer output provides data on the attenuation coefficient of each tissue element. These coefficients cannot be applied directly as inhomogeneity correction factors in the treatment plan as the therapy beam will be of a different photon energy but the data will enable the appropriate factors to be determined.

PART II

Radiotherapy—Oncology

13. The cancer problem

Nobody will need persuading of the importance and seriousness of cancer. It is the second most important cause of death in western countries—one in five deaths is due to cancer.

To put it another way, *a quarter of all infants born in the U.K. will develop cancer* at some stage of their lives. Eighty per cent of these will die of it, 20 per cent will be cured.

Considerable fear and prejudice, much of it born of ignorance, surrounds the public image of cancer. The lay mind tends to associate cancer with pain, incurability and misery and this makes its management even more difficult. There are, of course very good reasons for cancer's evil reputation, but the picture is not so black as it is often painted. There has been much progress in all aspects of the problem, so that we are now in a better position to cope with it than ever before.

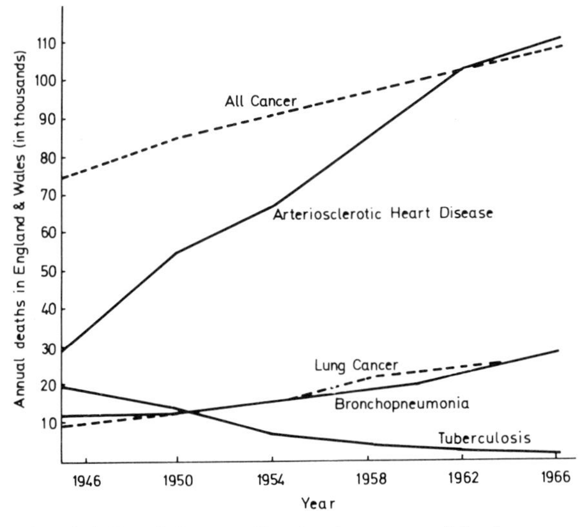

Fig. 13.1 Trends in mortality showing causes of death.

Cancer incidence

In spite of its importance as a cause of death, cancer is not a common disease. A general practitioner, with 3500 patients on his list, will sign on average 40 death certificates a year, of which eight will have cancer as the primary cause. In a large general hospital less than 5 per cent of admissions will be for cancer. Compare this with a radiotherapy ward where nearly every patient will have

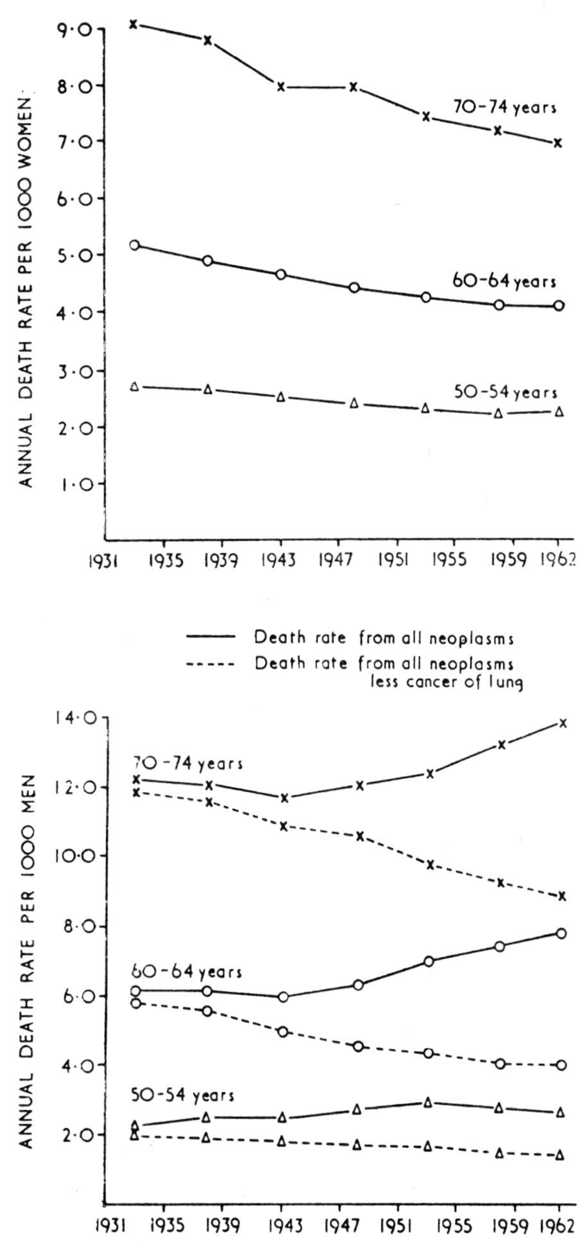

Fig. 13.2 Trends in cancer mortality (England and Wales) for women (*upper graph*) and men (*lower graph*) at different ages. Note the great difference due to lung cancer in man. (*Courtesy Prof. R. Doll*).

malignant disease. There is a relatively high incidence in geriatric and terminal care units too.

Overall about three new cases are seen annually per thousand of population, so that a city of one million will have some 3000 new cases per year. It is worth noting that rather *more than half of these will be treated by some form of radiotherapy* at some stage.

In particular age groups, especially 50–60 and 60–70, cancer mortality is not increasing. A notable exception is lung cancer. In some instances it has fallen—e.g. mouth, stomach, uterus. It is only the grand total that has increased and this is because the population is becoming bigger and living longer. Naturally this imposes a growing strain on nursing and medical services.

Figure 13.3 contrasts the current incidence of cancer at various common sites in the body. Although figures refer to this country, the picture is much the same in North America.

The management of cancer

Management means care. It also means prevention, early detection and education. When a case has been diagnosed, we have four specific modes of treatment at our disposal

1. *Surgery*
2. *Radiotherapy*
3. *Hormones*
4. *Cytotoxic drugs*

This book will not concern itself with surgical treatment, except to indicate its role in certain types of cancer.

(a) Surgery may be the treatment of choice.
(b) It may be combined with radiation, cytotoxic drugs, hormonal therapy.
(c) It may form an alternative to radiation, or merely serve to palliate.

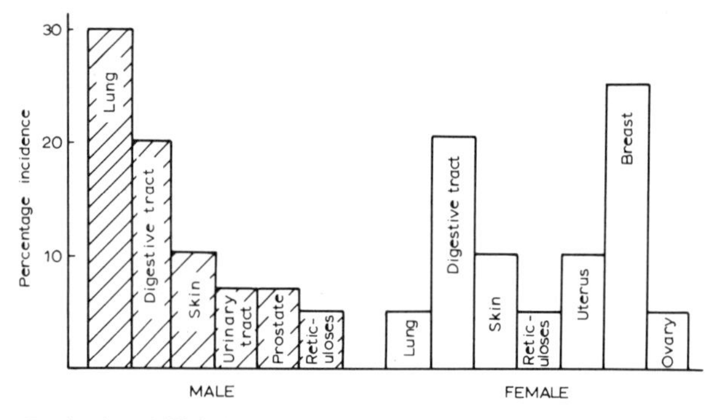

Fig. 13.3 Commonest cancers (England and Wales).

All of these except the reticuloses are *carcinomas*. Carcinomas account for 70–75 per cent of all cancers.

We shall deal with most of these groups in later chapters, but will have less to say about cancers of the gastrointestinal tract since here radiation has only a minor part to play. In this group, till recently, the commonest site was the stomach, but gastric cancer is now declining and colon cancer is now rather commoner. In fact, cancer of the large bowel is now the second commonest lethal cancer in Britain, only exceeded by lung cancer. Next comes cancer of the rectum, followed by the pancreas. In contrast, oesophageal carcinoma is rare.

The picture presented in Figure 13.3 is not a static one. The frequency of cancer and its relative incidence at different sites show considerable variations between countries, different social classes, different cultural and educational groups, different environments (urban, rural, industrial etc.) and therefore will change from decade to decade even in the same country. Compared with the early years of this century, we find more leukaemia and cancer of the lung and bladder in England today, but less of the stomach, bowel and mouth.

We shall concentrate on the other three agents, but nevertheless try to keep the total picture in focus. It is helpful and important to consider treatment as two kinds:

1. *Radical*—the attempt, heroic if necessary, to remove all the malignancy present.
2. *Palliative*—if radical treatment is thought to be impossible, the aim is to relieve symptoms.

At present, surgery and radiotherapy are the chief agents capable of radical treatment, but all four can be used for palliation.

Radiotherapy was a latecomer to the scene. It was only natural therefore and only reasonable that it should be used in the first instance for surgically hopeless cases. Its great palliative value was soon recognised—for relieving ulceration, bleeding, pain from bone secondaries etc. It was so successful, in fact, that its potential in radical treatment was overshadowed, and it became psychologically associated with inoperable, incurable cancer. Something of this reputation still lingers, but it should now be clearly recognised that *radiotherapy stands on its own feet as a curative agent comparable to surgery, and capable of giving results just as good (and often better) in properly selected cases.*

However there is sometimes a psychological advantage in the choice of surgery, evem when radiation offers a sound alternative, e.g. early cancer of the cervix. This is because radiation is closely associated in the minds of many patients with malignancy. Surgery on the other hand covers a much wider field and may be preferred when it is particularly important to conceal the truth from an apprehensive patient.

The relative value and uses of the various agents will be discussed later. In many sites, surgery is definitely better. In others radiotherapy has been found to yield superior results, and a summary account is given in Tables 17.4–17.8 pp. 184–5. In some cases it is advisable to use a combination of agents:

1. *Pre-operative radiation*
2. *Post-operative radiation*

} as described under breast cancer (Table 22.1, p. 217) and Wilms's tumour (p. 263)

In others, agents are best used in sequence:

a. *The primary growth*
b. *The secondary lymph-nodes*

For example, in seminoma of testis, the primary is first removed by orchidectomy and the secondary nodes are then irradiated. In cancer of the tongue, the primary is first treated by radiation, and secondary nodes in the neck then removed by block dissection.

Hormones are usually reserved until the resources of surgery and radiotherapy have been exhausted. Till recently the same applied to cytotoxics, but these have now acquired a predominant role in some types of cancer (malignant lymphoma, childhood tumours, leukaemia etc.) where they become the sole treatment or an essential part of it.

Prognosis in cancer: The concept of cure

We might have used the expression 'curative' instead of 'radical' treatment. 'Cure' seems simple enough. We know what we mean by cure of a simple fracture or of the common cold even though complications like sinusitis or pneumonia may follow. We mean restoration to health—more or less complete, and free from further trouble. But what do we mean by 'cure' of, say, diabetes or pernicious anaemia? We can usually hold them at bay for long periods of time, maybe till the patient dies a natural death from some other cause at a ripe old age. In other words we can usually control them well enough for the patient to lead a normal—or nearly normal—life. This often holds good for cancer, and *'control' is a better word than 'cure'*.

In many early cancer cases we can be fairly confident that we have eliminated all traces of malignancy, but we can never be sure. Local recurrence or distant metastasis* is always a possibility. A notorious example is breast cancer, where late recurrence or metastasis can occur after many years, even after many decades.

Indices of success. In measuring the success of treatment for cancer, a conventional 'yardstick' is the proportion of patients who survive for a certain number of years—usually five. *Five-year survival* figures give us valuable information about the results of different forms of treatment. But five-year survival is sometimes interpreted as 'cure' and this can be seriously misleading. In many cases, if a patient survives and is well after five years, his chances of remaining permantly free of malignancy are excellent, e.g. cancer of the lip. In other cases apparent cure can be followed by recurrent trouble at any time, as we have seen above.

The natural history of each particular cancer is clearly of paramount importance. Some grow so slowly that the patient can lead a comfortable and useful life for many years in the presence of the primary growth, even with secondaries too—e.g. some types of breast, parotid and thyroid cancers. *Ten-year* and *fiften-year* survival figures are even more valuable than five-year survivals, but it requires an elaborate follow-up organisation to provide these. Yet it is only from painstaking statistics of this kind that we can derive sound knowledge both of the natural history of cancer and the real effects of treatment.

Survival figures are essential for estimating the success of radical treatment, but are only of limited value in assessing the effects of palliative treatment. We may be unable to achieve complete eradication or permanent control of a cancer, yet treatment may lengthen life, control primary or secondary disease even for years, alleviate painful symptons and improve the quality of life generally. Radiotherapy, by and large, is more successful than surgery in this kind of therapy. It therefore provides a high proportion of the work of a radiotherapy department, and even apart from its success in the radical treatment of certain types of cancer, its value in palliation would amply justify the expense.

Clearly it is very difficult to be dogmatic about prognosis in cancer, even more to promise cure, and this makes for difficulty in discussion with the lay public. But we can at least be firm in our assurance that cancer is far from synonymous with 'an incurable disease'.

Results. If we take five-year survival as the index of success, we find the best results—i.e. an appreciable percentage of all cases surviving over five years—in cancers of the following sites:

Five-year survival rates (all cases)

Skin	75%	Uterus (cervix and body)	45%
Lip	75%	Breast	45%
Testis (seminoma)	70%	Thyroid	40%
Salivary glands	70%	Mouth	35%
Larynx (vocal cords)	50%	Bladder	35%

It is worth noting that many of these are included in the list of lesions for which radiation is the treatment of choice and that they are almost all *accessible*—i.e. they can be seen or felt in their early stages.

The worst results—less than 10 per cent alive at five years—are seen in

1. Pancreas
2. Oesophagus
3. Stomach
4. Lung
5. Leukaemia
6. Hypopharynx

Other cancers occupy intermediate positions, with survival rates in the region of 20 per cent.

In discussion of results, it should always be remembered that *cancer is not one disease but many*, and each separate site represents a different disease, a different natural history and a different problem. Generalisations therefore are of very limited value. It is probably true to say that about *one cancer in three is now 'curable'* in the five-year survival sense. But this is not much help in considering an individual cancer in a particular organ, since five-year rates can range from 2 per cent to 95 per cent. In any given case the prognosis depends on many variables. These we detail in later chapters.

At the present time the overall picture in the results of cancer treatment is not a cheerful one. Good results are obtainable in a few situations, especially for early cases of accessible growths, but in most of the common sites (e.g. lung, gastrointestinal tract, breast, bladder, ovary) the long-term results are poor. Unil the situation improves it would be wrong to be optimistic. It is improving now, albeit slowly, and will certainly improve further, perhaps much further, before the end of the century. But there is a great deal that can be done for almost all cancer patients, whether treatment is radical or palliative, and *good nursing and doctoring can make a very significant difference to the quality of life of* most of our patients. This is our best incentive and our best reward.

To help keep things in proper perspective, we may compare the prognosis in some other serious diseases (Fig. 13.4).

These disorders do not rouse the same alarm and mortal fear as cancer in the public mind, yet the outlook in some cancers, as shown above, is far more favourable. These facts are important in questions of health education.

Some important data are given for reference in Tables 13.1 and 13.2

Table 13.1 Causes of death in England and Wales—1973 (Total population = 55 000 000)

All deaths	550 000
Cardiovascular	250 000
Cancer	113 000
Respiratory	70 000
(for comparison)	
Motor accidents	6500
Suicide	3600
Special infections (tuberculosis, etc.)	3000

Table 13.2 Commonest cancers (England and Wales)

Males (%)	Site	Females (%)
30	Lung	5
	Breast	25
20	Digestive tract	20
	Uterus	10
	Ovary	5
10	Skin	10
7	Urinary tract	
7	Prostate	
5	Lymphomas	5

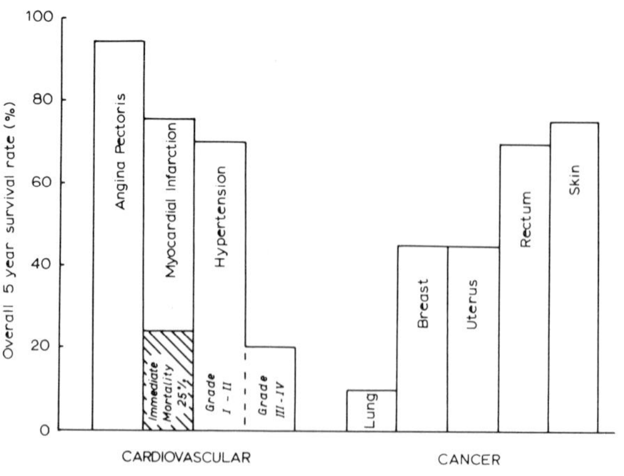

Fig. 13.4 Cardiovascular disease and cancer five-year survival rates.

14. Biological and pathological introduction

By 'radiotherapy' we mean the treatment of disease by means of radiations. The name could be applied to all forms of radiation used medically, but in practice is confined to *ionising radiation*. (Non-ionising radiations used medically include: (1) ultra-violet rays, (2) infra-red rays, (3) short-wave diathermy. These treatments are usually carried out in the physiotherapy department of a hospital.) We shall deal with (1) X-rays, (2) radioactive elements, such as radium and artificially produced isotopes, (3) subatomic particles such as electrons and neutrons.

Malignant Disease or Cancer will be our chief concern. The total of cancer victims is increasing in developed countries. This is due mainly to the increased expectation of life achieved by modern hygiene, nutrition etc., so that more and more people survive to middle and old age when cancer becomes more prevalent.

Cancer is a pathological variation of normal growth.

Growth is a fundamental property of living things, and the most remarkable feature of it is the beautiful way in which it is organised and controlled, from conception to maturity and beyond. Rates of growth vary greatly. Growth in the fetus is very rapid—more so than in any cancer—then gradually diminishes till adulthood. This holds good both for initial growth and for 'wear and tear' replacement and repair.

Growth rates vary widely between different tissues. At one extreme, nerve and muscle cells are all present at birth and if later destroyed can never be replaced. At the other extreme is bone marrow, where blood cells are in constant production. Since the red cell has a normal life span of only four months, over two million red cells must be produced every second. Besides the marrow, the fastest growing tissues are the lining epithelium*of the intestine (which is replaced every two days), the skin (epidermis*) and the male gonad (sperm cells). In fact, about 2 per cent of our body cells die and are replaced every day. Other organs have a much lower replacement rate, e.g. liver and kidney, but losses can be made good when necessary.

Control and differentiation. To develop into the various organs with widely different structure and function, cells must specialise. This process is called *differentiation*, and the more highly differentiated cells become, the less likely they are to divide.

Clearly there are delicate controlling mechanisms at work which determine the various degrees of differentiation, the size of organs and of the whole organism, the maintenance of a balance between normal destruction and regeneration. We know almost nothing about these mechanisms of control and differentiation, and this ignorance bedevils our whole insight into the problems of abnormal growth, especially cancer.

Neoplasia.* We must distinguish several varieties of abnormal growth.

1. *Hypertrophy* of organs, as in heart muscle when the heart has to work against increased resistance. Its constituent cells are enlarged, and some of their nuclei may also be enlarged and bizarre.
2. *Hyperplasia*—cells are more numerous and more tightly packed, with a high incidence of mitosis. It may precede cancer in epithelial surfaces and glands.
3. *Metaplasia**—replacement by a cell type not normally present in an organ—e.g. as a result of inflammation, transitional epithelium (see p. 238) of the urinary tract may change to squamous* and become keratinised.*
4. *Neoplasia*. Another general term often used is 'tumour', which means simply 'swelling' and was originally used for enlargements due to injury, inflammation, cancer or anything else. The word is now restricted to neoplastic swellings.

Until the true nature of neoplasia is understood, it will be impossible to offer a really satisfactory definition of a tumour. Meantime, that of R. A. Willis is as good as any—'a mass of abnormal tissue, the growth of which exceeds and is uncoordinated with that of the normal tissue from which it originates, and which persists in the same excessive manner after the stimuli which evoked the change have ceased'.

The abnormal cells give rise to a reparative response in the neighbouring tissues, in the form of connective tissue whose blood vessels supply the tumour cells with nourishment. This tissue is called the *stroma*;* it may be very scanty in a rapid growth, or very dense, fibrous and hard (scirrhous*).

Neoplasms or tumours are in two broad classes
1. *Benign or simple*
2. *Malignant or cancerous*

Benign growths are universal since they include the common birthmarks—everyone has at least a few, e.g.

moles (pigmented or not), simple melanomas,* papillomatous* warts, fibro-fatty tags, infantile angiomas.* They are harmless and usually of no more than cosmetic importance. Very rarely they may become malignant, e.g. a simple melanoma may become a malignant melanoma.

Their growth is strictly localised and never widespread, but they can be dangerous in certain situations, e.g. a simple meningioma may exert fatal pressure on the brain, whereas a similar kind of lesion* on a limb would be merely inconvenient.

Malignant growths are the *cancers*. Cancer is Latin for 'crab' and a picture of a crab is often used to symbolise cancer. The ancients used this name because swollen veins are often seen radiating from advanced growths, giving a picture resembling a crab's limbs. We cannot point to any single feature to define cancer, and the essential difference between the cancerous and the normal cell is quite unknown. Multiplication and growth are properties of most normal tissues, and the healing growth that repairs a cut finger is faster than that of a cancer. But when the wound is healed growth ceases. We tend to think of a cancer as rapidly growing. In general, the cells of a cancer grow no faster than their normal counterparts and in fact often more slowly. But all normal cell growth is under control, and addition of new cells is matched by loss of old cells, in accordance with the physiological needs of the organ. In a cancer these balancing constraints are lost, new cells pile up regardless of physiological need and thus *appear* to be growing rapidly.

What has happened when a normal cell starts to behave as a malignant cell? It is tempting to suppose it has acquired some new property, a mysterious faculty for unlimited growth. The truth seems to be that it has not gained anything at all, but has lost something. All cells possess—or in their developmental history once possessed—theoretically unlimited growth potential; but this is subordinated to the controlling process—of unknown nature—responsible for tissue organisation. The cancer cell has lost the capacity to respond to normal control mechanisms—it is a delinquent, unresponsive to civilising influences. It is easier to lose something than to acquire something—it is relatively easy for a normal cell to become cancerous, but virtually impossible for a cancer cell to become normal. It follows that treatment must aim at the destruction of every cancer cell, or failing that, the abolition of their capacity to reproduce. Otherwise residual cells will survive to form recurrent growths.

Though there is no universally acceptable definition of cancer, it has some essential behavioural qualities. *The important difference between a simple and a malignant new growth is that the latter possesses the power of 'invasion' while the former does not.* It may grow slowly or rapidly, but if not successfully treated will lead to death sooner or later, unless death from some other cause intervenes. The spread of cancer is dealt with in detail below.

The different characteristics of benign and malignant tumours are summarised in Figure 14.1 and Table 14.1

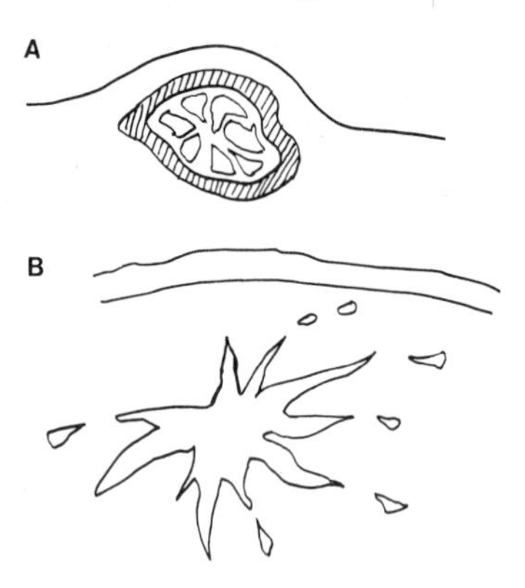

Fig. 14.1 Showing the difference between a benign tumour (A) contained by a definitive capsule and a malignant tumour (B) actively invading the tumour bed.

Table 14.1 Characteristics of Neoplasms

Benign	Malignant
Growth usually very slow, often ceases after a time.	Growth variable–may be very rapid.
Usually encapsulated.	Usually not encapsulated.
Growth by expansion.	Growth by infiltration.
Remain localised.	Metastasise (lymph or blood etc.).
Histological appearance similar to normal tissue.	Histology diverges more or less from the normal, but often widely.
Little or no destruction of normal tissue.	Tissue destruction eventually extensive if untreated.
Not fatal (except from mechanical pressure in special sites).	Always eventually fatal if not halted in time.

Carcinogenesis (i.e. cancer production)

If we paint a patch of skin in a mouse or a rabbit repeatedly with tar, a skin cancer can eventually be produced. The particular chemical responsible has been isolated, and many other carcinogens (i.e. 'cancer-producers') are now known. Their mode of action is still obscure, but it is at least interesting that many compounds, including hormones, which occur naturally in the human body, have related chemical structures.

This type of chemical *carcinogenesis* occurs in man, e.g. long years of work with tar are liable to produce skin cancers on exposed hands and faces. In 1775 Percival Pott in England described chimney sweeps' cancer—epithelioma* of the scrotum due to contamination with soot. There are many other industrial examples, e.g. mule-spinner's cancer from the oil used in treating yarn in spinning mills, and similar lesions from lubricating oils in machine tools. Internal cancer can also arise from chemicals—e.g. workers in the dye and rubber industries absorb and excrete in the urine compounds liable to cause bladder cancer. The most important instance of all is the production of lung cancer by tobacco smoke. It is also probable that inhaled tobacco products excreted in the urine contribute significantly to bladder cancer, which is on the increase.

In all these instances there seems to be a common factor like *chronic irritation* at work. We cannot pin-point the mechanism—all we can say is that it is a reaction to injurious agents, a constantly repeated cycle of insult and repair until the reparative processes become 'fatigued' and control of growth is lost. The result is a cancer, with production of changed cells often of peculiar shape and size and with a capacity for purposeless multiplication with no regard to the local or general economy of the body. However, it is *not just a matter of simple irritation in the mechanical sense*. There is no correlation between the mechanically irritative and the carcinogenic properties of materials. Coal dust is highly irritant mechanically but never produces cancer; whereas soot (from coal after combustion) is a powerful carcinogen. It is a *chemical factor* in the products of combustion that is the responsible agent.

Another important type of irritant factor is *radiation*, e.g. ultra-violet rays, invisible components of sunlight responsible for suntan, can cause skin cancer after prolonged exposure—as in farm-workers and sailors and in sun-blessed countries like Australia, Argentina and Southern U.S.A.

X-rays. Before the dangers of radiation were appreciated, many pioneer workers developed skin cancers, especially on the fingers. Other carcinogenic effects included leukaemia, and diagnostic radiology is a further potential source of danger (see p. 173 and Table 16.1) especially where growing tissue is involved. For example, some childhood cancers including leukaemia are thought to be due to pelvic radiography of the mother during pregnancy.

Radioactive mineral ores. Workers at certain mines in Bohemia (Germany)—where Marie Curie obtained the pitchblende from which she isolated radium—were for centuries susceptible to 'Mine Disease', now recognised to have been lung cancer. The atmosphere down the mines contained radon, and its various decay products were deposited in the lungs, giving off beta and gamma rays and the even more dangerous (since they are more intensely ionising) alpha particles (see Fig. 17.8, p.183). The average induction period was about 20 years.

Comparable effects resulted from luminous paint for watch dials, among the girls in a factory in New Jersey (U.S.A.). The paint contained small amounts of radioactive material (including radium) and when the workers pointed the brushes between their lips, traces of paint were ingested, absorbed and deposited mainly in bone where they continued to emit alphas, betas and gammas. Some died in a few years from aplastic* anaemia; others developed bone sarcomas after some 15 years.

Atomic bombs offer a striking example. The survivors at Hiroshima and Nagasaki suffered an increased incidence of cancer, especially leukaemia. This is an effect of acute rather than chronic irritation on the highly susceptible bone marrow which is normally the site of rapid cell proliferation. A clinically important case is the production of leukaemia (as well as other cancers) after X-ray treatment to large areas of the spine for ankylosing spondylitis (p. 269). These examples are enough to show why we are increasingly reluctant nowadays to use ionising radiation in treatment of non-malignant conditions.

Infection. The role of chronic infection is hard to isolate, but it is probably an important factor in cancer production in e.g. the mouth and also the cervix. In both these instances cancer is commoner as we descend the social scale, and seems to be associated with poor personal hygiene.

Hormones. We have already mentioned the chemical resemblance of some hormones to some known carcinogens. A milestone in experimental cancer research was passed when administration of natural ovarian hormone was shown to produce breast cancers in mice. This at once raised the possibility of physiological substances acting in the body as carcinogens, though the amounts of hormone needed experimentally were relatively huge and there may be no real parallel with natural events in humans. The carcinogenic action of hormones must be different from that of specific chemical carcinogens.

However, it is reasonable to postulate hormonal influences in breast cancer. The breast is an unstable organ, with cyclical growth changes, subsidence and regeneration, depending on a delicate balance of endocrine secretions. Breast cancer grows faster during pregnancy and lactation. There is good reason to believe that abnormalities of hormone patterns—i.e. of growth stimulus—may be common factors in initiating breast cancer (p. 215) and this also applies to other organs subject to hormonal influences, e.g. testis, uterus, ovary, thyroid, even kidney.

Even if hormones cannot be directly incriminated in the causation of human cancer, they are undoubtedly important in maintaining the growth of some tumours. These are the so-called *hormone-dependent* tumours. The best example is cancer of the prostate, followed by some breast cancers (see Ch. 30).

Worms and Cancer. There are several examples in animals of cancers caused by worm infestation. In some parts of the world, notably Egypt, there is good reason to believe that bilharziasis* is the external irritant factor responsible for the high incidence of bladder cancer, see p. 238.

Viruses and cancer. Viruses can unquestionably induce cancer in animals (and plants)—e.g. leukaemia in mice, sarcoma in chickens, skin tumours in rabbits, even breast cancer in mice. An immense amount of laboratory work has been done in this field and the virus theory of cancer has enjoyed great popularity—and still does. It obviously links up with immunological aspects discussed below. One speculative hypothesis is that there is a widespread virus lurking in human cells, normally harmless, but capable of malignant activation by various stimuli (physical, chemical, radiation etc.)

It seems reasonable enough to think that at least some human cancers are due to virus infection, but it remains true that so far there is no definite proof of this for any cancer in man. (For Burkitt's tumour see p. 284).

Immunological aspects

Antigen and antibody. 'Anti' is Latin for 'against, opposed to'; 'gen' is Greek for 'producing'. Anything 'foreign' to the body is liable to provoke the production of an opposing chemical substance called 'antibody' to neutralise it. The stimulant (bacterium, virus, vaccine etc.) is the antigen, the opponent produced (by the reticuloendothelial system) is the antibody.

In the nature of things, normal cells are not antigenic (otherwise they would not be normal!), but abnormal cells may be. Since many millions of cells are dividing by mitosis in the body every day, mutation, though rare, must occasionally occur and produce abnormal cells with some degree of antigenicity. These abnormal cells are likely to be promptly attacked and destroyed by lymphocytes etc. which are the body's 'watchdogs'.

Resistance to cancer. Resistance to infection (by antibodies etc.) is well recognised. The possibility of a viral origin of cancer, or of cancer cells carrying antigens, raises comparable possibilities of antigens, antibodies and antisera in treatment—but these are as yet little more than theoretical.

But some form of resistance to cancer does appear to exist, and to play a real part in determining whether a clinical cancer arises or not. It is even possible that the normal body is constantly producing microscopic cancers, which are prevented from developing by existing defence mechanisms, just as bacteria and viruses are constantly antagonised and neutralised by the reticuloendothelial system before clinical infection can occur.

We know that many cancers shed cells into the blood stream and they can be detected in the venous outflow at the time of surgical resection. But only a small proportion of these give rise to metastases—the cancer cells are presumably killed or incapacitated before they can 'take' at any site. Another suggestive feature is the local infiltration round some cancers of lymphocytes and other components of the lymphoreticular system. We know, for example, that breast cancers which have this infiltration carry a better prognosis than those that do not.

Immunological factors therefore appear to be important both in the rise and spread of cancer. There have been many reported instances of tumour regression after injections of vaccines or other foreign material. The mechanism is obscure—possibly it may trigger the body's immune defence system into treating the tumour as foreign tissue and rejecting it in the same way as transplanted organs. Vaccine is now proving useful in the treatment of some leukaemias (p. 249). Immune mechanisms seem to be the major factors in the control of choriocarcinoma (p. 283) and Burkitt's lymphoma (p. 284).

Whether the immunological front will achieve worthwhile success in the campaign against cancer still remains to be seen.

Heredity in cancer. 'Can cancer be inherited?' is a question commonly and anxiously asked by patients and relatives. The short answer is—No.

We can in-breed mice so that they all develop cancer, but this laboratory experiment has no possible counterpart in humans. This is not to say that inherited, i.e. genetic, factors have no importance. They undoubtedly have, but they are only one set of factors among many, and to determine a cancer there must be a combination of factors at work—*cancer is multi-factorial*. Hereditary factors may decide which particular organ is liable to develop cancer and there are certainly some families where several members develop e.g. breast cancer. But the cancer itself is not inherited, though the predisposition may be, and other factors must operate before the cancer can arise.

Further evidence comes from identical twins who develop cancers, If heredity is important, we would expect them to develop similar cancers at about the same time, but this is very uncommon (though commoner than in non-identical twins). Hereditary factors do have some importance here, but, as usual, environmental factors are far more important than genetic.

There is one genuine instance of genetically determined cancer—retinoblastoma (p. 264).

Infectivity of cancer. 'Is cancer catching?' is another frequent question. 'Is there any danger to the children in the house?' There is no evidence whatever that cancer can be caught in this way—there is not a single recorded instance—and complete reassurance can be given.

Even if a virus is ever proved to be involved in human cancer, it is unlikely that this statement will need changing for practical purposes.

Injury and cancer. It is popularly believed that physical trauma may cause cancer, and a patient may attribute say a breast tumour to a blow previously received. This belief often leads to claims for compensation if the injury was received at work. It is extremely unlikely that trauma alone can ever produce a cancer, but it may draw attention to a tumour previously unnoticed.

The incidence of injury is so very high and the cases where even a moderately plausible case for a cause and effect association with cancer can be made are so very few, that we are driven to conclude that no definite relationship exists.

Chronic injury e.g. from asbestos dust in the lungs—or tobacco smoke!—falls of course in a different category (see above).

The natural history of cancer

It is better not to speak of 'The Cause' of cancer but of multiple causes or causative factors. Though one factor may be the 'last straw', it is probable that in nearly all cases there must be a combination of causative factors at work, including genetic, hormonal, metabolic, inflammatory, physical, chemical, thermal and environmental generally.

The earliest stage of malignancy. This is clearly a subject of fundamental importance, since this is the stage where we can hope for complete eradication. In recent years some progress has been made in recognising these early changes in epithelial surfaces.

Pre-cancerous lesions may be recognised, a halfway house between the normal and the frankly malignant. Individual cells are abnormal, but the picture of invasive cancer is not present and the term *'pre-invasive cancer'* is sometimes used. In such cases experience shows that malignancy is very likely to follow, though it may be many years later, perhaps 10–20 years.

Examples of pre-cancerous situations where cancer is more liable to develop than in normal tissue:

1. Hyperkeratoses* of the skin—small warty growths; may be due to chronic exposure to tar, sunlight, X-rays etc.
2. Leucoplakia*—thickened whitish patches on muscosal surfaces, especially in the mouth
3. Papilloma* of the bladder—commonly multiple (papillomatosis, p. 238)
4. Stage 0 of carcinoma of cervix, pages 225–6
5. Chronic ulcers, especially if infected.
6. Plummer-Vinson syndrome in post-cricoid cancer (p. 204)
7. Paget's* disease of bone (p. 258)
8. Congenitally abnormal organs e.g. imperfectly descended testis (p. 242).

Although cancer may possibly arise from a change in a simple cell, it is normally due to a 'field' change in a whole area. It is not surprising to find that many foci may be abnormal in a tissue where cancer has arisen, either naked-eye or microscopically. The same factors responsible for a clinical cancer will often produce changes in the rest of the organ—e.g. multiple malignant or pre-malignant patches may be present in a breast removed for cancer. In other words the whole epithelium may be unstable and liable to degenerate into malignancy at multiple points. Thus a patient treated for cancer of the lip may later develop one on the tongue, then on the buccal cheek, then in the pharynx.

The spread of cancer

Growth may be (a) continuous, by local enlargement of the *primary* mass, or (b) discontinuous, by *secondary* growth elsewhere, near or far, with normal tissue intervening—this is also called *metastasis*.*

Spread may occur in several ways.

1. *Local invasion*
2. *By lymphatic vessels*
3. *By blood vessels*
4. *Across cavities*

1. *Local invasion.* Multiplication of cells results in a microscopic mass which will enlarge to become visible or palpable or produce other clinical effects. Cancer cells insinuate themselves between the normal cells and compete for the available nourishment. Normal tissue will eventually be disorganised and destroyed, depending on the blood supply and the available 'elbow room'.

The edge of the tumour is therefore ill-defined (Fig. 14.1) and complete excision by surgery correspondingly difficult and uncertain. A generous margin of apparently normal tissue should always be removed but even then microscopic examination—or the later clinical course—may show that some malignant cells have been left behind; if so, they will eventually produce a *local recurrence* if nothing further is done.

Ulceration of a surface, internal or external, may cause bleeding and secondary infection. Erosion of a large blood-vessel may cause serious, even fatal, haemorrhage. Cancer cells may be temporarily halted at a barrier, e.g. bone but even this will be invaded and eroded.

Functional changes may follow. A growth in the oesophagus or bowel may cause obstruction, damage to the brain may cause headache or fits, involvement of the liver or obstruction of the bile duct may lead to jaundice and liver failure, just as spread of cancer of the cervix to the ureters may result in renal failure and uraemia.* Unless they can be relieved by treatment (surgery or radiation) these changes are likely to be fatal, even in the absence of metastases.

Pain will occur when nerves are involved, but (contrary to popular belief) it is a late symptom, often very late.

2. *Lymphatic spread.* At some stage, local lymph capillaries may be involved and isolated cells or clumps of cells may be carried in the lymph flow to one or more of the regional lymph nodes. (They are still often called 'glands', a misnomer from the days when they were thought to be organs of secretion. It is time the old name was abandoned.)

If the cancer cells survive and multiply in the node, they form a secondary deposit (= metastasis) and the node may become palpably enlarged, e.g. secondary axillary nodes in breast cancer. If the whole node is involved, spread may occur beyond the node capsule, fixing it to surrounding tissues, and even ulcerating the overlying skin eventually. From one node, further deposits may arise in other nodes along the chain—e.g. spread to supraclavicular nodes may come from internal mammary or axillary nodes (Fig. 22.1 p. 216)

From lymph vessels and nodes cancer cells may eventually enter the general blood circulation, just as lymph itself does.

3. *Blood spread.* Cancer cells enter the blood stream by direct invasion of capillaries or the thin-walled small veins. Metastatic growth can arise at almost any site in the body, but there is a definite preference for certain organs—after regional nodes, the commonest are liver, lungs and bone.

Metastasis is the chief danger in any cancer and the cause of death in most fatal cases, from gross involvement of vital organs.

In general, carcinoma shows a marked tendency to early lymphatic spread, whereas sarcoma usually spreads via the blood stream.

A secondary deposit may give the first clinical evidence of cancer. For example, a tiny primary in the nasopharynx may give rise to much larger secondary nodes in the neck. Pain from bone secondaries—or even a pathological fracture—may be the presenting feature in lung or breast cancer etc. In some cases, secondaries may dominate the picture from the start and the primary may be so small and insignificant that it is not discovered till post-mortem examination (or not even then!).

Cavity spread

a. Cancer cells in the upper abdomen—e.g. from a gastric cancer—may be dislodged into the peritoneal cavity and come to rest in the ovary or pouch of Douglas where they will form a secondary deposit. Similarly an ovarian cancer in the pelvis may deposit seedlings on peritoneal or mesenteric surfaces.
b. A similar sequence may occur in the thorax, from one part of the pleural surface to another.
c. Cells from an intracranial growth—e.g. medulloblastoma of cerebellum—may be seeded into the cerebrospinal fluid and form deposits on other brain surfaces or in the spinal canal.
d. Cells from the upper urinary tract—e.g. renal pelvis—may be carried down in the urine to form a deposit in ureter or bladder.

As the word 'malignant' implies, death is the natural result of untreated—or unsuccessfully treated—cancer. It occurs from general exhaustion, liver or kidney failure, haemorrhage, terminal broncho-pneumonia, cerebral complications etc.

Staging of cancers

It is of the greatest practical importance to estimate the geographical extent to which any tumour has spread. We call this *staging*. There are various criteria and systems in use, but in general terms something like the following is adopted:

Stage 1. Tumour confined to organ of origin
Stage 2. Local lymph nodes invaded
Stage 3. Distant nodes invaded, or local spread beyond organ of origin
Stage 4. Blood-borne metastases present

Staging is based on the *clinical* findings (including radiological) before treatmeant is begun and is therefore not necessarily the same as the *pathological* staging based on histological examination. Staging enables us to group our cases and to make meaningful comparisons with cases at other centres.

The best example of staging is in carcinoma of cervix. This has been internationally accepted so that results of treatment all over the world can be profitably compared. Details are given on page 225. Breast cancer also lends itself to staging (see p. 216), but unfortunately there is so far no widely accepted international system. Another example is Hodgkin's disease (see p. 246).

TNM system

This is a system of staging proposed for international use and it will certainly be widely accepted before long.

T refers to local extension of primary Tumour
N refers to the condition of regional lymph Nodes
M refers to the presence of Metastases, beyond regional nodes

T1 means a relatively small primary, confined to the organ of origin
T2 means a relatively large primary, confined to the organ of origin
T3 means infiltration of neighbouring structures

N0: no nodes palpable
N1: movable nodes on same side as primary tumour
N1 may be further sub-divided:
N1a: nodes not considered to contain growth
N1b: nodes considered to contain growth
N2: movable nodes on opposite side, similarly sub-divided into N2a and N2b
N3: fixed nodes

(The system is modified in some sites where nodes cannot be examined clinically.)

M0—no evidence of distant metastases
M1—distant metastases present

Thus a very early cancer would be labelled T1N0M0 and a very late one T4N3M1.

An example of TNM staging is given for breast cancer—p. 217.

Histological grading: Differentiation

In tumour histology we are concerned not only to recognise in which organ a growth originates, but also to assess the degree of malignancy of the constituent cells, by studying the individual cells and noting how far they diverge from the normal for that particular tissue.

Development from the embryonic and primitive to the adult and specialised is called *differentiation*, and the microscopic appearance of an embryonic undifferentiated cell differs from that of a mature differentiated cell. Malignant growths may or may not grow faster than normal tissue —in fact they usually do not—but they generally contain a relatively high proportion of undifferentiated cells more concerned with growth and less with function. The malignancy of a tumour in fact, is shown by the relative proportion of poorly differentiated to well differentiated cells.

Histological grading has its uses, but is not nearly as important as Staging. We define three grades:

1. Well differentiated
2. Poorly differentiated
3. Undifferentiated—also called *anaplastic**

more likely that numbers of cells in foci become changed under some carcinogenic stimulus and it would hardly be worth arguing whether one or more cells were the true originators of the clinical cancer.

It is important to distinguish between *histological* cancer and *clinical* cancer. At the earliest recognisable stage, abnormal cells may be detected under the microscope—e.g. by abnormal staining or abnormal shape of the nucleus. This is the basis of the cytological detection of early or pre-invasive cancer of the cervix (p. 226). At a later stage cellular abnormalities—abnormal shapes and sizes of cells, bizarre nuclei etc.—and early local invasion would be found microscopically, though there might be no clinical evidence whatever—no tumour, no ulcer, no symptoms. In the earliest stages of cancer the histological diagnosis may be very difficult, for there is no sharp boundary between the normal and the pathological and two equally expert pathologists may disagree whether or not a cellular hyperplasia is malignant. Such doubt is common for instance after examining prostates removed for urinary obstruction or thyroids removed for nodular goitre.

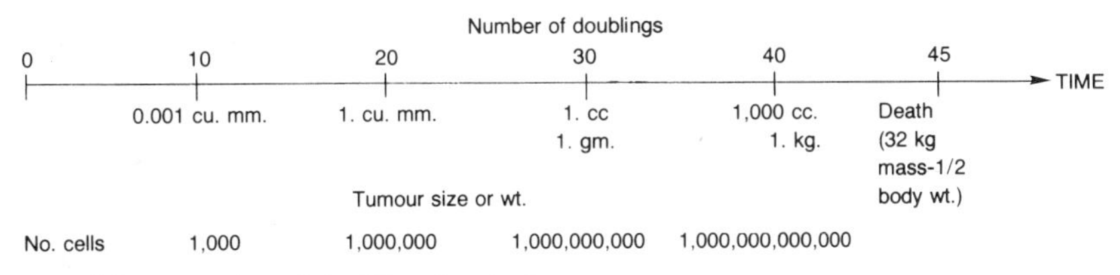

Fig. 14.2 Tumour growth by successive cell doublings (see text).

The more undifferentiated or anaplastic a growth appears, the more liable it is to throw off early metastases, and the worse therefore is the prognosis. On the other hand, it is likely to be more radiosensitive. Knowledge of the grading of a tumour can be one factor in deciding the best treatment of a case, and this is another reason for taking a biopsy.* For example, anaplasia may well be a point against surgery and in favour of radiation, since anaplastic growths are more liable to recur locally and to have silent metastases.

However, grading has its limitations. The degree of differentiation may not be the same in all parts of a tumour, and may not be the same in metastases as in the primary. The organ involved is of more importance than the grading—e.g. a Grade 1 (well-differentiated) cancer of the skin has an excellent prognosis, but the same grade in the lung has a very poor prognosis.

Growth rate of cancers

Theoretically a cancer could start as a single abnormal cell, which divides to form two cells, then four cells, then eight cells . . . and so on. And the same could occur with a single-celled metastasis. Single cells can produce cancers in laboratory animals, but under human conditions it is far

There will thus always be a latent interval from the inception of a cancer to its first histologically recognisable state, and a still longer interval before its first clinical sign—visible tumour, ulceration, bleeding etc. Clearly these intervals are matters of life and death, and the more rapid the growth, the shorter the latent interval.

It is instructive to consider these intervals in concrete terms (Figs. 14.2, 14.3). If a typical human cancer cell

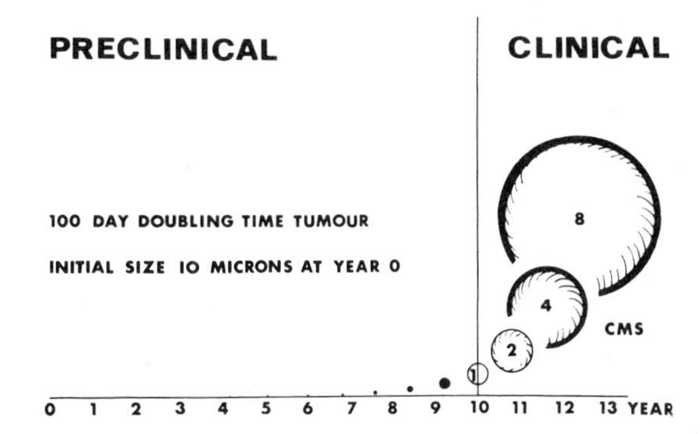

Fig. 14.3 Growth of tumour with a doubling time of 100 days. Note the long pre-clinical interval from inception to diagnosis. A clinical tumour can easily be ten years old when first detected. (Courtesy Mr W. H. Bond.)

divides after a certain time—so many hours or days to become two cells, the *doubling time*— and if the daughter cells keep up regular division in the same way, then simple calculation shows that it will take 20 doublings to produce a mass one cubic millimetre containing one million cells. This is about the smallest tumour mass one could ever hope to detect clinically. After 10 further doublings—i.e. 30 in all—there would be a mass of one cubic centimetre weighing one gram, containing about 1000 million cells. This would generally be considered an early lesion and we might well congratulate ourselves if we picked up, say, a breast cancer as small as this.

Ten more doubling times—i.e. 40 in all—would bring the mass to one kilogram, by which time of course there would be gross clinical changes, maybe even cachexia.* If a patient could possibly survive more doubling times—e.g. 45 in all—it would mean a mass of 30 kilograms or half the body weight—an unlikely event, but certainly fatal.

But let us note that the time for 30 doublings, i.e. from one cell to one gram, is at least twice the length of time from one gram to death. When we reflect that a cancer can rarely be diagnosed before the stage of one gram (= one cubic centimetre), it is clear that the greater part of its lifetime will already be behind it at the time of diagnosis, in fact about two-thirds of its total history, with all that this implies in terms of metastasis. *The clinical period is therefore only the visible tip of a pre-clinical 'iceberg'.*

These calculations are theoretical and may even seem fanciful at first, but there is good reason to believe that they are not far from the truth. Evidence from both animals and human cancers suggests that most tumours do grow at a fairly constant rate, i.e. with a fixed doubling time, which varies little for most of the tumour's life. Actual growth will depend on this doubling time. Our knowledge of doubling times is scanty, but it has been possible to measure it, for example by serial X-ray films of primary or secondary masses in lung fields, or mammograms of breast tumours. Not surprisingly there is a great variability in growth rates. In lung cancers, doubling times were found to range from 11 to 164 days, in breast cancers up to 450 days, i.e. 15 months.

These figures are of obvious importance. For example, with a doubling time of 60 days a tumour could take five years to grow to a diagnosable mass of one gram—and this is about the diagnostic boundary for a lung cancer on a chest film. With a doubling time of 100 days the pre-diagnostic interval would last 8–10 years (Fig. 14.3). They would also explain how secondaries can appear many years after removal of the primary growth. As we have seen, more than half a tumour's life—about 30 doubling times—has generally passed before it can be diagnosed and removed, and metastases could be shed at any stage in its pre-clinical history, from the first doubling to the thirtieth. With doubling times of a year or more, the appearance of secondaries after 20–30 years and even longer, following apparently successful removal of the primary—and this certainly does happen in some cases, especially in breast cancer—could easily be explained.

Although metastasis is usually the tragic element in cancer, it is only the final stages of secondary growth that are of clinical importance and cause symptoms. The early stages are silent. *If we can reduce a growth or metastasis to a sub-clinical level, the patient may be trouble-free for years, even though harbouring a silently growing lesion.*

It follows that 'cure' in the popular sense is not the only—not even the best—measure of success in cancer therapy.

Spontaneous regression of cancer

This sounds like a contradiction in terms, but is does seem possible that cancer can disappear spontaneously, that is without obvious treatment. We are rightly sceptical of all such claims and demand convincing proof, including expert histological confirmation. There are several hundred articles in medical journals describing alleged cases, and one whole book reviewing the subject and accepting the claim in some 180 cases.

As an example, there is a small series of bladder tumours confirmed as malignant in the usual way by biopsy. Urinary diversion was carried out by ureteric transplantation as a preliminary measure, and after a suitable interval the bladder was removed at cystectomy. Examination of the resected bladders showed no remaining cancer tissue. Presumably there had been a carcinogenic stimulus, probably a chemical factor in the urine, acting on the bladder epithelium. When this stimulus was removed by urinary diversion, the epithelium was able to recover its integrity. If this is really so, then it does appear possible that the earliest stages of carcinogenesis may sometimes be reversible and underlines the possibility of *resistance* to cancer.

Over half of the collected cases occurred in four types of cancer—carcinoma of kidney (hypernephroma), neuroblastoma, malignant melanoma and choriocarcinoma. In renal cancer and malignant melanoma, *hormonal* influences are known to be concerned and it seems cetrtain that they must have played some part in the process. Choriocarcinoma (p. 283) is a peculiar kind of cancer of different behaviour from most cancers and it is highly probable that *immunological* reactions are involved in its regression. In neuroblastoma (p. 262) spontaneous maturation into the simple adult type of ganglioneuroma* is known to occur in a small proportion of cases, about 1 per cent.

In other cases, bacterial infection seems to have played a part and possibly the antigen-antibody mechanism may have been responsible. Many years ago bacterial toxins were actually used in cancer therapy and seemed to be helpful at times. The present-day revival of interest in the immunology of cancer (see above) makes these cases particularly interesting.

The total number of these cases is relatively very small, so small that they might seem practically speaking unimportant. However the significance of these cases lies in the support they lend to the concept of biological controls even in cancer, and in reinforcing the hope that future research may yet give us better methods than surgery, radiation and poison, which are after all the very crude instruments we have at present.

Classification of neoplasms

Table 14.2 lists the chief types. The ending 'oma' is used to indicate a neoplasm, just as 'itis' indicates inflammation.

It took many centuries for the concept of cancer as a distinctive entity to evolve. The suffix 'oma' was applied to swellings of any nature. This relic of the past survives today in the use of names like *tuberculoma*, *haematoma* etc.

Most cancers are in one of these sub-divisions:

1. Carcinoma
2. Sarcoma
3. Reticulosis

Carcinoma is from the Greek for 'crab' but it is used in a more restricted sense than cancer and applied to *malignancy arising in surface (i.e. epithelial) tissues*. This is the largest and most important group of cancers—in fact 75 per cent of all cancers are carcinomas.

Squamous epithelium* lines the skin, where it is called epidermis, and the mucosal surfaces of oral and nasal cavities, accessory sinuses, pharynx, most of oesophagus, respiratory tract (larynx, trachea, bronchi) middle ear, vagina, cervix. The terms squamous carcinoma, squamous epithelioma, and simply epithelioma are used synonymously.

Glandular (secretory) epithelium may form a naked eye surface (as in the bowel) or remain microscopic (as in the breast). In the case of the latter it is continuous with an ordinary surface from which it is derived by a complicated downfolding process, the proximal part of which forms the duct (Fig. 14.4). The ductless endocrine glands lose their original ducts and their secretions pass into the blood stream.

This type of epithelium lines the whole of the alimentary tract from lower oesophagus to upper part of anal canal as well as all the glands opening into it including

Table 14.2 Types of neoplasms

Benign	Tissue of origin	Malignant
	Epithelium = surface tissue	General name = **Carcinoma**
Papilloma	Squamous epithelium	Squamous carcinoma or epithelioma
	Transitional cell epithelium	Transitional cell carcinoma
	Basal cells of epidermis	Basal cell carcinoma or Rodent ulcer
Adenoma	Glandular epithelium	Adenocarcinoma
	Supporting tissues	General name = **Sarcoma**
Fibroma	Fibrous tissue	Fibrosarcoma
Myoma	Muscle	Myosarcoma
Leiomyoma	Smooth muscle	Leiomyosarcoma
Rhabdomyoma	Striated muscle	Rhabdomyosarcoma
Lipoma	Fat	Liposarcoma
Chondroma	Cartilage	Chondrosarcoma
Osteoma	Bone	Osteosarcoma (osteogenic sarcoma)
Osteoclastoma		
Synovioma	Synovial membrane	Synovial sarcoma
	Lymphoreticular tissue	General name = **Reticulosis** or **Malignant Lymphoma**
Lymphoma (follicular lymphoma)	Lymphoid tissue – including spleen	Lymphosarcoma Reticulum cell (histiocytic) sarcoma Hodgkin's disease
	In skin	Mycosis fungoides
	Blood-forming tissues: – white cells	Leukaemia (myeloid, lymphatic)
	– red cells	Polycythaemia Rubra Vera
Solitary Myeloma	Plasma cell	Multiple myeloma (myelomatosis)

Table 14.2 *(Continued)*

Benign	Tissue of origin	Malignant
	Pigment cells	
Mole or naevus (benign melanoma)	Skin Mucosa Eye (choroid)	Malignant Melanoma (melanocarcinoma, melanosarcoma)
(Haem) angioma	**Blood vessels**	(Haem) angiosarcoma
	Intracranial	
	Supporting tissue of central nervous system	Glioma (this includes astrocytoma, glioblastoma, oligodendroglioma ependymoma, pinealoma etc.)
Meningioma	Meninges	
Pituitary Adenoma (chromophobe, eosinophil, basophil)	Pituitary gland	
Hydatidiform Mole	**Placenta** (fetal tissue)	Chorionepithelioma (Choriocarcinoma)
Dermoid Cyst Benign Teratoma	**Gonad** (germ cells)	Teratoma Seminoma (male) Dysgerminoma (female)
	In children	
	Kidney	Nephroblastoma (= Wilms's tumour or embryoma)
Ganglioneuroma	Sympathetic nerve tissue	Neuroblastoma
	Cerebellum	Medulloblastoma
	Retina	Retinoblastoma

N.B. This table is not complete. For rarer types, standard works on pathology should be consulted.

pancreas and liver. It also lines the sebaceous and sweat glands of the skin, the endometrium and mammary gland, the salivary and mucous glands of respiratory and upper digestive tract (including mouth and nose), the kidney, ovary and testis and all endocrine glands (pituitary, thyroid, parathyroid, suprarenal).

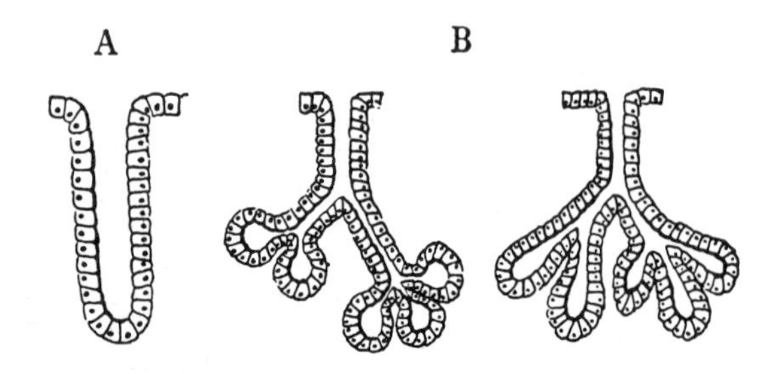

Fig. 14.4 Diagrams of simple and complex glands, showing their formation by infolding and down-growths of surface layers. The terminal parts become the secreting cells (e.g. milk-secreting cells of breast) and the initial parts form the main ducts along which the secretion passes. (*Whillis, Elementary Anatomy and Physiology, Churchill*)

*Adenocarcinomas** may retain some of the secretory function of the parent tissue, e.g. mucoid secretion in nasal cavity, hormonal secretion in thyroid. Cyst formation is common, e.g. thyroid, ovary (called cystadenoma and cystadenocarcinoma). Spread from carcinoma is commonly to regional lymph nodes and also via the blood to lungs, liver, brain etc.

Sarcoma is from the Greek for 'Flesh' after the fleshy appearance of many of these tumours. The commonest is the fibrosarcoma in the thigh or buttock. Spread of sarcoma is usually blood-borne to lungs and rarely to lymph nodes.

*Gliomas** arise from the specialised connective tissue of the central system (brain and spinal cord).

*Reticulosis** The reticuloenthothelial system is a widespread group which includes the blood-forming bone marrow, the lymph nodes, spleen, and parts of the liver and thymus. The group of reticuloses is a very mixed one, of variable malignancy, ranging from the semi-benign polycythaemia to the rapidly fatal acute leukaemia. They may arise at many foci simultaneously or spread successively from node to node—it is often difficult to be sure.

15. Public health aspects of cancer

Epidemiology of cancer

Cancer affects men of all races, all colours and all countries. It affects animals too, and even plants. The study of its incidence in space and time, its variations from country to country, between different races, in different circumstances of nutrition, hygiene, employment and all kinds of environmental factors, is the province of the epidemiologist and statistician. It was this kind of study that drew attention to the important association between smoking and lung cancer.

Cancer statistics and registration. It is clearly of prime importance to secure full and accurate figures in all cancer cases, with details of treatment and cause of death (which may or may not be cancer). This is true of all types of disease, but particularly cancer. Using these figures as a basis, we can extract the facts about its incidence at different sites; about variations in different parts of the country and between different social classes; the effects of treatment, length of survival etc. The Registrar-General in England publishes some of these details in his annual reports. Health Authorities also have cancer registers, and hospital treatment centres such as radiotherapy departments maintain their own records.

It is impossible to obtain complete coverage, for some cancers are never even reported and death certificates are by no means always reliable. But the figures are now good enough for most comparative purposes.

It is evident why we take such trouble to follow-up our cancer patients, for only from full and detailed statistics and comparisons can we assess the results of various methods of treatment—or lack of treatment.

Age. No age group is exempt. Cancers in childhood form a rather special group (see Chapter 28), but most cancers take many years to produce clinical symptoms. It is therefore not surprising to find an increased incidence in the older age groups. The longer we live, the greater the chance of developing cancer.

If cancer is on the increase, as it is in most countries, this is largely a reflection of the fact that people are living longer. This in turn is a tribute to the effectiveness of public health measures—nutrition, hygiene, housing, factory inspection etc.—rather than an indictment of the health professions.

Some two centuries ago the average expectation of life of a new-born infant was about 30 years in Britain. Today it is about 70, and old-age pensioners are even becoming a 'problem' in western countries. But it is still in the 30's in many parts of the world—Africa, India, South America etc. Even in the 19th century serious epidemics of cholera and typhoid occurred in England and it is this type of infection, as well as the parasitic diseases, both helped by malnutrition, which are still largely responsible for death in the less prosperous countries. There death rates will be high, but cancer incidence low, since potential victims will be killed off before cancer can develop.

Geographical, racial and environmental factors

Here are a few random findings:

1. Cancer of the stomach is rare in Indonesia (1 per cent of all cancer deaths) but very common in Japan (50 per cent)
2. Chinese are very susceptible to cancer of naso-pharynx
3. Liver cancer is common in Africa but uncommon elsewhere
4. Lung cancer is uncommon in Finland and Iceland, but common in England
5. Cancer of cervix is very uncommon in Jewesses, and even rarer in nuns
6. Cancer of penis never occurs in men circumcised in infancy
7. Breast cancer is common in the West, uncommon in Japan

Facts like these point forcefully to the determining or at least contributory influence of various internal (including genetic and hormonal) and external factors—for example, food deficiencies, special infections, marriage and weaning customs, chemicals in the soil, smoking habits etc. The epidemiologist tries to disentangle and pin-point possible factors by comparing and contrasting different groups in different settings. It is fascinating but difficult work beset by pitfalls. In establishing correlations, it is easy to confuse cause with effect. It does not follow, for example, that because there is a higher concentration of nurses in towns than in villages, the higher incidence of cancer in towns is somehow due to the nurses! Perhaps the greatest triumph in this field has come from the prolonged patient statistical work which finally removed all doubt about the cause-and-effect association between smoking and lung cancer.

Let us quote some further details for the light they shed on various aspects of the cancer problem.

Breast cancer is uncommon in Japan, its mortality only a sixth of that in England. Is this a built-in, i.e. genetic, immunity? When Japanese emigrate to America, the incidence of breast cancer is still low in the second generation—suggestive of inbuilt rather than environmental factors.

Various factors have been considered. Prolonged breast-feeding is common in Japan, but if this acts as a safeguard against breast cancer it fails to account for the equally low incidence in unmarried Japanese women. Dietetic differences are another serious possibility. Japan has one of the lowest fat consumptions in the world, western countries among the highest. Increased fat consumption by Japanese in America may well be relevant. Attractive as this proposition seems, it receives no support from the fact that in both world wars the low fat consumption of Central Europe did not lead to any decrease in breast cancer. Similarly cancers of colon, endometrium, ovary and prostate are all low risks in Japan, but take on the higher risk of the West after emigration to U.S.A.

Hormonal factors must also be considered. The breast, uterus, ovary and prostate are all subject to hormonal influence. Urinary hormone excretion has been investigated and the patterns in native Japanese women differ from those who have migrated to the U.S.A. Again a correlation with dietary fat seems plausible for we know that oestrogenic hormone tends to be retained in adipose tissue. It is very interesting to find that myocardial infarction is also uncommon in Japan, but becomes commoner after emigration, once again suggestive of dietary factors.

Table 13.2 (p. 148) gives much food for thought. Half of all the cancers in British men are seen to originate in only two tissues, the surface linings of the respiratory and gastrointestinal tracts. And about three-quarters in either sex arise at only five sites. Of these, the lung, bowel, skin and urinary tract are all subject to constant assaults from the environment with consequent repeated demands for repair—i.e. chronic irritation in a broad sense (see p. 151). It is no surprise to find that these tissues are liable to develop multiple cancers.

The cervix is another site where environmental stress is crucial. Cancer of cervix is virtually unknown in celibate women (e.g. nuns) and much commoner in the married than in the single. There is some association with multiparity and we used to think cancer was due to trauma in childbirth and could be prevented by good obstetric care, but this is not so. A far more important relation lies in sexual intercourse, particularly promiscuity. Age at commencement of intercourse is important—if it is under 20 years, the risk is double compared with over 20 years. The number of partners is also relevant—remarriage after widowhood or divorce increases the risk, while the incidence in prostitutes is six times the normal. Circumcision of the male may also be relevant. It has long been thought to have some association with the freedom of Jewesses from cervical cancer. In India the incidence in communities which practise circumcision (Moslems) is less than in those which do not (Hindus). It seems to be a question of standards of personal hygiene. Circumcision is part of this, but not an essential part as shown by the Parsees—a religious sect in India—who are not circumcised but have high hygienic standards and a low rate of cervical cancer. Poor hygiene probably accounts for the increasing incidence found as we descend the social scale.

The conclusion seems to be that social factors are of prime importance in cancer of the cervix and it is estimated that improvement in hygiene (both male and female) could abolish perhaps four-fifths of the present incidence. But the evidence from different races is still doubtful and inconclusive. Circumcision in itself is now thought not to be protective. Even poor Jewesses rarely develop cervical cancer—so there may after all be some genetic factors at work. Certainly the evidence does not justify advising universal circumcision.

The association between lung cancer and smoking, particularly cigarettes, is now notorious. Cigars and pipes, although not free from risk, are much less dangerous. It is an unquestionable statistical fact that the risk of developing lung cancer increases in proportion to the amount smoked daily. Among men smoking 25 or more cigarettes a day, one in eight will die of lung cancer. At 35 cigarettes a day, the risk of cancer is 45 times as high as in non-smokers. Yet it is still on the increase, is now the commonest and most serious cancer in men, and becoming more frequent in women.

Atmospheric pollution is probably a contributory cause, and may help to explain the higher incidence in urban areas. This explanation appeals to smokers and the tobacco trade, but if it was true, road workers and policemen would have a significantly higher incidence of lung cancer. In fact they do not, and smoking habits are a far more likely reason for the discrepancy.

Industrial respiratory irritants are often carcinogenic —e.g. arsenic, asbestos, chrome, nickel, radioactive materials. Another striking illustration is the high incidence of lung cancer in Mexican women due to smoke-filled kitchens from flueless stoves, and similarly in South African Bantu women from braziers.

Early detection and cancer screening

Next to prevention, early diagnosis—preferably before the stage of symptoms—gives the best chance of success. Too many patients do not report symptoms to their doctors as soon as they could. This is a problem in health education and is discussed below. Here we are concerned with pre-symptomatic diagnosis. Although we use the word 'early', we must remember that when a cancer becomes detectable, it is always 'late' in the sense that most of its natural life will already have passed and there may already be hidden distant metastases (see p. 155). 'Early' is therefore a relative concept, of clinical application. An early breast cancer that is only just palpable (or perhaps picked up on mammography and not yet palpable) is a very different problem from a late growth that has been allowed to fungate and invade the chest wall, even though the 'early' tumour may already have numerous microscopic secondaries lurking in liver, lung and bone, and the 'late' tumour may have none.

A simple test for malignancy would be enormously helpful—say, a serological test as for syphilis, or a biochemical blood estimation. There have in fact been many such claims advanced, but none, alas, has been confirmed.

Pre-symptomatic diagnosis clearly involves a positive search—getting at 'well' people who would not otherwise consult their doctors or feel the need to submit to any examination or test. Cancer screening programmes of this kind have been launched in many places in recent years and we are now accumulating some experience of their possibilities and value. They may cover whole communities or populations, or selected groups thought to be 'at risk', e.g. workers in a certain industry or factory.

Mass screening, to be worthwhile, must be acceptable to the population concerned. It must not be too troublesome for the individual nor too difficult to organise. It must be supported by educational campaigns and good follow-up with repeat examinations at appropriate intervals and reference of all detected and suspected cases for further investigation and treatment. Questions of expense and economic use of available resources must of course be carefully weighed.

Apart from simple self-examination, the available methods include:

1. Periodic medical examination
2. Cytological examination
3. Radiological examination
4. Highly specialised techniques—thermography, radioactive isotopes, biochemical and hormonal estimations and ultrasonics. These are mostly still experimental and not—or not yet—suitable for mass surveys.

Periodic medical examination. This has achieved a limited popularity, especially in U.S.A., applied either to selected groups such as business executives or offered commercially to those able to afford it. In addition to ordinary clinical examination, it includes a variety of laboratory investigations (blood and urine), chest film, possibly barium meal, sigmoidoscopy etc. It is unavoidably costly and therefore unsuited to large-scale surveys.

*Cytological** methods differ from histological*—the latter relate to cells organised in tissues, the former to cells isolated singly or in tiny clusters. Growing tumours shed—or 'exfoliate'—cells from their surface and these can be fixed and stained on a slide and examined microscopically. The expert can recognise abnormal features of cancer cells, e.g. nuclear irregularities.

This technique of exfoliative cytology was perfected by Papanicolaou in U.S.A. Its greatest merit is that it avoids the drawbacks of surgical biopsy.

(a) *Uterine cervix*—early cancer or pre-cancer can be detected by cervical smear (see p. 226). This is the most common application of cytological diagnosis.
(b) *Urinary tract*, especially bladder. After centrifugation, cells deposited from urine are examined for evidence of malignancy. This technique is useful for screening, e.g. workers in the dye and rubber industries who are at above-average risk. The method is more acceptable than routine cystoscopy, though less accurate.
(c) *Lung.* Cells from sputum may be detected as malignant and may make bronchoscopy superfluous.
(d) *Stomach.* Gastric washings may yield cancer cells.
(e) *Prostate.* Cells in the secretion obtained from prostatic massage may be malignant.

Radiological examination is also of great value. Examples are:

(a) *Chest films*—for detection of lung cancer in surveys comparable to mass miniature radiography.
(b) *Barium meal* investigation. In Japan, which has an exceptionally high incidence of stomach cancer, mobile teams with radiographic equipment have developed a comparable technique for mass miniature radiography.
(c) *Mammography* is a recent technique which is undoubtedly able to detect many breast cancers months and even years before they can be felt.

Thermography depends on the fact that cancers as well as other lesions (especially inflammatory) have a greater blood supply than their surroundings and consequently a raised temperature. A 'heat picture' can be taken, using the infra-red rays emitted from the body surface, and abnormal areas detected. Breast cancers may be picked up in this way. The technique is difficult and has not been widely adopted. It is much less acurate than mammography.

Evaluation of screen programmes. No one is likely to dispute that early diagnosis is a good thing. The widest possible mass surveys would therefore seem to be the merest common sense. But they are expensive and time-consuming. What then can be learned from experience so far? As usual we must consider different cancers separately.

Lung. In view of the appalling current rise in incidence

and mortality, this is the most urgent problem of all. In Britain the recorded deaths have increased thus:

1910: 1000 1946: 8000 1966: 27 000

Numbers have actually doubled since 1954 and are still rising. They are now in the region of 30 000 annually, and in men lung cancer is responsible for 1 in 12 of all deaths. One heavy smoker in eight will die of it. It actually accounts for 40 per cent of all male cancer deaths.

There have been several mass surveys, by radiography and/or sputum cytology, but the results of pre-symptomatic treatment have been little better than that given after symptoms have appeared. Mass radiography is relatively cheap, but sputum cytology is very costly in expert man-hours. The bitter fact is that 80 per cent of all patients undergoing surgical resection die with secondaries which usually precede radiologically detectable disease. In theory, earlier diagnosis should increase the small proportion without metastasis, but in practice it does not seem to reduce the ultimate mortality.

The conclusion so far must be that, until there is a marked improvement in the results of treatment, the case for routine mass radiography or sputum cytology is weak, both on medical and economic grounds. Exception should be made for middle-aged people with persistent cough and for heavy smokers—if only as a gentle reminder! Much better dividends could be expected from preventive measures if only they could be widely applied.

Cervix. Here the picture is brighter. The cervical smear is well established in many places as a routine procedure. With a specially shaped wooden spatula a surface scraping is made round the external os* (Fig. 23.1, p. 224) and the material transferred to a slide. Experience in several countries has confirmed that cervical cancers are detectable many years before the symptomatic stage and earlier treatment gives better survival figures.

'Carcinoma-in-situ' (see p. 226) is of special interest and importance. It is found in four or five out of every 1000 apparently normal women over the age of 25. It is generally believed to be pre-cancerous, but not always. There is fairly good evidence that it may turn into true invasive cancer, but the proportion which do is uncertain and estimates vary from a quarter to two-thirds. Some workers question the whole relationship of carcinoma-*in-situ* to invasive cancer and consider that, even if some do become invasive, most never pass through a pre-invasive phase. In other words, the two lesions may be quite distinct and not parts of a single process.

We must also appreciate that the discovery of early cervical (or any other) cancer by routine examination of 'well' people will still not completely prevent cancer deaths. For instance, in one series of early invasive cancers so discovered, almost one in six later died of cancer.

In spite of these doubts, there is now no question that mass screening can give results. In those centres where intensive screening programmes have been successfully maintained for years, the incidence of invasive cancer has fallen dramatically and advanced cancers are almost non-existent. Our most reliable evidence comes from British Columbia (Canada) where the incidence of clinical cancer of the cervix dropped by over a half in 11 years. But the acid test is, of course, the effect on mortality. Is the ultimate death rate from cancer any lower? So far in British Columbia it has not fallen; we hope it soon will and are anxiously awaiting later figures.

Our knowledge of the natural history of cervical cancer is still so poor that it is difficult to assess these facts and figures. We can at least say that early detection can and does mean a great reduction in suffering and benefit for the individual. But an elaborate organisation is required and the total cost is high. Estimates range from £300 to £1000 per cancer detected.

We must await further evidence and long-term results before we can be certain whether cervical cytology is justifiable on a national scale. Meanwhile we have 5000 new cases annually and 2500 deaths. Treatment is obviously much more successful than in lung cancer, but a mortality of 50 per cent is higher than that in most countries with comparable health services. Local experiments are justified.

Breast. This is an even more important cancer numerically speaking (Table 13.2, p. 148) with an annual incidence of about 14 000 new cases and 10 000 deaths—i.e. 5 per cent of all women will be affected at some time of their lives. The mortality from breast cancer, unlike most other cancers, is not falling; there is actually a slight rise.

Self-examination for a lump in the breast is the simplest and most obvious means of early detection. Intensive public education has been achieved, especially in parts of U.S.A., but initial enthusiasm and regular self-examination soon fade. Though many tumours can be and have been discovered early in this way, the ultimate results in terms of mortality have been frankly disappointing. In general the larger the growth the worse the prognosis. Yet in one series of really small breast cancers (about 1·5 cm, no bigger than a hazel nut) in which the results up to 10 years after operation were relatively good, more than half eventually died of their cancer. One-and-a-half centimetres is near the limit of ordinary clinical detection and we must therefore be very cautious in any claims we make.

Mammography has some definite advantages. It is of little value in the small fibrous breast, but very helpful especially in the large breast. It can reveal clinically impalpable cancers up to four years before they become clinically evident, including lesions less than a centimetre in diameter. It should be used as a supplement to clinical examination, not as a substitute, since the results of both combined are distinctly superior to either alone. An important large-scale survey is in progress in New York. Initial mass screening showed nearly three cancers in each

1000 women and at operation a lower incidence than usual of axillary node invasion. This was a very favourable feature which is certainly encouraging. On the other hand the false positive rate was high, since two-thirds of the suspected lesions submitted to biopsy proved non-malignant. It is too early to assess the long-term effect on mortality and the trial continues.

Looking at the possibilities on a national basis, we must bear in mind—quite apart from how many women could be persuaded to submit to it—the expense of X-ray equipment, radiographers, radiologists, doctors for clinical examination etc. There would be several negative biopsies for each cancer detected, and the cost would be about £1000 per cancer found. The resources for anything but small local experiments in Britain simply do not exist and are unlikely to exist for a long time yet. In a few years' time the New York survey should give us the answer to the key question—does early diagnosis improve survival? If it does, then large-scale surveys may be justifiable.

Even in New York, however, it has now been decided that the results do not justify continuing the programme on the same scale. Some useful lessons have been learned and some useful results achieved, but it is not 'the answer to the breast cancer problem.'

Stomach. In Japan mass miniature radiography has had some appreciable success and discovered early cancers which were successfully resected in two-thirds of cases, a remarkably high proportion. The method is justified in countries with an incidence of gastric cancer as high as that in Japan. In the West the incidence has declined markedly, perhaps due to improved diet.

Large intestine and rectum. Cancer of the large bowel is now the second commonest lethal cancer in Britain, exceeded only by lung cancer. Screening has been practised in some cancer detection centres in U.S.A., by clinical examination and sigmoidoscopy plus barium enema if indicated. Early cancers were discovered, but this type of examination is not likely to be acceptable on a large-scale—even if the resources were available, which is not the case at present.

Reliance must still be placed on immediate investigation of suspicious bowel disturbances or episodes of rectal bleeding, especially in middle age and after.

Prevention of cancer

Prevention is not the same thing as early detection, though the subjects overlap both with each other and with cancer education, so we may usefully consider them together.

At a recent cancer congress one authority asserted that 90 per cent of cancer could be prevented, at least in principle. We needed new organisation, application of new knowledge, a better liaison between public, research worker and doctor. This is a large and optimistic assertion. Even if we do not know 'the cause' of cancer, we actually know more about the causative mechanisms of some types of cancer than of many other diseases which are more or less controllable, e.g. diabetes. *Complete knowledge of causes is not essential for prevention or control.* For instance, we know that protecting the skin from contact with pitch will prevent pitch warts and cancers, and that abolition of cigarettes would prevent most lung cancer, even though we do not know precisely how these irritants work.

The known causes include various physical and chemical factors and some of these we have already mentioned—ultra-violet and ionising radiations, soot, lubricating oils, dyes, asbestos. Occupational exposure is known to be responsible for some cancers of skin, lung, bladder etc. Medical inspectors of factories are well aware of the potential dangers, and ancillaries employed in industrial work can play an important part in education and reassurance of workers, in medical examinations and protective measures (clothing, respirators etc.) and periodic checks (e.g. urine samples for detection of bladder cancer, see above). The number of people involved is relatively small, but there is good reason to suspect that environmental carcinogens essentially similar in action play a large part in cancer production, especially in modern industrialised urban conditions, which includes general atmospheric pollution. Occupational cancers form only a small proportion of human neoplasms, but their importance far exceeds their incidence. They are excellent models of chemical carcinogenesis and give us invaluable leads for prevention. New chemicals and industrial processes are pouring out, but their dangers are difficult to recognise because of the long latent interval before cancer production becomes apparent. It is only reasonable to assume that there are many unrecognised occupational hazards in our midst.

A host of other possible factors have been incriminated such as pesticides, fertilisers, creosote (which is carcinogenic for mice), though the evidence for many of these is slender. Dietetic factors are certainly important, either non-specific malnutrition and avitaminosis allowing absorption of environmental carcinogens (possibly responsible for the high incidence of liver cancer in Africa) or specific but unknown irritants. Suspicion has rested on cooking processes, overheating of fats, food additives (saccharin, cyclamates, preservatives, colouring and flavouring agents) and on smoked foods (the smoke contains the same type of carcinogen as tar and tobacco). Gastric cancer is rife in Iceland where consumption of smoked food is high. But for some obscure reason it has been declining in incidence for the past half century, both here and in U.S.A., while colon cancer has not changed. There is also some correlation between diet and hormonal patterns and changes in dietary fat intake may help to explain differences in breast cancer in Japanese women (see above).

Hygienic factors have been illustrated in cancer of the cervix and penis (see above). Improved oral and dental hygiene is probably responsible for the fall in cancer of the mouth. Septic and jagged teeth used to be associated with cancer of the tongue. Chronic alcoholism, now declining but far from extinct, is also accompanied by a high incidence of cancer of mouth, pharynx and oesophagus and in New York half the patients with oral and oesophageal cancer are alcoholics.

The contraceptive pill has also been a subject of speculation. Whether it has any effect either way—by increasing or decreasing cancer incidence (e.g. in breast or cervix or ovary) still remains to be seen.

These lists could be extended, but we clearly have enough knowledge already to reduce the influence of at least many known carcinogens if only the knowledge was properly applied.

Cancer education

There is a public 'image' of cancer as 'an incurable disease'. In fact the terms are commonly used interchangeably. The natural reaction to this is fear and 'escapism'—a refusal even to face the possibility—till forced by serious symptoms such as pain or heavy bleeding. This attitude is understandable since most families will have had experience of terminal cancer in relatives, friends or neighbours. Cancer's ugly reputation is bound to persist—after all, it is only about a century since medical science stood helpless before the problem, and even now the overall results of treatment are not brilliant.

It is difficult for the layman to appreciate what can now be done—that something like a third of all cancers can be either 'cured' in the popular sense or controlled well enough to give patients an expectation of life not far from normal, and that good palliation of symptoms can be achieved in most cases.

At this point it may be noted that even nurses and other paramedical workers are liable to undue pessimism. For in the nature of things we require their services to care for the failures of treatment rather than the success. Successful treatment is apt to go unsung, while failure cries out loud.

So great is the fear of cancer that doctors tend not to give the full details to patients to avoid causing them excessive worry. One result of this is that many patients successfully treated are unaware of the real nature of their trouble and cannot spread the good news. Only the failures are left to spread bad news. It is not generally realised that there are many thousands of cured cancer patients leading normal lives. And it is difficult to appreciate the difference between radical and palliative treatment, so that the inevitable eventual failure after palliative therapy is interpreted as a failure of attempted cure. This subject is difficult and controversial, but must be an everyday concern to workers in contact with these patients. It is impossible to lay down general rules. Cases will vary in maturity, personality, family background, economic circumstances etc. and each patient is a separate problem.

This fear of cancer is a potent factor in causing delay in diagnosis. If I have cancer—an incurable disease—I prefer not to know it or not to have it confirmed, so I won't go to a doctor. It may not be put consciously in these terms of course, but this sums up the psychological mechanism. For example, medico-social surveys have shown:

1. Most women are well aware that a lump in the breast might be cancer
2. Most patients, after detecting such a lump, delay many weeks or months (and even years) before seeking advice
3. Patients who knew the possible significance actually delayed longer than those who did not
4. The most important factor in delay was the belief that cancer is inevitably fatal

Cancer is seen as a threat to one's personal integrity in the most literal sense. It is our task to get the public to regard cancer as one serious disease among others, to be feared no more, yet no less. Paramedical workers have an important part to play in this campaign of enlightenment, whether inside or outside hospital. But we must realise that we are dealing with something that has deep roots in the human psyche. We are all subject to the illusion of immortality. And the evidence shows that even doctors do not usually get earlier treatment than their patients. Clearly it is not just a matter of simple ignorance; it is much more deep-seated. So let us temper science and rationalism with charity and realism, and appreciate that public education will not accomplish miracles. And we must make no extravagant claims which would only discredit our activities and make a bad situation worse.

There has been a good deal of educational activity, notably by the American Cancer Society, in an effort to enlighten the general public and encourage people to undergo routine checks (e.g. cervical smears, chest X-ray) and report leading symptoms at an early stage. Radio, television and films have been used as well as books, pamphlets and discussion groups (Women's Institutes etc.). Of all these media, the small group is best where individuals are not inhibited from asking questions freely, and can even discuss their fears and phobias. It is uphill work and its results are not easy to assess, but it is possible to lower the barriers of ignorance and prejudice among whole communities in this way. Some of the myths surrounding cancer can thus be dispelled, but cancerphobia is only one of the many and widely prevalent neuroses of our time, and even if we succeed in removing this particular worry from an individual he may merely switch to a cardiac neurosis instead.

Apart from pre-symptomatic checks, educational drives seek to induce people to come to their doctor with early symptoms or signs. It is customary to stress the significance of the following:

1. Unusual vaginal bleeding or discharge, especially intermenstrual or post-menopausal (cancer of uterus)

2. A lump or thickening in the breast—or neck, armpit, groin or elsewhere (breast cancer, Hodgkin's disease etc.)
3. A sore or ulcer that does not soon heal—on skin, lip, tongue etc. (epithelioma)
4. Unusual indigestion or difficulty in swallowing (cancer of stomach or oesophagus)
5. Blood in the urine or from the rectum (cancer of bladder or bowel)
6. Spitting or vomiting blood (cancer of lung or stomach)
7. Unusual cough or hoarseness (cancer of lung or larynx)
8. A change in regular bowel or bladder habits (cancer of colon, bladder, prostate)
9. Bleeding or a change in size or colour in a wart or mole (epithelioma, malignant melanoma)

It is most important to emphasise that *not one of these is at all diagnostic of cancer*. Blood from the rectum is commonly due to 'piles', a lump in the breast may be a simple cyst, vaginal bleeding may come from a benign polyp (or even an old forgotten pessary!). But any of them *might* be the first indication of cancer and should be investigated. Unfortunately early diagnosis does not guarantee cure, and it is foolish and misleading to make such a claim. We have already noted that a cancer is often present for years before even pre-symptomatic detection is possible. In any particular case of apparently early cancer we can rarely be sure whether metastasis has already occurred, but in a significant proportion of cases earlier diagnosis can mean the difference between life and death. This alone would justify our efforts. We can further claim that late diagnosis will worsen the chances of either cure or useful palliation, and will carry a high risk of increased suffering.

The object of public education is to change the climate of opinion, to dispel the belief that cancer is necessarily fatal and replace it by the knowledge that cancer is often curable or at least manageable. An informed public opinion is our best ally against cancer and it is worth stressing such points as the excellent chances in clinically early cancers of skin, larynx, cervix, rectum, tongue etc., and the prolongation of life and comfort (and even working capacity) in other cases where complete control cannot be achieved. Even if the cancer mortality figures are to be marginally improved—and this is the most we can expect in our present state of knowledge—the benefits of early diagnosis and treatment are still real and significant in many cases and of considerable value to the patients and families concerned.

'Cancer education' alone is not an ideal for psychological reasons. Ideally it should be incorporated in a wider programme of health education, beginning at school and including the elements of personal and public hygiene, with the stress on preventive measures and healthy living, fresh air, proper diet, exercise and improvements in our social environment. At the moment the biggest impact of all would result from an effective campaign to reduce smoking, especially cigarettes, but here the psychological problems are notorious. It is worth mentioning that the incidence of lung cancer in doctors is now falling—from 1954 to 1964 it fell by 30 per cent, while in the general population it rose by 25 per cent. Doctors are about the only section of the community where consumption of cigarettes is known to have decreased. Is the same true of nurses, one wonders, especially as cigarette smoking by women generally is increasing. In fact the rate of increase of lung cancer in females is now higher than that for males. We quoted above an estimate that 90 per cent of cancer is preventable. At the moment we should certainly be able to prevent about 40 per cent of cancer deaths in men (mainly lung) and 10 per cent in women (mainly cervix). Even though we are still ignorant of the intimate mechanism of cancer and of the way the body normally controls cellular proliferation, and though each type of cancer has still to be considered separately and empirically, there is good reason to believe that a considerable proportion of the remaining cancers should be ultimately preventable and that major advances will probably be made by the end of this century. Medical science can pin-point many causative factors and estimate the risks, but implementing an effective policy of prevention calls for governmental action and social and personal discipline. This means that public education on a broad front is essential and progress will be slow. It may be regrettable, but most people are reluctant to take good advice about health if it interferes with their pleasure, otherwise there would be far fewer sweets eaten and far fewer cigarettes smoked! (There may be some food for thought in the fact that among the religious sect of Seventh Day Adventists in the U.S.A.—who prohibit tobacco and alcohol and consume little meat, fish, coffee or tea—there is a low incidence of cancers of lung, larynx, oesophagus and cervix.)

Footnote. The health risk from lung cancer to our affluent cigarette-smoking society in the 20th century is comparable in scale to the ravages of tuberculosis in the slums of the industrial revolution in the 19th century. It is a melancholy fact that the revenue from tobacco is enough to finance a third of the National Health Service.

A recent writer has estimated that over 5000 hospital beds in this country are occupied by victims of cigarettes—chronic bronchitis, lung and other cancers, and cardiovascular disorders—i.e. 4·5 per cent of all hospital beds. In other words, in an 800-bed hospital, a ward of 36 would be filled by the results of cigarette smoking.

In U.S.A. it has been estimated that almost 20 per cent of working days lost between the ages of 17 and 64 may be ascribed to cigarettes.

16. Biological effects of radiation

Effects on cells

There have been considerable recent advances in our knowledge of cells, cell nuclei, chromosomes and genes. For our purpose the 'target' structure is the chromosome, and the 'target' molecule is *DNA*, i.e. deoxyribonucleic acid which together with protein goes to form the chromosome. Its two spiral threads twisted round each other form the famous 'double helix'. It is the DNA molecule which holds the genetic information (as genes) of the nucleus.

Mitosis and cell division are among the cell's most radiation-sensitive processes and though the action of radiation is completely indiscriminate and affects the cytoplasm and every other part of the cell, *it is the mitotic activity and its inhibition which are the critical features in growth*. These are therefore of cardinal importance to the radiotherapist.

At the molecular level, radiation has both *direct and indirect* actions. *Direct* absorption by complex molecules causes rupture of chemical bonds with damaging effects. Far more important (since most of living matter consists of water) is the *indirect* action on H_2 molecules by ionisation, with production of chemically active fragments such as H, OH, H_2O_2 and electrons. Most of these recombine immediately and harmlessly to form water again, but some react with nearby complex molecules and cause bond ruptures and chemical changes. At vital points like DNA molecules this can disrupt cell function grossly. Both nucleus and cytoplasm are vulnerable.

These changes will affect cell metabolism and produce observable biological effects. The most serious effects result from damage to chromosomes in the nucleus, on which the vitality and reproduction of the cell depend. *Immediate cell death* occurs only after huge doses of radiation, of the order of 100 000 rads (e.g. atom bombs) which have no place in clinical therapy. At lower dose levels the chromosomal damage does not reveal itself till the cell attempts to go into mitosis (Fig. 16.1, 16.2).

The cell is most vulnerable during actual mitosis, but since this is only a small fraction of the cell cycle, death from damage during mitosis plays only a minor part. In general, mitosis is delayed or suppressed for a shorter or longer time, depending on the dose, and then resumed. At low dosage, the damage may be almost completely repaired and the cell revert apparently to normal (but see below for mutations). At higher dosage, mitosis may appear microscopically normal or abnormal, the daughter cells may be normal or abnormal, and there may be one or several generations of daughter cells before the cell line dies out. Or mitosis may be permanently suppressed, so that the cell will continue to live and function, but cannot reproduce itself—a sort of premature old age.

In the absence of cell death, impairment of cytoplasmic function may follow—e.g. secretory cells may be more or less inhibited, such as sweat glands of skin, mucous and salivary glands of mouth, thyroid gland etc.

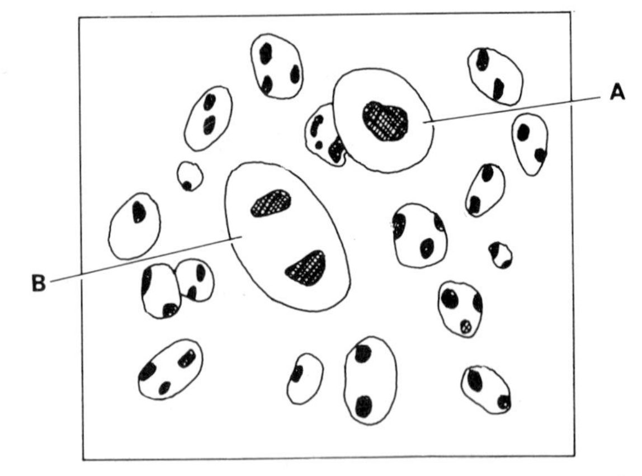

Fig. 16.1 Microscopic appearance of rapidly-growing cancer, showing some cells in process of division (mitosis). A, mass of chromosomes in middle of cell at commencement of activity. B, the twin daughter masses have moved to opposite poles of the cell prior to actual division.

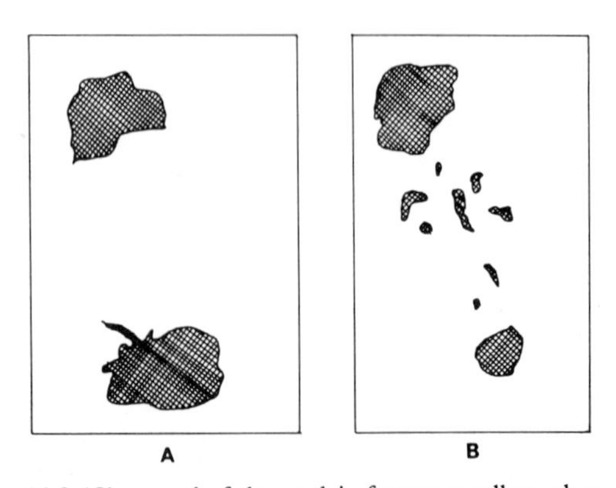

Fig. 16.2 'Close-ups' of the nuclei of tumour cells to show the effects of radiation. A, the daughter chromosome masses in the process of normal division. B, after radiation, showing the broken chromosome fragments left behind.

Effects on cancer cells. There is no essential difference between the effects of radiation on normal and on cancerous cells.

Till recently, the clinical effect of cancer destruction was attributed to the more rapid growth of cancer cells than of normals. We now know that this simple idea is false: cancer cells generally grow no faster than the cells of their tissue of origin—and often in fact grow more slowly. But in normal tissue growth is balanced by destruction (see p. 150). In a cancer more cells are formed than are lost, so there is an accumulation of cells, slow or fast, owing to the failure of balancing physiological controls. This is of the essence of cancer and in this sense cell loss is even more important than cell gain. In any particular cancer system the critical factor is the net surplus; both cancer cells and normal cells will be lost to some extent, but in the cancer the loss is not enough to balance the gain—hence the continuous growth and enlargement of the cancerous mass. How then can radiation ever destroy a cancer without also at the same time destroying its normal host tissue? The reason lies in *the good capacity for recovery and repair of normal tissues compared with the poor capacity of cancer tissues. At an appropriate dose level a cancer can be destroyed by abolishing its reparative powers, while still leaving enough in the host tissue to enable it to heal the gap. This is the basis of the radiotherapy of malignancy.*

The mechanisms discussed so far are not the whole explanation of the success of radiation in cancer. To achieve the death of all the cells of a tissue culture, huge doses are necessary—quite impractical for therapy, where much smaller doses suffice. Other mechanisms are also involved. Of these, the most important is the destruction of the small blood vessels on which a growth depends for its blood supply. Another indirect effect (this is more controversial) is a kind of inflammatory response of the reticuloendothelial defence-cell system; cells weakened by radiation are destroyed, just as bacteria are destroyed in a boil. The effects of radiation on a growth may therefore be summarised as follows:

1. *Direct*
    - (a) some few cells are killed early, before mitosis.
    - (b) some cells die later in attempted mitosis.
    - (c) some cells are sterilised by 'ageing' i.e. they *lose their reproductive ability* and never again go into mitosis. This is probably the most important effect of all.

2. *Indirect*
    - (a) loss of local blood supply causes local death.
    - (b) damaged cells fall victims to tissue defence-cells.

After destruction of the cancer cells, the resultant wound is healed just like a surgical or traumatic wound, by reparative growth of normal tissue in the form of a scar.

Effects of radiation on normal tissues

Since it is a patient that we treat, not just a tumour, the effects of the rays on other tissues are of obvious importance. Though we try to confine the rays as nearly as possible to the lesion*, normal tissues must always be more or less involved. There are definite limits to the amount of radiation they will tolerate, and doses in excess of these will have serious, even fatal, results.

Radiosensitivity. This means the (relative) vulnerability of cells to damage by ionising radiation. Different tissues, normal or cancerous, have differential sensitivities *dependent mainly on their different growth rates* (i.e. doubling time p. 156). This is the basis of the pioneer research summarised in the famous 'Law' of the French workers Bergonié and Tribondeau (1904)—'The biological action of Roentgen rays is greater, the higher the reproductive activity of the cell, the longer the period of its mitosis and the less the degree of differentiation' They studied the testis of the rat and found high sensitivity of the (rapidly growing) germ cells. It is no surprise therefore to find that the seminoma (a cancer of these germ cells) is also very radiosensitive, whereas a slow-growing osteosarcoma shares the low sensitivity of its parent bone tissue. Table 17.3 page 184 shows the importance of this scale of sensitivities. Radiosensitivity is discussed further on page 177.

The various tissues and organs can be placed in a scale of sensitivity from the highly sensitive to the relatively insensitive. The former are readily damaged by fairly low doses, the latter can withstand much larger doses without obvious ill effects.

The sensitive group includes:

1. The reproductive cells of the ovary and testis.
2. The cells in the bone marrow and elsewhere that produce the blood cells i.e. haemopoietic* tissue.
3. The epithelial lining (inner surface) of the alimentary tract.
4. The epithelium of the skin (epidermis).

Intermediate sensitivity group: Liver, kidney, many other glands (thyroid etc).

Low sensitivity group: Muscle, bone, connective tissue, nerve tissue.

The skin

This is a subject of prime importance since we are bound to irradiate the skin in most treatments, even if it is not the site of the tumour.

In superficial and deep therapy, the skin reaction is the limiting factor in dosage since the highest dose is on the actual skin surface and falls progressively in the deeper tissues. *Skin sensitivity* is taken as the 'yardstick' for measuring the sensitivity of other tissues. The characteristic radiation effect on skin is *erythema* (Greek for 'redness'), and in the days before definition of the roentgen and the rad, the 'erythema dose' was actually used as the measure of radiation dosage. In the orthovoltage era skin reactions were a very useful guide in actual treatment and could give warning signals to the therapist. Nowadays with the advent of megavoltage, which produces maximum dosage below the skin surface, skin reactions are not such limiting factors but are still of fundamental importance.

There is always an interval of time or *latent period* before a skin reaction appears, since it is an inflammatory reaction following the breakdown of damaged cells (basal cells of the epidermis which are the most actively growing cells of the skin). In a typical course of treatment, erythema begins in about two weeks.

Radio-Dermatitis. This may be acute or chronic.

1. *Acute*. This occurs typically in patients undergoing a course of treatment and receiving a large dose in a short time
2. *Chronic*. This includes late effects in:
   (a) Patients—after months or years.
   (b) Radiation workers (in industry etc.), or patients who receive small doses accumulating over a long period of time—usually many years—and do not pass through any acute phase.

The acute reaction may be divided into four stages or degrees of severity, with increasing dosage:

First degree. This is the *epilation* dose* due to the destructive effect on hair roots. The hairs become loose and may be pulled out painlessly or fall out spontaneously. This happens about 18 days after an epilation dose and the hair grows again after 2–3 months. The effect was used in treatment of ringworm of the scalp, but has now given way to modern antibiotics.

Second degree. This is the typical bright-red erythema, sharply localised to the irradiated area. Sweat glands are inhibited and may be permanently destroyed. The hair will fall out; it usually grows again, but may not. When the erythema subsides, there is often some residual pigmentation, like sunburn, which may be permanent. There is usually some itching and some temporary dry peeling (*dry desquamation**) as the superficial horny layer of the epidermis becomes thickened and flaky and scales off.

Third degree. The erythema is deeper and purplish. Blisters appear, raising the epidermis; these coalesce and burst, forming superficial ulcers exuding serum, i.e. *moist desquamation*. If left exposed, an adherent scab forms. Healing is usually rapid and complete in 2–3 weeks. Sweat glands are destroyed and loss of hair is permanent.

Fourth degree. The first three stages may be seen in normal treatment, but the fourth stage—the radiation 'burn' or acute necrosis—should not normally be seen at all. It represents an overdose due to a technical mistake or an error of judgement. The erythema is intensely dark, blisters go deeper than the epidermis to involve the whole skin or even deeper, and the whole surface eventually sloughs off. Like an ordinary heat burn, it may be very painful. Healing may occur, but is always a long and doubtful matter of weeks or months, chiefly because of the destruction of blood vessels.

At medium voltage and for small fields, reactions appear after single doses approximately as follows: 1st degree, 500 rad; 2nd degree, 1000 rad; 3rd degree, 1500–2000 rad; 4th degree, over 2500 rad.

Individual and local variations. People vary in their sensitivity to ionising radiation, just as they do to ultra-violet rays (sunburn). Blondes with fair hair and fair skin are more sensitive than brunettes.

Different parts of the body vary also. The more sensitive areas include those subject to moisture and friction (axilla, groin, vulva, anus) and those with a relatively poor blood supply (back of hand, back and sole of foot, midline of back, and areas overlying bone or cartilage, e.g. pinna).

Sequelae to skin reactions: Chronic radio-dermatitis

After treatment carried to high dosage, the skin will show some radiation effect for a long time, usually permanently, especially after superficial or deep therapy.

All the following changes are now seen far less frequently since the introduction of megavoltage with its skin-sparing effects.

Ischaemia.* Many of the late effects of therapeutic doses in the skin or any other organ are due to the destruction and choking of local blood vessels with consequent ischaemia, often associated with fibrosis. We shall have occasion to refer to this when discussing particular treatment, e.g. complications of radium treatment in cervical cancer (p. 232) and effects on brain, bone, bowel etc. (see below).

Pigmentation. This may vary from light to very dark brown and will show the size of the irradiated field. It may be distributed in a patchy manner, especially at the edges of a treated area, and may be mingled with whitish patches of depigmented atrophic skin.

Thickening. The skin may heal with considerable fibrosis of the dermis, giving a typical leathery feel with loss of elasticity.

Telangiectasia,* i.e. dilatation of terminal blood vessels. Destruction or narrowing of the small arteries of the skin may lead to compensatory dilatation of capillaries, which can be very disfiguring.

Late ulceration. An atrophic area is always vulnerable to injuries that would normally be of negligible importance. A scratch or burn, even years later, may lead to a persistent breakdown. This is late necrosis; it is very slow to heal and may need excision and grafting.

Malignancy. This is a good example of the carcinogenic effect of chronic irritation (p. 151). In pioneer X-ray and radium workers, before the dangers were realised, skin changes appeared especially on the fingers. The skin became dry, lost its elasticity, and erythema formed round the nails, which became fissured and irregular and might be shed. Later warts and fissures appeared on the skin and eventually, after some years, epitheliomatous degeneration occurred.

Similar changes also happened in some patients subjected to repeated courses of radiation, especially for non-malignant conditions pruritus, psoriasis etc. see p. 267). These dangers are now well appreciated in industrial and medical situations, and are—or should be—chiefly of historical interest.

Treatment of skin reactions

Explanation to patients. Patients should always be given a simple explanation of the likely effects of treatment on the skin—and any other organs likely to be affected such as bowel and bladder. It is useful to have a leaflet to hand out at the start, telling them what to expect and what precautions to take—(Fig. 16.3)—otherwise reactions may cause needless alarm. If moist desquamation is expected, they should be told of the probable breakdown and discharge, crusting and eventual healing after two weeks or more. They should be assured that these are normal reactions, not 'burns'.

In the milder reactions, little or no special treatment may be necessary—e.g. in dry desquamation—unless the part is exposed to friction when a simple covering may be useful until the skin has healed. Even with a moist desquamation, if it is only a small area (e.g. a small epithelioma of the face), it is often quite satisfactory to leave it alone, allow it to crust over and leave healing to proceed until the crust drops off the new epidermis. If infection underneath the crust is suspected, it may be removed with forceps. If infection is present or threatens during the course of treatment, an antiseptic cream such as Hibitane or Cetavlex (cetrimide) can be used.

In first and second degree reactions, the chief complaint is usually of simple irritation or itching. An ordinary dusting or talcum powder may be used, but since most of these contain a heavy metal (zinc or bismuth) they should not be put on before treatment has finished, because the metal when irradiated gives rise to secondary radiation, which increases the skin dose and therefore the severity of the reaction. Instead a simple starch or baby powder should be used. For the same reason zinc oxide adhesive strapping should be avoided and sellotape or micropore used instead. In areas of friction, lanolin or tulle gras may be applied. When the full course of treatment is over, creams or ointments containing metals may be used freely, e.g. zinc and castor oil.

The patient should be cautioned against all forms of irritation to the treated area. It is usual to advise against washing the part at all until reactions are over. But it is not the (lukewarm) water or toilet soap that is to be feared, but the vigorous towelling likely to be applied after washing that really aggravates reactions. If the patient can be trusted, very gentle washing and gentle drying by dabbing are permissible.

If skin marks have been outlined on the patient for the radiographer's guidance, they must not be washed off. They tend to be removed by sweat and friction, and may need frequent reinforcing by marking pencil or gentian violet.

Shaving with ordinary razors is forbidden, as the trauma increases the severity of reactions. Electric shavers may be cautiously allowed. On the neck, tight or stiff collars should be discouraged because of chafing. Collars anyhow should be unbuttoned and a silk scarf makes an acceptable neckwear. Tight brassieres, corsets etc.

INSTRUCTIONS TO PATIENTS TREATED BY X-RAYS OR RADIUM

Care of the Skin

Some redness and tenderness of the skin is to be expected about two to three weeks after beginning treatment.

It is important to observe the following instructions in order to keep the skin healthy and avoid trouble now or in the future:

1. Keep the area dry.
2. Do not wash the part (unless permitted by the doctor).
3. Do not apply ointments or lotions (unless prescribed by the doctor).
4. Do not apply hot fomentations, either during the treatment or afterwards.
5. Avoid direct sunshine or cold (winds, etc.) on the part.

Fig. 16.3. Pamphlet of instruction given to patients at the start of a course of treatment.

are also bad as are counter-irritants like Thermogene wool.

In short all forms of irritation, mechanical, thermal or chemical—including hot water bottles—are to be avoided since they increase the severity of reactions, just as in sunburn.

Moist reactions usually need dressings. Non-adhesive siliconised dressings are useful, or Melolin (which is expensive). An antiseptic may be required to prevent infection, for sepsis will delay healing and worsen the cosmetic result. Every department has a favourite prescription. One is an antiseptic dye such as gentian violet (1% aqueous); a single layer of gauze is applied to the surface and the dye is painted over it several times daily and allowed to dry. No attempt is made to remove it at subsequent paintings. It has a coagulant effect and forms a protective coat which is left on until it eventually separates from the healed surface. If frequent dressings are needed, an oily preparation is useful like acriflavine emulsion, or flavine and paraffin.

Patients are fond of applying Vaseline, but this tends to hinder discharges and promotes infection. If the moist area is large or the patient's condition poor, hospital care is advisable with nursing as for an ordinary burn.

Fourth degree reactions, i.e. acute necrosis, need similar but more prolonged treatment. For chronic ulcers, Aserbine cream is useful. If ulceration persists, excision and possibly grafting may be needed.

At the end of the course of radiotherapy, if skin reactions are present or anticipated, the patient should always be told what he should do and not do. Gentle washing is usually allowable after two to three weeks with lukewarm water. No special attempt should be made to rub off the skin markings.

Reactions on mucous membranes

The gastrointestinal mucosa is more sensitive than the skin since it renews itself at an even faster rate. Cancericidal doses would usually involve intolerable ulceration, diarrhoea, bleeding etc. with high morbidity and even mortality. This is why common cancers of stomach and colon are not treatable by radiation, and even palliative treatment is seldom rewarding.

Mucosal reactions are comparable to skin reactions, allowing for structural differences. Instead of hairs and sweat glands, mucous and salivary glands are inhibited. There is no horny layer to flake off, but at the stage of superficial ulceration (third degree) the exuded fluid coagulates and forms a whitish membrane composed of fibrin—the fibrinous (or membranous) reaction—which can be mistaken for thrush. This membrane becomes yellowish and gradually decreases in size as healing proceeds.

In the mouth and pharynx, these reactions may cause unpleasant dryness, loss of taste, sore throat and dysphagia. In the oesophagus, which is bound to be involved in the treatment of lung and mediastinum, there may be soreness, painful spasm and dysphagia.

In the bowel, when abdomen or pelvis is treated, there may be spasm and diarrhoea which can lead to dehydration, and also bleeding. When cancer of the cervix is treated, the rectum (immediately behind the vagina) receives a considerable dose and some degree of proctitis* is usual with irritation, tenesmus,* passage of mucus and possibly blood. In the bladder, reactions may cause dysuria with pain and frequency.

Late effects on mucosal surfaces may appear after weeks, months or years. There may be adhesions, fibrosis and stenosis,* leading to obstruction. There may be ulceration, fistulae* and bleeding. Surgical intervention may be needed for any of these. Another possibility is a malabsorption syndrome.

Treatment of mucosal reactions. As a prophylactic measure, dental treatment should be carried out where necessary in head and neck cases—see p. 192

When mouth and throat are involved, the diet should be light. Foods of high calorie value (e.g. Complan) are helpful. Mouth washes should be used liberally to remove viscid secretions. When the reaction is at its height, generally towards the end of the course or soon after, eating may be painful. Aspirin mucilage or a local anaesthetic before meals (e.g. Mucaine) will help. Hot or spiced foods (vinegar, pickles etc.) should not be given. If possible smoking and alcohol should be stopped.

Bowel reactions are common and important in the treatment of abdominal and pelvic lesions. Drugs may be required to control vomiting, spasm or diarrhoea—chlorpromazine (Largactil), codeine phosphate, opiates etc. Reactions may set in early and may at times be so marked that treatment will have to be interrupted for a few days, or in extreme cases stopped entirely.

Effects on blood-forming tissues

Haemopoietic tissue—mainly bone marrow and lymphoid tissue—is highly radiosensitive. The most marked effects are on the parent cells of the leucocytes, lymphocytes and platelets. Red cells are much less radiosensitive, as their life-cycle is much longer—about four months, as compared with a day or less for most white cells.

In patients, the effect on the blood count is very variable. It depends on many factors, particularly the size of the area treated and the amount of bone marrow irradiated. There is a fall in total white cells—*leucopenia**—and in platelets—*thrombocytopenia** or thrombopenia—but red cells may hardly be affected at all. If only a very small part of the body is under treatment, the effect on the blood will be negligible and in superficial therapy—e.g. for skin cancer—we do not bother about blood counts. But the larger the field and the more penetrating the radiation, the greater will be the effect on haemopoietic tissue. During most courses of therapy, full blood counts are usually made weekly and often more frequently—sometimes even daily.

In really large field therapy, e.g. abdominal baths, the white cell and/or platelet counts may be the limiting factor. If counts drop too far it will be necessary to suspend treatment for a few days (or even stop it completely), otherwise the patient will be dangerously exposed to infection because of leucopenia, and haemorrhage because of thrombopenia. As to the lower limit of safety, this is a matter of medical judgement and will vary from case to case. As a rough guide we may say that counts of less than 2000 white cells or 80000 platelets (normal = 250000 per cubic mm) are danger signals.

Just as for skin effects, marrow damage occurred in pioneer radiation workers who developed aplastic* anaemia or even leukaemia as a result. This is another example of the carcinogenic effect of chronic irritation.

Effects on reproductive organs

The gonads (ovary and testis) contain two separate types of tissue:

1. Reproductive, for formation of germ cells (ova and sperm). They are among the most radiosensitive in the body.
2. Endocrine, for production of sex hormones (oestrogens* and androgens*).

Males. Sperm production can be temporarily halted by quite low doses, e.g. 50 rad. Permanent sterility occurs after about 1000 rad. Androgenic* hormone production is much more radioresistant and this effect is rarely of clinical importance.

Females. Sterility can similarly be induced by radiation and depends on physiological age. Hormonal effects are more obvious and of much greater clinical importance than those in the male. Production of ovarian hormone can be reduced or abolished with temporary or permanent suppression of menstruation. This effect is used in inducing an artificial menopause (p. 268).

Gene mutations. Very low doses of radiation, too low to have any obvious effect on mitosis, can still affect the genes from which chromosomes are made and so produce genetic effects in the offspring. These mutations (Latin for 'changes') are almost always detrimental. Once produced they are permanent and transmitted to any offspring. When we treat the pelvis of a pre-menopausal patient to high dosage, e.g. for cancer of cervix, sterility will be inevitable and there will be no problem of gene mutations, since there will be no further children. In any event, the alternative surgical treatment would achieve the same result. But irradiation of the gonads, male or female, in the younger age group is only justified for life-saving purposes or for deliberately producing sterility, as in the artificial menopause.

Years ago, ovarian radiation at low dosage was used, e.g. for treatment of dysmenorrhoca* or sterility. This would be frowned on today, as it would mean inevitable 'contamination of the stream of human life', even though the results might not be apparent until later generations.

Radiation in pregnancy. This is particularly undesirable. Damage may be done either to the mother's ovaries or to the fetus. Gene mutations in the ovarian germ cells (ova) may cause changes in later children of the mother or in later generations. The fetus itself is highly vulnerable—not surprisingly, in view of the relatively enormous growth rate and the extreme immaturity of all its tissues. The first three months are the most dangerous and even low dose radiation then can produce such defects as hare lip, cleft palate, mental deficiency etc. Larger doses will kill the fetus and lead to abortion.

We now have evidence that fetal irradiation carries a definite risk of childhood leukaemia. Because of this, diagnostic X-ray departments, before taking films of the pelvis, enquire routinely into the menstrual history, to avoid exposure in women of reproductive age if there is a possibility of pregnancy.

Effects on other organs

Every organ in the body is affected to a greater or lesser extent i.e. they have varying radiosensitivities. This is largely a function of growth rate—e.g. mature nerve cells never divide, so radiation damage does not show itself, whereas the stem cells of the blood are in continual active growth.

Eye. The conjuctiva is rather more sensitive than the skin, but most of the eye is relatively insensitive. Damage to the cornea, especially if accompanied by infection, can lead to corneal opacities with some degree of blindness. The lens is very liable to post-radiation changes—typically a few years after treatment. A dose of about 750–1000 rads can cause *cataract,** depending on the patient's age.

The lacrimal ducts, which begin at the inner ends of the eyelids, are at risk when lesions such as rodent ulcers of the inner canthus are treated. The radiation reaction in the duct can lead to blockage and if the obstruction is not relieved, it may be permanent. Tears will then be unable to escape by their normal channel to the back of the nose and will overflow on to the cheek (epiphora). To avoid this, it is good practice to send the patient to the eye department after the height of the reaction is over, for the duct to be syringed once or twice. This ensures its patency.

When treating cancers near the eye, e.g. nasal cavity, antrum, naso-pharynx, we try to minimise the dosage to the eye itself by specially arranged radiation fields and lead protection. But in some cases, e.g. a growth that has invaded the orbit, it is essential to include all or most of the eye in the high-dose region and we must accept the consequences if curative treatment is our aim. With megavoltage, the surface-sparing effect is valuable in lessening the actual dose which conjuctiva and cornea receive, and hence the severity of the inevitable conjunctivitis. When the reaction begins, antiseptic eye-drops or ointment should be used freely, e.g. sulphacetamide (Albucid) or hydrocortisone-neomycin.

Apart from cataract, degeneration of the globe is liable to occur at high dosage, with complete loss of vision. The eye will be painful and may have to be enucleated. This may occur about 3 years after radical dosage.

See also page 188 for precautions taken in treatment of the eyelid.

Kidney. Radiation can damage renal blood vessels and lead to degenerative changes. The possible consequences are increased blood pressure (benign or malignant hypertension), and acute or chronic nephritis with raised blood urea,* renal failure and uraemia.*

The borderline of safety is about 2000 rads, and if the kidneys are included in the radiation field they must be shielded by lead blocks when this dose is reached, after being localised either by a plain X-ray film of the abdomen or by IVP.

Brain and nerve tissue. Although relatively resistant (since adult nerve cells never go into mitosis) serious effects can follow high dosage. This is due to damage to the blood supply of the brain or spinal cord, which in turn leads to degenerative changes. After treatment of a brain tumour, there may be late symptoms suggestive of recurrence, but these may be due to radiation damage alone. In the spinal cord, radiation myelitis* can cause paralysis etc. even years later and be mistakenly attributed to metastasis.

When treatments are planned, e.g. to pharynx, oesophagus etc., we always take account of the dose to the central nervous system to ensure that tolerance is not exceeded.

For tolerance levels see p. 254.

Bone and cartilage. High dosage may result in devitalisation, necrosis (early or late), and fracture. Vascular damage is again the chief cause. In rare cases, malignant change (osteosarcoma or fibrosarcoma) has developed years afterwards. In growing bone, damage to the epiphysis* can retard growth and cause, for example, shortening of a limb or spinal scoliosis. This must be remembered whenever radiation is used in children.

Cartilage necrosis can occur in, for example, the outer ear (pinna and external auditory meatus), nose (ala nasi) and larynx.

All these changes are aggravated, or precipitated by trauma and infection. Dental caries is an example we have already discussed.

Lungs. The inflammatory reaction in lung tissue—radiation pneumonitis—may cause serious scarring (fibrosis) which prevents the lung expanding properly and so reduces vital capacity and makes the patient more liable to pulmonary infection. We therefore try to avoid lung tissue. For example, in treating the chest wall in breast cancer, special techniques are used to minimise damage to underlying lung (Fig. 22.4, p. 220).

Radiation sickness

This is a general reaction which is liable to occur during a course of treatment. Its severity depends on the part of the body and the volume of tissue which is irradiated. If we treat a small epithelioma on the skin, there will be no general reaction at all. But if we use large-field or 'bath' type of therapy for a deep tumour, especially if the upper abdomen is included, there may be marked general upset.

The clinical picture resembles sea-sickness—tiredness and weakness, headache and nausea, maybe vomiting. In very sensitive or debilitated patients there may be prostration and treatment may have to be interrupted.

The cause of radiation sickness is obscure. Radiation has a destructive effect on tissue, especially rapidly dividing cells. Abnormal breakdown products may act as toxic foreign protein, which the body is not accustomed to deal with, and cause a minor or major degree of shock. The nervous type of patient is more liable to sickness than the stoic and much good can be done by simple explanation before the course begins, with reassurance that some degree of tiredness etc. is to be expected and is not abnormal.

If drugs are indicated, a simple sedative may be enough or Vitamin B_6 (pyridoxin). There is no specific treatment for the sickness, but anti-sickness drugs may help, e.g. Avomine, Stemetil, Fentazine, chlorpromazine (Largactil). Adequate fluid intake should be maintained—4–5 pints daily—to dilute and eliminate any offending toxins.

Radiation sickness is much less common than it used to be. This is due to better patient care, improved radiation techniques, and megavoltage.

Whole body radiation

As a therapeutic procedure this is rarely used now (e.g. polycythaemia). However, chronic exposure is natural and unavoidable and affects all living beings on earth. We are subject to radiation from (a) natural sources and (b) man-made appliances etc.

Natural radiation arises from (a) cosmic rays and (b) naturally-occurring radioactive elements of the earth (terrestrial).

Cosmic radiation, from the sun and stars, is a mixture of penetrating radiation, subatomic particles (protons, electrons, neutrons, mesons) and nuclei of light elements (carbon, helium etc.). Most of it is filtered off by the earth's atmosphere. The small fraction reaching us is highly penetrating and irradiates the whole body uniformly.

1. Terrestrial radiation comes from the families of thorium and uranium, which are distributed over almost the whole earth in minute quantities in rocks and soils, especially granitic rock. Gamma-rays arise from them, and building brick and stone are therefore liable to be sources of radiation. Radon gas, a decay product of uranium, diffuses from the earth into the atmosphere, where it is normally present in very low concentration. The human body itself contains traces of radioactive carbon and potassium, as well as of radium absorbed from food and water and deposited in bone.

Of the total natural radiation, the terrestial provides about half and the cosmic and body fractions a quarter each.

In considering the effect of natural radiation on mankind, it is probably the dose to the gonads that is of most importance. The average dose from all the above sources has been estimated at roughly 0.1 rad per year, or about 3 rad per generation of 30 years.

Genetic effects of radiation. Gene mutations are of prime importance, as they provide the 'raw material' on which the evolutionary processes of natural selection may work. Mutations occur naturally and continually in genes at a definite but very low rate; all genes are subject to possible mutation.

The causes of mutation are not clear; genes are complex molecules, and disturbances in structure can arise from random chemical influences as well as ionising radiation. Natural background radiation is responsible for only an uncertain fraction of ordinary mutations, perhaps about 10 per cent. There is no reason to suppose that radiation, natural or otherwise, produces any novel type of change; *its effect is simply to increase the rate or frequency of change.*

The effect of radiation on genes differs from the usual biological effects (skin changes, etc.) in some important respects. *There is no threshold level of dosage and the increase in mutation rate is directly proportional to the dosage received; in other words, the smallest exposure has a definite even if minute effect.* Moreover, there is *no recovery effect;* all doses, even the tiniest, add up inexorably, and repeated low doses produce the same result as one or more large doses if the final totals are the same.

Man-made sources of ionising radiation. These include the increasing uses of radiation in industry and medicine. *In medicine*, diagnostic X-rays contribute a far greater dose to the population as a whole than radiotherapy. *In industry*, radiography is widely used for e.g., examination of castings. Physicists, chemists, engineers, and others engaged in the production of X-ray tubes or radioactive isotopes are similarly exposed. Isotopes are employed in a wide variety of industrial processes for calibration, accurate control etc. The development of *atomic energy* plants has added a new sector on the industrial front and here radiation hazards involve many thousands of workers. Other minor sources are luminous clock and watch dials and unshielded rectifying valves in television sets.

There are many *historic* examples of the dangers involved and some have already been described—skin cancers in pioneer X-ray and radium workers, anaemia and leukaemia, sterility, cataract after some industrial accidents, bone sarcoma in luminous dial painters, lung cancers in miners of radioactive ores—see page 151.

The acute radiation syndrome

This is exemplified by the *atom bomb*. At Hiroshima and Nagasaki, where the bombs fell in 1945, most of the casualties were due to blast and fire, but 15–20 per cent were from gamma and neutron radiation. The effective range for serious injury from radiation was about one mile, compared with three miles for flash burns and five miles for blast.

The radiation doses have been estimated and the survivors carefully followed up and observed. Other episodes have been: unintentional exposure in the Marshall Islands in the Pacific in 1954 in the course of bomb testing, affecting nearly 300 people, over a dozen accidents involving high exposure in industrial plants, laboratories and hospitals. All these, plus experimental animal work, have given us a detailed picture of the effects and mechanisms of acute radiation damage.

1. Doses of about 700 rad—i.e. to the whole body at once—leave few if any survivors.
2. Doses of 450 rad make all victims very ill, and 50 per cent die in 2–3 months.
3. 150 rad cause illness in about half.
4. 100 rad cause sickness in about one-seventh, with very few deaths.
5. 50 rad rarely cause sickness.

The effects may be summarised under three headings:

1. The haematological syndrome
2. The gastrointestinal syndrome
3. The central nervous system (CNS) syndrome.

The typical sequence of events after a dose of about 400 rad is as follows. Nausea and vomiting within the first few hours, then a latent period of several days. Then sickness again, fever and weight loss. Intestinal ulceration now dominates the picture with diarrhoea, fluid loss, dehydration and prostration. Bleeding occurs from bowel, nose etc. Deaths begin in the third week, till 50 per cent succumb by the end of the third month. The rest recover very slowly and suffer premature ageing later and increased risk of cataract,* leukaemia, amenorrhoea,* sterility etc.

At this level the chief target tissue is the bone marrow, where the parent cells of the peripheral blood cells are inhibited or killed. Lymphocytes are affected most, then the other white cells, then the megakaryocytes.* The red cell precursors are also sensitive, but since the life-span of the adult red cell is about four months, anaemia sets in later then leucopenia. Thrombopenia* leads to haemorrhages, leucopenia to bacterial infection, especially from the bowel.

There is some direct effect on gastrointestinal mucosa,* which is one of the fastest-growing tissues in the body and normally renews itself every two days. This is slowed down or prevented, with resultant ulceration and bacterial invasion leading to septicaemia.

At higher dose levels, over 1000 rad, the gastrointestinal effect becomes even more important, with death in a few days from fluid loss and massive bacterial invasion of the blood-stream, in addition to the marrow damage.

At above 2000 rad the CNS syndrome sets in, with headache, coma and convulsions. It is always fatal.

In those who survived the bomb a special hazard involved pregnant women. Abortions and stillbirths (i.e. birth of dead babies) were brought about. If the child in the womb survived, its brain development was liable to be retarded and it might be born with an abnormally small head and mental deficiency.

Late hazards of atom bombs etc. The newer (thermonuclear) bombs are thousands of times as powerful. In addition to the late hazards already mentioned (ageing, leukaemia etc. there are some special dangers from the '*fall-out*'. This is the radioactive dust containing fission products (p. 15) carried to great heights in the atmosphere with the vaporized material from the bomb. The fine particles are carried in air currents, settling down gradually to earth. These dust clouds can travel enormous distances, even encircling the earth many times over, so that *no part of the world is immune*. A bomb exploded e.g. in Nevada (U.S.A.) produces fall-out detectable in England after about five days. If the firing of test bombs continues at its present rate, the gonad dose involved will be quite low, only about 1 per cent of the natural background; but any marked increase in the number of bombs could soon become significant genetically to the future world population. As it is, the fall-out from a bomb can have fatal effects over an area of 7000 square miles and be dangerous for tens of thousands of square miles.

A special danger from fall-out concerns radioactive strontium, a fission product. This is soluble and is absorbed by plants, cattle, etc., and thereby reaches humans in food and drink. It behaves chemically like calcium and is deposited in bone tissue after absorption. It is therefore potentially capable of producing late effects comparable to those in the luminous-dial painters mentioned above. So far the danger is negligible but is one of the indices that have to be watched.

Taking all the man-made sources together (radiography, industry, bombs, etc.) the total gonad dose is estimated to amount to about 20 per cent of the natural background, or rather less than 1 rad in a generation of 30 years. This is a figure that need cause no present alarm. Constant watch is now kept on the dosage to the population from industry, atomic energy plants, fall-out, etc., and our knowledge of the potential hazards is extensive enough to give us warning of any serious risk. Complacency would be foolish, especially as regards any increased firing of atomic bombs, but pessimism would be equally unreasonable.

Table 16.1 summarises the present situation. Radiotherapy is not listed separately; its contribution is almost insignificant. It involves large individual doses to a few but this is of very minor importance to the population as a whole and in any case the patients are nearly all beyond reproductive age.

Table 16.1 Radiation to whole population

Source of radiation	Dose as % of natural background
Natural background	100%
Diagnostic radiology	14.1%
Occupational exposure (medical, industrial, atomic energy)	0.5%
Fall-out from nuclear explosions	3.7%

Relative importance of radiation hazards. It should be realised that radiation forms one of the smallest hazards of the world, even in industrial societies. The main health problems, especially in less developed areas, remain those of infant mortality, malnutrition and communicable disease; and in the so-called 'advanced' countries, those of degenerative disease (e.g. arteriosclerosis), malignancy, respiratory disease and accidents. The number of deaths and injuries that can reasonably be attributed to the effects of radiation in ordinary life is very small—in fact it is not easy to find them at all. If we compare this with over 7000 people killed annually in Britain through road accidents, it will help us to keep things in proper perspective.

17. Principles of treatment and dosage

General assessment and choice of treatment

Once a diagnosis of malignancy has been reached, some crucial decisions must be made:

1. To treat or not to treat
2. To give radical or palliative treatment
3. Which line of treatment to use—surgery, radiation, hormones, cytotoxic drugs—or a combination of these

Very exceptionally, the decision may be to give no treatment beyond sedatives and ordinary nursing care—e.g. if the growth is hopelessly advanced, or the type of cancer is known to be unresponsive. In such cases it is psychologically better not to attempt specific treatment.

Assessment before treatment

In every case we need to gather all relevant information about the patient in general and the growth in particular:

1. General condition—age, coincident disease etc.
2. Extent—how localised is it? How much invasion has occurred? Are there any lymph-node or distant metastases? This information is conveniently summarised where possible by clinical staging (see p. 154).

General condition. It is important to assess the patient's fitness for any drastic procedure. Old age, coincident diabetes, cardiac failure, hypertension, obesity etc. may rule out a hysterectomy for cancer of body of uterus, or radical radiation for bladder cancer.

Extent. Local, regional or distant spread is assessed by:

1. Clinical examination—especially for accessible cancers, such as skin, mouth, cervix, breast. This includes palpation of regional lymph-node areas to detect enlargement, e.g. in the neck for mouth and throat cases, in axilla and above the clavicle for breast cases etc.
2. Radiography—X-ray films of chest for secondaries in lungs, skeletal survey for secondaries in bone, intravenous pyelography (IVP) for investigation of kidneys and ureters, lymphography for lymph-node involvement in pelvic and paraortic areas (see p. 245)
3. Instrumental endoscopy*—bronchoscopy for lung, cystoscopy for bladder, sigmoidoscopy for colon etc.
4. Isotope scans—radioactive tracer studies may detect primary or secondary growths in bone, brain, liver etc. (see pp. 112–13)

Type. Histological examination may be needed for diagnosis in a doubtful case. If there is no clinical doubt e.g. in many basal cell carcinomas of skin, biopsy is often omitted. Sometimes biopsy before treatment is not practicable and histology must wait for the surgical specimen, e.g. after nephrectomy.

Histological grading may provide valuable information about the degree of differentiation of a growth and its liability to metastasis; its possible importance in choice of treatment has been mentioned (p. 155).

When the relevant investigations on these lines have been completed, we have to consider whether treatment is to be radical or palliative.

Radical and palliative treatment

1. *Radical treatment* means the attempt to remove all the malignancy present. Staging of a growth is a logical approach in deciding whether treatment should be radical. In general, only stages 1 and 2 are suitable for radical surgery. Early stages are also the most suitable for radical radiation, but later stages are not ruled out, e.g. a Stage 3 cancer of cervix, with malignancy confined to the pelvis, is unsuitable for radical surgery but often suitable for radical radiation.

With very few exceptions, radical treatment means either primary surgery or primary radiation, or both, with radiation given either before or after surgery. Occasionally cytotoxic drugs may be included as part of radical treatment.

For surgery the growth must of course be operable—it must be considered possible to remove the whole of the primary mass with a safe margin. In some sites it is also possible to remove in the same block of tissue the lymphatic vessels and regional lymph-nodes. This will include any early lymphatic metastases, e.g. in radical mastectomy, the breast is removed with underlying muscle and the axillary lymphatic contents; in radical (Wertheim's) hysterectomy, the uterus and appendages are removed with as much as possible of the pelvic lymphatic tissue.

Radical radiotherapy similarly attempts to destroy the whole of the primary growth and may include the regional nodes.

2. *Palliative treatment* is used if radical treatment is considered impracticable because of age, debility or extent of spread—expecially if there are known to be distant metastases. In general, these cases will be in Stage 3 or 4. The aims of palliative treatment are:

(a) To relieve symptoms—e.g. pain, bleeding, cough—arising from primary or secondary growth
(b) To forestall or heal ulceration or fungation or pathological fracture of bone
(c) In general—to improve the quality of the rest of the patient's life

In radical treatment heroic measures are justified, even if the strain on the patient is severe, as it often will be, for the stakes are life and death. But in palliative treatment we are trying, not primarily to prolong life, but to make it more bearable and we are not justified to impose any severe strain. Treatment should be reasonably short, radiation reactions minimal. There should be no general upset.

Any of the four agents can be used for palliation. *Surgery* may be required for:

(a) Obstruction, e.g. transurethral* resection in prostatic cancer, colostomy* in bowel cancer
(b) Pain, e.g. nerve block, section of posterior roots or sensory tract in the spinal cord, amputation of a painful bone sarcoma on a limb
(c) Paraplegia*—laminectomy* for relief of pressure on the spinal cord by a tumour mass
(d) Fixation of a pathological fracture by pinning
(e) Local removal of a fungating mass, as a toilet procedure—e.g. residual breast cancer

Radiation is even more widely useful for palliation. Here are a few examples—others will be described in later chapters:

(a) Relief of symptoms in lung cancer—haemoptysis,* cough, pain
(b) Control of haematuria in bladder cancer
(c) Bone secondaries—relief of pain and local healing
(d) Cerebral secondaries—improvement in neurological symptoms
(e) Improvement or healing of ulceration or fungation in cancer of skin, breast etc.

Hormones and *cytotoxic drugs* are now used extensively for palliation in advanced cases. These are described in Chapters 30 and 31.

Surgery or radiotherapy?

The choice between these may be easy or difficult. The following factors will influence the decision:

1. Site
2. Operability
3. Radiosensitivity
4. Histology
5. Clinical experience

These factors overlap, and some have already been discussed above.

1. *Site.* Some lesions are definitely unsuited to primary radiotherapy, e.g. all the gastrointestinal tract from stomach to rectum; operable growths of breast, ovary, kidney, thyroid, lung; intracranial gliomas, meningiomas (see Tables 17.4–17.8 below)
2. *Operability.* Some lesions are clearly inoperable, e.g. breast cancer with deep invasion of chest wall
3. *Radiosensitivity* (see Table 17.3 below)
   (a) Low sensitivity of the growth. As a rule some lesions are so unresponsive to radiation that surgery is the only real hope of cure—e.g. osteosarcoma, fibrosarcoma
   (b) High sensitivity of normal tissue. This may be the limiting factor—e.g. if it were not for the great sensitivity of normal intestinal epithelium, some bowel cancers would be treatable by radiation.
4. *Histology.* The degree of differentiation is relevant. Very undifferentiated (anaplastic) growths are more liable to invade and metastasise than well-differentiated growths. This factor should weigh in the balance, though not necessarily be decisive in itself. For example, if biopsy of a Stage 2 cancer of cervix shows high anaplasia, surgery is unlikely to be curative and radiation would be preferable
5. *Clinical experience.* This is really the most important criterion of all and it incorporates all the above factors. Knowledge accumulates on the basis of recorded and sifted experience. Hence the great importance of cancer registration, follow-up of patients, and accurate recording of results—including side-effects, complications and survival. For example an early mass of Hodgkin's disease or lymphosarcoma is easily removable by surgery, but new regional or distant lesions are virtually inevitable sooner or later, so that even enthusiastic surgeons rarely advocate this. In contrast, the results of radiation for malignant melanoma are usually so poor that surgery should always be the method of choice if possible.

All this is not to say that the choice between surgery and radiotherapy is always clearcut—far from it. Other factors may be involved. For example, in rodent ulcers of the face (the commonest site) the cosmetic results of radiation may be poor (they are always rather unpredictable) and this may determine the choice of surgery, especially in women.

In some cases, good results can be obtained by either method, e.g. in early (Stage 1) cancer of the cervix. Similarly with early cancer of the vocal cord of the larynx, excellent results follow both surgery and radiation, but radiation is usually preferred as it leaves the patient with a better voice.

In general, treatment policy varies between different countries and different individuals. There is no one method which is invariably 'best' and different methods give equally good results. Local resources and experience will often determine the choice. If radiotherapy is not readily available—it may be hundreds of miles away —surgery or cytotoxic drugs may be preferred. Individual doctors may have special experience in certain techniques and come to advocate them—not without justification, since any technique will give poor results in incompetent hands. For example, the criteria of operability in lung cancer may vary considerably and some thoracic surgeons will operate on border-line cases where others would not. Some chest clinics refer a large proportion of lung cancers for radiotherapy, others only a small percentage. It is the person that counts, not the scalpel or the machine.

Ideally patients should be seen at joint clinics where all relevant specialties are represented—surgeon, radiotherapist and (where indicated) gynaecologist, urologist, haematologist, dermatologist etc. In this way the best treatment can be jointly decided for the individual patient and clinical trials can be conducted. The ultimate test is the actual result. This is easy to say, not so easy to evaluate. Survival figures are the most obvious criterion of success, but not the only one, as we have already discussed.

Principles of radiation treatment

General medical principles

The radiotherapist is a physician first, a technician specialist second. If he is dealing with a small 'rodent ulcer' on the cheek of an otherwise healthy patient, radiotherapy is a comparatively simple matter—a few brief out-patient attendances suffice and though mistakes and complications can occur, ordinary care and routine thoroughness should ensure a smooth passage and complete success. Most patients however are not in this happy position, and though some may be able to receive treatment as out-patients others will need hospital care. Those treated by radioactive sources (radium etc.) will almost always have to be in-patients, if only for reasons of radiation protection.

The patient's general condition must always be considered first. He may need fluid replacement by intravenous drip, blood transfusion after haemorrhage, antibiotics for infection, concentrated nourishment and vitamins for malnutrition, diuretics* for oedema, drugs for pain, spasm, cough etc. etc.

He may present as a medical or surgical emergency—cardiac failure, laryngeal obstruction, intestinal obstruction, compression paraplegia, pathological fracture etc.—and require the attention of the ENT, orthopaedic or neurological surgeon.

A course of radiotherapy, especially for radical treatment, is usually a strenuous affair for any patient, comparable to the strain of a major surgical operation. One or more of the various radiation effects described in the last chapter are bound to afford at least some discomfort and they may last for many weeks and sometimes even months before healing. The patient will need all the physical and moral support we can give him—not to mention his family also.

Technical factors in treatment

In any radiation treatment there are always two fundamental factors to be decided:

1. *Dose*—what total dosage (rads) should be given?
2. *Time*—how is this dose to be distributed in time?

Should it be delivered in a quick 'single shot', or a slow continuous course, or in successive fractions, and over what overall length of time?

Of necessity, treatment methods grew up in the early 20th century on an empirical basis, by trial and error. Later, scientific (as opposed to clinical) methods were used to gain deeper insight into the mode of action of radiation and to find a rational basis for dosage and timing, i.e. what factors give best results in cancer destruction without sacrificing tissue healing. This is the science of *radiobiology*, which involves laboratory work with tissue cultures, bacteria, small organisms etc. The results have to be applied with great caution to human therapy—a human cancer differs in many ways from a pure tissue culture or a bacterial colony. But the findings have given us valuable insights and are now beginning to affect practical therapy regimes (see below—Cell survival curves).

Radiosensitivity. Some preliminary discussion has already been given (p. 167). Till recently we believed that different types of cells had inherently different sensitivities, and that this accounted for the rapid clinical shrinkage of for example, a mass of lymphosarcoma and the slow shrinkage (if any) of, for example, a fibrosarcoma. Modern radiobiological work has shown that most cells (normal or cancerous) are about equally radiosensitive—which is surprising, as it seems at first to conflict with our ideas of malignancy. But the damage will not become clinically apparent till later mitosis, which may be soon (hours or days) or late (months or years)—or never. And when cells do die, they must be removed from the scene before the tumour can be seen or felt to shrink.

To the clinician a growth is radiosensitive if it regresses fairly rapidly and does not recur locally. But this is not the same as '*radiocurable*', since rapidly growing radiosensitive tumours are liable to metastasise early. It is pleasing to patient and doctor to see a tumour regress during a course of treatment, but this is largely a measure of the rate of cell death and removal of dead cells. It does not guarantee success; the tumour may not recur locally, but the prognosis depends on the biological behaviour of the growth, whether it has already metastasised or is of multifocal type (e.g. most lymphoreticular growths).

Bergonie and Tribondeau's law (p. 167) stresses the period of mitosis and the degree of differentiation. All cells are more vulnerable to radiation during the actual process of mitosis, and anaplastic growths will have a higher percentage of cells in division at any given time than well-differentiated growths. It follows that anaplastic tumours are likely to be more radiosensitive—which is borne out by clinical experience.

This is superior both in safety and in curative effect. It is better tolerated by the tissues and takes advantage of the difference in recuperative ability (see below) between cancerous and normal tissue. The sequence of events is pictured on Figure 17.1. The cancer cells suffer progressive damage as the course proceeds and the gap between them and the normal cells widens. Eventually all the tumour cells are irreversibly damaged while normal cells can still recover. Figure 17.2 shows what occurs when dosage is too high or too low.

Fractionation is clearly of great importance in any scheme of therapy (Figure 17.3). In the early days methods developed empirically by trial and error, and only recently have attempts been made to place it on a scientific basis, by incorporating the results of modern radiobiological research. Ideally we should match the dose

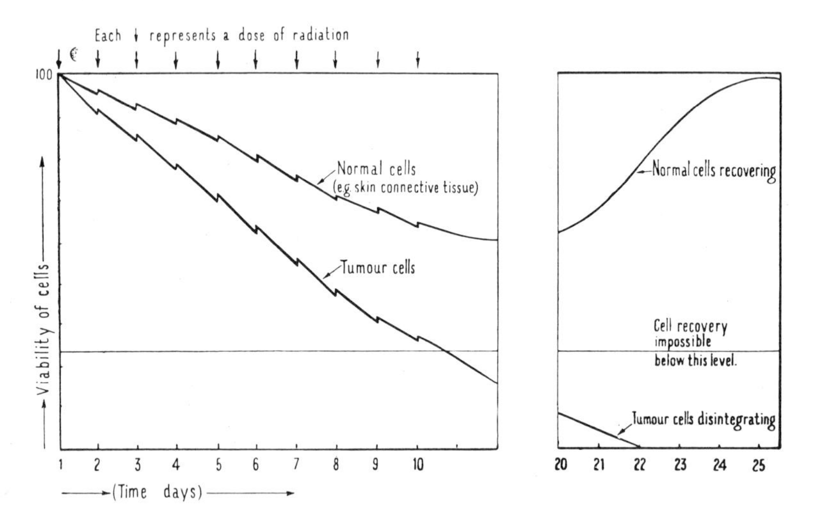

Fig. 17.1 Diagrammatic representation of a successful course of radiation treatment. For explanation see text.

A dose of 2000 rad may suffice for a small (1 cm) skin cancer if delivered in a single treatment session, but if delivered in five daily fractions (5 × 400 rad) it would be insufficient. To achieve a comparable curative result, it would have to be 5 × 600 = 3000 rad. If we took two weeks (10 treatment sessions) the dose would have to be 10 × 450 = 4500 rad. This is a general rule—the longer the overall time, the higher the dosage needed to achieve a comparable result (Figure 17.3).

It would of course be very convenient to use single shot treatments, but this is possible only for very small superficial lesions and even then the cosmetic results are liable to be inferior and the risk of necrosis higher. It was once thought that a single large dose was optimal for cancers generally, even in internal organs, but the unfortunate effects soon led to the use of smaller and smaller repeated doses over longer periods of time. This is the basis of modern technique, i.e. *protracted fractionation*, usually over several weeks.

and its distribution in time to the individual growth, but we simply lack the knowledge to do this. Many patterns of fractionation have been tried, and many attempts made to work out rational treatment plans, including mathematical formulae to cover variables such as number of fractions and overall treatment time. But there is still considerable doubt about the clinical application of laboratory work. We are still feeling our way cautiously and using regimes tested by hard experience. Modifications are being tried out clinically, but it is usually only after several years that results of changes can be properly assessed. The matter is further discussed under 'Cell survival curves' below.

Daily treatments—Monday to Friday—are usual. This began as an administrative convenience, though there is no reason why it should remain the only method or the best method. Some centres now treat many cases three times a week—Monday, Wednesday, Friday. Results are equally good, possibly better in some lesions, and the strain on the patient is likely to be less. With this regime, the overall total dose must be reduced by 10–15 per cent from the conventional total with daily fractions.

Split course. This is another variation—e.g. a fortnight's treatment, after which the patient is rested for two to three weeks, then the treatment resumed for a further week or two; there may be a third course after a further rest. This is particularly useful for elderly patients. It imposes less strain on them and enables us to avoid troublesome radiation reactions, keep the treatment within tolerance limits, and often achieves better results than a standard course.

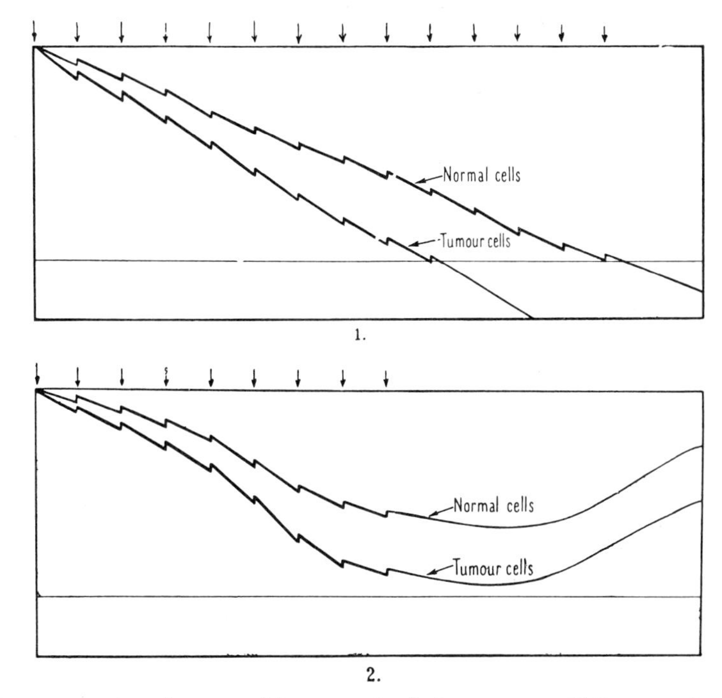

Fig. 17.2 Diagrammatic representation of unsuccessful courses of radiation treatment. (1) Course too long, resultant permanent injury to normal tissues and inability to heal. (2) Course too short, some tumour cells remain active to form later recurrence.

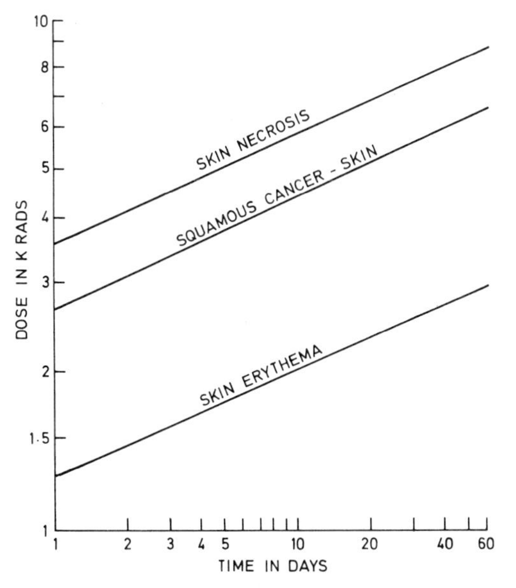

Fig. 17.3 Iso-effect curves relating the total dose and overall treatment for skin necrosis, cure of squamous carcinoma and erythema. (After Strandqvist, 1944. Courtesy of Duncan and Nias, *Clinical Radiology*, Churchill Livingstone.)

The oxygen effect. Another of the findings of modern radiobiology is the importance of oxygen. It has long been known from experience that tumours with a good blood supply tend to respond better than those less vascular. The presence of oxygen enhances the physico-chemical reactions mentioned on p. 166, and this explains the clinical findings, since more blood means more oxygen carried to the tumour.

The radiosensitivity of cells increases rapidly as the oxygen concentration in their immediate neighbourhood during radiation treatment increases, up to a critical level. Above this level, which is about that of the ordinary atmosphere (21 per cent) sensitivity does not increase appreciably (Figure 17.4). The absence of oxygen seems to interfere with the intracellular biochemical events that follow radiation. This 'oxygen effect' is so important that it is clearly desirable to increase the concentration in a tumour during treatment. Most growths tend to outstrip the available blood supply because of the cell accumulation and consequent compression of surrounding blood vessels, leading to slowing of the local circulation and partial necrosis with oxygen deficiency. Thus, the outermost parts of a tumour may be well supplied with blood vessels and oxygen while the centre is liable to be much poorer in both—i.e. relatively *anoxic*. Such anoxic areas will be less radiosensitive than the rest of the tumour—in fact, *an anoxic cell is only one third as radiosensitive as a well oxygenated (aerobic) cell*. Most tumours contain some anoxic, or nearly anoxic, cells which will be foci of radioresistance and therefore liable to give rise to later recurrences after radiation treatment. The larger the tumour, the more likely is the centre to be anoxic and the greater the bulk of anoxic cells. This helps to explain why large tumour masses are more resistant and require higher doses for eradication than small masses. Even a small proportion of anoxic cells can have a great influence on the success of therapy (see Figure 17.7).

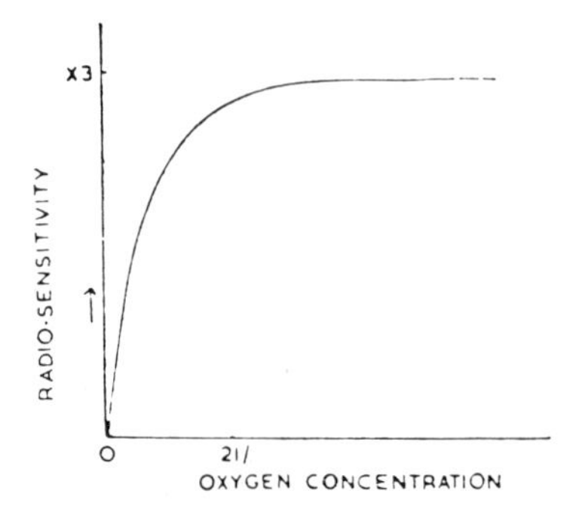

Fig. 17.4 Graph showing change in radiosensitivity with variation in oxygen concentration. Ordinary atmospheric oxygen concentration = 21%.

Other factors may also decrease the oxygenation of tumour cells, e.g. destruction or fibrotic choking of vessels after previous surgery or radiation, or infection (especially syphilis). Anaemia from any cause will have a similar effect.

Hyperbaric oxygen therapy*. It is a logical application of this knowledge to try to increase the oxygen supply to a tumour during actual irradiation. The patient lies in a hyperbaric oxygen tank and oxygen pressure is gradually raised to three times atmospheric pressure, which is maintained during the period of irradiation and then gradually lowered back to normal. The oxygenation of the whole body is thus increased, including—it is hoped—the tumour. The radiosensitivity of normal tissues will be very little if at all increased, since it is normally at its maximum, but any increase in the tumour will be valuable.

This difficult and time-consuming technique has been practised in several centres for several years. Some useful results have been found, e.g. improved results in secondary nodes of neck and in advanced cancer of cervix. Not all reports are encouraging, but clinical work and controlled trials are still in progress. There is some doubt whether there is a significant increase in oxygen where it most matters, i.e. in the anoxic tumour centre. At present the case is 'not proven', but it is too early for final assessment.

Cell survival curves. The laboratory approach is illustrated in Figures 17.5, 17.6, 17.7. They show the results of experimental irradiation of cells (normal or cancerous) cultured in colonies, in the manner of bacterial cultures, in the laboratory. In this way we can study the effects of radiation on mammalian cells directly rather than having to depend on plants or fruit flies (drosophila). Even single cells, like single bacteria, can be grown on the surface of a suitable culture medium and the number of resultant 'colonies' counted.

The higher the dose, the fewer are the cells surviving with reproductive integrity. The steeper the slope of the curve, the greater is the fractional kill, i.e. the higher the radiosensitivity. Any given dose will kill a certain percentage of cells. The remaining cells, after temporary inhibition of mitosis, will recover; if the same dose is applied again, the same percentage of cells will again be

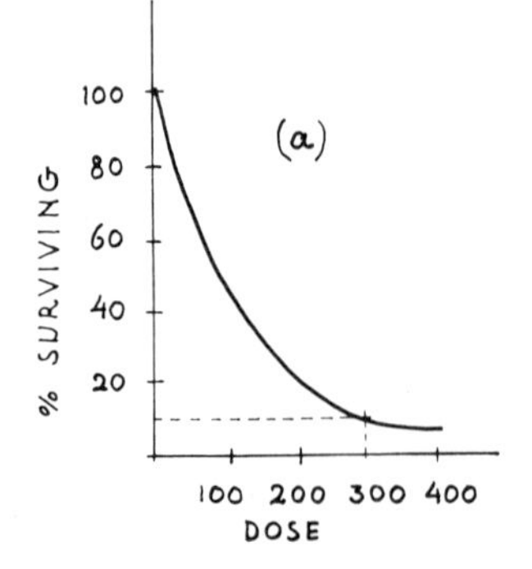

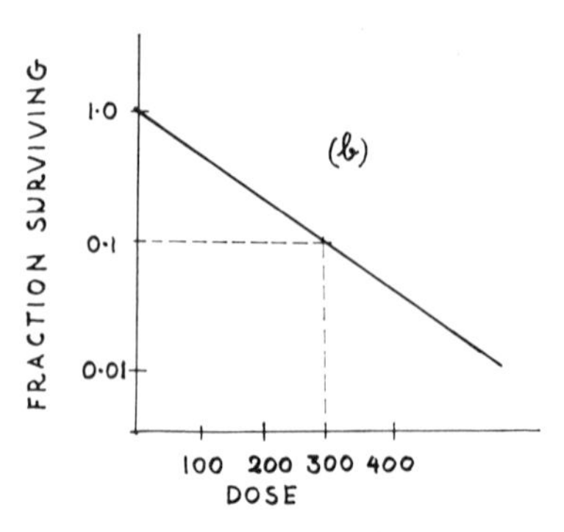

Fig. 17.5 To show the basis of cell survival curves, survival plotted against dose. Dose is plotted in linear fashion in both graphs, i.e. double the distance represents double the dose. Survival is plotted linearly in (a) but logarithmically in (b) i.e. double the distance represents double the percentage survival in (a) but ten-fold in (b). A plot such as (b) is therefore called 'semi-logarithmic', and the effect is to convert the exponential curve of (a) into the straight line of (b) which is simpler to use. The dotted lines show an example such as might be found in an experiment—a dose of 300 rad yields a survival of 10 per cent = (0.1).

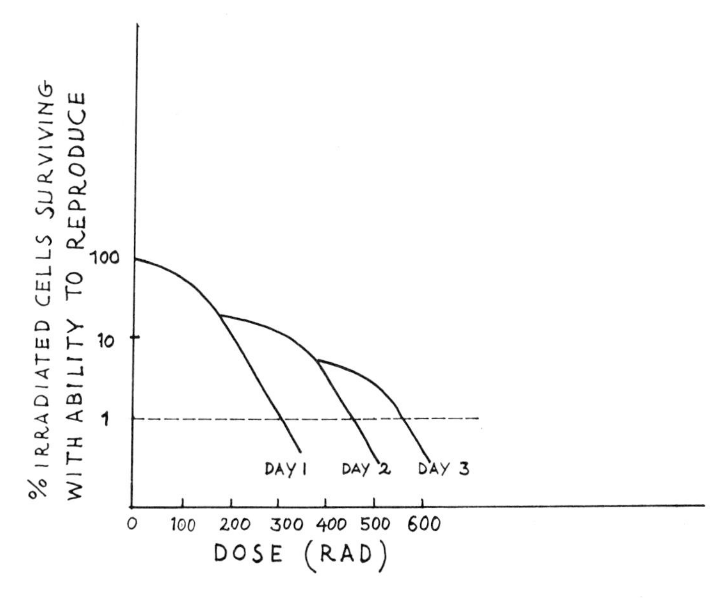

Fig. 17.6 Survival curves after fractionated irradiation of a cell culture on successive days. The initial 'shoulder', before each curve becomes a straight line, represents an early partial recovery effect. The dotted line is at the 1 per cent survival level. A single dose of 300 rad is seen to yield 1 per cent survival, i.e. to kill 99 per cent. To achieve the same result by two fractions is seen to require 450 rad (2 × 225), and three fractions require a total of 550 (3 × 185). Clinical fractionation is believed to resemble such experimental laboratory findings.

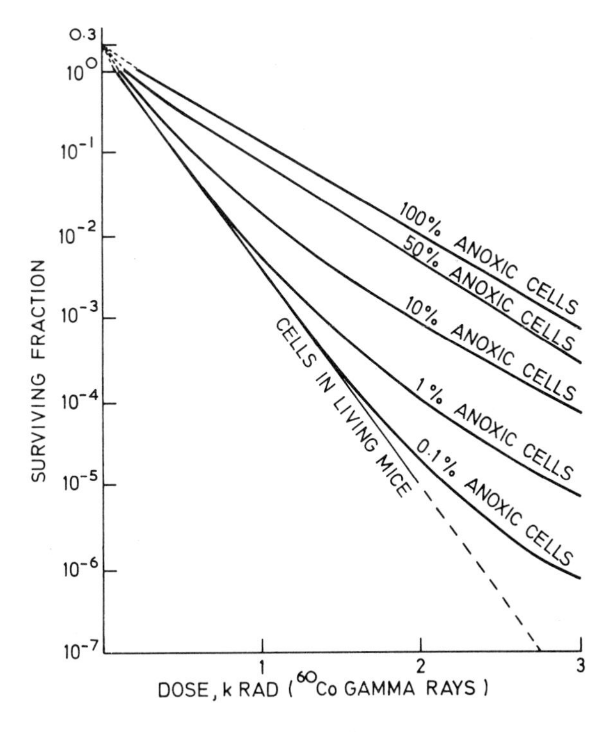

Fig. 17.7 Survival curves for cell populations with various proportions of anoxic cells. The lowest curve applies to well-oxygenated cells, in which a dose of 2000 rad can be seen to give a survival of only 10^{-5} (i.e. 1 in 100 000). In the presence of 1 per cent anoxic cells, survival is 10^{-4} (1 in 10 000) and with 10 per cent anoxic cells it rises to 10^{-3} (1 in 1000). (From Hewitt and Wilson. Courtesy of Duncan and Nias, *Clinical Radiobiology*, Churchill Livingstone.)

killed—and so on. We can thus determine experimentally the effects of variations of dose and fractionation.

After moderate doses there is a small 'shoulder' before the curve becomes a straight line (exponential). This is due to *sublethal damage and cell recovery*, i.e. a proportion of the cells recover rapidly in the course of the next 24 hours. The shoulder is reproduced in subsequent doses of the same order.

Clearly, the most efficient method of cell 'killing' is the single large dose. This is excellent in the laboratory, but not for radiotherapy in humans, as we have noted above. The single large dose was soon replaced, for example, by serial smaller daily doses and the typical clinical course is illustrated in Figure 17.6. Reference to the curves shows that a single dose of, say, 300 rad will kill 99 per cent of the cells. To achieve the same result in two daily fractions needs 450 rad, and in three days 550 rad.

These curves provide a basis for planning new treatment regimes e.g. longer intervals and larger fractions, to obtain equivalent effects. The importance of cell recovery is clear. If there were no recovery, then fractionating the dose would make no difference: 2 × 250 or 5 × 100 would produce the same effect as 1 × 500 rad. Moreover—and this is fundamental to the cancer therapist—the capacity for recovery in normal cells is greater as a rule than in cancer cells, and this accounts for the divergent curves of Figures 17.1, 17.2. This is the really important reason why protracted low dose fractionation is so much more effective and better tolerated than one or a few large doses. Normal tissues have a strong natural tendency to organise themselves again by means of normal repair mechanisms, but the cancer cells have no such mechanisms (or at any rate very impaired powers) of self-organisation—in other words, cancer is distinguished by lack of 'self-control'. The technical term for this is 'homeostasis'. Fractionation takes advantage of this and permits of the maximum damage to the cancer cells with minimum damage to normal cells.

Other factors also come into play. Slow and gradual shrinkage of a tumour mass tends to improve the blood and oxygen supply to the anoxic centre, by relieving the pressure on the local blood vessels. There will also be increased blood in the area owing to the mild inflammatory reaction provoked by radiation. Further, the more frequently the doses are delivered, the more likely is any particular cell to be in process of mitosis, which is the most vulnerable part of the cell cycle. These are all important advantages, but it does not follow that a course of radiation should be prolonged indefinitely. Apart from the strain on the patient (and on hospital resources), repair mechanisms will lead, before many weeks have passed, to some degree of fibrotic scarring and impairment of blood supply, and so to decreased oxygen and radiosensitivity. Local recurrence at a site previously irradiated to radical dosage is not amenable to a second radiation attack, which would be liable to achieve only local necrosis because of the poor blood supply.

The importance of the oxygen effect is shown in Figure 17.7.

Oxygen enhancement ratio (OER). If a dose D is required to cause a certain amount of damage in fully oxygenated (aerobic) conditions, then a larger dose xD is required to cause the same damage in anoxic conditions. The ratio of xD to D, i.e. x, is the OER. Cell survival curves are the best means of measuring it. For X and gamma rays the OER is about 2·5 – 3·0.

Neutrons and OER. (See also p. 135). Fast neutrons have an OER much lower than that of X and gamma rays—about 1·6, as well as high LET and RBE (see below). The oxygen effect is therefore of much less importance, which means that for the same amount of damage to normal (well-oxygenated) tissue, neutrons cause greater damage to anoxic tissue, which is the important danger area in a tumour, than X or gamma rays. This is probably the most significant advantage of neutrons, and trials are now in progress to test it clinically. Early results are very promising.

Tissue tolerance and the time factor. A small area or volume can tolerate a larger dose than a large area or volume, since the damage can be more easily repaired. Spreading out the dose in time, e.g. fractionation, brings the recovery powers of the tissues into play and increases tolerance. It is of obvious importance to know the limits of tissue tolerance. Table 17.1 gives figures of skin tolerance derived from experience. They refer to 200 kV and would need adjustment for other voltages in accordance with RBE (see below).

They should not be taken in an absolute sense, but only as rough guides. Tolerance varies in different parts of the body as noted in Chap. 16. It also varies with natural pigmentation, for a brunette will tolerate more than a blonde (just as for ultra-violet rays in sun-bathing). The

Table 17.1 Tolerance doses in rad (deep X-ray therapy)

Size of field in cm	Single dose	One week	Two weeks	Four weeks	Six weeks
6 × 4	2000	3000	4500	5500	6000
10 × 8	1700	2500	3500	4500	5000
15 × 10	1500	2000	3000	4000	4500
20 × 15	1000	1700	2700	3500	4000

figures apply primarily to skin, but they are also applicable to many other tissues and form an approximate guide to the order of suitable dosage in treating many malignant growths.

Figure 17.3 shows a useful graph based on these principles, for general guidance in treatment.

The quality factor: LET and RBE

In radiotherapy we make use of a whole series of 'qualities' of rays, ranging from the very soft Grenz rays to the very hard gamma rays of radium or megavoltage X-rays. The effects of the various wavelengths, though similar in general, differ in degree. We may illustrate this by listing the doses in rads needed to produce similar erythema effects on the skin (Table 17.2).

We see that the higher the voltage, i.e. the shorter the wavelength, the bigger must be the dose in rads to produce a comparable effect on the skin (and on most other tissues also).

Table 17.2

KV	Erythema dose in rads
50	200
100	270
200	700
1000	1000

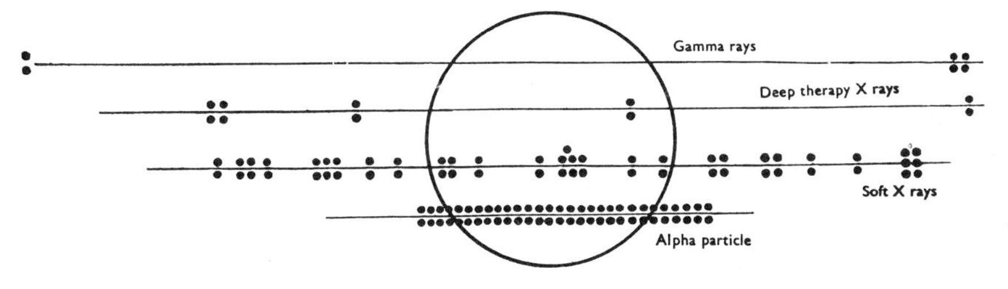

Fig. 17.8 Diagram showing comparative ionization densities along the tracks of various qualities of radiation. (The diameter of the circle is 27 millionths of 1 mm.) [*Gray*, "*Brit. Med. Bull.*".]

Figure 17.8 represents the basis of Table 17.2. As ionising radiation passes through tissue, it gives up energy along its track by setting electrons in motion. The amount of energy given up per micron of tissue (1 million microns = 1 mm) is called the *Linear Energy Transfer* or *LET*. High LET value (e.g. alpha particles, neutrons, soft X-rays) means densely ionising, dissipating more energy per micron of its path than less densely ionising radiation (e.g. gamma rays). The higher spatial concentration of energy loss accounts for the greater biological effect reflected in Table 17.2.

High LET radiation is therefore more efficient than low. This factor is measured by the *Relative Biological Effectiveness or RBE*. This is a useful concept to summarise comparisons between the biological effects of the different qualities of radiation used in clinical practice. It is defined as the ratio of the (absorbed) doses which produce the same biological effects. Since we are most familiar, for historical reasons, with effects at 200–250 kV, we use this as our base line:

$$\text{RBE} = \frac{\text{Absorbed dose for biological effect at 250 kV}}{\text{Absorbed dose for same effect at quality in question}}$$

Thus, if we find the same effect from a dose of 100 rad of a given quality of radiation, as from 200 rad at 250 kV then the RBE of the given radiation is 200/100 = 2; if 400 rad are needed, then the RBE is 200/400 = 0·5.

If we designate the RBE at 250 kV as 1.0, then the RBE's of some important qualities are:

Cobalt–60	0.85–0.90
4 MV	0.85
20 MV	0.80

The facts about relative ionisation density and linear energy transfer (see above) are relevant in explaining these differences in RBE.

The REM (= Radiation Equivalent Man)—see details on page 121—takes the RBE into account. It is the absorbed dose which produces the same biological effect as 1 rad of X-rays at 250 kV.

Dose in rems = dose in rads × RBE value.

Other considerations may also influence our choice. For instance, if we are treating a lesion adjacent to bone or cartilage we prefer the harder rays, for they produce much less secondary radiation from the atoms of comparatively high atomic number contained in the hard tissue (chiefly calcium, see Fig. 5.18, p. 57). Consequently the harder rays are less liable to cause tissue necrosis at a later date in such cases. Calcium tends also to be deposited in cartilage after middle age and its presence in such cartilages as those of the larynx and ear means greater local energy absorption under long wave irradiation than under short waves, with consequent greater liability to late necrosis. This is why gamma-rays are often preferred to X-rays in treating these parts.

Radiosensitivity of tumours

This is another factor of obviously fundamental importance. Tumours like normal tissues vary in their vulnerability to radiation and this depends chiefly on their rate of growth.

Table 17.3 lists some of the most important growths in the order of their radiosensitivity. It is a useful approximation, but there are exceptions—e.g. most lymphosarcomas melt away very rapidly, but the occasional case will prove unresponsive even to high dosage. Most cancers of rectum or melanomas are fairly insensitive, but a few prove to be sensitive.

The most important factor of all is the biological nature and behaviour of the particular cell type of the particular growth. Growths, like people, are never twice the same. Unexpected successes and failures occasionally occur for no apparent reason and there is no way at present of accurately assessing in advance the responsiveness of a particular tumour.

It is important to draw a distinction between *radiosensitivity* and *radiocurability*—see above, pages 177–8.

The highest proportion of successes is achieved in the epithelial cancers (especially squamous) of moderate sensitivity such as skin, mouth, cervix etc.

The treatment of choice

It will be useful to list the chief sites at which the various methods have been found to achieve the best results. The following classification (Tables 17.4–17.8) would not be universally accepted. In fact, no classification can hope to be, since there are quite wide divergencies in resources, experience and judgement. It should be taken as a broad generalisation with considerable room for flexibility.

Table 17.3 Tumour radiosensitivity

Highly sensitive	
Reticuloses—e.g. lymphosarcoma leukaemia, Hodgkin's	complete local disappearance to be expected
Seminoma of testis	
Medulloblastoma neuroblastoma, nephroblastoma (Wilms)	
Sensitive	
Basal cell carcinoma	complete local disappearance to be expected
Epithelial carcinoma—skin, mouth, cervix, bladder etc.	
Carcinoma of breast	complete local disappearance possible, but not usually to be expected
Carcinoma of lung	
Carcinoma of ovary	
Poorly sensitive	
Sarcoma—osteosarcoma, fibrosarcoma	only temporary growth restraint usually to be expected
Malignant melanoma (most cases)	
Glioma	
Carcinoma of rectum	
Carcinoma of kidney	

Table 17.4 Surgery—the treatment of choice

Stomach, intestine, colon, rectum, pancreas, kidney, prostate
Wilms's tumour (in association with radiation and chemotherapy)
Bone—most primary tumours
Fibrosarcoma
Melanoma
Testis—orchidectomy for primary mass
Radiation may be useful for palliation (e.g. rectum) or pre-operatively (e.g. osteosarcoma)

Table 17.5 Radiation—the treatment of choice

Mouth—including lip, tongue, cheek, alveolus, fauces
Intra-nasal, accessory sinuses (antrum etc.), middle ear, naso-pharynx, oro-pharynx, hypopharynx
Skin—except melanoma
Uterus—cervix
Bladder—all except stage 1 growths
Larynx—vocal cords
Testis—seminoma (for secondaries)
Medulloblastoma—after surgical exploration
Lymphomas—Hodgkin's and others (cytototics very often useful as alternative, especially in late stages and leukaemias)

Table 17.6 Surgery—the first line

In operable cases, but radiation is justified even in early cases if surgery is contra-indicated. Useful results from palliative radiotherapy

Breast	Body of uterus
Lung	Ovary
Brain	Oesophagus
Urethra	Thyroid
Testis—teratoma (secondaries)	
Salivary glands (parotid etc.)	

Secondary carcinoma in lymph-nodes

Table 17.7 Radiation—the first line

Though results usually poor with any form of treatment.

- Hypopharynx (epiglottis, piriform fossa, arytenoids)
- Post-cricoid
- Naso-pharynx
- Neuroblastoma (in association with radiation and cytotoxics)

Table 17.8 Cytotoxic drugs—the treatment of choice

Acute leukaemia
Choriocarcinoma
Burkitt's lymphoma
Lymphomas (Hodgkin's etc.) —late stages

Hormones are rarely the treatment of choice in the first instance. They are useful in selected cases of prostate, breast, thyroid, body of uterus and kidney (see Ch. 30).

18. Skin, lip, penis, vulva

Skin tumours are the commonest of all neoplasms and skin cancers are of great importance to the radiotherapist. They are the most accessible of cancers and their limits generally easy to define. The patient usually seeks advice before growth becomes extensive and diagnosis therefore tends to be earlier than in inaccessible organs. This, plus the fact that metastasis—if it occurs at all—is generally later, gives most skin cancers an excellent prognosis.

Causative factors. The skin is an outstanding instance where carcinogenic 'irritation' can be seen at work.

1. Ultra-violet radiation in sunshine is the most important factor of all, accounting, for example, for the very common basal cell carcinoma in Australia etc. Exposure to sun, wind and cold leads to similar changes in farmers, sailors etc.
   Individuals and races with fair hair and skin are more susceptible than the dark-skinned, since the skin pigment acts protectively by absorbing ultra-violet.
2. Over-exposure to X-rays and radium etc. in the pioneer days produced radiation dermatitis, especially in fingers and hands.
   Manufacturers, laboratory workers, doctors etc. were affected and there were many cancer deaths.
3. Chemical carcinogenesis is clearly at work after prolonged exposure to many substances—pitch and tar, soot, mineral oils (shale industry), lubricating oils (mule spinners, machine tool workers). Arsenic, which used to be prescribed medicinally in low dosage for long periods, can lead to dermatitis and later malignant degeneration.
4. Old scar tissue, especially after wounds, burns or tuberculosis
5. Chronic irritation from infected discharging sinuses and ulcers

Pathology and clinical features

Pre-malignant conditions. There are a number of common tumours which can occasionally become malignant. They include:

*Hyperkeratosis**—small warty nodules; may be due to sunlight, tar, X-rays etc.

*Papilloma**

*Benign Melanoma**—the universal flat or raised pigmented patches—see under Malignant Melanoma below.

Malignant lesions

1. *Basal cell carcinoma* or *rodent ulcer*. This forms about 80 per cent of all skin cancer. It arises from the cells at the base of the epidermis. The commonest sites are on the face—nose, cheek, temple, eyelid, scalp. There are several clinical types—a raised pearly nodule, a scaly patch, a depressed cicatrising* lesion healing in the centre and creeping outward at the edges. There is also a less common cystic type, important because it is less responsive to radiation than the others.

Growth is slow and they may exist for years with only very gradual increase in size, but superficial ulceration, scabbing and slight bleeding are usually early. A history of a year or two is common. They do not metastasise, but at a late stage they can turn into squamous epithelioma and then give rise to secondaries. Typically they are 1 to 3 cm in diameter; exceptionally, e.g. in tramps or mental defectives, advice may be delayed till they have done extensive damage by their 'rodent' activity, e.g. destroyed an eye or eaten deeply into underlying bone; death may eventually occur from haemorrhage (erosion of a large blood vessel) or meningitis (erosion of skull).

2. *Squamous carcinoma* or *epithelioma* forms most of the other 20 per cent. They grow more quickly than basal cell lesions and the history is usually shorter. Lymph node metastasis occurs, especially in anaplastic lesions.

A mixed type also occurs, with features of both basal and squamous cells—basi-squamous carcinoma.

3. *Malignant Melanoma* or melanocarcinoma. This is the rarest of the important skin cancers. It has special features and is separately considered below.

4. *Other skin tumours.*

*Bowen's** disease or intra-epidermal carcinoma—superficial brownish plaques, often multiple, on the trunk and elsewhere. Similar to basal cell carcinoma and easily treatable on similar lines.

*Kerato-acanthoma**—a projecting dome-shaped tumour with cheesy centre and a typical history of a few weeks of rapid growth. It resembles squamous epithelioma but is benign and often self-healing, or heals after simple curettage.* Even biopsy* may not distinguish it from carcinoma, and if there is doubt it should be treated as epithelioma by full radiation or excision.

Skin glands—sebaceous and sweat glands may give rise to simple tumours (adenoma) or malignancy (adenocarcinoma). They are rare and best excised.

*Reticulosis.** Skin lesions (lymphoma cutis) may be early manifestation of lymphoreticular disease. The most serious is *mycosis fungoides* (p. 248).

Secondary carcinoma often appears as nodules just beneath the epidermis, especially from a lung primary (it may even be the first clinical evidence of disease) or in breast cancer. Local X-ray treatment is useful to prevent fungation.

Diagnosis is usually obvious on clinical appearances alone. In case of doubt, biopsy is done (unless melanoma is suspected—see below). If the lesion is small, complete excision is often the simplest means of both diagnosis and treatment.

Treatment

Both surgery and radiotherapy yield good results in most cases. The choice can therefore be governed by comparative functional and cosmetic results, as well as by the overall length of time for treatment.

Surgery is preferable in the following circumstances:

1. Excision is speedier than a course of radiation involving multiple attendances
2. Where tissue preservation is not important, e.g. on the trunk or thigh
3. Some sites tolerate radiation poorly and are liable to late necrosis, e.g. back of hand, sole of foot. Skin subject to friction and moisture is also vulnerable, e.g. perineum
4. If bone is invaded, surgery will usually be needed, though preliminary radiation is helpful by reducing tumour bulk. Involvement of cartilage, actual or potential, especially on ear or nose, may weigh in favour of surgery. The avascularity of cartilage may lead to poor healing and necrosis. But if cartilage invasion is only slight, radiation may be successful—and surgery can be kept in reserve in case of failure
5. If the lesion is on the basis of unhealthy skin—e.g. radiation dermatitis, scar tissue, chronic infection—excision of the whole area, malignant and pre-malignant, is advisable
6. Radiation failures or recurrences—which should be infrequent—can usually be rescued by surgery
7. Cosmetic considerations may be decisive, because of the uncertain post-radiation pigmentary changes (p. 168). This will be more important in, for example, a 45 year old actress than a 65 year old farmer.

Small lesions are easily excised, but on the face recurrence is common, as the surgeon naturally tends to be conservative and preserve as much tissue as possible. On eyelids and nose excision may involve difficult plastic repair, which is not often justified, as radiation usually gives good results. Most cases of basal cell and squamous carcinoma are in practice treated by radiation at most centres.

Radiation techniques. Many varieties are available, including X-rays (any voltage from Grenz-rays for very superficial, to DXR or cobalt beam for thick lesions—Figure 18.1; radium (or cobalt etc.) needles or tubes as implants or surface moulds; gold grain implants or moulds. Radium moulds had a long vogue, but are little used now, as they involve expensive equipment, hospitalisation of the patient and unavoidable radiation exposure of nurses and other staff. Some workers still prefer moulds, e.g. for back of hand (Fig. 18.2), claiming it to be safer (for late necrosis) and cosmetically superior to X-rays. Most therapists are content to rely on X-rays.

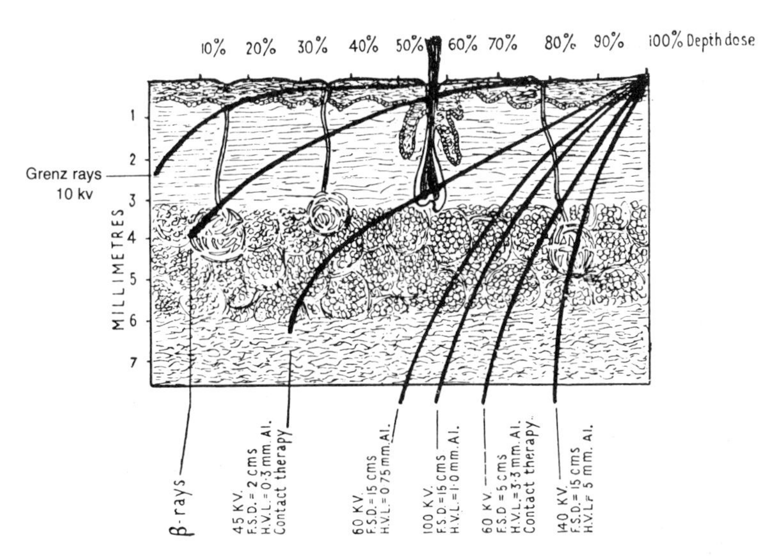

Fig. 18.1 Depth-dose absorption curves for radiation of various qualities, superimposed on a cross-section of the skin.

Superficial X-rays in the 60–100 kV region are by far the most popular and important technique. Open-ended applicators are used on, e.g. a KX10 machine at 15 cm FSD. Lead-glass walled applicators are helpful aids for visual control. Larger fields can be treated with a wider applicator at 25 cm FSD.

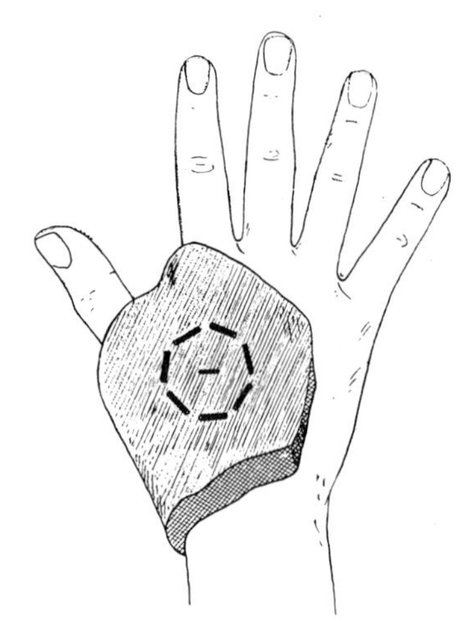

Fig. 18.2 Diagram of radium (or cobalt etc.) sources, mounted on a surface mould, for treating an epithelioma on the back of the hand. (*Paterson, Treatment of Malignant Disease, 2nd edition, Arnold*)

A series of lead cut-outs, 0.5 mm thick, should be available in all the standard sizes—both circles and ovals. A special cut-out can easily be made for an unusual shape. The field should include a margin of 0.5–1.0 cm beyond the apparent edge of the lesion.

Lesions on the *eyelids* are common, usually basal cell carcinoma, e.g. at the medial ends near the bridge of the nose (inner canthus). When X-ray treatment is given it is important to protect the cornea and lens. Eyeshields are used, like contact lenses, incorporating 1 mm thick lead. They should have a perfectly smooth surface, to avoid scratching cornea or conjunctiva. They may be kept in an antiseptic solution and rinsed before use in sterile water. They should be regularly inspected for damage and cracks. The surface of the eye is anaesthetised and the shield inserted gently under both eyelids, or under one lid and over the other. After each treatment the eye is covered with a pad, which the patient is told to retain for a few hours, in case particles of dust or grit enter the eye and damage the cornea while it is still insensitive.

If the ala* nasi is treated, some protection can be given to the inside of the nose by a narrow lead strip inserted in the nostril.

Dosage. In a typical case, for a lesion about 1 to 3 cm diameter, out-patient treatment is given daily, or three times a week, at 500 rad per session, to a total of 4000–4500 rad. For larger and thicker lesions, higher voltage, e.g. 140 kV, and higher dosage, e.g. 5000–5500 rad, may be given, in smaller fractions of 300–400 rad. Protracted fractionation is particularly important on the cartilaginous areas of ears and nose, to minimise the risk of late necrosis.

It is possible to destroy small lesions, up to about 2·5 cm, by a *single-shot* technique giving 1750–2000 rad. This is very useful, e.g. for a senile patient whom it would be unreasonable to subject to multiple visits. The method, however, is objectionable both because the cosmetic results are likely to be poorer (this may well be unimportant) and because the risk of late breakdown of the treated skin after minor trauma is higher than after fractionated doses.

For really large lesions DXR may be preferable—or cobalt beam if bone or cartilage is invaded or underlying, with bolus to bring the maximum dose to the surface. Fractions of 200–250 rad are used, up to 5500–6000 rad in about five weeks, or three weeks for smaller growths. Single fields may suffice, or a parallel pair; a wedge pair is often useful.

Reactions. Patients should always be warned what to expect. Reactions will be more marked in those with fair complexions. After a single dose of 2000 rad erythema appears in 7–10 days; in a 10–15 day course it often appears towards the end. The colour deepens for a week then fades, or goes on to moist desquamation. After moist peeling a scab usually forms, healing proceeds underneath it, and the scab falls when healing is complete. Healing is generally rapid, and complete in a few weeks; it may be followed by pigmentation, especially in a dark skin, but this usually fades in a few weeks or months. Some residual pallor may also persist, more noticeable against the background of a dark skin, but it is often difficult, at first glance, to detect any changes after treatment of a small field.

Large areas will inevitably show some late effects, with pigmentary changes, a degree of atrophy and maybe telangiectasia* (see p. 168). Irradiated skin is always more vulnerable to damage than normal skin, and patients should be warned against exposure to sun, cold winds and trauma, both at the time of treatment and in later years.

Late necrosis is more liable in some areas—see above.

Treatment of reactions. If possible, the area should be left alone and undisturbed, uncovered by any dressing or application. This is practicable for small areas or if the patient is in bed and infection can be avoided. Otherwise simple tulle gras covered by a plain dressing is enough. Greasy substances like vaseline are bad as they prevent natural discharges and encourage infection. Other useful applications are: calamine cream; Cetrimide or neomycin ointments for infection (see also pp. 169–70).

Recurrences may appear, especially at the edges after a 'geographic miss'. In the spreading type of basal cell lesion it may be very difficult to tell the exact extent of spread, and generous margins of treatment fields should be allowed. Recurrences are best treated surgically. For very superficial recurrent patches, good results are sometimes achieved by local application of an ointment containing a cytotoxic drug (5-fluorouracil—see p. 280).

Prognosis. If diagnosed early, skin cancer has an excellent prognosis. Radiation of basal cell carcinoma should yield 90–95 per cent cures. Squamous carcinoma can metastasise and therefore carries a somewhat worse prognosis unless it is really early. Basal cell lesions can be discharged from follow-up after a few weeks or months, but squamous lesions should be observed for at least three years. Secondaries in nodes worsen the prognosis considerably, as usual, but are very rare after three years.

Malignant melanoma

Melanomas* contain the pigment melanin. Benign melanoma is almost universal; most people have 10–20 brownish spots or moles, usually birthmarks. They lie just beneath the epidermis; the surface may be smooth, warty or hairy. The name 'naevus'* is also used for them, but this is unfortunate as the same word is also applied to the angiomas.

Melanoma is of relatively high malignancy, but forms only a small percentage of skin cancers. It is rare before puberty. Some common sites on the skin are—sole of foot (especially negroes), leg, nail-bed. Apart from the skin, it also occurs in the retina, meninges and some mucosal surfaces (anus, nose, gastrointestinal tract).

Pathology and natural history. Malignancy can occasionally arise in a benign melanoma. Irritation by repeated trauma is a factor in inducing malignancy, e.g. on the sole of the foot. Most lesions, however, are new formations of unknown causation. Danger signs are—increase in size, darkening in colour, ulceration and bleeding. Pigmentation varies from light brown to nearly black, but some contain no pigment (amelanotic melanoma) and there may be non-pigmented secondaries from a pigmented primary or vice versa.

Growth may be very slow, over years, or rapidly fatal in months. Metastases can show up years after removal of the primary lesion. Spread may be local, with pigment seeping into the skin around, and there may be satellite nodules nearby. Lymph spread is first to regional nodes, blood spread is to liver, lungs etc.

Some melanomas are subject to hormonal influence, e.g. there may be growth stimulation in adolescence or pregnancy. This is of no practical value in treatment. Immunological factors may also be at work, and account for spontaneous regression which is relatively common.

Removal of a piece for biopsy should never be done, but only complete excision with a wide margin; otherwise, malignancy and spread may be precipitated.

Treatment. Adequate surgery is the only satisfactory method. Some advise 'prophylactic' block dissection of regional nodes (inguinal, axillary etc.), i.e. to remove micrometastases even though there is no suspicious node enlargement. Most prefer to observe at monthly follow-up, and operate only for clinically suspicious nodes.

Radiation is only a poor second-best, as most melanomas are not radio-responsive. There are occasional exceptions, but not enough to make radiation a serious rival to surgery. It may be used for inoperable or recurrent lesions, primary or secondary. Occasionally the response is better than expected, with useful palliation at high tolerance-dosage levels.

Cytotoxic therapy has had a limited success in palliation. The best drug is melphalan, since drugs of the phenylalanine group are selectively metabolised by pigmented tumours. It can be given either intravenously, or intra-arterially by infusion or (better) perfusion (see p. 281).

Some success has been claimed for immunotherapy.

Prognosis. For small lesions adequately excised, the five year survival is about 60 per cent. But the general prognosis is poor, as dissemination is so common.

Cancer of the lip

Cancer of the lip is far commoner in males than females, rare under 50 years of age, and nearly always on the lower lip. Causative factors are similar to those of skin cancer—exposure to sun and wind—plus hot pipes and cigarettes.

Pathology and clinical features. Almost all are squamous carcinoma, usually well-differentiated and slow-growing. Local spread may involve the whole of the lip, the skin below or the oral mucosa, and eventually the mandible. Node metastasis is unusual, unless the primary is anaplastic; spread is to submental* and to cervical nodes.

There may be a premalignant basis, e.g. superficial keratosis* or leucoplakia.* The typical growth is on the red margin, has an indurated base and may be ulcerated. Growths at the angle of the mouth may invade the buccal* mucosa; this makes treatment more difficult and node spread more likely.

Treatment. If the lip epithelium is clearly unstable or leucoplakic, surgery is advisable, to remove not only the growth but the whole abnormal area, to forestall later malignant degeneration. If the growth is large, with much loss of tissue, or if bone is invaded, surgery will probably be needed, either as sole treatment or after pre-operative radiation. Plastic repair will be necessary.

For most lesions, up to a few centimetres in size, surgery and radiation are both satisfactory. Radiation is commonly preferred and avoids admission to hospital.

Preliminary attention should be given to oral hygiene and dental treatment as necessary.

Radiation technique. This is on similar lines to that for skin cancer. Superficial X-rays (100–140 kV) are adequate for small lesions, with the methods and doses described above. The standard lead cut-outs can be used, or a special one made with a 1 cm margin beyond the lesion. During treatment the mouth is protected by a 2 mm thick lead insert. DXR is good for thicker growths, with appropriate lead protection; 5000 rad incident can be given in three weeks.

Interstitial methods (radium needles etc.) were previously used by many and still are by some. They have no advantages over X-rays, and involve hospital admission and general anaesthesia.

A double 'sandwich' radium mould was also in use—elegant and successful, but expensive, time-consuming and involving radiation exposure of staff—5500 rad to the centre in 8 days, and 7000 rad maximum to the mucosa.

For really large tumours, if surgery is not to be used at once, external radiation is given by megavoltage or cobalt beam, with a wedge pair or parallel fields—comparable to the larynx techniques (see p. 203).

Lymph node metastases are treated on the lines laid down on pages 194–5.

Radiation reactions are the same as described above, plus mucositis and membrane formation on the inner surface. Lanolin is useful and soothing as a local application. Antibiotic ointments may be useful to control or prevent infection.

Prognosis. Small lesions, up to 2 or 3 cm should do well, with 80 per cent five-year survival. Large bulk, or anaplasia, especially with node secondaries, worsens the prospects as usual—about 30 per cent.

Cancer of the penis

Penile cancer affects the older age group. It is never seen where circumcision has been performed in infancy or early childhood (e.g. Jews, Muslims) but later circumcision provides incomplete protection. Like cancer of cervix it is associated with poverty and poor personal hygiene. Phimosis* is commonly present.

Pathology and clinical features. Almost all are squamous carcinoma, either a warty growth or indurated ulcer on the glans or the sulcus at the base of the glans. The first sign may be an infected or bloody discharge from beneath the prepuce. If the lesion is visible, diagnosis is usually obvious, but if phimosis hides it the glans must be exposed by incising and peeling back the prepuce (dorsal slit) or by complete circumcision, under anaesthesia. A biopsy is taken at the same time.

Growth is superficial at first, then by invasion of the penile shaft. *Lymphatic spread* to inguinal nodes is common and may be early. But enlarged nodes may be due merely to infection. From inguinal nodes further spread is upwards via iliac nodes to the abdomen. Blood spread can occur later.

Treatment. Surgery is curative in localised lesions, but since it involves partial or complete amputation of the penis is obviously objectionable, especially for the younger patient. Radiation is a good alternative for most cases. But if there is deep invasion of the shaft, radiation has poor chances of success and surgery is preferable. Invasion of the urethra also favours surgery, since post-radiation fibrotic stricture is very liable.

For small superficial lesions, treatment can be on the same lines as for skin cancer elsewhere, with superficial X-rays. Interstitial methods have little or no place.

Larger growths are treated by a parallel pair. DXR is satisfactory, and the scrotum and testes can be protected by lead. A wax or plastic block with a cylindrical cavity can be used for convenience of set-up. 5000 rad are given in four weeks. Cobalt beam can be used in the same manner, to 5000–5500 rad.

Cylindrical radium moulds were formerly popular, but are subject to the same objections as lip moulds (see above). However, superior cosmetic results are claimed, at 5500 rad in seven days.

Secondary *inguinal nodes* are treated surgically if practicable. If inoperable, or in patients unfit for surgery, external radiation may be used, by DXR or (preferably) cobalt beam, to 4000–4500 rad in four weeks. The inguinal region, subject as it is to moisture and friction, is intolerant of high dosage, and treatment is essentially palliative, for growth restraint.

Radiation reactions. These are like skin reactions elsewhere but somewhat more marked, and moist desquamation is commoner. Local infection can be kept at bay by gentian violet or an antibiotic ointment. Urethral reactions can cause discomfort and dysuria. Necrosis is a rare late complication, after treatment of a large tumour. Urethral stricture has already been mentioned.

Prognosis. The five-year survival for early lesions is high, about 80 per cent. The overall figure, however, is much lower, since the lesions are treacherous and node secondaries common.

Cancer of the vulva

Cancer of the vulva occurs in older women. There is commonly a basis of atrophy, dryness, thickening, pigmentary changes; maybe venereal disease, obesity, diabetes; leucoplakia is a frequent forerunner.

Almost all are squamous, with an exceptional adenocarcinoma.

Tumours may be warty and proliferative, or ulcerative.

Local spread is to surrounding skin, perineum, vagina and urethra.

Lymphatic spread is to inguinal nodes; it is common and may be bilateral. Blood spread may occur later.

Treatment should always be surgical if possible. The standard operation is radical vulvectomy, i.e. removal of the whole of the vulva and adjacent skin. An alternative is thorough destruction by diathermy coagulation.

For small lesions local excision may be preferred, with careful follow-up to detect further degeneration.

Radiation should be used only if surgery is contra-indicated by poor general condition, age or inoperability, as the vulva is notoriously intolerant and liable to late necrosis. Small implants of needles or gold grains can be quite successful. Two-plane implants are possible, but more difficult and hazardous. Superficial X-rays, as for skin elsewhere, are also possible, but difficult to apply.

Large growths may be treated by DXR or (preferably) cobalt beam, by a single field, parallel pair or wedge pair. Worthwhile results are often obtainable.

Radiation reactions are relatively sharp in this area subject to moisture and friction, but settle with proper antiseptic and nursing care.

Inguinal nodes. The remarks under cancer of penis (above) are also applicable here, including the bad effect on prognosis.

19. Mouth, secondary nodes of neck, tonsil, nasopharynx, paranasal sinuses, middle ear, salivary glands

Cancers of the mouth fall into several anatomical groups with general similarities and some individual differences—tongue, floor of mouth, alveolus,* hard and soft palate, buccal cheek.

Males are much more commonly affected than females. Environmental irritants predispose—tobacco (especially chewing), dental sepsis, badly fitting dentures, carious and jagged teeth, alcohol—plus syphilis and leucoplakia. Improved hygiene, nutrition and dental care, with decreased syphilis, have lowered the incidence in many countries. It is often the commonest site of cancer in parts of Asia, where causative factors include chewing of the betel nut (with tobacco and caustic lime).

Pathology. Over 90 per cent are squamous. A few are adenocarcinoma or malignant mixed tumours from mucous glands or minor salivary submucosal glands.

The degree of differentiation varies, but the more highly differentiated—and therefore less aggressive—growths usually occur in the front part of the mouth, while the anaplastic and aggressive, as well as the malignant lymphomas, tend to occur at the back.

Early *spread* is local, e.g. to gum, palate, cheek. Bone may be invaded, especially the mandible. There is a rich lymphatic network and invasion of regional nodes of neck follows. Blood spread is late and uncommon in practice.

Clinical picture. Most lesions present as ulcers with more or less deep induration; others as heaped up or papillary growths or fissured forms. Infection may follow causing halitosis, and excessive salivation. Pain may be prominent in later stages and may radiate to the ears. The tongue may become fixed, with difficulty in speaking or eating.

Diagnosis is usually straightforward on inspection, but thorough local examination and palpation are necessary to reveal the full extent of invasion. The neck is palpated for enlarged nodes. Biopsy should be taken for confirmation and to determine the histological grading, which may be useful in subsequent management (see p. 155).

Preliminary dental treatment. Many patients are without teeth at the start, others obviously require complete clearance of their remaining septic stumps. Treatment should be delayed a short time if necessary for this. Dental sepsis will not only increase discomfort, but may even prejudice the success of radiotherapy. At the least, scaling and cleaning should be done, with liberal mouth washes. If healthy teeth are to be included in the high-dose volume in radical radiation, they are bound to suffer impaired vitality because of the radiation ischaemia. Later decay will therefore be more difficult to deal with owing to poor healing and liability to infection. If later extractions become necessary, osteomyelitis* may follow, especially in the mandible, and even bone necrosis. Some therapists therefore advise prophylactic removal of teeth in the danger area, to insure against this unpleasant possibility.

Treatment. Surgery is usually undesirable because loss of tissue here is mutilating and likely to impose functional disturbance (speech, swallowing). Successful radiotherapy usually avoids both these drawbacks and is therefore the first choice in the great majority of cases.

Radical radiation of head and neck tumours should be by small beam-directed fields (megavoltage or cobalt beam). Immobilisation is best achieved by a full head mould (see p. 69); alternatively sandbags or headclamps are used, but great care is needed since small movements of the head can lead to serious errors of dosage. In general, tumour doses of 5500—6500 rad in five to six weeks are adequate, depending on the size of fields. Every case needs careful individual planning.

Cancer of the tongue. This is the commonest cancer of the mouth and the commonest site is the lateral aspect of the anterior two-thirds (Fig. 19.1). The posterior third differs, for growth here tends to be more anaplastic (and therefore more liable to metastasise) and include lymphoepithelioma (see p. 195) and malignant lymphoma.

Late syphilitic disease of the tongue is a bugbear, for the inflammatory fibrotic induration may easily be mistaken for malignancy. Syphilis is in any case a premalignant lesion in the tongue, and syphilitic glossitis* and cancer may co-exist. Radiation should not be used in such cases, since syphilis itself causes ischaemia from narrowing of blood vessels (endarteritis),* and added radiation usually precipitates necrosis.

Node secondaries may be apparent when the patient is first seen, especially with growths in the posterior third.

Surgery involves partial or total glossectomy, but is the treatment of choice in some circumstances:

1. A small lesion at the tip of the tongue. It is very difficult to implant this small volume, with its curved surfaces, satisfactorily, whereas local excision is simple and adequate, with no functional disabilities.

2. Growths on a syphilitic basis, as noted above.
3. If there are associated leucoplakic patches on the tongue or elsewhere in the mouth, all the lesions, malignant and premalignant, are best excised. They are in any case radioresistant.
4. Post-radiation recurrences.
5. If bone is invaded, radiation is unlikely to succeed. An alternative is to use radiation first and be prepared to resort to surgery in case of failure.

The radiation technique depends on the site and the size of the growth and also on the presence of node secondaries.

Interstitial technique. Small or medium lesions (up to about 4 cm) are suitable for treatment by implanted radium needles (or a radium substitute—see p. 89). The anterior two-thirds of the tongue is one of the few sites in the body where interstitial methods still retain their place as the treatment of choice. Figure 19.1 shows a single-plane implant, which will irradiate effectively a slab of tissue 1 centimetre thick. Two-plane and volume implants are possible, but it is technically more difficult to obtain a good lay-out without bunching or angulation of the needles. Lesions too large for a single plane are better treated by external radiation.

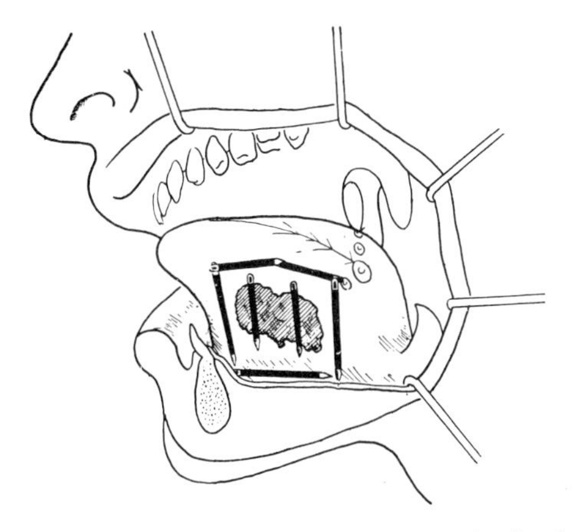

Fig. 19.1 Diagram of radium needles for single-plan implant in the tongue. (*Paterson*)

Growths which would be otherwise suitable for needle implant should not be implanted—(a) if there is a syphilitic avascular base—see above; surgery is better, and (b) if there are inoperable node secondaries in the neck; tongue and neck should be treated together by external radiation.

The radium insertion is done under general anaesthesia in the operating theatre. The therapist will have made a preliminary calculation, based on the size of the lesion. In the example shown (Fig. 19.1) a single plane of needles will be adequate, covering a rectangular area. Needles of 1 and 2 milligrams of radium are generally of most use in this type of case. Precise details will be given by using a system such as that of Paterson and Parker (see p. 86). In this case the dose will be worked out at half a centimetre from the plane of needles, i.e. it will suffice to dose adequately a slab of tissue 1 cm thick, which should include all the malignancy. A typical dose is 6000 rad in six days.

Based on his calculations, the operator will have ordered the requisite types and numbers of needles. He inserts the needles at the base of the tumour, one by one, carefully measuring the distances and separations with a ruler so as to be sure of securing the correct lay-out as calculated. The implant should cover the whole growth with a safety margin of about 1 cm all round. The needles are held in special grooved long-handled forcepts and pushed home with special 'pushers'. Each is stitched in with catgut. At the end of the operation all the threads are gathered together and brought outside the mouth, where they are securely strapped to the cheek—so that even if a needle were to work loose it could not be swallowed. At the end of the operation the position of the needles is checked—in the theatre if possible—by fluoroscopy and/or an X-ray film from a portable unit. If any serious errors of positioning are found, they can be corrected before the patient leaves the theatre for the ward.

Needles are removed at the end of the treatment time with the help of the attached threads; no anaesthetic is necessary and it can be done on the ward. The needles are checked, cleaned and finally returned to the radium safe. While the needles are in position the patient remains on the ward, and talking is discouraged. He is usually quite comfortable, but the diet is one of fluids and slops, as a minimum of tongue movement is desirable. Soon after the end of treatment the radiation reaction begins, on the tongue itself and also to some extent on adjacent surfaces (floor of mouth etc.). The successive changes of the typical mucosal reaction will appear, leading to membrane formation which gradually shrinks as the underlying ulceration heals. Healing is complete in a few weeks.

Instead of needles, gold grains can be used as permanent implants. They are useful particularly for old people, as they allow free tongue movement and so lessen the risk of post-operative pneumonia.

External radiation. For tongue lesions unsuitable for implantation, megavoltage or cobalt beam is used, and the precise technique will depend on the size and position of the growth. A simple parallel pair or a wedge pair will be satisfactory for most cases, to give a tumour dose of about 5500–6000 rad in about five weeks.

Growths of the *posterior third* of the tongue present different problems. One type is a localised mass posterolaterally, near the anterior faucial pillar, or involving the pillar—the typical faucio-lingual carcinoma of elderly men. Adjacent tissues may also be invaded—soft palate, alveolus, buccal cheek etc. Such a growth may be treated on the same lines as a tonsillar mass (see below).

The posterior third is also the site of the more anaplastic

types, including the lymphoepithelioma (described under Tonsil). The posterior surface of the tongue is also an anatomical part of Waldeyer's* ring of lymphoid tissue (p. 195) and may be involved by malignant lymphomas. All these are liable to metastasise early to cervical nodes, and large-field regional radiation is indicated. The fields should include all the lymphatic tissues of head and neck, from base of skull to clavicles (and it may sometimes be advisable to include the upper mediastinum also). The technique is described under Tonsil below.

Floor of mouth. Early growths were formerly treated by radium needles or radon seeds or surface radium moulds (intra-oral or combined intra and extra-oral) in preference to external orthovoltage 200 kV beams with their damaging effects on the mandible. Results were good in most cases and the techniques are still valid, with gold grains replacing radon seeds, but radium has now been largely superseded by megavoltage or cobalt beam. If bone is invaded and it is decided not to use primary surgery, megavoltage has no rival. For small growths a wedge pair is elegant and satisfactory (Fig. 19.2). For larger growths a parallel pair may be preferable, especially if they cross the midline.

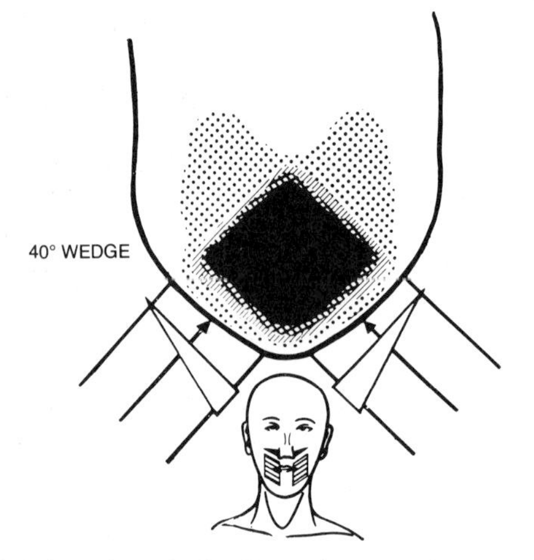

Fig. 19.2 A wedge-pair for floor of mouth (from Paterson, *Treatment of Malignant Disease*, 2nd edn., Arnold.)

Alveolus and hard palate. Surface radium moulds were formerly favoured, but have given way to external radiation (Fig. 19.3).

Soft palate. This is a very difficult site to implant successfully. Gold grains may occasionally be suitable, but most cases are best managed by external beams.

Buccal cheek. Small growths are often best excised; otherwise they can be adequately treated by a single-plane radium implant. Leucoplakic patches, especially with frankly malignant degeneration, should be excised. Larger growths, especially at the back, are difficult to implant and are best treated by external radiation, e.g. a single DXR or cobalt field, or a wedge pair.

Radiation reactions. Smoking and alcohol should be forbidden if possible during the reaction period. About the third week of the course the mucosal reaction will begin (see p. 170) and last several weeks. It is always a strain on any patient, with soreness, loss of taste and dysphagia. Patients need considerable moral support and reassurance from the treatment staff, and must be encouraged to take adequate nourishment and fluid. It is often advisable to admit to hospital during the acute phase, so that the nursing staff can ensure proper hygiene and nutrition. Mouth washes should be freely used, e.g. glycerine and thymol, though none are better than plain sodium bicarbonate (a teaspoonful to a pint of warm water). Local anaesthetic suspensions may be useful before meals.

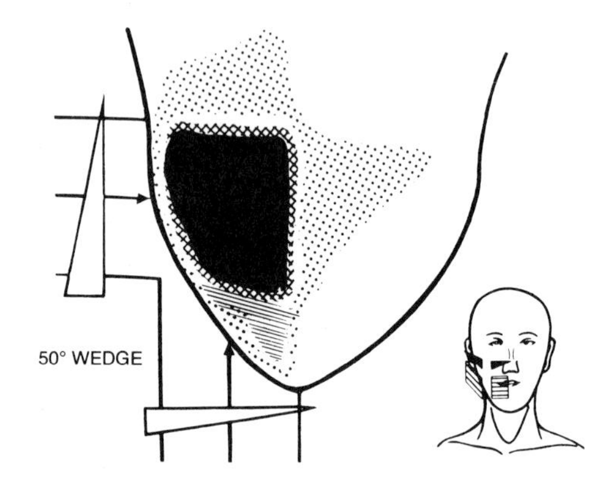

Fig. 19.3 A wedge-pair for alveolus (from Paterson, *Treatment of Malignant Disease*, 2nd edn., Arnold.)

After the acute reaction has passed, late effects follow, which may be troublesome. Some degree of dryness of mouth and throat is unavoidable, owing to inhibition of mucous glands. This will be particularly marked if the major salivary glands, especially the parotids, have been in the high-dose region. It is important to leave at least one parotid gland in a low-dose area if possible, and it is worthwhile to adjust field lay-outs with this in mind. Loss of taste sensation is another unpleasant effect, but usually recovers.

Skin changes, including epilation, are generally mild with megavoltage and give no real trouble. Submental* (i.e. below the chin) oedema sometimes occurs, due to lymphatic obstruction, giving a 'dewlap' appearance, but is seldom of sinister importance.

Late dental complications have been discussed above. Bone necrosis, to which the mandible is much more liable than the maxilla, used to be a bugbear in the orthovoltage era, but is rare now unless bone has been invaded by malignancy. Surgical removal of dead bone usually allows healing.

Lymph node metastases in the neck

Metastasis to the neck is the most serious aspect of mouth

(and throat) cancer. Prognosis usually depends more on the presence or absence of secondaries in cervical nodes than on the primary growth.

As a generalisation, it may be said that *operable nodes in squamous carcinoma are best treated by surgery*, usually radical block dissection, while inoperable nodes must rely on radiation as a poorer second best.

But if the primary is anaplastic, surgery is likely to fail and radiation is preferable, especially if nodes are bilateral. In many of these cases the lesion is sufficiently radiosensitive to allow the whole of the neck as well as the primary to be treated in one block to a cancericidal level of dosage—see under Tonsil below.

In squamous carcinoma, three clinical situations have to be discussed:

1. *No nodes palpable*. At one time 'prophylactic' removal of nodes on the affected side was advised, to remove early secondaries. The operation is drastic, with removal of as much lymph node tissue as possible. This involves removal of sterno-mastoid muscle and internal jugular vein also—all in one block and hence called 'block dissection'. However, nine out of ten cases proved to be free of secondaries, and it seemed unjustifiable to penalise the many for the sake of the few. Most therapists now prefer a 'wait and see' policy, with regular follow-up at monthly intervals. If nodes become palpably enlarged, surgery is usually advised. Radical surgery is, of course, unsuitable for patients who are elderly or in poor condition.

2. *Nodes palpable and operable*. If this is the situation at the time of diagnosis of the primary, there are several possibilities, each with its advocates. Some (the majority) irradiate the primary first, then after healing excise the nodes. Others remove the nodes first, then irradiate the primary. Others prefer to remove primary and nodes together in one extensive surgical attack.

Bilateral blocks, the second side at a later stage than the first, are possible, but it is dangerous to remove the internal jugular vein on both sides. If surgery is contra-indicated. radiation is used.

Operable nodes on one side, in squamous carcinoma, discovered at follow up after the primary is healed are usually treated surgically by block dissection, while radiation is reserved for inoperable nodes. This rule is still accepted by most workers, and was certainly valid in the orthovoltage era. However, it is now challenged by modern megavoltage therapy, which is claimed to be capable of controlling growth in secondary nodes if not too advanced. Figure 19.4 shows the technique. The fields cover the whole of one side of the neck just lateral to larynx and spinal cord, and extend from mastoid to clavicle. Care must be taken to avoid overlap with fields already irradiated in treatment of the primary in the mouth etc. The treated volume is of irregular shape; to secure uniform dosage it can be built up with bolus, though this sacrifices the skin-sparing effect. Dosage must be high, 5500–6000 rad in five to six weeks. If the opposite side is suspect, similar treatment can be given to 5000 rad in 5 weeks on a 'prophylactic' basis.

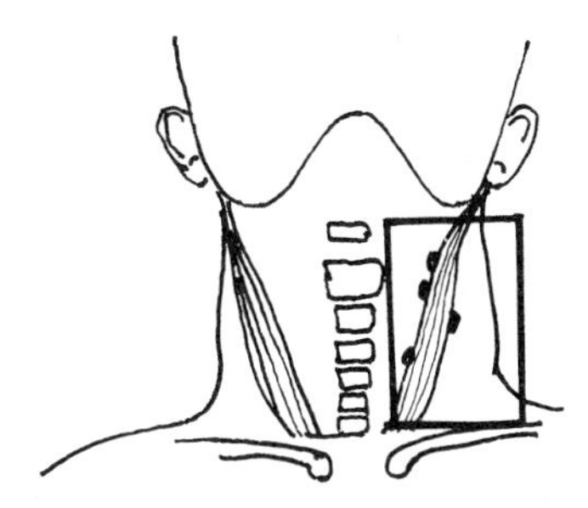

Fig. 19.4 Anterior field for secondary nodes of neck. A parallel posterior field can be added. The hyoid, thyroid, cricoid and upper tracheal cartilages are shown.

There is less experience with radical radiation than with radical surgery, and opinions on their comparative merits will continue to differ at present, until long-term results provide firm answers.

3. *Nodes inoperable* (fixed). Palliative radiation is usually adopted, e.g. single direct fields with cobalt beam to about 4000 rad incident in three weeks. Or an anterior-posterior pair can be used as in Figure 19.4. Occasionally it may be possible to use limited surgery after this.

Some workers attempt to be more aggressive in selected cases—e.g. multiple split courses, external radiation combined with small implants (needles or gold grains), and hyperbaric oxygen, for fixed nodes, and claim some success.

Prognosis. The relevant factors have already been discussed. For an early moderatedly well-differentiated squamous carcinoma, e.g. of the tongue, the five-year survival is about 65 per cent. Development of secondary nodes lowers this to less than 20 per cent. Anaplastic growths, as usual, do worse, but lymphoepithelioma if adequately treated gives a good and lasting response.

Cancer of the tonsil

There is a protective ring (Waldeyer's* ring), more or less complete, of lymphoid tissue at the entrance to the pharynx (Fig. 19.5). It includes the two lateral tonsils, the nasopharyngeal tonsil (adenoids) on roof and posterior wall of nasopharynx, and lymphoid tissue on the posterior surface of the tongue (lingual tonsil).

Tumours in this region are of two main types—*squamous carcinoma* from the epithelium, of varying degrees of anaplasia, and *lymphoma* (lymphosarcoma etc.) from the lymphoid tissue. There is also the badly-named *lymphoepithelioma*, which contains both epithelial and lymphoid elements but is really an anaplastic squamous carcinoma with an unusually large infiltration of lymphocytes.

Pathology and clinical features. Each tonsil lies in a recess between the anterior and posterior faucial pillars. Two-thirds of tumours are squamous carcinoma, with a high proportion anaplastic, and the rest lymphoreticular. Symptoms are minimal at the start, with slight sore throat; in later stages, dysphagia and pain may be prominent. Often it is the secondary mass in the neck which brings the patient to the doctor, and the primary is only then discovered. Local spread is early, beginning with anterior and posterior faucial pillars and soft palate. The lower end of the anterior pillar extends to the postero-lateral corner of the tongue, which may be invaded. Spread backwards involves the lateral wall of the oropharynx. The extent of local spread must be carefully assessed, since the technique and succcess of local treatment depend on it.

Treatment is almost always by external radiation. Surgery is possible in very early cases, which are rare; it usually involves unacceptable mutilation, and the results are anyhow poor, especially in view of the high proportion of node secondaries.

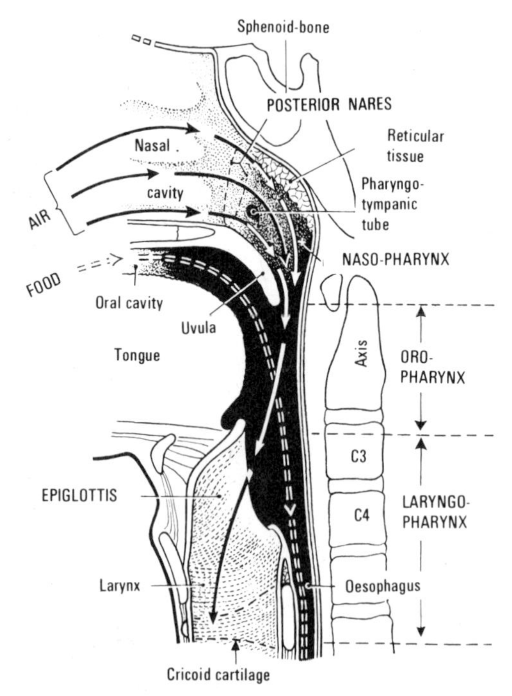

Fig. 19.5 Section of the pharynx, from front to back. (From Roper, *Man's Anatomy, Physiology, Health and Environment*, 4th edn., Churchill Livingstone.)

For the moderately *early case*, with limited local spread, radical treatment should be attempted, e.g. with a head shell and small beam-directed fields, or a wedge pair. A three-field technique is useful—anterior and posterior obliques on the affected side and a third from the opposite side either in front of or behind the parotid (Fig. 19.6). Each case needs individual assessment and planning, to cover the full extent of spread with a safe margin, and avoid spinal cord, orbits and at least one parotid gland. The tumour dose should be about 6000 rad in five weeks, with megavoltage or cobalt beam.

For the more *advanced cases*, especially with spread beyond the midline, a simple parallel pair may be as good as anything, in spite of the effect on both parotids. Special beam direction is not necessary. Treatment is essentially palliative, and dosage should not be pushed above 4500–5000 rad in four or five weeks. Good results can at times be obtained even in unpromising cases.

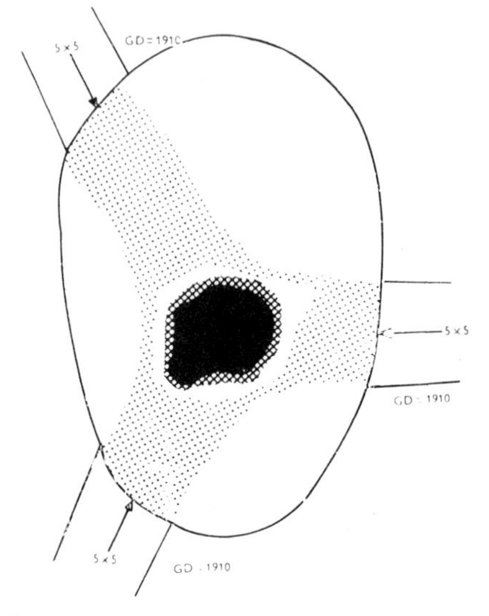

Fig. 19.6 Three-field scheme for e.g. tonsil. (From Paterson, *Treatment of Malignant Disease*, 2nd edn., Arnold.)

For *lymphoreticular growths* the approach is different, as small fields and high dosage are not indicated. They are nearly always of radiosensitive type, and *wide-field regional therapy* should be used, to include all the lymphoid tissue from base of skull to clavicles. There are two chief alternative techniques—(a) lateral parallel opposed fields reaching high enough to include the nasopharynx, and (b) antero-posterior fields, making allowance for the irregular contour and varying thickness of head and neck, either by using bolus on the neck or leading off the laryngeal region for part of the course. Dosage should not exceed 4000 rad in four weeks, otherwise the tolerance of the cervical spinal cord may be exceeded, with risk of radiation myelitis (see p. 172). The primary lesion in the mouth etc. can, if necessary, be raised to higher dosage locally by supplementary small fields after the main course, to give an extra 500—1000 rad; these must avoid the spinal cord and should be checked on the simulator or by films.

Radiation reactions with mucositis, especially in radical treatments, are inevitably troublesome, with sore throat, dysphagia, dryness and loss of taste. It is often desirable to admit to hospital towards the end of the course for nursing care, mouth hygiene and nutrition, especially fluid intake. An aspirin-mucilage mixture before meals is helpful, to make swallowing more comfortable.

Secondary nodes of neck. The management has been discussed above (p. 195) and most of it is applicable here. The situation is similar to posterior third of tongue, with a high proportion of anaplastic growths, and involvement of nodes of neck from the start. In such cases treatment should be directed to the primary and the whole of the neck, as for lymphoma and lymphoepithelioma, to 4000 rad in four weeks and possible supplementary fields as outlined.

Prognosis for growths with no evident secondaries is fairly good, with five-year survival about 75 per cent, falling to 40 per cent in the presence of secondary nodes.

Cancer of nasopharynx (Post-nasal space)

This is similar in many ways to tonsillar growths, as would be expected. For reasons unknown it is extraordinarily common in the Chinese, outnumbering any other cancer.

Squamous carcinoma is commonest, then lymphosarcoma and lymphoepithelioma. Adenocarcinomas of salivary type also occur as rarities. Growth is insidious, and the first indication is often an enlarged node in the neck just below the lobe of the ear. If so, suspicion should centre on the nasopharynx which should be examined by a mirror or in the theatre under anaesthesia.

The mode of spread is important (Fig. 19.5) and accounts for the great variety of presenting symptoms—to posterior part of nose (nasal obstruction, bleeding), Eustachian* tube (deafness), upper cervical vertebrae (pain), base of skull and intracranially (cranial nerve palsies). Any of these can be the first indication. Tumour may be visible in the nose or bulging the soft palate in the mouth.

Investigation includes soft tissue lateral films to show the air cavities and irregularities of the nasopharyngeal walls, and special projections to examine the base of the skull for bone erosion.

Treatment. Surgery is quite impracticable, and external radiation is the best hope. The fields must cover the routes of spread, especially the base of the skull. A head shell may be used or a free set-up. The chief fields are two laterals, and their coverage should be carefully checked by films on the simulator, diagnostic set or treatment machine. An additional anterior field is valuable, especially if the nasal cavity is invaded; this lies between the eyes, and homogeneity of dosage is secured by giving part of the lateral dosage by wedged fields; the scheme resembles Figure 19.10. Dosage is about 5500–6000 rad in five to six weeks.

Anaplastic and lymphomatous growths, as well as secondary cervical nodes, are managed on the lines laid down above for mouth and tonsil.

Complications. In field planning care is taken to minimise dosage to the eyes, especially the lens, brain and spinal cord. Late cataract is not too high a price for cure; but it may not be possible to avoid risking damaging dosage to brain stem and spinal cord; the dilemma must be faced, in the knowledge that late effects on central nervous tissue can be devastating and even fatal.

Prognosis. For squamous carcinoma, five-year survival is about 30 per cent. Lymphosarcoma and lymphoepithelioma are somewhat more favourable—about 50 percent.

The paranasal sinuses

These include the maxillary antrum, ethmoid, sphenoid and frontal sinuses. Growths in any of them, or in the nasal cavity itself, are uncommon, but the commonest site among them is the antrum, followed by the ethmoid. Causative factors are obscure, but there is a relatively high incidence in wood-finishing workers (furniture etc.) presumably due to some locally irritant factor.

Cancer of maxillary antrum

Pathology. Squamous carcinoma is the usual type, though adenocarcinoma, lymphoma and various sarcomas (osteo, fibro etc.) can occur. Growth usually begins in the lining membrane of the cavity and can therefore be silent for a long time. Figure 19.7 shows the various routes of spread after bony invasion occurs—to nasal cavity, cheek, hard palate, upper alveolus, alveolo-buccal sulcus, ethmoid sinus, orbit; also to pharynx and adjoining areas. Spread to nodes of neck is late.

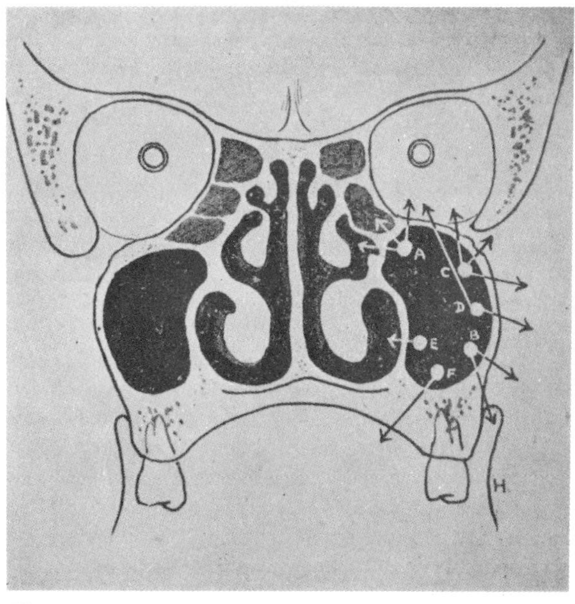

Fig. 19.7 Cross-section of the head, from side to side, through the nasal and other cavities. The arrows show the various possible directions of spread of a growth beginning in the antrum. (Windeyer, *British Practice in Radiotherapy*, Butterworth)

Clinical picture. Symptoms arise eventually from local spread—nasal blockage, discharge and bleeding, bulging of the cheek, bulging or ulceration in the mouth. If the orbit is invaded the eye may be pushed upwards or forwards (proptosis)* and there may be double vision (diplopia).

Pain may at first be mistaken for toothache. Radiographs, including tomograms, will show the extent of bone destruction of the antral walls, including floor of orbit.

Treatment. Surgery is technically feasible and can be curative, but involves gross mutilation, with removal of half the upper jaw and any invaded structures. If the orbit is involved, removal of orbital contents including the eye, will be necessary. Such heroic procedures are usually unacceptable to the patient or contra-indicated by age or general condition. Since modern radiation offers prospects at least as good, with better cosmetic results, it is now the preferred method. Limited supplementary surgery may be advisable after radiation.

Before radiation begins, it is important to provide adequate drainage of the antral cavity because—(a) it is always more or less infected from the start, and (b) as radiation proceeds, fragments of growth, dead bone etc. will become foci of further infection, leading to pent-up discharges, necrosis and pain. Surgical drainage is carried out preferably by palatal antrostomy, i.e. removal of part of the hard palate and upper alveolus, thus permanently exposing the interior of the antrum. The surgeon removes any loose and decayed teeth, dead bone and necrotic debris and takes a biopsy. The opening has another advantage—the cavity can be inspected later for evidence of recurrence. The gap is closed by an individually made obturator worn like a denture, so that the patient can eat and speak normally.

Radiation technique. The region is awkward for the therapist because of the irregular contours and the proximity of the eyes, both of which must be taken into account in the treatment plan. A full head mould with beam direction is desirable. If there is no suspicion of invasion of the orbit, the fields need go no higher than the inferior orbital margin and so avoid both eyes. A simple wedge pair is satisfactory, one field anterior, the other lateral, as in Figure 19.8. The anterior field should cross the midline to include the ipsilateral* nasal cavity. If the overlying skin of the cheek is involved by growth, wax bolus is used to bring the high dose (from cobalt beam etc.) to the skin surface, as shown in Figure 19.8. The lower borders of the fields enclose the hard palate, but the tongue can be kept out of the high-dose volume to some extent by a mouth block, made at the same time as the head shell—as in Figure 19.9.

If the orbit is invaded by growth it must be included in the treated volume. Some therapists include it routinely, to be on the safe side, even if there is no evidence of invasion. The situation must be fully and carefully explained to the patient, so that he understands the vital necessity of irradiating the eye and the likelihood of deterioration of vision in later years. The fields should extend to the roof of the orbit and the anterior field should cross the midline to include the ipsilateral ethmoid sinus, which is very liable to be invaded. No effort is made to spare the eye on the involved side, but every effort to avoid the other eye. A wedge pair with vertical hinge (i.e. anterior and lateral fields—Fig. 19.8) is the commonest

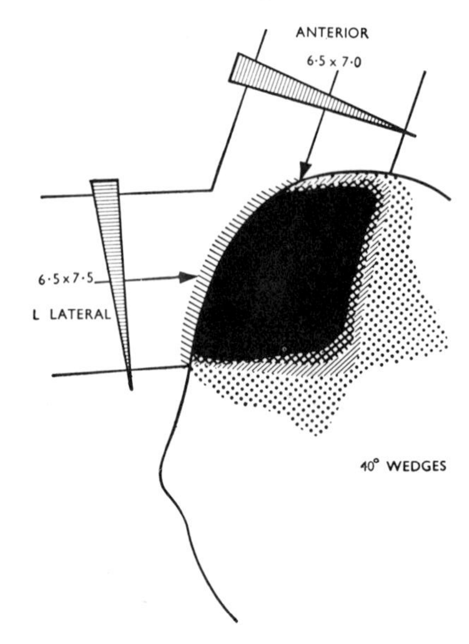

Fig. 19.8 Wedge pair for treatment of carcinoma of antrum at 4 MV. (Paterson, *Treatment of Malignant Disease*, 2nd edn., Arnold)

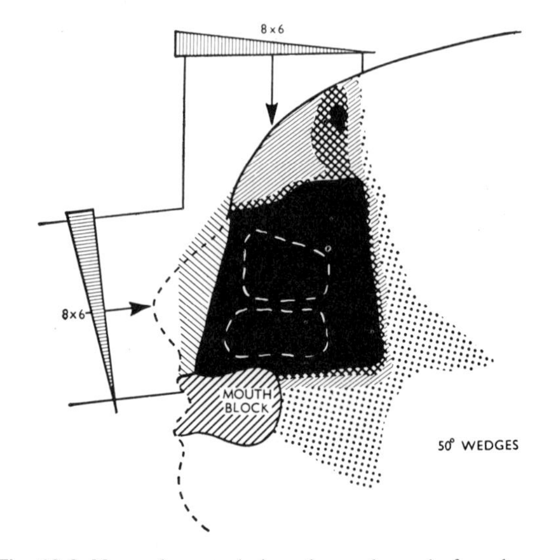

Fig. 19.9 Megavoltage technique by wedge pair for advanced carcinoma antrum, including orbit. Note the mouth block to decrease dosage to tongue etc. Dose symbols as in Fig. 20.5 (Paterson, *Treatment of Malignant Disease*, 2nd edn., Arnold)

technique. The lateral field is angled backwards to avoid the opposite eye as much as possible. An alternative technique, which completely avoids the opposite eye, is a wedge pair with horizontal hinge, one field anterior the other superior (Fig. 19.9). This involves irradiating the frontal area of the brain, but dosage is within acceptable limits. Whenever the orbit has to be included, the patient should be instructed to keep the eyes open during the radiation exposure and look up into the X-ray beam, thus making use of the surface-sparing effect of megavoltage. If wax bolus is used, cylindrical 'peep-holes' 1.5 cm diameter should be cut for this purpose. This minimises damage to the cornea.

A tumour dose of 5500–6000 rad is aimed at in five weeks.

Some departments explore the antrum surgically six weeks after the end of radiation and take biopsies from any suspicious areas. If active growth is found microscopically the patches can be destroyed by electro-fulguration or additional radiation given by an intracavitary applicator carrying radium sources.

Palliative treatment for advanced cases can be on similar lines with lower dosage, It is much simpler and often adequate to apply a single direct megavoltage field, giving, for example, 3500–4000 rad incident in two or three weeks.

Secondary nodes of neck are uncommon, and associated with advanced primaries. Radical treatment by surgery or radiation may be possible, but usually palliation by radiation is all that is called for.

Reactions. Skin and mucosal reactions develop in the usual way and are managed on routine lines.

Irradiation of the eye leads to conjunctivitis beginning in the later part of the course. Antibacterial eye drops or ointment should be instilled daily to counteract infection which might damage the cornea.

Late effects on the eye have been described on p. 171. If full dosage has been given, degenerative effects on the lens (cataract)* and retina are virtually certain, commonly about three or four years later. Vision will be impaired and may be lost entirely. Sometimes the eye becomes painful and has to be removed surgically.

Cancer of ethmoid sinus

The ethmoids lie between the orbits and are liable to be involved not only primarily but secondarily to growths of maxillary antrum and nasal cavity. It is often difficult to say exactly where a growth originated.

It may present as a swelling at the side of the bridge of the nose, but more often as a nasal tumour with blockage. The orbit is very liable to invasion and the eye may be pushed laterally. X-ray studies including tomograms should be made, to determine the extent of bone destruction, especially posteriorly.

Surgical treatment is hardly ever practicable. Radiation is the usual method and the general principles are similar to those for antral cancer. As described above, the ipsilateral ethmoid is commonly included in the treated volume for cancer of antrum. Figure 19.10 shows a technique for an early growth, using mainly an anterior field, with supplementary dosage from two wedged lateral fields directed behind the lens of each eye, to boost the dosage in the posterior part of the sinus.

Prognosis. The outlook generally in growths of the paranasal sinuses is not good, because of the late stage at which they are usually discovered. The average five-year survival is about 25 per cent. Prolonged useful palliation is often achieved in advanced cases.

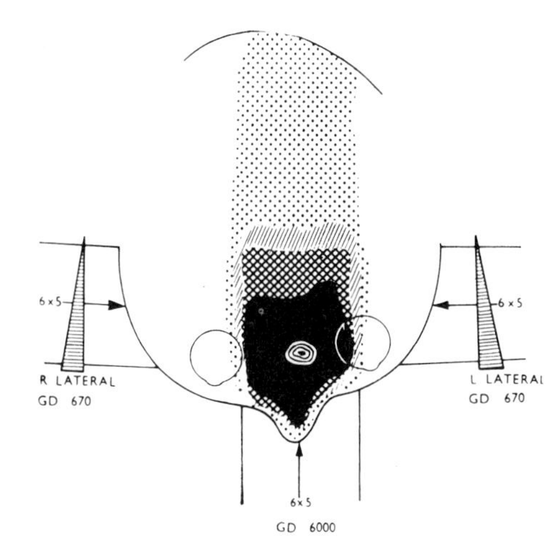

Fig. 19.10 Three-field scheme for ethmoid sinus (From Paterson, *Treatment of Malignant Disease*, 2nd edn., Arnold.)

Cancer of the middle ear

This is an uncommon and very serious lesion. Squamous carcinoma is the usual type. There is an association with chronic middle ear infection (otitis media) and a discharging ear in most cases. Because of this, malignant change is insidious and apt to be overlooked.

Local spread, with accompanying sepsis, occurs early. It may be backward to the mastoid bone, forward towards the nasopharynx, outward to the external auditory meatus, upward to the cranial cavity, downward to the jugular region, inward to the internal ear and petrous* part of the temporal bone. The facial nerve is often invaded, causing facial palsy (weakness or paralysis of the muscles of the face). The temporo-mandibular joint and parotid gland may also be invaded.

The clinical picture is usually of infection, with eventual blood-stained discharge. Examination may show granulation* tissue and polypoid fragments. Pain is absent at the start. Later there will be facial palsy, deafness and other symptoms, in conformity with the mode of spread. When suspicion is aroused, one or more biopsies confirm the diagnosis.

Skull X-rays, including tomograms, reveal the extent of bone destruction, including the petrous temporal bone.

Treatment. Surgery is disappointing. In the rare event of early diagnosis, radical mastoidectomy is carried out, i.e. removal of all involved tissue of the middle ear and cavities of the mastoid process (which are in continuity with the middle ear). This can be curative, but additional X-ray therapy is advisable. Heroic surgery, with excision of part or even all of the temporal bone, is technically possible, but still yields very poor results.

As a rule, external megavoltage is the best hope, usually following surgical exploration to assess the extent of spread and provide drainage. A full head mould is made, with careful planning by a wedge pair to produce a wedge-shaped zone of high dosage (Fig. 19.11) including the temporal bone, but not dosing the mid-brain to excess. 6000 rad tumour dose is given in five weeks.

Results are poor even in expert hands, and the best five-year figures are not above 25 per cent.

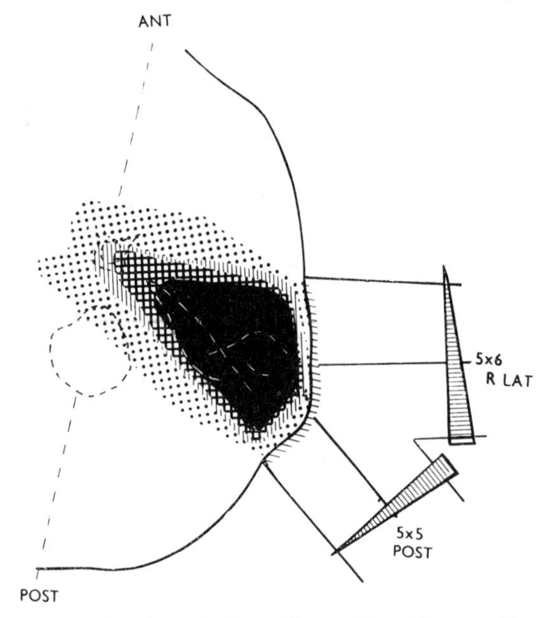

Fig. 19.11 A wedge-pair for middle ear (From Paterson, *Treatment of Malignant Disease*, 2nd edn., Arnold.)

Salivary tumours

Tumours of salivary glands form a difficult subject. They occur not only in the major glands (parotid, submaxillary, sublingual) but also in the many small mucous glands of the mouth and upper air passages including nasopharynx, trachea and bronchi. But 80 per cent are in the parotid.

Some are benign, some semi-malignant, some malignant, with no sharp divisions separating them. Very few are radiosensitive.

The commonest benign tumour is the *salivary adenoma*, often called 'mixed parotid tumour' though it is not restricted to the parotid. It is of epithelial origin, but contains material that stains like cartilage and was therefore thought to arise from more than one type of tissue—hence the misleading name 'mixed'. It is very slow-growing and presents as a firm painless swelling, usually near the angle of the jaw, and there may be a history of up to 40 years. The best treatment is generous local excision, but recurrence is common. There may be repeated excisions and recurrences, and 25 per cent eventually become malignant after many years. Post-operative radiation has often been practised, both interstitial and external methods, but its effectiveness is debatable. When malignancy sets in, growth becomes more rapid, skin and muscle are invaded and also the facial nerve, whose course is closely associated with the parotid gland, with resultant palsy. Pain may be severe.

Other malignant types are—adenocarcinoma, squamous carcinoma and cylindroma (or adenoid cystic carcinoma—columns of epithelial cells undergo cystic mucoid change). Metastasis occurs to nodes of neck, and later to lungs and liver.

The best *treatment* for salivary cancer is complete surgical removal. In the submaxillary and sublingual glands this is often practicable, but in the case of the parotid (the commonest site) it is a hazardous operation, for radical parotidectomy may involve sacrifice of the facial nerve with resultant palsy. A more limited operation is therefore often carried out, and followed by radical radiation. An alternative is to irradiate first and excise any residue later.

Radiation technique. Cobalt beam is very suitable, either a single direct field or a wedge pair, giving a tumour dose of 5500–6000 rad in five weeks.

Secondary nodes of neck are treated on the same lines as for growths of the mouth (see above).

The *prognosis* in malignant tumours is poor in general, though patients may survive with disease for many years.

20. Lower pharynx, larynx, post-cricoid, thyroid

It is convenient to deal with these sites together. The lower pharynx (hypopharynx or laryngopharynx) extends from the tip of the epiglottis (just above the level of the hyoid bone) to the lower end of the cricoid cartilage, at the junction of hypopharynx and oesophagus. The anatomy is complicated; details are given in Figures 20.1, 20.2.

The **larynx** is best considered as consisting of three sections:

1. The glottis—the gap framed by the vocal cords
2. Above the glottis, or supraglottic
3. Below the glottis, or subglottic

Tumours on aryepiglottic* folds and posterior surface of epiglottis are sometimes labelled supraglottic but are best classified as *hypopharyngeal*. It can be seen that the boundaries of the various compartments are not sharply delimited, and it is often difficult to say where a growth really originated.

Pathology. Almost all growths here are squamous carcinoma. Adenocarcinoma and sarcoma can occur but are very rare.

Premalignant lesions may be seen, especially in the larynx—hyperkeratosis,* papillomas* and general thickening of the vocal cords.

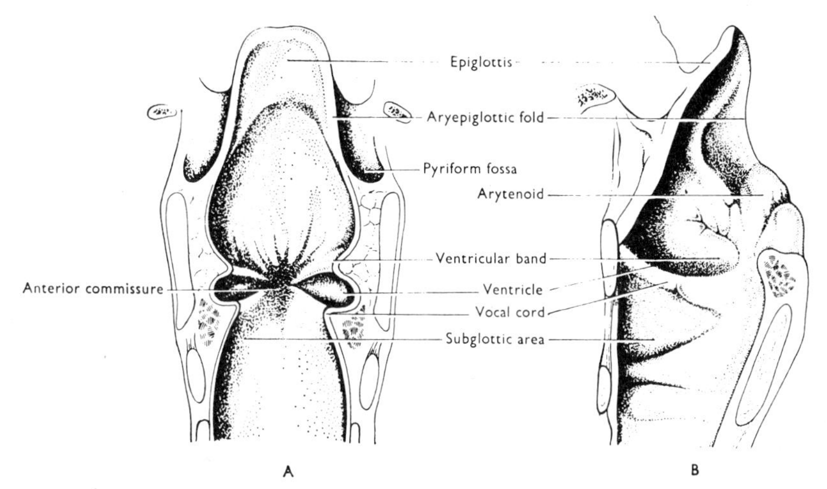

Fig. 20.1 Landmarks of the normal larynx and pharynx from (a) the posterior (b) the lateral aspect. (From Robinson, *Surgery*, Longmans)

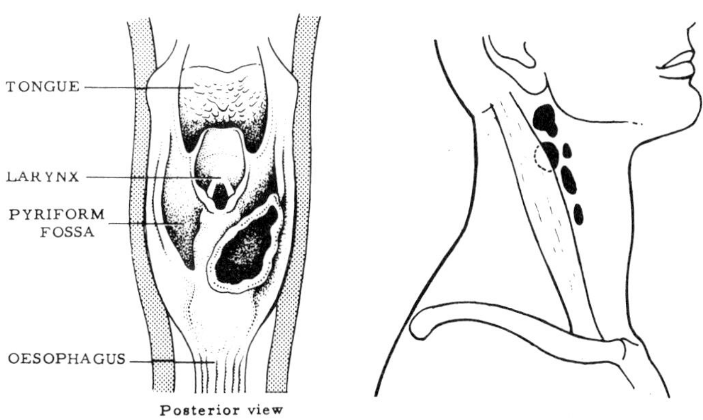

Fig. 20.2 Carcinoma of the hypopharynx (right piriform fossa). Metastasis occurs to many lymph nodes. (From Macfarlane, *Textbook of Surgery*, Churchill Livingstone.)

On the vocal cord the lesion commonly begins near the centre and spreads along the cord, later involving the junction of the cords (anterior commissure—Fig. 20.1) and then spreading across to the other cord. It is usually well-differentiated and slow-growing, and confined to the cord for a long time. The cords move (apposition and separation) in normal respiration and phonation, and their mobility is maintained till spread impairs it and causes some degree of fixation. Late local spread involves the supraglottic area and laryngeal cartilages; subglottic invasion is less frequent. The cord itself is poor in lymphatic vessels and node metastasis is rare unless local growth goes beyond the cord.

Supraglottic and hypopharyngeal growths are more commonly anaplastic than cord tumours, with earlier local spread and node invasion.

Subglottic growths are the least common, and node secondaries late.

Clinical features. There is an association between smoking and throat cancer. The same applies to alcohol, and these two factors almost certainly play a part in causation. Other predisposing conditions are: chronic laryngitis, dusty occupations and singing. Growths of the vocal cord are in a special category, since even tiny irregularities of their margins cause enough changes in the vibrating air column to produce early hoarseness. Prompt attention to this favours early diagnosis. Advanced growths will narrow the breathing 'chink' of the glottis and produce stridor, i.e. audible wheezing on inspiration. Neglect results in increasing dyspnoea, and impending suffocation may have to be relieved by an emergency operation (tracheostomy*) whereby an opening is made in the trachea lower down in the neck, and a tube inserted through which the patient can breathe.

Growths in the rest of larynx or pharynx are less favourable for early symptoms, and tend to remain silent for many months. A mass may reach considerable size—3 or 4 cm—before causing really troublesome symptoms. Lesions in the pyriform* fossa are particularly liable to long latency. There may be intermittent vague discomfort or sore throat at first, followed later by interference with speaking (hoarseness, dysphonia), swallowing (dysphagia) or breathing (dyspnoea)—depending on the exact site of the growth and extent of local invasion. By this time—or very often before—there may be secondary cervical nodes, and a lump in the neck is all too commonly the presenting complaint, leading to the discovery of the primary in the throat.

Investigation and diagnosis. Preliminary clinical examination is by indirect endoscopy* (laryngoscopy) using a long-handled mirror (Fig. 20.3). Except for very early growths, especially on the cords, it is not usually possible to see the whole of a tumour, but only the upper part, or the edge of an ulcer, or an abnormal bulge.

Further detail, to outline the full extent of the mass, is obtainable by radiography. Soft tissue lateral films of the neck are taken, together with tomograms which will reveal the cords and any masses distorting the normal air spaces.

In most cases it is desirable to carry out *direct endoscopy* under general anaesthesia in the theatre, to inspect the whole of the growth if possible and to take a biopsy.

Treatment. Preliminary attention is to the patient's general condition, nutrition, correction of anaemia etc. Oral hygiene is important, and any dental sepsis should be dealt with at the start.

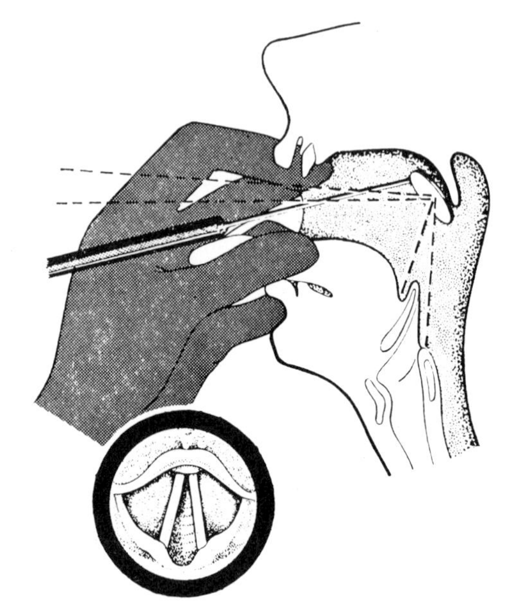

Fig. 20.3 Indirect laryngoscopy, and the mirror view of the larynx showing vocal cords, ary-epiglottic folds, etc. (From Robinson, Surgery, Longmans.)

In general, the best treatment of small growths is radical radiation, and of very advanced growths with secondary nodes, palliative radiation. For intermediate growths, especially if laryngeal cartilage is invaded, surgery may be superior, with pre-operative radiation perhaps assisting. Malignant cartilage invasion is unlikely to be controlled by radiation and dangerous cartilage necrosis is very liable to be caused thereby. Radical radiation failures can sometimes be rescued by subsequent surgery, but recurrences after surgery cannot be rescued by radiation. Subglottic growths are not good subjects for radiation, and surgery should be seriously considered.

If tracheostomy has been carried out, the metal tube must be replaced by a plastic tube before the start of radiation, otherwise the soft secondary rays from the metal will cause excessive reaction in the adjacent skin.

For early growths of the middle of the cord, both surgery and radiation give excellent results. Excision is by *laryngofissure*, i.e. splitting open the larynx in midline and retracting the two halves laterally to permit removal of the growth, including all or most of the cord. An impaired voice is however inevitable, whereas successful radiotherapy leaves a normal or near-normal voice and is clearly superior. Apart from laryngofissure, radical operations have to be drastic and mutilating, with complete removal of the larynx and part of the pharynx, plastic reconstruction of the pharynx, and permanent tracheostomy. As a primary procedure this is clearly objectionable if there is any reasonable alternative; it is ruled out, on both physical and mental grounds, in the elderly. The emphasis nowadays is therefore on radiation, with the exceptions noted above.

Laryngectomy does not necessarily involve loss of voice production and speech. With the help of a speech therapist a patient can learn to speak with an 'oesophageal voice', using swallowed instead of inspired air.

Radiation technique. In all radical techniques the general principles apply—small fields, high dosage, accurate beam direction. Dosage to spinal cord must be borne in mind, and its tolerance must not be exceeded for the consequences can be catastrophic (see p. 172).

The best results are achieved in early growths of the vocal cords, because the volume involved is small and regional nodes need not be included, since invasion is late.

Accurate localisation is easily achieved by radiography, which shows the laryngeal and other cartilages, and the air-spaces, epiglottis, spinal canal etc. Free set-up on the treatment unit is quite possible, but if facilities are available it is better to make a head-and-neck shell for accurate daily reproduction of fields. Excellent results are obtainable with DXR, but megavoltage/cobalt beam is preferable and the cosmetic results in the skin are definitely superior.

Antero-posterior and lateral radiographs of the neck are taken while the patient wears the shell, the tumour is then outlined on the film and related to radio-opaque markers on the surface of the shell and to anatomical landmarks (epiglottis, cricoid cartilage etc). Field planning and isodose distribution proceed in the usual way (Ch. 6). Windows are cut in the shell to preserve the skin-sparing effect of megavoltage.

The simplest technique is a parallel opposed lateral pair (Fig. 20.4). Wedge compensators (p. 63) can be used to avoid excessive dosage anteriorly while retaining the skin-sparing effect. A good alternative is an anterior wedge pair (Fig. 20.5) which gives an even better localised high-dose distribution limited to the larynx. Field size should be not less than 5 × 5 cm nor more than 7 × 7 cm. A tumour dose of 6000 rad in five weeks (25 treatments) is normally adequate.

Small supraglottic lesions can be treated in a similar way. Supraglottic and pharyngeal growths, however, are liable to be locally invasive and metastasise early to regional nodes. It is wise therefore to use larger fields, not less and 8 × 6 cm or up to 10 × 8 cm and give about 5500–6000 rad in five weeks. Lateral opposed fields are advisable, to include node areas. Care must be taken that the posterior edge of the field is clear of the spinal canal; this should be controlled on the simulator or by check films.

For growths too advanced for radical radiation—and node secondaries will most likely be present—if surgery is not applicable, palliative radiation may be used. Fields should be of generous size, covering most of the neck, part or all of the way from mastoids to clavicles. A dose of 4000 rad in four weeks is adequate.

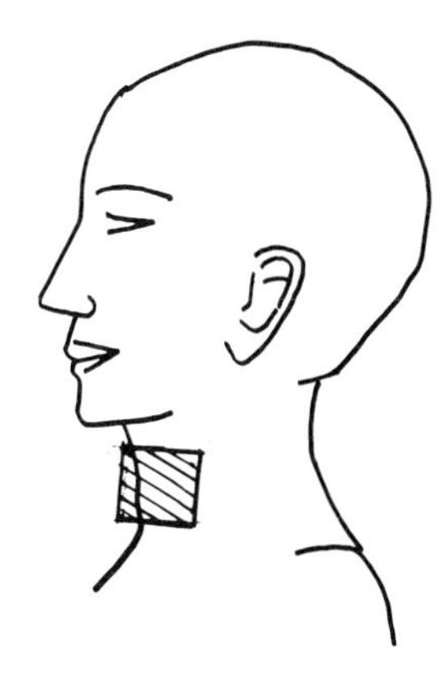

Fig. 20.4 Field for larynx, using a parallel opposed pair.

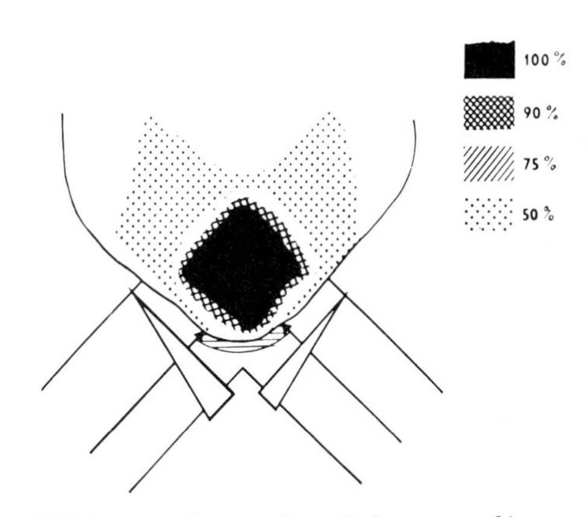

Fig. 20.5 A megavoltage wedge pair for cancer of larynx.

If a tracheostomy has already been performed, it will not be practicable to make a shell, but with megavoltage radical radiation is still feasible and it is often possible to close the tracheostomy later after healing of the growth.

Secondary nodes of neck appearing after radical radiation are managed surgically if possible. Further radiation will not usually be possible.

Radiation reactions. These are on the same lines as for the mouth, with similar management (p. 194). Huskiness may persist for months before improving. Occasionally, especially with large growths, there may be reactionary oedema very early on, even after the first or second dose, which can be dangerous if the airway is already narrowed.

Acute obstruction may be precipitated, calling for emergency tracheostomy to prevent suffocation. If this possibility is anticipated, dosage should begin at a reduced daily rate and be gradually increased.

Following radical radiation there may be persistent oedema of the laryngeal region and maybe patches of ulceration with infection. It is difficult to distinguish from persistent or recurrent growth, and even biopsies may not do so with certainty. Such cases may require laryngectomy to rescue the patient from an intolerable existence, even if there is no residual malignancy.

Late cartilage necrosis. Necrosis of laryngeal cartilages is a serious reaction, developing several months after radical radiation. It is rarely seen after megavoltage, but occurred in the orthovoltage era, since 200 kV radiation deposits much more energy in calcified cartilage. It is a painful condition, and infection is likely to follow, with dyspnoea and dysphagia. Tracheostomy may be necessary or even laryngectomy. The danger of necrosis is a good reason for preferring surgery in cases where cartilage is already invaded by growth.

Prognosis. Early cases of vocal cord cancer, with the cord mobile, yield 80–90 per cent permanent successes, after either surgery or radiation. If the cords are fixed, the five-year survival falls to about 40 per cent.

Other growths of larynx and hypopharynx do less well, as their long clinical silence enables local spread and even node metastasis to develop before diagnosis. The five-year survival seldom exceeds 20 per cent.

Post-cricoid carcinoma

This is a special variety of hypopharyngeal growth, of the food passage behind the larynx, just above the oesophagus. The typical victim is an elderly woman suffering from chronic anaemia, with a smooth tongue (superficial glossitis) and hollow spoon-shaped finger nails (koilonychia*). This is known after the workers who described it, as the Plummer-Vinson syndrome in Britain or the Patterson-Brown-Kelly syndrome in America. The whole epithelium of the upper digestive tract is atrophic, unstable and premalignant, and there may be malignant degeneration in the mouth, pharynx or oesophagus at any time.

The cardinal symptom is dysphagia. Mirror examination (Fig. 20.3) may show only slight oedema or a suspicious pool of mucus behind the larynx. Soft tissue films of the neck and fluoroscopic examination of a barium swallow help to outline it. Direct endoscopy should be done, to see the full extent and for biopsy.

The anaemia should first be corrected, by vigorous medical treatment, or blood transfusion.

Radiation presents some technical difficulties, since the growth is usually so low in the neck that parallel opposed lateral fields are not possible. Lateral oblique wedged fields are generally practicable on the same lines as for thyroid growths (Fig. 20.7) or a three field technique as for oesophageal cancer (Fig. 21.1 p. 210). Meticulous localisation (with the help of a barium swallow), neck mould and small beam-directed fields are advisable. Dosage to spinal cord must be kept low. 6000 rad are given in five weeks if possible.

The results are poor, with a 5-year survival in the region of 10 per cent.

Cancer of the thyroid gland

This is relatively rare, but of great interest to endocrinologist, surgeon and radiotherapist. It affects females more than males and any age including childhood.

Causative factors are obscure, but it is commoner in a goitre* (enlarged thyroid) than in a normal gland. Thyroid secretion is controlled by TSH (thyroid stimulating hormone) produced by the pituitary.* Iodine deficiency probably plays a large part in producing goitre, and it is reasonable to assume that the pituitary plays a part in production of some thyroid cancers by stimulation of an abnormal gland.

Another predisposing factor is radiation. In former days children's necks were irradiated for enlargements of the thymus* and a significant number developed thyroid cancer later in life. Most thyroid cancers in children are, in fact, preceded by external radiation to the neck and retrospective investigation showed that 200 rad to a child's neck can cause cancer in 3 per cent of cases. In the light of this it is natural to suspect that the therapeutic use of radioactive iodine for thyrotoxicosis (p. 269) might have carcinogenic effects, but after more than 30 years' experience there is no evidence of this.

Pathology and associated clinical features. Simple adenomas, in the form of nodules, are common and are further discussed below.

The chief malignant types are: *papillary*, *follicular* and *anaplastic*. Table 20.1. and Figure 20.6 give their chief differential characters.

1. *Papillary* carcinoma* is so called because the microscopic picture is of papillary projections into cystic spaces. It is the commonest type and accounts for almost all cases in childhood and adolescence. Its growth is remarkably slow, over very many years. The typical presentation is as a small hard slow-growing nodule in one lobe of the thyroid, but careful examination of the rest of the gland will often show other foci. Frequently, the correct diagnosis is only made by the pathologist when he examines microscopic sections after an operation for removal of what was thought to be a benign nodule.

An enlarged secondary node in the neck may be the first clinical sign.

2. *Follicular* carcinoma* (also called alveolar*) is somewhat more aggressive and metastasises earlier via lymphatics and blood-stream. The microscopic appearance resemble normal thyroid tissue very closely. There is often a

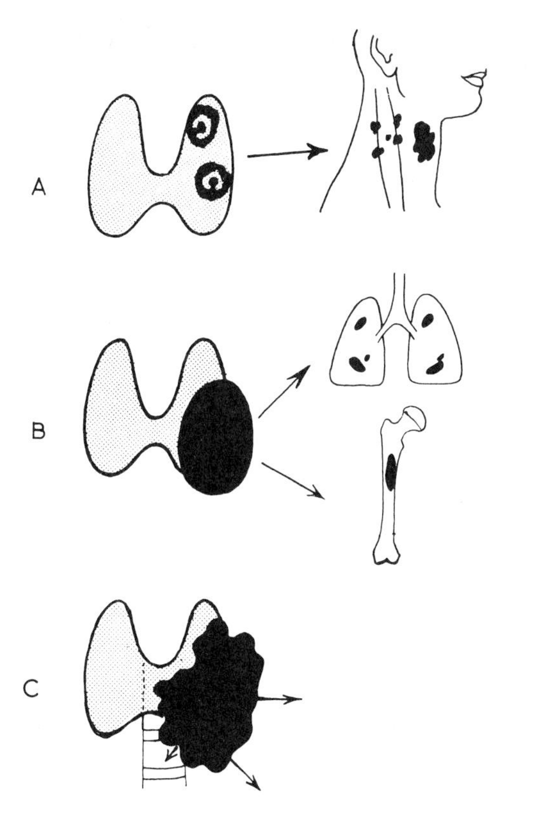

Fig. 20.6 A, Papillary, metastases to lymph nodes. B, Follicular, metastases to lungs and bones. C, Anaplastic, local infiltration. (From Green, in Macfarlane, *Textbook of Surgery*, Churchill Livingstone.)

nodule in the neck, present for some years, with recent enlargement. Most cases already show metastases in lung, bone or nodes at the time of diagnosis. One in three cases presents with signs due to secondaries, e.g. pathological fracture or dyspnoea from lung deposits.

An important and surprising feature is that some (about half) are capable of synthesising and secreting thyroid hormones. Even more striking is the fact that this capacity is shown by their metastatic deposits also. As in normal thyroid tissue, the hormones are built up from iodine removed from the blood and this enables radioactive iodine to be used in treatment of these particular growths. *Follicular carcinoma is unique in this respect.*

3. *Anaplastic carcinoma* is typical of the elderly, throws off early metastases and is usually inoperable when first seen. Its rapid growth may compress the trachea and cause dyspnoea and stridor. The oesophagus may also be involved, with dysphagia. Invasion of nerves may cause pain, also vocal cord palsy with hoarseness.

A rarer type is the medullary* carcinoma. This is slow-growing and best treated by surgery. Radiation, including ^{131}I, has little or no value, though some workers believe it to be moderately sensitive and worth treating by external radiation, especially post-operative residues.

The above classification is neat and satisfying, but in practice most differentiated growths (papillary and follicular) show a mixture of histological patterns and they are classified by the predominant type. But the three main types tend to follow different clinical courses and require different treatment. They must therefore be considered separately.

Table 20.1 Clinical features of thyroid cancer

	Papillary	Follicular	Anaplastic
Typical age	Under 40 incl.children and adolescents	40–50	60+
Growth rate	Slow—may be many years	Intermediate	Rapid
Presenting sign	Small slow-growing nodule in one lobe. Or secondary node in neck	Thyroid nodule for years, recent enlargement or secondary, e.g. pathological fracture	Large diffuse fixed swelling
Spread	Nodes of neck Blood spread rare	By blood, e.g. lung, bone. Nodes of neck—late	Lymphatic and blood
Primary treatment of choice	Total thyroidectomy or lobectomy	Total thyroidectomy	External radiation
Thyroxin therapy	Full (for TSH suppression)	Full (for TSH suppression)	Full (for replacement)
Radiosensitivity	Moderate	Low	High
Radioactive iodine therapy	No uptake	Ablation doses for residual thyroid tissue and functioning secondaries	No uptake
Treatment of secondary nodes of neck	Operable—surgery Inoperable—external radiation	If ^{131}I uptake—ablation dose. If no uptake—surgery if operable; external radiation if inoperable	External radiation
Prognosis and survival	Good. Can be >20 years even if tumour present (local or metastatic)	Intermediate	Poor. 1–2 years

Investigation and treatment. Management depends very much on the pathological type, so that a histological diagnosis is virtually essential for rational treatment. As a general rule, one of the first steps in all thyroid masses is a radioactive iodine uptake test and scintiscan of the neck (p. 268). 'Cold' areas are suspicious of cancer, but not diagnostic; 'hot' areas rarely prove cancerous (see Nodules below).

There is least diagnostic difficulty in rapidly-growing anaplastic tumours of the elderly, where the diagnosis is usually obvious. The only lesion with which they might be confused is one of the forms of thyroiditis (associated with the names of Riedel and Hashimoto); biochemical blood tests will distinguish these. A biopsy should be obtained if possible. If the patient is in good condition, surgical removal of as much of the growth as is practicable is worth while, and forestalls possible early compression of trachea and oesophagus. The mainstay of treatment in most cases is external radiation.

Nodules in the thyroid. Clinically solitary nodules are common (though half prove to be multiple). Malignancy may or may not be suspected, but one in eight prove to be malignant, and the rest simple adenomas. An isotope scan is unfortunately not very helpful; a cold spot may mean cancer, but may be merely a simple adenoma that has undergone degenerative or cystic change. The risk of cancer usually justifies surgical exploration; if facilities for immediate frozen section are available in the theatre, a biopsy is examined. If carcinoma is confirmed, or if the surgeon is in no reasonable doubt that he is faced with malignancy, total thyroidectomy should be carried out. Thyroid histology is difficult and the pathologist may be unwilling to commit himself to a firm diagnosis before examining further sections at leisure, especially as follicular adenoma and carcinoma can resemble each other closely. Failing a firm diagnosis in the theatre, most surgeons carry out only a partial thyroidectomy e.g. removal of the affected lobe.

Often enough, the diagnosis of malignancy and its type has to be awaited after partial thyroidectomy. It may even come as a surprise. If it is of *papillary* type, a decision must be made whether to proceed to a further operation for complete removal of the thyroid. In favour of this is the knowledge that multiple foci are common. Against it is the fact that papillary carcinomas grow very slowly and metastasise late; also, second operations are always more difficult and carry increased danger of damage to laryngeal nerves and parathyroid glands. Practice varies; some are content to leave well alone and observe, with further treatment only in the event of later recurrence or metastasis; others proceed to external radiation to deal with any remaining malignancy; a minority advise further surgery.

If the nodule proves to be *follicular* carcinoma, the case for complete removal of the thyroid is stronger, since this type, in addition to being possibly multifocal, is more aggressive and more liable to metastasise, as well as being even less radiosensitive. A second operation is therefore fully justified. But if the decision is against a second operation, thyroid ablation* by ^{131}I is carried out as described below, to remove any remaining normal thyroid tissue.

Thyroxin therapy. The activity of the thyroid gland, and therefore the production of thyroid hormone, is dependent on pituitary TSH (thyroid stimulating hormone). The amount of TSH produced by the pituitary is itself regulated by the amount of thyroid hormone produced by the thyroid, i.e. the two hormones are in mutual adjustment by a 'feed-back' mechanism. Low circulatory thyroid hormone provokes more TSH and so more thyroid activity, and high circulatory thyroid hormone leads to less TSH and so reduces thyroid activity.

All patients, after any form of thyroidectomy for malignancy (or even failing operation in anaplastic growths) should be given thyroid hormone as thyroxin tablets by mouth for the rest of their lives. This is for two good reasons. (1.) After removal of the thyroid, replacement of the missing hormone is necessary, otherwise the patient would suffer all the effects of hypothyroidism (see p. 268). (2.) If there is any thyroid tissue remaining, it will be subject to stimulation by TSH, which will extend to cancer cells also. Thyroxin in large doses removes the normal stimulus to the pituitary to secrete TSH; low TSH results in low activity of thyroid cells, normal and malignant, and thus favours quiescence of any residual foci of tumour tissue in the neck or elsewhere.

In the case of anaplastic growths, thyroxin sufficient for replacement therapy only should be given, since this type of cell is unresponsive to hormonal factors.

Metastatic nodes of neck. Here again, treatment depends on the histological type. In anaplastic growths surgery is not helpful even if the nodes are operable, since distant secondaries are virtually certain to be present. External radiation is therefore best.

In papillary carcinoma, operable nodes may be removed by block dissection. Inoperable nodes are treated by radiation.

In follicular carcinoma, treatment will depend on the ability of the nodes to take up ^{131}I as described below. If they cannot be induced to do so, block dissection is justified for operable nodes, otherwise external radiation.

Indications for external radiation of the thyroid

1. All anaplastic growths (regardless of any surgery performed)
2. Inoperable growths (even if partial thyroidectomy has been done)
3. After incomplete thyroidectomy for papillary or follicular carcinoma. (The alternative is a second operation to complete the thyroid removal).

External radiation is also useful for palliation of secondaries in the neck or elsewhere. Papillary carcinoma is more responsive than follicular; anaplastic growths are the most sensitive.

Radiation technique. External radiation of the thyroid region presents technical difficulties. The contours are irregular, with rapidly changing dimensions from point to point. The spinal cord is close enough to be at risk and its tolerance must always be borne in mind, to avoid any danger of late radiation myelitis.

As a rule, the regional cervical nodes should be included in the treated volume. If spread to the mediastinum is known to have occurred, the fields must be extended to include at least the upper mediastinum. For anaplastic growths the fields should extend all the way from mastoid to mediastinum.

Figure 20.7 shows a technique for treatment of the neck—two fields angled down towards the chest, with vertical wedge filters to compensate for the angle. The posterior edges must be anterior to the spinal cord, and this must be checked by films or on the simulator. A shell is desirable, even if not essential. Alternatively, two lateral fields can be used, with bolus to even out the dose distribution, but the high dose will not cover the lowest part of the neck so well. A tumour dose of 5500 rad is given in five weeks.

Figure 20.8 shows a technique which can be extended to cover the upper mediastinum also—an anterior field with two wedged antero-lateral obliques, with a tumour dose of 5000–5500 rad in five weeks. The spinal cord receives about half this dose, well within the safety margin.

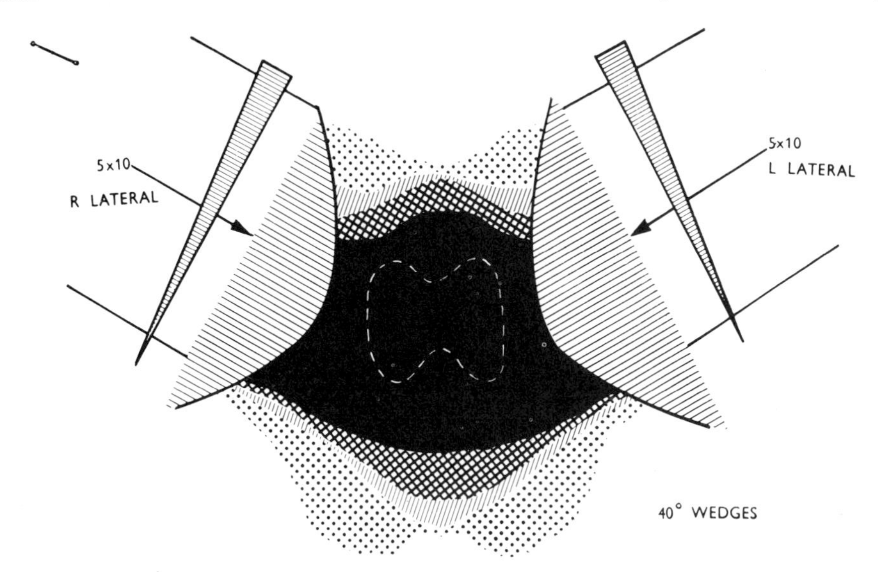

Fig. 20.7 Two lateral oblique wedged fields for thyroid cancer. (From Paterson, *Treatment of Malignant Disease*, Arnold,)

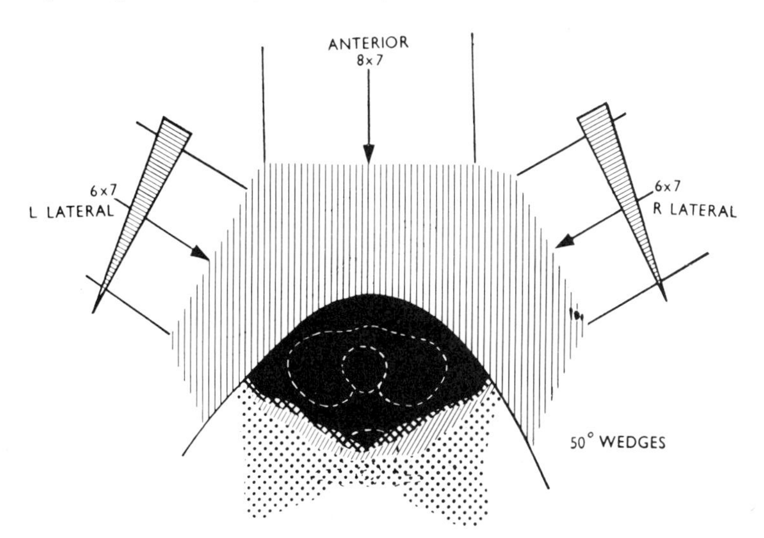

Fig. 20.8 Three-field technique for thyroid cancer. (From Paterson, *Treatment of Malignant Disease*, Arnold.)

For large anaplastic tumour masses, large fields must be used, to include all or most of the neck, and usually the upper mediastinum also. Simple anterior and posterior fields are satisfactory, with a downward tongue for the mediastinum. The spinal cord is inevitably included and dosage should not exceed 3500 – 4000 rad in four weeks.

Metastases. Isolated deposits, especially in bone, may be treated by external radiation, with doses of about 2500 rad in five treatments. Pain relief is usually possible. The treatment of secondaries which take up iodine is described below.

Radiation reactions. The trachea and oesophagus are unavoidably in the treated volume and will suffer the usual inflammatory reactions with associated discomfort. If the trachea is narrowed by the growth, dosage should be low for the first few days, to avoid oedema which might cause acute respiratory obstruction and even call for emergency tracheostomy.

Radioactive iodine and follicular carcinoma

The differential take-up of iodine by the thyroid is described on page 110. Radioactive iodine, being clinically indistinguishable from stable iodine, is taken up in the same way, and thus is an ideal tool for internal—even intracellular—radiation. Well-differentiated follicular carcinoma closely resembles normal thyroid tissue, not only microscopically but functionally, i.e. it takes up iodine and synthesises thyroid hormone which is then discharged into the circulation. The best treatment of follicular carcinoma, as described above, is total thyroidectomy, to remove all functioning thyroid tissue. In practice this is very difficult to achieve and some residual tissue in the neck is usually found on post-operative scan. Following operation, a tracer dose of ^{131}I is given and a scintiscan taken (see p. 268), which will reveal any functioning thyroid tissue still present. The scan can be extended to the rest of the body, and a 'profile' count obtained, with peaks opposite areas of above-normal uptake elsewhere, i.e. any functioning metastases.

Any remaining thyroid tissue after thyroidectomy of any degree, as revealed by the scan, should be eliminated as a potential danger. A *therapeutic ablation* dose of ^{131}I is given, about 50–75 mCi. The patient should be in a separate side-ward and full precautions taken against radiation hazards (see Ch. 11). This will destroy remaining thyroid tissue. About 50 per cent of follicular carcinomas will also take up iodine in the same way, but if normal thyroid tissue is also present, then, as a rule, the normal tissue will take it up preferentially from the circulation, as it will be the successful competitor for whatever iodine is available in the circulation. The purpose of the ablation dose is to remove this competition; when all normal thyroid tissue has been destroyed by the ^{131}I, the cancer tissue, stimulated by TSH, begins to function, to replace the missing thyroid. Following the first ablation dose, follow-up tracer scans should be made at intervals of three or four months, to detect any functioning tissue in the neck or elsewhere. Such tissue will be foci of recurrence or metastases, and their uptake can be measured. If they take up worthwhile amounts of ^{131}I, they can now be attacked and destroyed by large therapeutic ablation doses, about 100 mc. Particular attention is paid to any part of the body where a metastasis is suspected, and the ^{131}I pick-up is estimated after a tracer dose. As long as adequate pick-up is demonstrated, ablation therapy is repeated, unless there is any contra-indication, e.g. excessive depression of the white cell count.

Reactions to radioactive iodine. These large ablation doses can produce local and general reactions but they are usually mild. In the gland itself there may be an inflammatory reaction with painful swelling and tenderness the next day, subsiding in the following week. There is also some excretion via the salivary glands (parotid etc.) and a similar reaction may occur there. General effects, radiation sickness, lassitude etc. are seldom troublesome and leucopenia* is short-lived. The gonads (ovaries and testes) will be affected, with menstrual disturbances and temporary inhibition of sperm formation; these are unlikely to be of practical importance. Pregnancy is an absolute contra-indication to isotope therapy because of the risk of radiation damage to the fetus.

When large amounts of ^{131}I have been used in successive therapeutic attacks on a thyroid cancer, late aplastic* anaemia and leukaemia have been known to occur. This is a hazard that is justifiable in an otherwise fatal disease.

Prognosis. Survival times are extremely variable, as shown in Table 20.1. The anaplastic types fare worst, the papillary best, with strikingly long survival in many cases, even in the presence of secondaries. Follicular carcinoma is not nearly as favourable, but ^{131}I is a new tool which can offer excellent palliation, in a way which nothing else can rival, in those cases which take it up in adequate concentration—including secondary deposits. This was a brilliant success for nuclear medicine, which 'hit the headlines' and seemed to some enthusiasts to herald the dawn of successful cancer therapy in general—at last. But it was not to be, and there is no other comparable example of such selective uptake in cancer tissue.

21. Oesophagus, gastro-intestinal tract, lung

Cancer of the oesophagus

The oesophagus continues the digestive tract from the post-cricoid hypopharynx to the stomach. It is subject to much the same kind of environmental influences as the higher parts of the tract and there is some association between oesophageal growths and previous growths of the mouth and throat, i.e. similar causative factors, including tobacco and alcohol, are at work.

It is a disease of the elderly, usually over 60, men more than women. It is relatively common in Japan and African negroes and in several other parts of the world, for unknown reasons.

Pathology. It is convenient to divide the oesophagus into thirds—the upper third in the neck, the middle third (the commonest site) in the middle of the chest at about the level of the aortic arch and the lower third above the stomach. Most are squamous carcinoma of varying anaplasia, but 10 per cent are adenocarcinoma, mainly in the lower third where they may actually arise from the proximal or cardiac end of the stomach and spread to the oesophagus.

The growth may be of proliferative or ulcerative type, and obstruction soon occurs. Spread is by extension upwards and downwards. The oesophageal wall is thin, and extra-oesophageal spread to aorta, trachea, pleura, pericardium and even vertebral bodies, may soon follow. Lymph-node metastasis is usually early—from upper third to cervical and supraclavicular nodes; from middle third to paratracheal, paraortic and mediastinal nodes; from lower third to adjacent mediastinal and upper abdominal nodes near the stomach. Most cases prove to have node secondaries at the start of treatment. Blood spread may also be early, to liver, lungs etc.

Clinical features. The cardinal sympton is progressive *dysphagia*—first for solids, later for fluids also. There may be regurgitation and vomiting. Discomfort and chest pain may follow, from extra-oesophageal spread. Malnutrition, loss of weight and dehydration will become serious if not relieved.

The first *investigation* is by radiography, with a barium swallow, which will show the level of obstruction and narrowing. The upper level of the growth can be determined in this way, but the lower end may be difficult to make out. Next, instrumental oesophagoscopy is carried out under general anaesthesia in the theatre, the tumour is inspected and a biopsy taken. It is sometimes possible to obtain temporary improvement by gentle dilatation to widen the lumen.*

Treatment. The situation is similar to that in lung cancer—surgery is possible for a minority, but many of these prove to be more advanced than was suspected, and inoperable because of node secondaries or spread to vital structures. Apart from age, poor general condition and evidence of spread beyond the oesophagus, surgery is contra-indicated if the growth is anaplastic or if its length is judged to be over 5 cm, since experience has shown that the chances of finding a genuinely localised growth are remote. If excision is feasible, continuity is restored by bringing up part of the stomach or bowel through the diaphragm. The mortality and morbidity of radical surgery are still high, and few patients are fit enough to withstand it. The success rate after surgical resection is low—from 5 to 20 per cent.

In the cervical part of the oesophagus, surgery is very difficult and usually involves removal of the larynx also; not surprisingly, it is rarely attempted.

Surgery is however the best method for adenocarcinoma of the lower third. These are anyhow radioresistant.

Palliation by surgery

1. *Intubation.* A very useful method when radical treatment is contra-indicated is to insert a tube to maintain a free passage past the growth and so ensure adequate swallowing for the remainder of life. Various types of tube are available, including Mousseau-Barbin plastic or (better) a soft rubber Celestin tube with a flanged upper end. It is necessary to open the stomach to enable the tube to be drawn down into the stomach from above the growth. An older type is the Souttar tube, of flexible metal coils, inserted simply from above, but rather liable to be dislodged into the stomach.

2. *Gastrostomy*, is the most obvious method of surgical palliation. The stomach is opened and feeding is through a tube brought out on the anterior abdominal wall. It may be valuable as a temporary measure while radical radiotherapy is carried out, but as a permanent measure it is clearly objectionable. If it is to be considered, the pros and cons should be fully discussed with the patient's family in the light of the prognosis.

For the really advanced case, especially with secondaries in the liver etc., treatment should be confined to nursing care and symptomatic relief by drugs, with terminal heroin if necessary. Attempts at specific therapy in such cases are fruitless and unkind.

If radical surgery is ruled out—which is the usual situation—radiation is the treatment of choice. Here also a decision has to be made for either radical or palliative radiation. Radical radiation is a strenuous affair for the patient, who must be in reasonably good condition for it. Preliminary attention must be given to nutrition, correction of dehydration and anaemia, by blood transfusion if necessary. If obstruction is severe, intubation or even (temporary) gastrostomy should be considered, though such patients are doubtful candidates for a radical course.

Radiation technique. The typical site is the middle third, near the centre of the thorax, and presents difficult problems to the radiotherapist. The distance from the body surface is relatively great, and meticulous localisation and beam direction are of crucial importance. The dosage to spinal cord and lung tissue must be borne in mind. Another complicating factor is the low density of lung tissue (because of contained air) and hence the increased transmission of radiation. A beam through lung may give 30–40 per cent higher radiation dose to the oesophagus than the same beam through solid tissue, and a correction factor must be used to avoid overdosage.

Localisation is by radiography, with swallowed barium outlining the lesion. The patient should be in the actual treatment position—either supine or seated—when films are taken or on screening on the simulator. Radio-opaque skin markers are placed over the centre of the growth (as near as can be judged) at front and back, and then antero-posterior and lateral films taken. The body contour is then outlined (see p. 68) and transferred to paper. The position of spinal cord, heart and lungs are easily recorded on the drawing, with the aid of the films.

Megavoltage is best and adequate dosage can be achieved by three fields. (In orthovoltage days, six beams were needed for satisfactory tumour dose). Figure 21.1 shows a typical plot. The length of fields should not be less than 15 cm, as it is impossible to determine the exact extent of spread up and down the oesophagus, so a generous margin should be allowed at each end. The width should be at least 6 cm; some prefer 8 cm, to include adjacent nodes liable to invasion.

Beam direction is by pin-and-arc, front-pointer or back-pointer. Some workers make a shell for the patient, which he wears during the radiographic localisation and at each treatment session, in conjunction with, for example, a back-pointer.

An alternative technique is rotational therapy, as the oesophagus is in the centre of the body—as in Figure 24.3 page 240—with the posterior 50° omitted, to spare the spinal cord. This involves more radiation to lung tissue.

For radical treatment a tumour dose of 5000 rad is given in four weeks. For palliative therapy (poor general condition or known secondaries) a similar technique may be used, or a simple antero-posterior pair needing no elaborate preliminaries or beam direction. 4000 rad in four or five weeks often gives good improvement of dysphagia.

Growths of the upper third (neck and thoracic inlet) can be treated with a similar three-field technique. Because of the awkward contours, build-up with wax is advisable. Since the upper oesophagus is relatively near the anterior surface, an alternative is to use anterior fields only, as for the thyroid (see Fig. 20.8, p. 207).

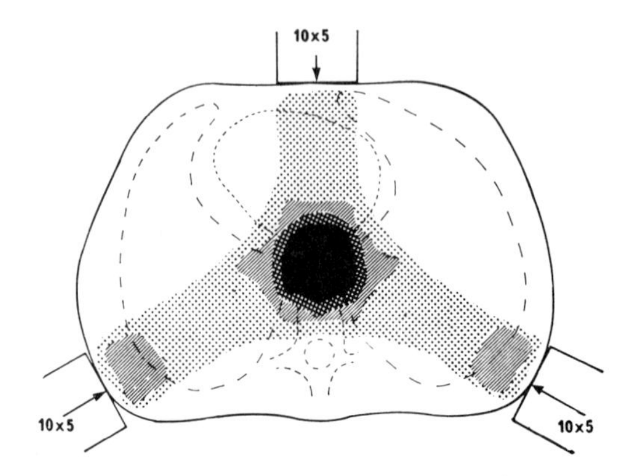

Fig. 21.1 Three-field megavoltage technique for cancer of oesophagus.

Pre-operative radiation has been used in Japan, where the disease is common.

Radiation reactions. Radical radiation can be a severe strain, and the patient should be in hospital. General reactions with nausea etc. may occur and even interrupt treatment for a few days. Adequate fluid intake and nutrition must be watched. Radiation oesophagitis is unavoidable during the course and makes feeding even more difficult. The patient will need much support and encouragement, with soft diet and sedatives.

Because of the thinness of the oesophageal wall, infiltration by the growth in depth can soon occur and lead to perforation. This may even be precipitated by destruction of the growth under radiation. Infective mediastinitis will follow, usually fatal. Similarly, a major blood-vessel may be eroded, with rapidly fatal haemorrhage.

Fibrosis of the lung is a late possibility, but rare.

Prognosis. In general, the outlook is very poor, as in lung cancer. In any event, more than a third are too advanced or in too poor condition for any form of radical treatment. However, post-mortem studies have proved that radiation is quite capable of controlling primary growths. The poor prognosis is due mainly to early spread to nodes and elsewhere.

The overall five-year survival is about 5–10 per cent; for growths in the neck, 10–20 per cent. The lower third fares better, with about 30 per cent after surgery.

Some of the best results come from Japan, after an aggressive policy of pre-operative radiation plus surgery.

Cancer of the gastro-intestinal tract

Growths of stomach and bowel, especially colon, are among the commonest cancers of all. With few exceptions they are adenocarcinoma. In general, surgery is the treatment of choice. Cancericidal doses of radiation are not tolerated, and attempts at radical treatment lead to ulceration, perforation and obstruction. Damage to kidneys is another limiting factor.

Malignant lymphoma may occur, especially in the stomach, and local node involvement is usual. Diagnosis is usually made on laparotomy* or even only after gastrectomy has been performed. If removal appears complete and the nodes are free, no further treatment is necessary. Usually, however, additional therapy is indicated, to include the whole of the stomach (if still present) and regional nodes. External radiation is given with large fields (localised with the help of barium), which must of necessity include the left kidney and cross the midline to the right, but the right kidney must be avoided. A tumour dose of 3000–3500 rad is given in four weeks.

Cancer of the rectum is of interest to the radiotherapist especially since the advent of megavoltage. Surgery is still the first choice in operable cases, but local recurrence is common, causing pain in sacral and perineal areas, maybe radiating down lower limbs. For advanced inoperable or recurrent cases radiation can often give worthwhile palliation. Simple antero-posterior pelvic fields of about 14 × 12 cm are used, to give a depth dose of about 4500 rad in four weeks.

Small perineal recurrences may receive a direct field on cobalt beam, with applied bolus.

Pre-operative radiation of about 2000–3000 rad before the radical operation for removal of the rectum seems to be of real value in improving the long-term results. It is now under trial and if confirmed will be widely adopted as a routine.

Cancer of the anus. Anal cancers are mostly of squamous type and show the same kind of radiosensitivity as squamous carcinoma elsewhere. Growths of the anal canal are usually adenocarcinoma and similar to rectal growths. Small lesions may be successfully treated either by surgery or radiation. The best radiation technique is a needle implant, single or double plane according to thickness of the lesion. The superficial end may be crossed, but not the deep end; to compensate for this, 'Indian club' type needles are useful (see p. 87). It is often desirable to perform a colostomy* prior to radiation; this can be closed later if treatment is successful.

Larger growths, if unsuitable for surgery, can be treated by external radiation, using single fields, wedge pairs or other appropriate techniques.

Cancer of the lung (bronchus)

Lung cancer (bronchogenic* carcinoma) is now the most important of all, at least in many developed countries. It is particularly common in for example Britain and Finland, but only half as common in U.S.A. Its incidence has been, and still is, increasing far more than that of any other cancer; this cannot be due merely to improved techniques of diagnosis. In England and Wales it now kills 25 000 men annually and 6000 women. In sum, it is responsible for nearly a third of all cancer deaths. It is by far the commonest of the fatal cancers of men. In women its rate of increase is even faster, so that it has recently jumped to second place after breast cancer.

Causative factors. In the early part of this century lung cancer was relatively rare. Environmental factors are clearly at work in industrial situations, e.g. mining and factories with exposure to asbestos, nickel, chrome etc. A classic example is the 'Mine disease' in Bohemia, known for centuries—see p. 151.

The association with smoking, particularly cigarettes, is now notorious. Cigars and pipes, though not free from risk, are much less dangerous. It is a statistical fact that the risk of developing cancer increases in proportion to the amount smoked daily. Even light smokers, e.g. 10 a day, have a higher incidence than non-smokers. Of men smoking 25 or more cigarettes a day, one in eight will die of lung cancer. The latent period is, as usual, very long. One estimate is that, on average, 20 cigarettes a day for 30 years will produce a cancer. To a youngster this is almost infinity, but if he or she starts before age 20, the cancer will present before 50.

Atmospheric pollution is probably a contributory cause, and may help to explain the higher incidence in town than country.

Pathology. The commonest type is squamous carcinoma, 60 per cent, with varying degrees of anaplasia. A special variety is the oat-cell tumour, 30 per cent, highly anaplastic and so called from the elongated appearance of the cells resembling oat grains. The remaining 10 per cent are adenocarcinoma.

Most arise centrally near the lung root, close to a major bronchus (Fig. 21.2). If the bronchus becomes blocked, part of the lung will be collapsed and liable to infection.

Spread is by local extension, and lymphatic and bloodstream invasion. Direct extension can occur to lung tissue, pericardium and heart, oesophagus and chest wall including ribs. Pleural involvement may cause a pleural effusion, often blood-stained. Lymphatic spread is to adjacent nodes (Fig. 21.2) and later to nodes above the clavicle and maybe the axilla. Blood spread may be early and wide, to liver, skin (as isolated or multiple nodules), skeleton, brain.

Clinical picture. Males predominate, six to one, and the highest age incidence is 45–65. The growth may be silent for years and be discovered only accidentally, e.g. when a chest film is taken for some other reason, or at mass radiography. Symptoms arise in various ways:

1. *Respiratory*. The commonest symptom is persistent cough. When growth ulcerates the bronchial wall, the sputum becomes blood-stained or there is frank haemoptysis.* Shortness of breath (dyspnoea)* arises when enough lung tissue is out of action. Infection behind a blockage will aggravate symptoms; a diagnosis of pneumonia may be made at first, the truth only emerging when the pneumonia fails to resolve completely after antibiotic therapy.

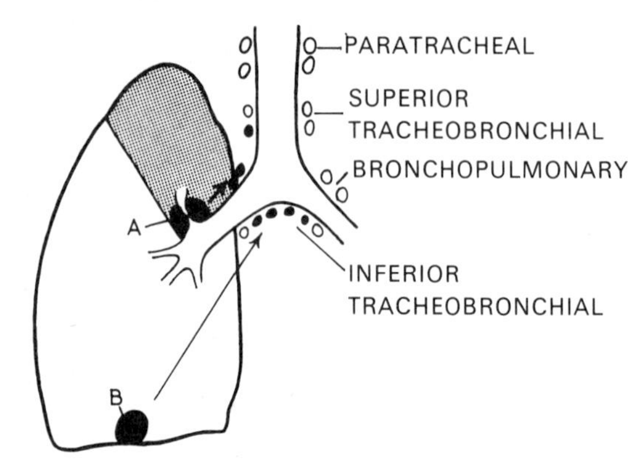

Fig. 21.2 Carcinoma of the lung, with lymph node distribution. A, Central tumour causing bronchial occlusion and peripheral pulmonary collapse. B, Peripheral tumour. (From Jayne, in Macfarlane, *Textbook of Surgery*, Churchill Livingstone.)

2. *Invasion of other structures*. Pain may arise from invasion of mediastinum or chest wall. Growths at the apex of the lung may cause pain radiating to shoulder and arm, from involvement of the brachial nerve plexus. One clinical variety is known as 'superior pulmonary sulcus' or 'Pancoast'* tumour at the apex. Involvement of recurrent laryngeal nerve may cause vocal cord palsy with hoarseness. Pressure on the great veins in upper mediastinum, including superior vena cava, leads to *mediastinal obstruction*, with swelling of head, neck and upper limbs, and engorgement of veins of neck and chest wall. Pressure on the oesophagus can cause dysphagia.

3. *Outside the chest*. The first sign may come from a metastasis while the primary lung lesion is silent. A deposit in the brain may have any of the effects of a primary brain tumour. A bone deposit may cause a pathological fracture. An enlarged liver may appear, maybe with jaundice. Nodes of neck or axilla may be present.

4. *General effects*. There may be anaemia, lassitude, loss of weight. Curious hormonal effects may appear, as some growths revert to primitive hormone production, with disturbances of metabolism, sex hormone changes etc. Nerves and muscles may also be pathologically involved, for reasons not understood.

Diagnosis. After full clinical examination, radiography almost always provides suggestive evidence. In addition to the abnormal shadow of the growth, there may be pleural effusion, a raised diaphragm (from phrenic nerve involvement) or erosion of ribs. Tomography often yields further information of value in doubtful cases.

Bronchoscopy under general anaesthesia is the next step, and if growth is visible biopsy is obtained. Even if the growth is not visible, there may be narrowing or distortion of a bronchus from extrinsic pressure.

Microscopy of the sputum is another diagnostic measure, and malignant cells may be recognisable.

Treatment. After full assessment, two vital decisions must be made:

1. Is the growth operable or inoperable?
2. Is treatment to be radical or palliative?

Surgery is the best hope, but is applicable only to a minority. Most patients show one or more contra-indications to surgery, i.e. removal of the affected lobe (lobectomy) or the whole lung (pneumonectomy). They include age, poor general condition, invasion of mediastinum or chest wall, pleural effusion. Histological anaplasia including oat cell type, even for growths otherwise operable, has been found by experience to make tumours unsuitable for surgery.

Radical radiation can be considered for a growth that appears truly localised and is not too large—e.g. if surgery is contra-indicated or is refused.

Palliative radiation is usually the best means of relief in the inoperable or advanced case, and in many centres is the treatment given at some stage to most patients. It is a very valuable agent and gives great, if temporary, relief to large numbers for whom no other means is available. Its use is for palliation of symptoms, but if symptoms are not really troublesome, it is generally wiser to withhold treatment until they do become troublesome. This is paradoxical and contrary to sound general principles of early treatment. It is certainly tempting to treat earlier rather than later, as likely to achieve more benefit, but in practice it does not prolong life in the incurable case, and may preclude effective palliation later on when symptoms are serious. Moreover, the patient is liable to attribute his later deterioration to the previous radiation. Relief is usually obtainable for cough, dyspnoea, haemoptysis and pain. A blocked bronchus may be re-opened, with drainage and relief of infection.

In advanced cases—e.g. very large tumour masses or multiple metastases—it may be better to give no specific therapy, apart from symptomatic drugs and nursing care. Pulmonary tuberculosis, especially if active, is a contra-indication to radiotherapy, which may cause serious exacerbation of the infection.

Radiation technique. In the small number of cases where radical radiation seems justifiable (e.g. an operable case where surgery is refused), careful planning and accurate beam direction are essential, as neighbouring vital structures in the chest are at risk. The general condition must be good, and anaemia, infection and nutrition controlled at the start. Since the aim is to deliver a cancericidal dose, the target volume should not be large, and field size should not need to exceed about 8 cm. If larger fields are needed, the case is unsuitable for radical radiation.

Localisation is done on the simulator or by films. Tomography may be helpful in delimiting the size and depth of the tumour. In the absence of a simulator, check films should be taken on the actual treatment unit. Planning proceeds in the usual way (see Ch. 6) with an outline drawing on which the tumour, heart, lungs and spinal cord are plotted. Megavoltage is indicated. Simple antero-posterior fields are rarely possible, as the cord would be dangerously overdosed. A three-field technique is usually best, one anterior and two postero-laterals (Fig. 21.3). Other three-field arrangements may also be useful, including one from the side, though this will irradiate all the intervening lung.

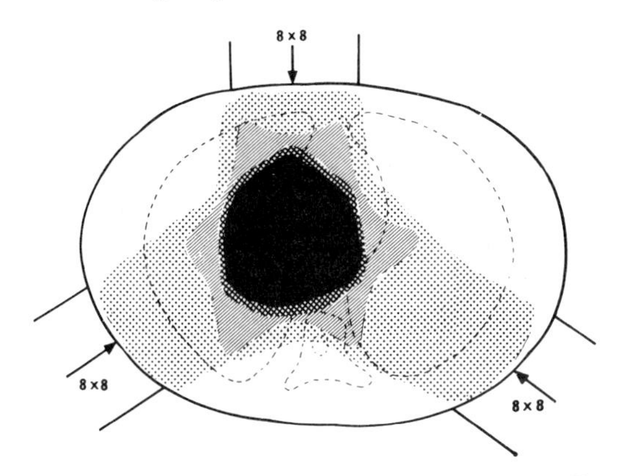

Fig. 21.3 Three-field technique for lung cancer. (From Paterson, *Treatment of Malignant Disease*, Arnold.)

Rotation therapy has its attractions, on grounds of simplicity, but involves more lung radiation than fixed fields.

The presence of air in normal lung is a complicating factor in dosage calculation, since depth-dose tables and isodose curves are based on solid tissue. The presence of air means decreased absorption and scatter and therefore greater depth dosage than would otherwise be the case. If a beam passes through much aerated lung tissue, a correction just be made to arrive at the true tumour dosage. In Figure 21.3 for the two postero-oblique fields on megavoltage about 15 per cent should be added to the conventional figure. For beams passing mainly through mediastinum, no correction is necessary.

Tumour dose should be 4500–5000 rad in four to five weeks, depending on field size. Local disease can often be controlled at this dosage, but attempts to push it higher result merely in increased late fibrosis (see below).

Palliative radiation. Technique can be simple—an anterior and posterior parallel pair suffices, with fields as large as needed to cover the primary and mediastinal nodes, up to 15 cm or more. Localisation on the simulator is most convenient. 3000–3500 rad can be given in 10 treatments, or up to 4000 rad in three or four weeks, depending on size of field.

If the patient cannot lie flat because of dyspnoea, treatment can be given in the sitting position. He will need adequate support during the treatment, e.g. on a chair or a fixture on the couch.

Mediastinal obstruction is usually relieved by similar dosage directed to the mediastinum. In emergencies, e.g. if radiation is not immediately available, relief may be obtainable by cytotoxic drugs, e.g. intravenous mustine (p. 278).

Radiation reactions. General upset is possible with large fields. There may be some oesophagitis during or after treatment. The diet should be soft, and an aspirin-mucilage emulsion before meals is helpful. A more serious reaction, after attempts at radical radiation, is late pneumonitis leading to permanent and irreversible *fibrosis* of lung tissue. This is visible in a chest film and may begin within a few months. Lung mobility and function are impaired, the patient suffers from intractable non-productive cough, and eventual shortness of breath. Intercurrent infection is liable to increase the damage. Such symptoms and radiological findings may easily be mistaken for recurrence of malignancy. Further radiation is not likely to be possible in any event, and would only make a bad situation worse.

Another serious late complication is narrowing of blood vessels at the lung roots, causing hypertension (i.e. high blood pressure) in the pulmonary circulation, and secondary heart failure.

Metastases are very common. Useful palliation is generally possible. Examples:

1. Bone—see pages 259–60
2. Intracranial—see page 256
3. Spinal—paraplegia—see page 256
4. Skin nodules. Superficial therapy may be useful, as for breast cancer (p. 222)

Pleural effusion is common, from involvement of pleural surfaces. External radiation rarely helps, but following removal of most of the fluid by paracentesis,* a cytotoxic drug can be instilled into the pleural cavity (e.g. mustine). This may control effusion for weeks or months and can be repeated. An alternative method of intra-cavitary treatment is by a radioactive isotope (phosphorus, gold, yttrium—see p. 98).

Cytotoxic drugs are useful in some cases, especially the oat-cell type. The most commonly used are mustine and cyclophosphamide. They can be combined with radiation, e.g. a short course of intravenous cyclophosphamide may be given first, then radiation, then maintenance therapy afterwards.

Prognosis. Most patients do not survive more than one or two years. *Surgery* is the best hope of success, but the overall results are lamentable. Of all patients seen by the surgeon, a third are clearly inoperable at the outset, another third are found to be inoperable when the chest is opened in the theatre, while the remaining third are apparently suitable for excision. Actually the operability rate is only 10 per cent of all cases and the five-year survival after surgery only 25 per cent. The sad fact is that at least 75 per cent of resected cases die with secondaries, which usually precede diagnosable disease. In sum, less than 5 per cent of all cases are successfully treated by surgery.

Radical radiation is not a serious rival to surgery, but is capable of destroying a growth locally, with very occasional long-term survival.

Exceptionally, long remissions are achieved even after treatment intended only as palliative, e.g. for oat-cell carcinoma. There is, however, no question about the *palliative value of radiation*, especially for mediastinal obstruction, where it can certainly prolong life. In most cases its value is to improve the quality, not the quantity, of the remainder of the patient's existence, by relieving symptoms.

Secondary cancer in lungs

This is very common indeed, usually multiple, and heralds the beginning of the end, with short life expectancy. But there are exceptions, and a patient may survive in good condition for several years after secondaries become visible on a chest film—e.g. in some salivary gland or thyroid cancers, and some slow-growing sarcomas. On rare occasions, a genuinely solitary secondary in the lung has been removed by lobectomy and the patient cured, especially in carcinoma of kidney.

Exceptionally, for a limited mass, palliative radiation may be worthwhile, on the lines laid down above for palliation of lung cancer.

In the vast majority of cases, however, neither surgery nor radiotherapy has anything to offer. In appropriate cases, hormones may cause temporary improvement, and cytotoxic drugs may also have a measure of success. If radiation is used it will usually be in some form of chest 'bath', and to give useful results the tumour must be more radiosensitive than the surrounding normal lung tissue. Otherwise, effective doses would cause severe radiation pneumonitis, itself probably fatal if the patient survived long enough. However, there are a few instances where growths are so radiosensitive that adequate dosage can be given without excessive lung damage—some lymphomas, seminoma testis, disgerminoma, chorionepithelioma, nephroblastoma (Wilms's tumour) and neuroblastoma. In these, radiation chest baths can even be curative at times.

Dosage to one lung should not exceed 2500 rad in four weeks; or 2000 rad to both lungs.

22. Breast

Breast cancer is the commonest form of malignancy in women in western countries: 20–25 per cent of all female cancer deaths (Table 13.2, p. 148). Its incidence and mortality are not decreasing; 7 per cent of all women develop breast cancer and the mortality has stayed the same for the past 30 years. Its causes are unknown.

Hormone factors. Breast tissue is not subject to environvental assaults comparable to, for example, skin or bowel. But, along with the uterus, it shows the greatest responsiveness in the body to hormonal influences. Apart from enormous growth in pregnancy and lactation, it is subject to repeated monthly cyclical changes under hormonal stimulation. This causes temporary growth of the milk-secreting glandular tissue, in preparation for possible pregnancy. The relatively higher incidence in single than married women, and the association with lowered fertility, difficulty in breast feeding, early onset of menstruation and late menopause etc., also presumably have a hormonal basis. The lower incidence in tropical and eastern countries—e.g. Japan, where breast cancer is only a tenth as common as in the west—are still unexplained. Differences in weaning customs, especially prolonged breast feeding, have been suggested, but this is little more than a guess.

The relation of these hormonal influences and variations to development of malignancy is not understood. Much research has gone into attempts to identify them in the hope of finding cause-and-effect mechanisms—e.g. urinary hormone excretion patterns vary, and there seem to be possibilities of picking out susceptible groups of women for detailed study, with a view to early diagnosis or even prevention. Whether this will be a fruitful line remains to be seen. At least, hormones have given us an important new tool for clinical management of established malignancy (Ch. 30).

Natural history. Breast cancer shows a very wide range of behaviour, with great differences in rate of growth and tendency to metastasis. A few are lethal in less than a year, despite energetic treatment. At the other extreme, notably in elderly women, growth may be almost imperceptibly slow, over 15 years or more. Most grow slowly for some years before causing serious trouble. Rate of growth depends on the doubling time of the cells, i.e. the interval between successive mitoses, and there is generally a long preclinical silent phase of months or years before any palpable mass appears (Fig. 14.3, p. 155). This gives considerable opportunity for metastasis to take place before the clinical period begins.

Even without any treatment, the average survival time is three years. Almost one in five (18 per cent) survive five years, and one in twenty five (4 per cent) live ten years. The conventional five-year survival figure is therefore of very limited value in assessing long-term results. For the untreated, life becomes miserable indeed, with ulceration and pain etc. Treatment, apart from usually prolonging life, relieves the misery of existence, clears up ulcers, controls pain and other symptoms, enabling the patient as a rule to enjoy most of her remaining life in reasonable comfort.

A typical sequence of events in a neglected case is as follows. A woman of 55 notices a painless lump 2 or 3 cm across. She ignores it, out of fear or 'because it didn't hurt'. It doubles in size in six months, when a small node appears in the lower axilla. Six more months and there is some dimpling of the overlying breast skin, or oedema, advancing later to 'peau d'orange'.* Then the growth shows deep fixation to underlying muscle. By this time more nodes are palpable higher in the axilla; later, also above the clavicle. Purplish skin nodules may appear on the breast and spread to the chest wall. Skin involvement, with more or less ulceration, may become confluent, and extend over a large part of the chest wall, even round to the back—called 'cancer en cuirasse'.* Symptoms of mestasis soon arise—cough and dyspnoea from spread to lung or pleura, and maybe pleural effusion; backache from vertebral deposits; liver involvement with jaundice; cerebral secondaries causing headache and vomiting; bone deposits with pathological fracture. The end comes with loss of appetite, weight and strength, and terminal pneumonia.

Sometimes the primary remains small and unnoticed, and the first symptom may come from a secondary deposit—e.g. pathological fracture (bone), headache (brain), dyspnoea (pleural fluid).

Pathology. There are many varieties of benign tumours, nodules and cysts. Of all the breast masses brought to the doctor, only one in five prove to be malignant. It is often possible to say at once, on clinical examination, that a mass is e.g. a simple cyst. In other cases the diagnosis of malignancy is obvious at the start. If there is the slightest doubt—a common situation—then excisional biopsy is imperative. This should be done in the operating theatre, with the pathologist in attendance. The surgeon removes the lump locally; the pathologist makes a rapid microscopic examination in an adjoining room by the procedure of 'frozen section' which takes only a matter of minutes. He will usually be able to pronounce at once with confidence whether the mass is simple or malignant. If malignant, the surgeon can proceed at once to full operation—or, if there are no facilities for frozen section, the surgeon closes the small wound and awaits the full report. If malignancy is confirmed, he can proceed to mastectomy a few days later.

Almost all breast cancers arise from glandular epithelium, i.e. are adenocarcinoma. A tiny proportion are sarcoma; rarely a deposit proves to be for example Hodgkin's disease.

The pathologist will report on the histological* *grading* of the tumour, whether anaplastic or well differentiated (see p. 155). This information is useful in assessing prognosis and optimal management. Anaplastic growths grow faster and spread sooner locally and generally.

Breast cancers contain a variable amount of fibrous tissue, not itself malignant but a reaction of invaded tissue to the 'irritation' or stimulus of the cancer cells—rather like a scar. It may be so dense as to produce a fibrotic hard tumour, called 'scirrhus'.* The fibrous tissue tends to contract, for reasons unknown, and cause retraction of the nipple or skin dimpling, and later stony fixation to the chest wall. These scirrhous* types are typical of elderly women.

Paget's disease* of the nipple is another type. It resembles simple eczema at first, but goes with an underlying breast cancer.

Spread may be very early or very late—anything from a few weeks to a great many years after the growth arises. As usual, there are three main varieties:

1. *Local*—to soft tissue and overlying skin, with nodules and ulceration, fixation to muscle and chest wall.

2. *Lymphatic* (Fig. 22.1)—usually first to *axillary* nodes. Over 60 per cent of cases present with palpable axillary nodes, and clinical staging (see below) is based on their presence or absence. But there is a considerable degree of error, when corrected by microscopic examination—a quarter of impalpable nodes are found involved, and nearly a third of palpably enlarged nodes are found free of growth.

From medial or central parts of the breast, invasion is to *internal mammary (parasternal)* nodes, 3 cm lateral to midline and 2 cm deep to skin surface. Parasternal nodes cannot be detected unless so enlarged as to bulge externally, which is rare. But we know they are invaded in almost a third of operable cases, in two-thirds when the primary is in the medial part of the breast, and in three-quarters when axillary nodes are invaded.

Supraclavicular nodes are invaded later, from the axilla or from mediastinal or parasternal nodes. Nodes of the opposite axilla may also be involved.

3. *Blood* vessels are invaded directly in the breast and detached cells settle and grow in almost any organ—especially bone, lung, liver, brain. Pleural and peritoneal membranes and cavities may be involved, by blood or lymphatic invasion.

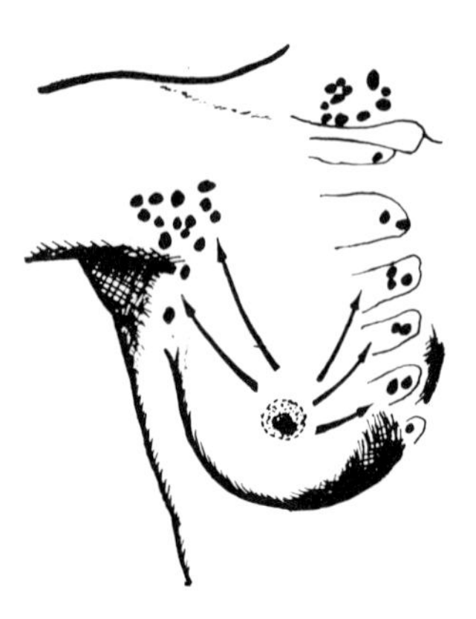

Fig. 22.1 Lymphatic drainage of the breast, showing axillary, parasternal (internal mammary) and supraclavicular lymph-nodes. Arrows show chief routes of spread.

Staging is most important, based on clinical examination when the patient is first seen. The size and mobility of the tumour are noted, and any involvement of skin, muscle etc. Regional node sites are examined—axillary and supraclavicular on both sides, as well as the other breast. The abdomen is palpated for liver enlargement. The chest is examined for abnormalities of lung or pleural fluid. Radiographs are taken of chest and skeleton (especially spine and pelvis).

Staging must be as accurate as possible, since choice of treatment depends on it. Just as for cancer of cervix, it is valuable for comparison of results of different methods of treatment. But, unlide the cervix, the breast has no internationally agreed system, and various systems, some very complex and confusing, are in use. The following is a simple classification:

Stage 1. Freely movable (on underlying muscle). No suspicious nodes.

Stage 2. As stage 1, but mobile axillary node(s) palpable on same side.

Stage 3. Primary more extensive than Stage 1—e.g. skin invaded wide of primary mass; or fixation to muscle.
Axillary nodes, if present, are fixed. Or supraclavicular nodes involved.

Stage 4. Extension beyond chest area—e.g. opposite breast or axilla involved. Or distant metastases.

The TNM system (p. 154) is now gaining acceptance internationally and its application to breast cancer is as follows:

T—Primary tumour

- T1 = Small tumour (less than 2 cm) confined to breast.
- T2 = Moderate tumour (less than 5 cm) limited or with slight extension beyond breast.
- T3 = Large tumour (less than 10 cm) with local infiltration of neighbouring structures.
- T4 = Tumour more than 10 cm, or widespread invasion of neighbouring structures.

N—Regional lymph nodes

- NO = No palpable nodes.
- N1 = Mobile ipsilateral* nodes.
- N2 = Fixed ipsilateral nodes.
- N3 = Ipsilateral supraclavicular nodes, movable or fixed.

M—Distant Metastases

- M0 = No distant metastases.
- M = Distant metastases including opposite breast or nodes.

These classifications are fairly simple, but are bound to let us down on some important points, particularly the significance of clinical axillary findings, as noted above. Most patients present in Stage 2, but this includes a wide range of development—the primary may be small or large, nodes single or multiple, skin invasion may be slight dimpling or actual ulceration, etc. We have no means of knowing the state of internal mammary nodes; only surgical removal or later developments can tell us this. Additional methods of detecting silent spread:

1. *Mammography* of the other breast can show up another primary mass or a secondary—it may be difficult to say which.
2. *Isotope scanning* of the skeleton (p. 113) can show up bony secondaries about 18 months before X-ray films can. This is particularly valuable in the presence of suspicious pain and negative films. Some workers are using it in early cases and silent deposits are sometimes detected quite unexpectedly; this would obviously lead to re-appraisal of treatment, especially the extent of surgery.

Scanning of other organs (liver, lung, brain) is also used in appropriate cases.

These refined and expensive techniques may not be available in smaller departments.

Treatment. Table 22.1 summarises the commonest lines of therapy currently advised. It will be seen that there are alternative approaches in most cases. In general, treatment depends on the staging arrived at by the clinician after complete investigation. The student should appreciate that breast cancer is the most controversial subject in all oncology, and there are divergent views about optimal treatment in the various stages. As a rule, patients are first seen by the surgeon, who will have the chief voice in decision. Ideally there should be joint clinics where both surgeon and radiotherapist assess each patient and decide on policy after discussion, since both surgery and radiotherapy have essential roles to play in management of most cases.

Some generalisations can be made at the outset. For early cases (Stage 1 and most of Stage 2) some form of surgery is advisable as primary treatment. For most of Stage 3 and all Stage 4, radiation and/or hormones and/or cytotoxic drugs will be preferable. Most workers would agree so far—but it begs some important questions. In marginal cases, opinions will differ about 'operability'—e.g. for a large primary mass or multiple axillary

Table 22.1 Management of breast cancer

	Aggressive	Conservative
Stages 1–2	Radical mastectomy + post-operative RT, especially if tumour anaplastic or in medial half or axillary nodes positive Pre-operative RT used by minority. Adjuvant cytotoxics on trial (Stage 2)	Local excision or simple (or extended simple) mastectomy + post-operative RT (routinely or as in 'aggressive')
Stages 3–4	RT to primary and/or painful secondaries. Simple mastectomy (early or late—e.g. 'toilet') Post-operative RT (± pre-operative RT) Hormones as initial treatment, or for residues etc. Cytotoxic drugs.	

RT = Radiotherapy

nodes. It depends on the experience (and temperament!) of the surgeon, and likewise of the radiotherapist. Competent, experienced and honest doctors can arrive at different conclusions—for it is, in the end, a matter of judgement. If surgery is agreed, the extent of the operation has to be determined—and here again there are decidely different schools of thought. The timing of radiotherapy —pre- or post-operative—and the necessity for radiation in the light of the histological findings, are further matters for divergent opinions. Wide variations in belief and practice exist, from the most conservative to the most aggressive—hence the (perhaps bewildering) divergencies of Table 22.1.

It is reasonably certain that *in most cases the ultimate result is already predetermined at the time of diagnosis*, and depends on the biological nature of the individual growth, whether aggressive (with a short doubling time and high probability of early metastasis) or slow (long doubling time and late metastasis). This would account for the lack of improvement in mortality figures in the past half century, but still leaves room for heated argument on policy in particular groups or the individual case.

The role of surgery. Apart from rapidly-growing anaplastic tumours, breast cancers are only moderately radiosensitive (less than squamous carcinoma) and as a rule cannot be reliably destroyed even by the largest tolerable doses of radiation. It is widely agreed, then, that for 'operable' cases surgery is the more reliable method of eradicating local disease. How extensive should the surgical procedure be? There are several possibilities:

1. Local excision of the tumour. (The monstrous word 'lumpectomy' has been coined—a barbarism which cannot be too strongly condemned.)
2. Simple (local) mastectomy—removal of breast alone without disturbing the axilla.
3. Extended simple mastectomy—easily accessible nodes in lower axilla are also removed, but pectoral muscles left intact.
4. Radical mastectomy—pectoral muscles removed to expose whole axilla, which is cleared of nodes as thoroughly as possible.
5. Super-radical (extended radical) mastectomy—the radical operation is extended to include internal mammary and supraclavicular nodes.

Pros and cons of surgery. Prior to 1950 the standard was radical mastectomy. This was based on sound logical principles of cancer surgery—to remove the primary and regional nodes in one block of tissue. The extended radical carries the good logic even farther, but obviously increases morbidity (post-operative debility) and occasional mortality, and has never been widely practised. The chief drawbacks of radical operations are:

1. The cosmetic results in chest wall and shoulder regions are unsightly (evening dress and bathing suits are unwearable) and therefore psychologically undesirable. Nevertheless a low price to pay for cure.
2. Post-operative oedema of the arm is common—usually not very troublesome, but sometimes an appreciable handicap. Again generally acceptable in the interests of cure.

The radical operation is still widely practised (especially in U.S.A.) but recent years have seen a strong reaction against it, in favour of more conservative operations, especially simple mastectomy. There are several arguments in favour of simple mastectomy:

1. The psychological advantage of less mutilation and better cosmetic appearance.
2. Very much lower risk of oedema of the arm.
3. If there actually are no axillary secondaries, the radical is unnecessary. If there are involved nodes, modern radiotherapy is now known to be capable of eliminating them. The best results after radical operation are in those cases shown (by microscopic examination) to be free of axillary secondaries, i.e. where the radical proved to be superfluous.
4. The long-term results of radical operations are not definitely superior.

This last point is of course the most important of all. Long years of experience and analysis of results, even after campaigns for earlier diagnosis and treatment, yielded a bitter harvest—the ultimate mortality rate obstinately refused to improve. The number of breast cancer deaths is as high now as ever.

Why does radical surgery so often fail, even in clinical Stage 1 cases? The reason is to be found in the natural history of the disease and the liability to early metastasis, beyond the reach of any form of local treatment, as indicated above. This danger is even greater if there are secondary deposits in regional nodes, for if regional nodes are involved, distant spread will already have occurred in most cases. This accounts for the sharp decrease in survival when axillary secondaries are found. *Five-year survival:*

Axillary nodes not involved — 70 per cent
Axillary nodes involved — 30 per cent

This means that *the cases where growth is genuinely confined to breast and axilla only, are relatively few.*

Disappointment in the long-term results of radical mastectomy led to a reappraisal. Trials began of *less radical operations, combined with radiation of regional nodes*. Follow-up results showed that 10-year survival rates were at least as good—Figure 22.2. Hence the diminishing popularity of radical mastectomy in favour of more limited operations, especially simple (or extended simple) mastectomy.

Distant spread may be clinically silent for years. But if it has occurred, then any form of local treatment, however drastic and extensive, is merely 'shutting the stable door after the horse has bolted'. Most deaths are due to blood spread, which has usually taken place by the time the patient is first seen, especially if regional nodes are invaded, though it may not become evident for years.

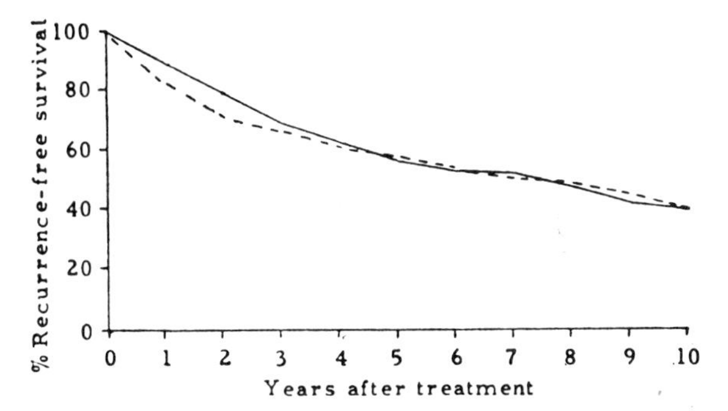

Fig. 22.2 Survival in breast cancer. Comparison of simple mastectomy with post-operative radiation (continuous line) and extended radical mastectomy (dotted line).

Here then is our dilemma—how do we know in any given case whether disease is truly localised or has already spread beyond regional nodes? If it is localised, then radical local treatment could be life-saving. But in practice the number of such cases is low. Is then the policy of treating the many for the sake of the few still justifiable? It is a genuine dilemma and there is room for difference of opinion. It is not surprising that it forms one of the major controversies currently exercising the minds of surgeons and radiotherapists. Feeling is apt to run high on the virtues or otherwise of radical mastectomy, especially in U.S.A. Some surgeons passionately contend that statistics (e.g. Fig. 22.2) are all very well in their place, but of no practical value for the individual surgeon faced with the individual patient. How can he be sure that radical operation would not save *this particular patient?* If so, would it not be unethical to deny her the possible marginal advantage, even if statistically insignificant? A patient, after all, is more than a statistic! At this point the argument tends to become philosophical and emotional rather than scientific.

The new techniques of detecting early and silent secondaries (see above) will help us to be more selective and save a proportion of patients from unnecessary and futile procedures. But our experience of these is still very limited, and the harvest has yet to come.

Local excision of the breast tumour alone is obviously attractive to the patient for its cosmetic and psychological advantages. Unfortunately it is unsafe, since we know from experience that microscopic foci of early malignancy are often present in the remainder of the breast in addition to the obvious clinical mass. However, the local procedure has its advocates on occasion, e.g. when wider operations are refused. With good fortune, perhaps with the help of post-operative radiation, it can be successful.

The role of radiation in early cases, stages 1 and 2.
In the early case, radiation is normally no rival to surgery, as we have seen. However, radiosensitivity varies with the degree of differentiation, and anaplastic growths can disappear rapidly under radiation. For this reason, if a growth is known to be histologically undifferentiated (e.g. after biopsy) or is growing rapidly with a short clinical history—which signifies anaplasia—primary radiation is preferable to primary surgery, especially if the growth is in the inner half of the breast or axillary nodes are palpable.

The chief role of radiation is supplementary to surgery, usually following it. Experience shows that mastectomy alone is followed by *local recurrence* in or near the scar in about 10 per cent of cases. Often the situation in the regional nodes cannot be known with any certainty from clinical examination—especially as axillary nodes can be felt in a third of normal women. This has been discussed above.

The purpose of *post-operative radiation* is to destroy any residual cancer cells in the chest wall or regional nodes, and so keep this whole area free of growth. This is sometimes called 'prophylactic' i.e. preventive. Local recurrence, with nodules in the skin flaps, is common. They may arise from actual residues of growth or from cells implanted in the wound by the surgeon's scalpel if it has cut through cancer tissue. To keep the chest wall free of growth is a worthwhile objective, since the presence of local disease—e.g. cancer-en-cuirasse* on the chest wall—is liable to cause more suffering, especially mental suffering, than most distant metastases.

Pre-operative radiation is rarely indicated in Stage 1, but may be used in Stage 2 in cases of rapid enlargement signifying anaplasia. Mastectomy is carried out four to six weeks after the end of the course. The object is to destroy the more undifferentiated parts of the growth, which are likely to be at the edges, i.e. those cells which are most liable to give rise to recurrence and metastasis—and so reduce the risk of residual cells being able to 'take' in the skin flaps. Occasionally, both pre-operative and post-operative radiation is used.

In many centres, the combination of mastectomy (radical or simple, but especially simple) followed by radiation became almost routine, particularly when axillary nodes proved to be invaded. Figure 22.2 may be cited as evidence of the value of post-operative radiation. However, a *retreat from routine radiation is now evident*, comparable to the retreat from radical surgery. It is clearly evident after radical mastectomy and now appearing after simple mastectomy. Just as hard experience has shown that radical operations often do more harm than good and do not seem to improve the chances of survival, so we are now realising that the same *may* apply to routine radiation.

Late cases—stages 3 and 4.
Treatment is essentially palliative, and primary surgery usually contra-indicated. In fact if it is used at the start it will probably do more harm than good, as surgery involves cutting through lymphatic and other tissue, containing cancer cells and, by the inevitable manipulation of tissue, helps to promote both local and distant spread. If surgery is contemplated, pre-operative radiation is a useful precaution.

If the mass is bulky, or if the breast is of the big fatty pendulous type, which tolerates radiation poorly, local mastectomy should be considered, with pre- or post-operative radiation. This will be a matter of surgical judgement.

Some advanced cases respond surprisingly well to radiation, with prolonged local control. This is another instance of the variability in the natural history and radiation response of breast cancers.

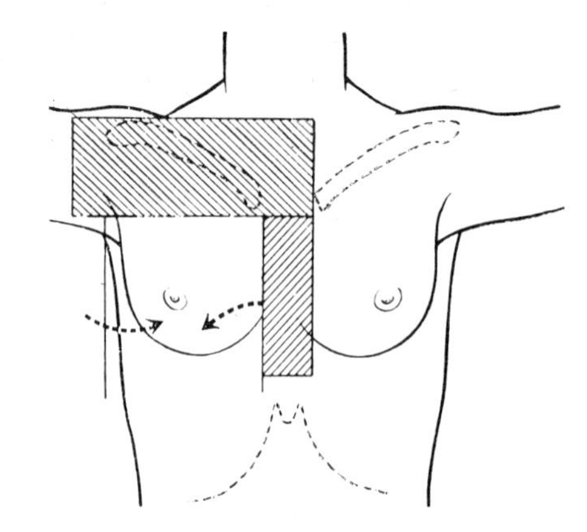

Fig. 22.3 Technique of X-ray treatment in breast cancer. Four fields are shown—1. Axillary-supraclavicular. 2. Parasternal. 3. and 4. Chest wall tangential (the breast may or may not be present). A posterior axillary field is usually added. (*Jones, in Robinson's Surgery, Longmans*)

The overall picture of breast cancer, as it presents itself in clinical practice, varies with the type of patient (social class, education etc.) and is also influenced by health education, propaganda for early diagnosis and so on. But in this country and many others, we can make the following rough generalisations:

1. In any 100 consecutive unselected patients 30 are inoperable when first seen.
2. Of 70 operated on, only 35 are apparently free of disease after five years.
3. Of all the 100 cases therefore, at least two-thirds (30 + 35 = 65) are either inoperable or not cured by surgery in the limited sense of five-year freedom.

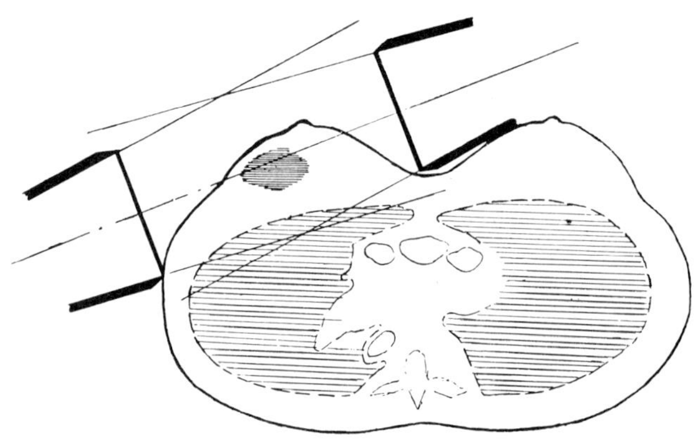

Fig. 22.4 The two tangential fields of Fig. 22.3. This technique spares the normal lung tissue.

It follows that the role of surgery in treatment is inevitably restricted, while the role of other agencies—radiation, hormones and cytotoxic drugs—is increasing.

Radiation technique. The purpose of post-operative radiotherapy is to sterilise any malignant cells residual in the chest wall and peripheral regional nodes. A large block of tissue must therefore be treated. The size and geometry of the area, with awkward shapes and curves, plus the necessity to minimise radiation to the lung, make it a very difficult subject for the radiotherapist. There is no really good technique available. Ideally the whole target volume should be included in a single set-up, without junctional fields which always involve risk of overdosage or underdosage at junctional margins. Some of the methods used are shown in Figures 22.3, 22.4, 22.5.

Internal mammary nodes are treated by a direct field from jugular notch above to upper end of xiphisternum below—about 15 cm long and 6 cm wide, from midline or just across.

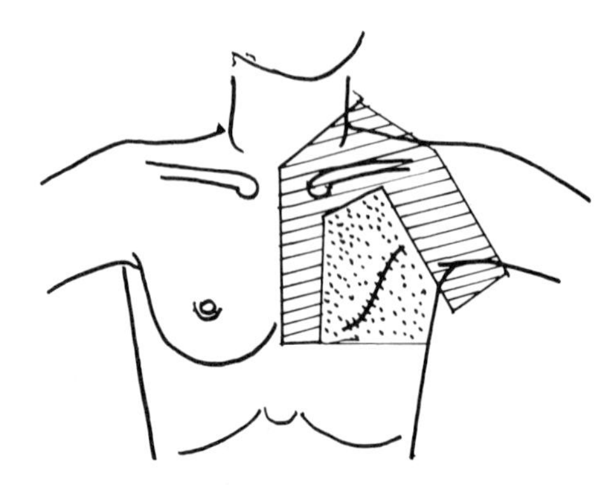

Fig. 22.5 Fields for breast cancer (post-operative). The cross-hatched area is a single megavoltage field, shaped by beam blocks. The stippled area is treated by a single field (or two adjoining) at low voltage (e.g. 100 kV).

Axillary apex and supraclavicular nodes are covered by a pair of antero-posterior parallel opposed fields, about 20 × 10 cm.

The chest wall, including the breast if still present, is treated by a pair of tangent fields, to reduce dosage to lungs. 15–20 cm length is needed to cover the whole of the scar. The medial field adjoins the lateral edge of the parasternal field; the lateral field is at or a little in front of the mid-axillary line. When the tangent fields are being irradiated, the anterior axillary field should be protected from additional dosage by lead rubber.

Variations on these basic techniques will depend on the equipment available and the preferences of the therapist.

1. DXR (deep X-rays, orthovoltage, 200–250 kV) may be used throughout, and was the standard technique before the advent of megavoltage. A maximum dose of 4500 rad is given in four weeks (20 treatments). The parasternal field is given 4000–4250 rad, since it will receive a small additional dose from the lateral tangent. A jig is useful for adjusting the angulation of the tangent fields—a pair of parallel plates sliding on a rod, to ensure that the applicator surfaces are really parallel. Full bolus is applied with the tangent fields, to secure homogeneity of dose distribution and enable standard depth-dose tables to be used. The central dose between the tangents is usually low and may be supplemented by a narrow direct field, e.g. 15 × 6 cm to bring it up towards 4500. A posterior parallel axillary-supraclavicular field completes the scheme.

The chief drawback of this method is the relatively low dosage to the axillary apex—about 3500 rad. But it is still preferred by some workers, even if megavoltage is available.

2. MXR (megavoltage or cobalt beam) is now widely used, for its better depth dosage. Parasternal and supraclavicular fields may be treated separately or (preferably, to avoid junctions) by a single inverted L-shaped field with the aid of beam blocks—the whole of the cross-hatched area of Figure 22.5. Increased FSD or SSD will be needed for a field of this size.

MXR can also be used for the tangent pair. A 1 cm layer of bolus is applied to the chest wall, to bring the maximum dose up to the skin surface. The skin-sparing effect of megavoltage would be a positive disadvantage, since tumour cells are liable to be present in the skin itself.

3. MXR may be used for peripheral node fields and DXR for the tangential chest wall fields. DXR for the tangents is technically simpler, gives full skin dosage and allows for easier protection of the clavicular field adjoining.

4. Instead of a tangent pair, superficial X-rays can be used by direct fields on the chest wall (Fig. 22.5); 100 kV is adequate and avoids lung damage. If the field is reasonably flat, a 25 cm diameter applicator can be used to cover the whole area. If the surface is curved, two separate fields are used. All junctions must be carefully demarcated by lead rubber. 4000–4250 rad is adequate in four weeks.

5. Megavoltage can be used in a single-block technique, a tangential pair including peripheral nodes (parasternal, axillary and supraclavicular) and chest wall. It extends from mid-neck to below the scar. The medial edge is across midline, so that parasternal nodes are included in the high-dose zone. A special jig is desirable for accuracy of set-up. A maximum dose of 5000 rad is given in four weeks. Theoretically this technique should be ideal—only two fields to set up and no junctions to worry about. But there are drawbacks. Dosage is liable to be too low, especially above the clavicle. In the attempt to include both parasternal and axillary nodes, the medial edge has to be beyond midline, and the lateral edge far enough back on the lateral chest wall to cover the apex of the axilla; there is danger of too high dosage to too much lung tissue. For these reasons some workers do not favour the technique in spite of its attractions.

6. The axillary-supraclavicular fields so far described are at a right angle to the parasternal field. This means that only the upper part of the axilla is included. Bringing the inferior border of the field lower would include too much of the lung. It is easily possible to angulate the field, to follow the lymphatic chain, bringing the outer corner downwards so as to include the lower axilla also. But this complicates the arrangement of a tangent chest wall pair, with awkward overlaps on any parasternal field. An alternative is to use direct superficial radiation on chest wall as in (4) above.

Figure 22.5 shows a useful combination of cobalt beam and low voltage. The peripheral field is single, at increased distance, shaped by beam blocks. The incident dose to the peripheral field is 5000 rad, and 4000 to chest wall in four weeks. This gives adequate dosage to supraclavicular and parasternal nodes. Axillary dosage is boosted by a small supplementary posterior field about 10 × 8 cm giving for example 5 × 150 = 750 rad incident towards the end of the course.

For chest wall, electron therapy is an alternative.

At the patient's first visit, the fields are drawn on the skin with a marker pen. The patient is instructed not to wash them off. Treatments are usually daily, five sessions per week; alternatively three times per week, Monday Wednesday and Friday. In the latter case, dosage must be adjusted to compensate for the larger fractions, and the doses quoted above for daily treatments will need to be reduced by about 10 per cent.

Successful treatment depends on attention to detail. Some important points are:

1. See that the patient lies in a comfortable and relaxed position, with supporting pillows etc. If not, she will tend to relax and sag during actual treatment, and a glancing field, for example, may come to miss the chest wall almost completely.

2. See that the arm is out at about a right angle, and the head turned to the opposite side. For the posterior field, see that the shoulder does not sag—support it with pillow or bags, to bring the axilla as close as possible to applicator or beam.
3. Take particular care against overdosage or underdosage at field junctions.
4. The head of the humerus and the shoulder joint should be protected by shaped fields or lead rubber.
5. Skin marks tend to be removed by friction of clothing and perspiration. They may need occasional renewal.
6. Use adequate bolus. Err on the side of generosity.

Reactions. The principles and details on page 168 are applicable. With DXR, reactions are brisk, usually going to moist desquamation, beginning in the axilla. With MXR, mild to moderate erythema and dry desquamation are the rule:

Late complications of treatment. Wherever radiation is given in tolerance dosage, a small risk of damage must be accepted—just as there are always risks attached to surgery. The possibilities include:

1. Skin scarring and thickening of skin and subcutaneous soft tissue (fibrosis), and the other effects detailed on page 168.
2. Scarring round the shoulder joint with limitation of movement.
3. Brachial neuritis from fibrotic strangulation of the brachial nerve plexus, causing pain and weakness of arm and hand.
4. Bone—late necrosis and fracture of ribs, clavicle and upper end of humerus.
5. Choking of blood and lymphatic vessels in the axilla may cause oedema of the arm, as fluid cannot drain freely. This complication is common after radical mastectomy (without radiation) since the surgeon removes as much lymphatic tissue as possible. But it can be caused or aggravated by radiation, and especially if surgery and radiation are both used.
6. Lung—transient post-radiation pneumonitis, especially at the apex, where axillary-supraclavicular fields cannot avoid underlying lung, is very common. Rarely it leads to permanent fibrosis, with dyspnoea and cough. Scarring of lung tissue decreases respiratory reserves and makes later infection more serious.

This is a formidable list, but modern equipment and technique have made serious trouble rare.

Palliative treatment

In many cases, radical treatment of any kind is clearly unrealistic—e.g. very advanced primary, distant secondaries, old age or poor condition. Surgery has usually little to offer, but radiation can be extremely useful in palliation and this sphere of treatment is one of the triumphs of radiotherapy.

Primary growth. There may be an advanced ulcerated bleeding growth, with infection, discharge and maybe pain. Even today such late cases are still seen, due to ignorance and fear. Local radiation will usually lead to considerable improvement, with cessation of haemorrhage and discharge, shrinkage of the bulk and even healing of an ulcer.

Technique need only be simple—e.g. a straight-on field at 200 kV, or 140 kV or a tangent pair. High dosage is not essential—e.g. 3000 rad in 10 treatments. In selected cases it is possible to give tolerance dosage—5000 rad in four weeks—with excellent local effect. Even a single dose may be useful—1000–1500 rad.

Local recurrence. At any time after operation, recurrence may take the form of small nodules in the skin of the chest wall, generally in or near the scar. Treatment should aim at a larger area than the obvious nodule, since neighbouring nodules are liable to appear before long. Treatment can be on the same lines as for skin cancer, with superficial X-rays. A 'chasing' technique may be needed, treating each group of nodules as they appear. If the skin has received previous high dosage, further treatment may be ruled out, or possible for one or two small patches only.

Bone secondaries and pathological fracture—see page 259

Cerebral secondaries—see page 256

Effusions and intracavitary treatment

Exudation of fluid from serous* surfaces (pleural or peritoneal) involved by growth is common (also in lung, ovarian and other cancers). It causes dyspnoea, abdominal swelling and discomfort etc. Paracentesis* gives relief, but re-accumulation may be rapid, and frequent tappings are inconvenient and exhausting, especially as considerable loss of protein occurs. External radiation hardly ever helps. The simplest method of therapy is intracavitary (intra-pleural or intra-peritoneal) injection of a cytotoxic drug (see pages 278, 281). If this fails, a radioactive isotope in colloidal* solution may be effective. The commonest in use are radiogold and radioyttrium (p. 98).

The deeper tissues of lung or bowel etc. will receive virtually no beta radiation, but only the small fraction of gamma rays (if gold is used) which give a dose insufficient to have noticeable effects. The radioactive particles are deposited on the serous surfaces and irradiate them. Yttrium is preferable to gold, since its betas are more penetrating, and the absence of gamma reduces radiation hazard. The method is essentially a form of internal superficial 'bath' therapy. Superficial malignant cells will be inhibited, surface capillaries will be obliterated by the radiation reaction, and a protective layer of scar tissue may form in the surfaces. But if there are large masses of growth, both cytotoxic and isotopic therapy will fail; the betas cannot penetrate to influence anything at a depth of more than about 2 mm. The procedure is carried out in the theatre, with full aseptic precautions. The bulk of the fluid is first removed and the radioactive solution is instilled by gravity feed from a lead container. For a case of ascites,* 100–200 mCi of radiogold are injected (or 75 mCi of yttrium); in a pleural cavity, 75–100 mCi gold (or 50 mc yttrium). These methods are helpful in at least half the cases.

Other metastases may occur in almost any organ, especially lungs and liver. Occasionally local radiation may be worthwhile if the general condition is good, but at this stage of dissemination hormones and cytotoxic drugs are more likely to be used.

The role of cytotoxic drugs—see Chapter 31, especially page 283

Breast cancer and pregnancy

After treatment, younger patients are often advised to avoid further pregnancy. However, after a year's freedom from recurrence or metastasis, this policy is not necessary. In fact, the evidence goes to show that subsequent pregnancy, if anything, does more good than harm, especially in those under 35.

Breast cancer discovered during pregnancy can present difficult problems. The interests of both mother and child must be considered, and they may at times conflict. Tumours tend to be more anaplastic in pregnancy and the prognosis rather worse than average, but by no means hopeless.

The normal principles apply as far as possible.

In the first half of pregnancy, mastectomy (radical or simple) should be performed. On post-operative radiation opinions differ. The risks to the mother from residual malignancy have to be balanced against the risk to the child from scattered radiation. At this early stage of development the child's tissues are extremely vulnerable even to low doses of radiation. On the whole, radiation is better avoided; if there is residual malignancy, the prognosis for the mother is poor in any case.

In the second half of pregnancy, if the growth seems early and not enlarging aggressively, it may be left and treated after the delivery. If growth is rapid or in Stage 3, the interests of the mother should prevail—pregnancy should be terminated and the lesion treated. If the child is viable, Caesarian section should be performed, followed by mastectomy.

Breast cancer appearing in lactation is apt to grow rapidly. Lactation should be suppressed (by drugs) and treatment carried out on the usual lines.

Prognosis. In its early stages, breast cancer is curable. But overall figures are of little value, because of the great variations in natural history (p. 155). Some important factors have already been discussed, e.g. the poorer prospect if axillary node invasion has occurred. Clinical staging is obviously relevant, but the most important single factor is not the actual stage but the biological potential of the cell type—which includes the histological grade or degree of anaplasia which in turns governs the rate of growth and liability to metastasise.

Anaplastic growths are commoner in the younger age group, while slow scirrhous* growths are common in the elderly. Growths in the medial half of the breast are more likely to metastasise early to parasternal nodes, and thus carry a worse prognosis.

Five-year overall survival figures are of little value. Ten-year figures are better, 15-year better still, but even then it is not safe to consider a patient 'cured' since recurrence or metastasis can still occur 25 years or more after initial treatment. It is still a controversial point, but after 15 years the expectation of life (statistically) tends to be the same as that of the general population, and in this sense it may be reasonable to regard these patients as 'cured'.

Some representative figures are given in Table 22.2

Breast cancer in males. Only one percent of breast cancer occurs in males—about the same ratio as the amount of mammary tissue. The basic principles of staging, grading and treatment are the same. Since the distance to underlying tissues of the chest wall is so much shorter, fixation often occurs before the mass is noticed. The stage is therefore liable to be more advanced and the prognosis correspondingly worse, than in women. Similar principles of hormonal and cytotoxic therapy apply in later management—see Chs. 30, 31.

Table 22.2 Results in breast cancer

Stage	Histological type	Survival per cent	
		5 years	10 years
1	Differentiated	85	55
	Anaplastic	55	37
2	Differentiated	75	45
	Anaplastic	25	7
3	Differentiated	70	40
	Anaplastic	15	5
4			Low survival

Overall 15-year 'cures' = 20%. Note: the markedly higher survival of differentiated than anaplastic types

23. Cervix, body of uterus, ovary, vagina

Cancers of the genital tract are among the commonest causes of death in women (Table 13.2 p. 148). Figures 23.1, 23.2 give the anatomical landmarks.

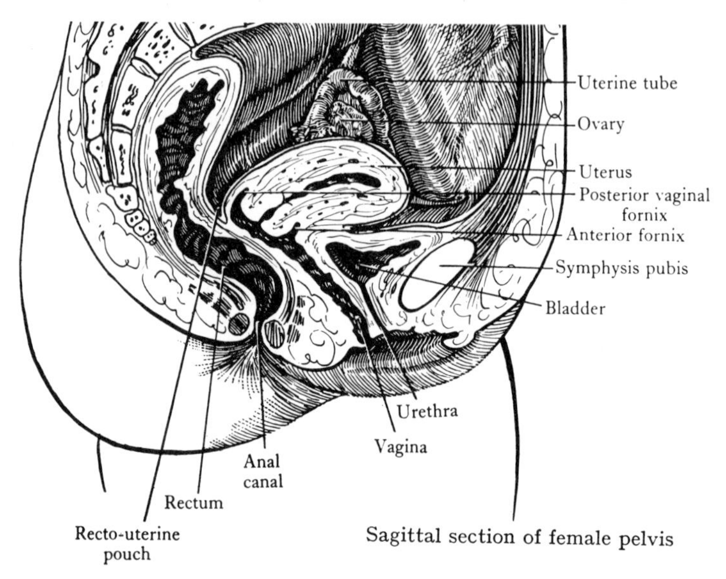

Fig. 23.1 The female pelvis—front to back.

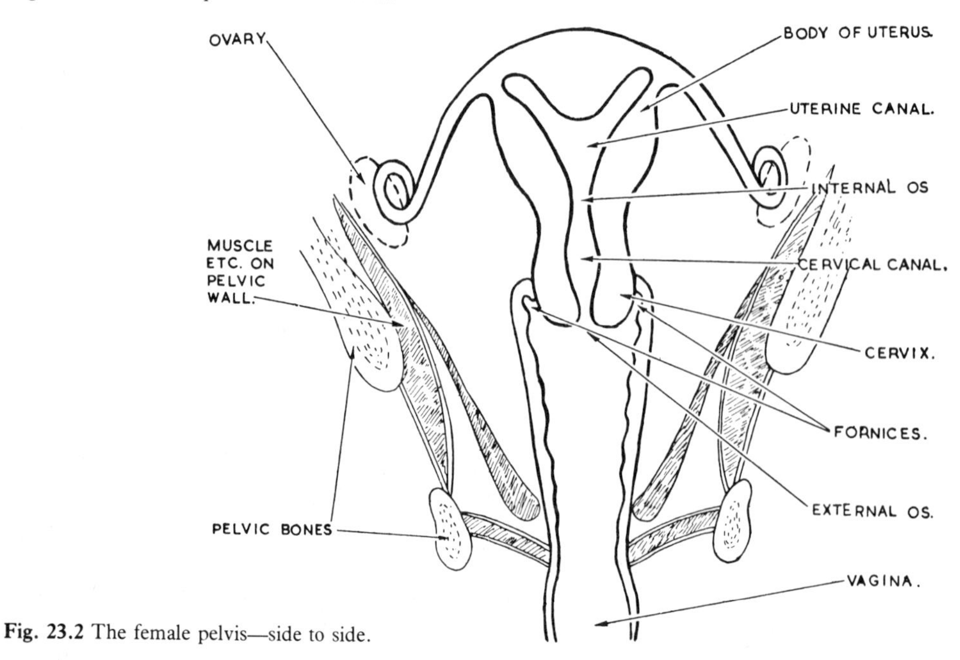

Fig. 23.2 The female pelvis—side to side.

Cancer of the uterus

Two types of cancer occur in the uterus, with different characteristics—growths of (a) the body or 'corpus' which are almost all adenocarcinoma, and (b) the neck or 'cervix' which are typically squamous, with a small proportion of adenocarcinoma. Growths of the *cervix* outnumber those of the body by two or three to one; the ratio a few decades ago was five to one. The reason for this change is unknown, but cancer of the cervix has been declining in frequency in recent years in many parts of the world. Cancer of cervix occurs in the child-bearing and menopausal years and even in the 20s and 30s, whereas cancer of corpus typically occurs during and after the menopause, though 25 per cent are pre-menopausal. Another difference is that growths of the body are relatively common in infertile women (e.g. nuns) and in the higher economic/social classes; the reverse holds for the cervix.

Cancer of the cervix

Local spread is to adjoining tissues (Fig. 23.3)—vaginal vault, fornices,* upward to corpus, laterally to parametria;* later to bladder and rectum.

Lymphatic spread is to adjacent nodes in parametria, then to nodes on the pelvic walls and from these upwards to abdominal nodes alongside the aorta (paraortic nodes).

Staging (Fig. 23.3) There is an internationally accepted classification:

Stage 0: Carcinoma-*in-situ;** also called pre-invasive or intraepithelial carcinoma.

Stage 1: Growth confined to cervix—but upward spread to corpus does not change the classification.

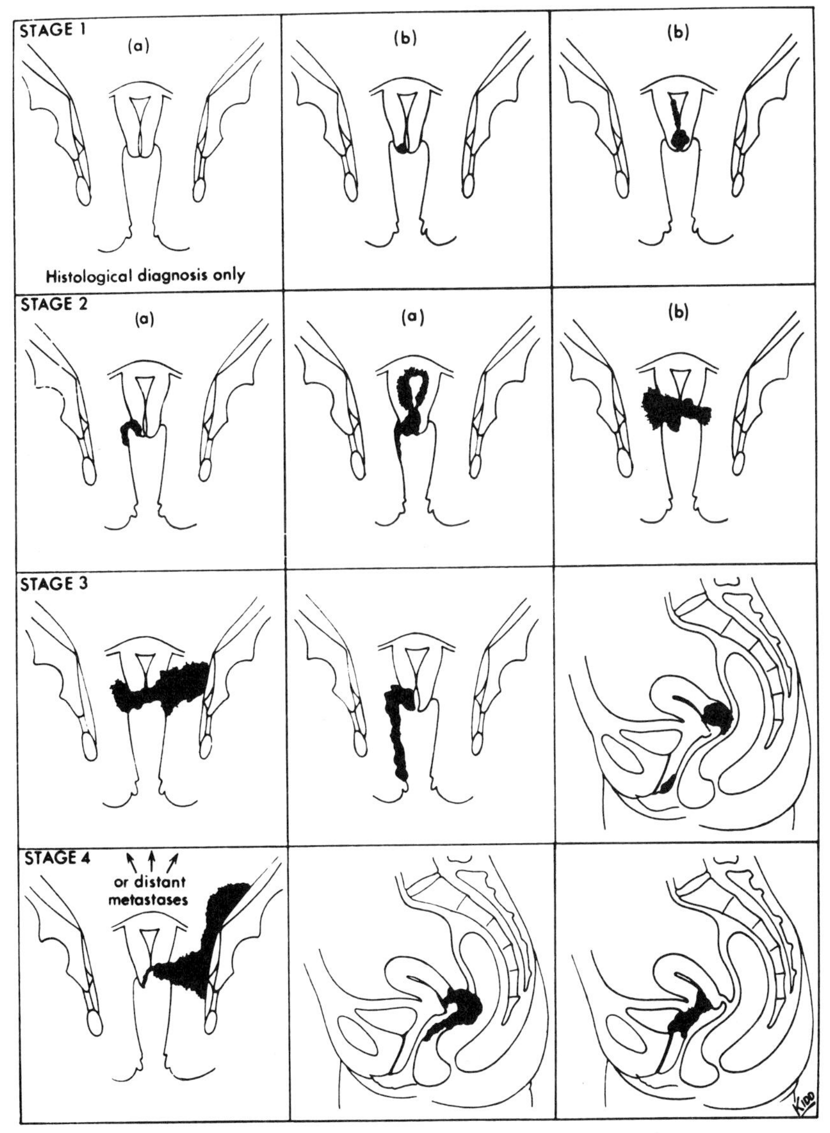

Fig. 23.3 Illustrations of the stages of cancer of the cervix. For details see text. (*Jeffcoate, 'Principles of Gynaecology', Butterworths*)

Stage 2: Spread beyond cervix—either to upper vagina or parametrium, but not as far as lateral pelvic wall.

Stage 3: Spread has reached lateral pelvic wall or lower third of vagina.

Stage 4: Growth has involved mucosa of bladder or rectum, or has spread outside the true pelvis.

Carcinoma-in-situ is a most important lesion as it may (but not inevitably) be the precursor of invasive cancer, and can be detected by *cervical smears* ('Pap' smear—see p. 161) in 'well' women. The smear may reveal either frankly cancerous cells or an intermediate stage towards malignancy, with e.g. bizarre and atypical nuclei. Immediate biopsy of the cervix is indicated, and this will reveal an abnormal epithelium, with similar malignant or semi-malignant cells which have not yet broken through the basement membrane separating the epithelium from deeper tissue to give the full picture of invasive cancer—i.e. a pre-invasive phase. Its place in screening programmes and early detection is discussed on page 162.

Lymphography is now used routinely in some departments, and can demonstrate node invasion in the pelvic and paraortic areas (compare p. 245). The findings however do not affect the staging detailed above (though they probably will at a future revision).

Symptoms and natural history. The cardinal symptom is bleeding, or blood-stained discharge—abnormal, irregular, between menstrual periods, maybe post-coital or post-menopausal. Unfortunately, irregularities are common at the time of the menopause, and pathological bleeding may be mistaken for 'only the change of life' and lead to delay. The growth may be obvious on clinical examination (palpation and visual inspection), but should be confirmed by biopsy which will also establish the histological grade.

As in breast cancer, pain is a late symptom and indicates considerable spread beyond the cervix. Later invasion of parametrial tissues and then pelvic nerves will cause vague lumbar aching, then pain radiating to hip and thigh. In ulcerated lesions there will be associated infection, which will aggravate symptoms. Invasion of the base of the bladder will cause dysuria with frequency and pain; destruction of tissue between base of bladder and vagina will cause a fistula* and urine will leak straight from bladder to vagina (vesico-vaginal fistula). The ureters pass through the parametria close to the cervix (Fig. 23.2) and ureteric blockage by compression will lead to back-pressure on the kidney and renal failure with uraemia.* This is the common terminal stage in neglected or failed cases.

Treatment. Full clinical and pathological assessment must be made before final choice of treatment. Blood count may show anaemia which may need correction by transfusion. In addition to routine 'bedside' examination (abdomen, pelvis, vagina, rectum etc.), cystoscopy* is desirable (this may be done at the first theatre session). If symptoms point to possible bowel involvement, instrumental and radiological examination of bowel are indicated. Since ureteric obstruction is potentially so serious, IVP (intravenous pyelography*) should be performed and may show unsuspected hydroureter*/hydronephrosis* on one or both sides. Blood chemistry is also investigated; impaired renal function is shown by a rise in the concentration of urea* in the blood.

Infection of the uterine cavity may be found even when not suspected (pyometra*); this should be drained via the cervical canal before further treatment.

Lymphography is a recent addition to pre-treatment assessment and has been mentioned above. If invasion is detected, this must be taken into account in the plan of treatment.

Surgery or radiotherapy?

Cancer of the cervix represents one of the greatest triumphs of radiotherapy; in earlier days the patient's sole chance lay in a severe major operation. The best-known operation is associated with the name of its German author Wertheim—an extensive removal of uterus, ovaries and lymphatic tissues of the pelvis—Wertheim's hysterectomy. Only the earlier cases—chiefly Stages 1 and 2—were suitable for the operation; for the others only palliative treatment was possible. But radiotherapy has completely altered the picture—it can accomplish much the same as surgery for the 'operable' cases, while it has a great deal to offer in the later stages where surgery is unsuitable. Moreover, the patient is spared the major operation with its associated risks and discomforts. The acid test, of course, is to see what the actual results are, i.e. what proportion of cases survive and do well; comparison of survival statistics shows that the results of radiotherapy are in practice generally better.

Radiation has, therefore, widely replaced surgery as the primary line of treatment. There remain, however, some cases (even in Stage 1) that radiation cannot cure—some show little response to treatment and obvious persistence of growth, others recur after apparent healing. Some of these can still be saved by prompt surgery.

Stage 0. This is still a puzzling situation, reflecting our lack of precise knowledge of the natural history of cervical cancer. The optimal management is still doubtful and controversial. Follow-up studies have shown that while many cases proceed to invasive cancer many others do not, and the suspicious changes can regress spontaneously. It may therefore sometimes be justifiable to wait and rely on close follow-up observation to see whether the appearances persist or improve. However, natural anxiety to be on the safe side usually dictates the policy of surgical removal of the potential danger area. The extent of the operation will depend on the patient's age, and especially on whether she wishes to retain the possibility of further children. For the older woman, simple hysterectomy will be the method of choice; likewise for the younger woman whose family is complete. For the rest, the operation of 'conisation' is advised, i.e. removal of a cone of cervical tissue (Fig. 23.4) carrying the dangerous area round the external os*. Regular follow-up observation is essential, with repeated cytological smears to detect any further malignant degeneration at an early stage.

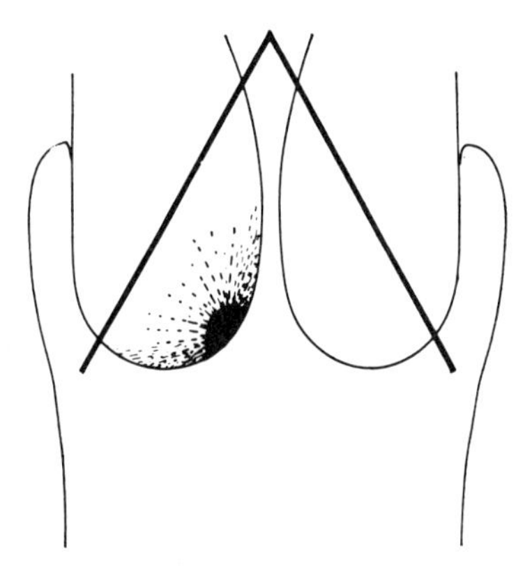

Fig. 23.4 To show the area of cervix removed by the operation of conisation. (*Jeffcoate, Principles of Gynaecology, Butterworths*)

Stage 1. Surgery and radiation are both curative in most cases. Some workers prefer surgery for (a) adenocarcinoma and (b) endocervical growths causing barrel-shaped expansion of the cervix, as they believe the response to radiation tends to be unsatisfactory; other workers disagree. Another argument for surgery is the psychological advantage of avoiding all reference to radiation and its implications in an apprehensive patient.

The chief controversial point—where opinions still differ—is the ability of the two methods to control lymph node secondaries. This is a crucial factor, since we know that one in five of all clinical Stage 1 cases have undetected secondaries in pelvic nodes. Surgeons usually claim that surgery is more reliable here, but we now have proof that radiation is also capable of eliminating pelvic node metastases. The statistical test of experience is the best available. On the basis of long years of observed results, most gynaecologists are satisfied that on balance, with average surgical skill and average radiotherapeutic resources, the results of radiation are superior to those of surgery—and this is the simple reason why most cases are now managed by radiotherapy alone.

The mortality from radiotherapy is almost nil, and the morbidity very low. Surgical mortality and morbidity have declined over the years, but are not negligible. The dangers of for example damage to ureters and urinary fistulas are still harrowing surgical possibilities. Some workers have tried to get the best of both worlds by giving full radiation followed by radical surgery; this is excellent in theory, but carries the possible complications of both. If surgery is used as principal treatment, a single pre-operative radium insertion to moderate dosage is advisable; this gives good protection to the vaginal vault against local recurrence.

Occasionally, a radiation failure, which can occur even in Stage 1, can still be rescued by operation.

Stage 2. There is less controversy than for Stage 1, though still a place for surgery in cases with minimal extension beyond the cervix, especially for the exceptions noted under Stage 1. The standard treatment is intracavitary radium (usually two insertions) followed by external megavoltage.

Stage 3. Radiation is almost the only treatment possible. One intracavitary insertion (or two) may be followed by external radiation, alternatively external radiation alone. Good palliation is usually achieved, and a proportion of lasting results. Some of the latter may be due to the fact that palpable parametrial induration may be more of inflammatory than malignant nature—clinical examination cannot distinguish.

The scope of surgery is restricted to heroic efforts necessitating removal of bladder and/or rectum also (with diversion of bowel and/or ureter)—called 'pelvic exenteration',* e.g. in a desperate attempt to salvage a radiation failure. Morbidity and mortality are inevitably high.

Stage 4. Palliation is the aim here, and clinical judgment may suggest that it is kinder to withhold radiation, and keep the patient as comfortable as possible by nursing care and sedative drugs, especially in the elderly. Radiation can be useful, e.g. in controlling bleeding and discharge or pain. Attempts at intracavitary treatment are liable to precipitate fistulas, by rapid local destruction of growth precariously plugging bladder or rectum, if not by mechanical trauma.

Surgical exenteration can be considered if the patient is fit enough.

Table 23.1 lists the usual methods of treatment for the various stages.

Contra-indications to intracavitary treatment

In addition to those already mentioned, the following situations make surgery preferable.

1. Pelvic infection, e.g. salpingitis,* past or present. Radiation may light up a seemingly quiescent infection
2. Pelvic adhesions from old or recent infection or surgery in the pelvis. A loop of bowel readily becomes adherent to pelvic structures (uterus etc.). The bowel's normal mobility gives it considerable protection from radiation damage—to which it is much more vulnerable than uterus or bladder. An adherent or trapped loop of bowel can receive damaging dosage, especially from an adjacent intracavitary source, with resultant ulceration,

haemorrhage, diarrhoea, even perforation, and later fibrotic constriction and obstruction

3. Uterine fibroids (benign fibroblastic tumours of uterine wall) may distort the cavity and make insertion difficult. Radium may also cause acute necrosis in the fibroid, leading to a surgical emergency.

choice of applicators of varied shapes and sizes, single or multiple.

A tubular radium container, about 6 cm long (but varying with the actual length of canal found) is inserted with the aid of long forceps. The tube is enclosed in a thin rubber sheath, or encased in plastic, and a silk cord is attached to the lower end, long enough to reach outside the vagina and enable easy withdrawal of the container at the end of the treatment time. Vaginal applicators are next inserted at the vault. Various models are illustrated in Figures 23.5, 23.7, 23.11. Figure 23.5 shows a pair of oval applicators or 'ovoids' of the *Manchester* type, which are made in three sizes, for small or large vaults, usually of hard rubber (or metal or plastic). Each contains radium in the central cavity. Ovoids were designed so that the surface coincides with the shape of the isodose curves of the central radium tube, and the surface doses are therefore equal all over (compare Fig. 8.3 p. 84). Rotation of the ovoid therefore does not cause any significant dose change. The two containers are held apart by a 'spacer' which serves to keep them forced outwards, with some distension of the vaginal vault. This brings the radium as near as possible to the pelvic walls and so gives as high a dose as possible to the path of spread of the growth. If the vault is not wide enough to hold them separated by a thick spacer, a narrow spacer may have to be used instead. Each container and spacer has an attached thread, for withdrawal.

When the applicators have been inserted, the operation is completed by plugging the vagina with gauze. This serves to hold them in place and prevent them slipping out of position. When inserting the gauze pack, care is taken to begin by packing as much as possible behind the vaginal

Table 23.1 Treatment of carcinoma of cervix

	Usual treatment	Exceptional treatment	Point	Dosage (rad) Radium	MXR	Ra + MXR
Stage 0	Conisation or Hysterectomy	Observation				
Stage 1	Radium with or without MXR	Hysterectomy	A	6000	1500	7500
			B	1500 2 insertions	3500	5000
Stage 2	Radium plus MXR	Hysterectomy	A	6000	1500	7500
			B	1500 2 insertions	3500	5000
Stage 3	Ra. plus MXR or MXR alone	Exenteration	A	4000	3000	7000
			B	1000 1 insertion	4000	5000
Stage 4	No specific treatment or MXR	Exenteration	Treatment according to case. Usually MXR. Pelvis up to 5000			

Overall treatment time 4½ – 5 weeks

Radiation technique—intracavitary

In most cases we use both intracavitary *radium*—or *caesium* (see Table 8.5 p. 89)—and external megavoltage.

Radium insertion is carried out in the operating theatre under general anaesthesia. The vagina, cervix and pelvis are assessed by visual and digital examination. Some workers do routine cystoscopy. A biopsy is taken for microscopy, if not previously done. The cervical canal normally allows the passage of a very narrow 'sound' of about 3 mm diameter; to hold a radium tube it has to be widened. This is done gently and gradually, by passing a series of long narrow metal 'dilators' of increasing width, till the canal is wide enough. Occasionally a growth begins inside the canal (endocervical) and there may be nothing obvious at first, until the canal is widened enough to give the operator access. An average uterine canal (cervix and body) is about 6 cm long.

The general principle is to insert radium applicators so as to cross-fire in and around the growth, while minimising dosage to critical neighbouring structures, especially bladder and rectum. This is achieved by placing radium (a) within the central canal and (b) across the vaginal vault, up against the surface of the cervix and fornices (Fig. 23.5). Figure 23.6 shows the kind of dosage distribution resulting. A narrow tubular holder is the appropriate type for the canal; for the vaginal vault there is a wide ovoids, in order to increase the separation of the radium from the rectum. The rectum lies just behind the vagina, separated only be a very thin layer of tissue (Fig. 23.2). It is bound to receive a considerable dose, especially from the ovoids, and since its radiation tolerance is lower than that of the vagina, it is important to keep the rectal dose as low as possible.

Various sizes and loadings should be available—e.g. uterine tubes holding 35 mg (15 + 10 + 10 as in Fig. 23.5) or 25 mg (15 + 10); ovoids are large, medium or small, holding 25, 20 or 15 mg. The aim is to insert the longest tube and the largest ovoids that the cavities allow. Isodose curves are seen to be pear-shaped (Fig. 23.6). Their shape

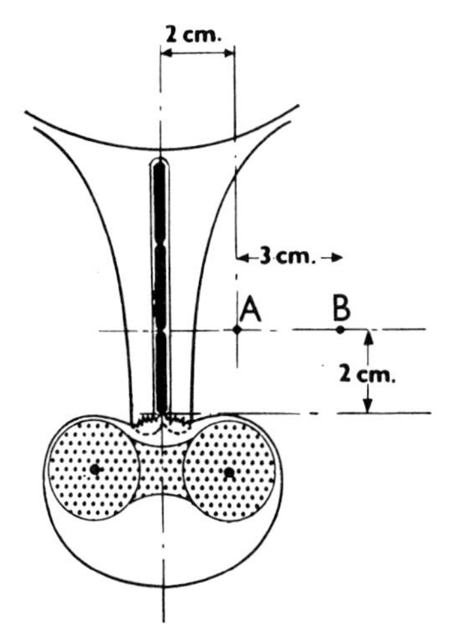

Fig. 23.5 Diagrammatic uterus and vagina, showing intra-uterine tube, cross-sections of ovoids in lateral fornices, separated by spacer, and points A and B. (*Paterson, Treatment of Malignant Disease, Arnold*)

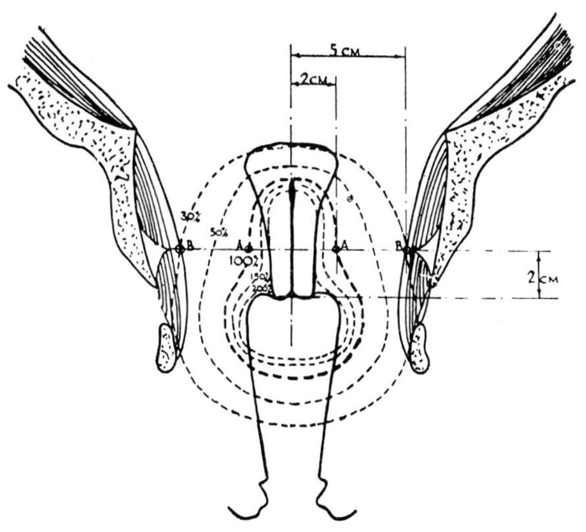

Fig. 23.6 Isodose curves in the pelvis for a typical radium distribution. The position of points A and B is shown. (*Tod, British Practice in Radiotherapy, Butterworths*)

depends on the relative amounts of radium in the uterine and vaginal sources. If the amount of radium in the uterine canal were increased, the isodose curves would become more ovoid in shape. The relative amounts are balanced to obtain the isodose distribution shown. Smaller ovoids in a narrower vault will be balanced by less intra-uterine radium. The Manchester loadings are designed to give about three fifths of the dose at Point A (see below) from the uterine source and two fifths from the vaginal.

A standard technique is used if possible, but treatment must be individualised to suit the individual patient. The chief cause of difficulty lies in the variable state of affairs found at the vaginal vault. For example, the vaginal part of the cervix may have disappeared, eroded by the growth, or a bulky growth may make it difficult to position the vaginal sources satisfactorily, or the fornices may be so contracted that there is simply not enough room at the vault for the applicators, e.g. it may hold only a single ovoid.

Intra-uterine tubes are of similar design in all techniques, but there is more variety in types of vaginal applicators used in different departments. Figures 23.7 and 23.8 show the *Sheffield* type. At the vault are kidney-shaped holders—one or two according to the space available—made of transparent plastic, mounted by sliding on a metal tube. They are made in three sizes, and the appropriate one can be chosen to suit the individual case. Tungsten inserts of 8 mm maximum thickness screen off about 50 per cent of the gamma rays posteriorly, to reduce the rectal dose. In a typical case, the uterine tube (total length 5.5 cm) contains 45 mg (25 above, 20 below), the upper vaginal holder 2 × 20 mg and the lower 2 × 10 mg. At the second insertion, a week after the first, the lower vaginal

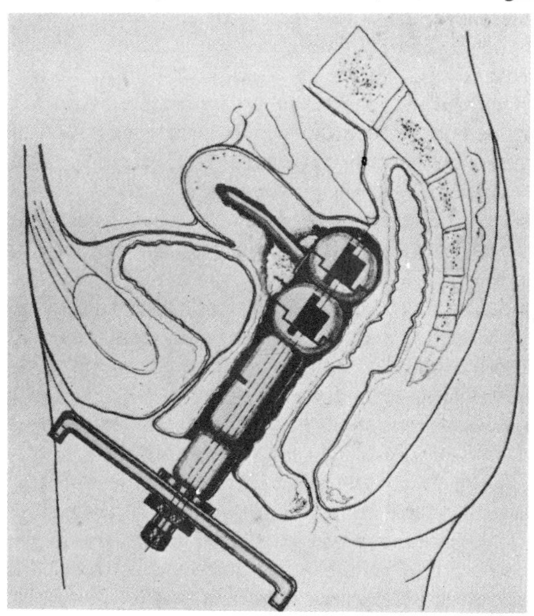

Fig. 23.7 The Sheffield applicator in position, with two radium-holding 'kidneys' and uterine tube.

holder is usually omitted. No packing is needed, and the plastic is less irritant than gauze to the vagina. The apparatus is held in position by attachment to a belt strapped round the waist.

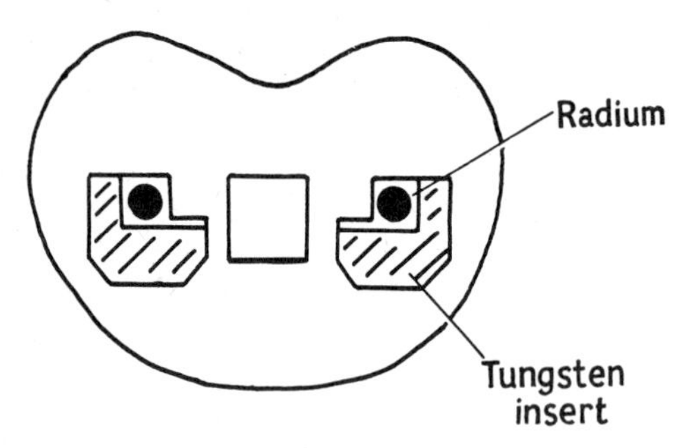

Fig. 23.8 Cross-section of kidney-shaped vaginal radium holder, showing position of radium tubes and removable tungsten inserts. If tungsten protection is not required, plastic inserts are used instead. The radium tubes are then inserted on each side and the applicator closed by the cap.

Figures 23.11, 23.12 show another type of applicator, flat silver boxes each holding 20 mg in four small 5 mg tubes.

After the operation, routine radiographs of the pelvis are taken to show the exact position of the applicators. This is best done in the theatre itself, with a portable X-ray machine. Occasionally their position is found to be unsatisfactory; the most serious mishap is that the radium tube should have slipped out and come to lie in the vagina alongside the vaginal applicators. This would mean an inevitable overdose to the rectum, leading to ulceration of rectal wall and probable fistula.* If positioning is seen to be bad, it can be corrected before the patient leaves the theatre.

The patient is returned to the ward and remains flat on her back in bed for the period of insertion—typically about a day and a half—to prevent displacement of the containers. All nursing procedures are carried out in bed, and full use is made of mobile lead screens 2.5 cm thick at the bedside to protect nursing staff etc. For removal at the appointed time, no anaesthetic is required. The pack (if any) is removed first, then the vaginal containers withdrawn by the cords, then the intra-uterine holder. They are placed at once in a lead box in a trolley which is wheeled away to the isotope safe.

A second insertion is usually one week after the first.

Dosage. Points A and B. In the pioneer days earlier in the century, the optimal length of time and frequency of insertions were cautiously worked out by trial and error. This resulted in the classical techniques of two schools: (a) *Stockholm*—two or three short intensive divided doses at intervals of one or two weeks—the basis of most current techniques and (b) *Paris*—less intensive continuous application for five days. Dosage was expressed in milligram-hours, i.e. number of milligrams of radium multiplied by number of hours it was left in. When dosage later became scientific, with development of instruments and concepts of roentgen and rad, detailed knowledge of dose distribution throughout the pelvis became possible.

In Figures 23.5 and 23.6 are two reference points labelled A and B. These were devised by the Manchester school as an aid to dosage calculations and comparisons. Point A is 2 cm above the vault and 2 cm lateral to midline. Point B is 3 cm farther out, i.e. 5 cm from midline, on or close to pelvic wall. Point A is approximately where the ureter lies, at the side of the cervix on its way to the bladder. It is regarded as critical for the minimal cancericidal dose level for a localised growth of the cervix and is therefore designated as being on the 100 per cent or base line of the isodose curves (Fig. 23.6). Point B is roughly at the site of the soft tissue (muscle, lymph nodes etc.) on the lateral pelvic wall, and therefore critical for cancericidal levels furthest from midline.

We can deliver cancericidal doses at A from central sources with safety—e.g. 7500 rad. This would give B about one fifth or 1500 rad, much too low to deal effectively with secondaries. To achieve adequate dosage at B, e.g. 5000 rad, from the same sources would take about 10 days and involve 25 000 rad at A. The result would be acute necrosis, early fistulas of rectum and bladder, and probable death from sepsis.

It is obvious from Figure 23.6 that the dose at the surface of the cervix and immediately adjoining is 200 per cent, i.e. about 15 000 rad—a very high dose, but this region is, fortunately, unusually tolerant and so permits an effective dose as far out as Point A. If this were not so, intracavitary treatment would be impracticable.

It is possible to make accurate dosage calculation from the radiographs taken after insertions, and some departments have this done routinely. It is very laborious, and most prefer to rely on simple inspection of the films, in the light of experience, to assure themselves that the radium lay-out is reasonable.

If the technique is of Manchester type, with reliance on packing to hold applicators away from the rectum, inadequate packing may leave the radium too near the recto-vaginal septum, and this cannot easily be appreciated from inspection of the films. The Sheffield technique, with a known fixed distance between radium and vaginal surface, avoids this difficulty.

A possible aid in avoiding rectal overdosage is a scintillation counter at the end of a narrow probe. This can be inserted in the rectum at the end of the operation and the maximum dose rate at the rectal mucosa obtained by a series of readings at various distances along the rectal wall. If the dose is found excessive, the radium must be re-adjusted. It is also possible to insert the probe via the uretha into the bladder and measure the dose rate at the bladder base. This method is attractive in principle, but many have found it less useful than expected.

Departments usually have their own dose charts, giving dose rates at critical points—A & B, base of bladder and anterior rectal wall—for all the standard insertions normally used with their own holdings of radium. Dose-rates from intra-uterine and vaginal sources are also itemised separately, so that dosage from any combination of sources can be easily calculated.

As Figure 23.6 shows, dosage around the sources changes rapidly and markedly from point to point, and it is not easy to express tissue dose in a meaningful way. Because of this, some workers have now gone back to the older system of dose expression in milligram-hours.

Afterloading techniques. To reduce radiation exposure of staff, methods have been developed in which suitable empty tubes are first inserted in uterus and vagina. This involves no radiation exposure whatever, and the position of the tubes can be checked by radiography and corrected if necessary. The radioactive sources are then inserted within the prearranged tubes by *remote control*, with virtually no radiation exposure of working personnel.

This afterloading technique is in use in a number of departments for growths of the uterus. The sources are several hundred times more active than in the standard techniques. No anaesthetic is usually required, and treatments can be given on an out-patient basis. With the catheters inserted, the best position for the treatment sources can be estimated from rectal dose readings from small monitoring sources. The applications are made in a specially protected room (Fig. 23.9). Each treatment is a matter of minutes instead of the usual hours. The very high dose rate is an unusual radiobiological factor, which must be taken into account. A few high-dose applications have been found to be biologically equivalent to the usual continuous low-dose method, if the same total dose and overall time are used.

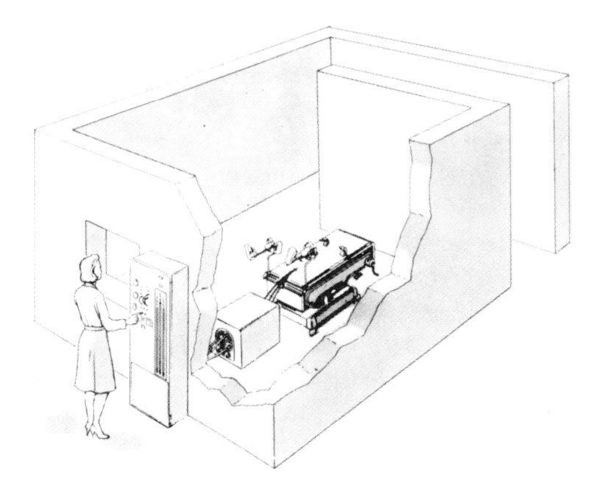

Fig. 23.9 After-loading technique. The patient is in a radiation-protected room. Cobalt sources are inserted, by remote control from outside the room, into tubes previously positioned in the patient. At the end of each treatment the sources are withdrawn into the safe. (The Cathetron, *courtesy T.E.M. Instruments Ltd.)*

The same type of afterloading method can also be used, e.g. for surface applicators on the skin etc.

External radiation. For a genuine Stage 1 case, i.e. pathological Stage 1 as well as clinical, radium alone is adequate, and capable of achieving a cure-rate of 90 per cent or more. But, as we know, node metastases can hardly ever be ruled out with confidence. Most workers therefore supplement radium with external (preferably megavoltage) radiation delivered to the lateral parts of the pelvis, to bring the dosage there up to tumoricidal level.

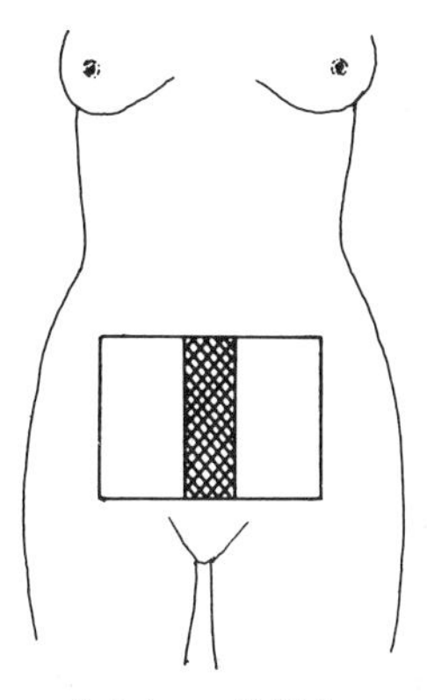

Fig. 23.10 X-ray or cobalt-beam field for supplementary radiation to pelvis. The central strip filter cuts out most of the radiation to mid-line structures (cervix, rectum, bladder). A similar field is applied at the back.

Treatment can begin the day after withdrawal of the radium. Figure 23.10 shows the usual technique, simple antero-posterior fields divided by a central filter strip about 4 cm wide to protect midline structures (cervix, bladder, rectum) which have already received considerable dosage from radium. Two pairs of anteroposterior fields could of course be used instead, with a midline gap of 4 or 5 cm. A typical size of field is 16 × 14 cm—from upper end of sacrum above to lower border of symphysis pubis below; lateral borders at the bony margins of the pelvis. The fields are planned on the simulator; or checked by films taken in the diagnostic department or (better) on the actual treatment machine. If invasion of paraortic nodes has been shown by lymphography, the fields can be continued upwards in inverted-T shape to the requisite length—comparable to the technique for seminoma of testis (Fig. 24.7, p. 243). Since the contrast medium will be visible in the nodes, the simulator is invaluable in arranging the fields front and back; 8 cm width is generally adequate. The objective is to give the desired dose at Point B, additional to the dose already given by the radium, while giving a much lower dose to A which is already heavily dosed. There will be some megavoltage dosage to A, if only from scattered radiation. The dose to A can be varied while leaving the dose at B unchanged, by using different thicknesses of central filter, and changing them, if need be, during the course. In this way, the doses can be arranged to summate to the required levels.

In Stages 1 and 2, two radium insertions in successive weeks are the rule, followed by external radiation for about three weeks—making an overall treatment time of about five weeks. In Stage 3, one insertion may be preferred to two, then external radiation.

Table 23.1 summarises treatments and doses.

If paraortic nodes are also treated, dosage should not exceed 4500 rad in four weeks in that section of the field.

The external radiation, though normally given after the insertions, can be given between insertions. Treatment is sometimes begun with external radiation, e.g. if there is a bulky mass at the vaginal vault making application of radium sources difficult. A preliminary fortnight's course to about 2000 rad should decrease the bulk and make insertion practicable.

Early radiation reactions. The cervix and body of uterus are unusually tolerant of high dosage (see above). The more vulnerable organs that are potential sources of trouble are: rectum, bladder and any loops of bowel trapped or adherent in the pelvis. As we have noted, special precautions are taken routinely for rectal protection—packing or heavy metal filter. A minor degree of rectal irritation with perhaps slight diarrhoea, is usual after a radium insertion, and settles rapidly. During external radiation symptoms may reappear and be quite troublesome. Treatment may then have to be slowed down or suspended for a few days.

Bladder irritation, with dysuria and frequency, is another possibility, but less common.

Skin reactions are unlikely to give trouble if megavoltage is used.

Late reactions and complications

1. *Infection*. An ulcerated growth is always more or less infected. The manipulations involved in treatment, including dilatation of the uterine canal, may spread infection; hence the importance of full surgical antiseptic precautions. The fundus* of the uterus may be perforated by the exploring uterine sound, with danger of pelvic and peritoneal infection. The patient's temperature is observed after insertion; an unusual rise points to pelvic infection, which is treated with antibiotics. It may be advisable to remove the radium before the prescribed time rather than risk spreading infection.

2. *Bowel*. Overdosage to rectum or other parts of the bowel can lead to late haemorrhage, ulceration and fistula. Fibrotic narrowing and obstruction can also occur. If conservative treatment does not lead to healing, surgical intervention will become necessary.

3. *Bladder*. The bladder is similarly vulnerable, lying just in front of the anterior vaginal wall. A possible late reaction is telangiectasia* at the bladder base, months or even years after treatment; these may bleed and cause blood in the urine (haematuria). This may be alarming but is rarely dangerous and soon settles. Cystoscopy, however, is advisable, to exclude a more serious cause of bleeding. *Vesico-vaginal fistula* (communication direct from bladder to vagina with constant leakage of urine) is almost always due to advanced growth rather than radiation damage.

4. *Fractured neck of femur*. In the orthovoltage era, especially when lateral fields were added to build up dosage to pelvic walls, the high energy absorption in bone sometimes resulted in weakening and fracture of the neck of the femur. This danger is avoided by megavoltage; lateral fields are not necessary, and the femoral neck can be kept out of the radiation fields.

Cancer of the cervical stump. Subtotal hysterectomy, i.e. removal of the uterus above the level of the cervix (e.g. for fibroids) was a common operative procedure in former times, but now largely given up. Carcinoma may develop in the remaining stump. Radium treatment is difficult, as the length of canal left, about 2 cm, is rarely enough to hold a radium tube. Vaginal applications can be made, supplemented—or replaced entirely—by external radiation.

Cervical cancer in pregnancy

This is fortunately rare—about one per cent of cases, one in several thousand pregnancies. The prognosis is less favourable, because the hormonal and vascular changes predispose to rapid growth and spread, while the pregnancy itself is a hindrance to treatment. Modern opinion favours full radiation in most cases, regardless of the pregnancy.

In the first six months, with the fetus non-viable, external megavoltage should be used, giving 1000 rad weekly to the pelvis for four weeks. The fetus will be killed, and after two or three weeks abortion will usually occur. At the end of the course, the uterus will be sufficiently involuted for treatment to be continued by intracavitary radium to usual dose levels. If abortion does not occur, hysterectomy should be performed.

In the last two or three months, with a viable fetus, Caesarean delivery should be performed, followed after a few days by external radiation and completed by intracavitary radium as above. Some prefer to apply intravaginal radium first, regardless of radiation effects on the fetus, which usually survives. This is followed by Caesarean delivery or hysterectomy.

It is never advisable to induce abortion or allow natural delivery before giving radiation, as this promotes rapid dissemination and the worst results of all.

With skilled treatment, five-year survival is 40–50 per cent.

Occasionally, in the sixth month or so, a decision may

have to be taken whether or not to wait a short time for viability, in the hope of a live child. This is an ethical problem of balancing the interests of mother against child. The wishes of the family should be taken into account after full discussion and explanation.

Prognosis. As usual, histological grading is some guide, and anaplastic growths tend to fare worse than the well-differentiated.

Over the years, results have steadily improved, owing to better technique, concentration of treatment in skilled hands at large centres, and earlier diagnosis and treatment. Thirty years ago the overall five-year survival was 30 per cent, now it is at least 50 per cent.

In Stage 1, radiotherapy alone gives a five-year survival of 70 per cent, and at the best clinics it is 85–90 per cent. The comparable figure for Wertheim's hysterectomy in Stage 1 is 65 per cent, and in the hands of a few very experienced surgeons 80–90 per cent.

Experience of combined radiation and surgery is not yet extensive, but results may be an improvement on either alone, especially in Stage 2.

The conclusion is clear—with average facilities and skill, radiotherapy offers the best chance in most cases.

Cancer of the body of uterus

(Carcinoma corporis uteri)

Pathology. The internal surface lining of the uterine cavity is called the 'endometrium'.* It is glandular epithelium and growths arising from it are therefore adenocarcinomas. They are the commonest cancers of the corpus (body). Sarcomas are rare (5 per cent of corpus cancers), usually malignant degeneration of the common fibroid tumours of the muscular uterine walls.

Hormones play a dominant role in the profound cyclical changes of the endometrium (menstruation, pregnancy, menopause), just as in breast tissue. Excessive stimulation by oestrogenic* hormones is related to endometrial overgrowth and malignancy—e.g. irregular menstrual cycles, late menopause, oestrogen-secreting ovarian tumours (see p. 235). Multiple hormonal disturbance is also probably the basis of the commonly associated features of obesity, hirsutism (hairiness), diabetes, hypertension (high blood pressure).

Spread may be downwards to endocervix, and it may be impossible to say whether a growth arose in cervix or body. Such cases should be regarded as primary growths of cervix. Or it may invade the uterine wall deeply and even penetrate into parametrium etc. Secondary deposits in the ovaries are common, likewise in the lower vagina. Lymph-node metastases are later and less frequent than with cervical growths, to pelvic and then paraortic nodes. Blood spread, to lungs, liver, bone etc. is relatively common in late stages.

Symptoms. Irregular bleeding, especially after the menopause, is the cardinal symptom. It is occasionally detected from a vaginal smear, from malignant cells cast off and lodging in the vagina.

Investigation is by dilatation and curettage* (D and C) under general anaesthesia. The uterine canal is dilated, as for radium insertion, and the cavity explored by a curette, to remove fragments of the lining tissue. The diagnosis may be immediately obvious, or only when confirmed by microscopy.

Treatment. The standard treatment is surgical, i.e. total hysterectomy (removal of uterus, ovaries and tubes), in early operable cases where there are no contra-indications such as hypertension, diabetes, obesity. Inoperable cases e.g. appreciable extension outside the uterus, or cases technically operable but unsuitable for surgery, are treated by radiation—either intracavity radium, external radiation or a combination of both.

Hysterectomy alone is often (10–15 per cent of cases) followed by recurrence at the vaginal vault, probably from cells implanted during the operation. The best combination is *pre-operative utero-vaginal radium*, usually a single insertion, and surgery can follow in a week or two, or up to six weeks later. This devitalises the malignant cells sufficiently to have marked effect in reducing the incidence of vault recurrences. Others prefer *post-operative radium*, with applicators inserted at the vaginal vault to deal with any residual cells in the suture area.

Pre-operative radiation can also be given by external radiation; this is preferable to radium if the growth is anaplastic or bulky, or if ovarian involvement is suspected. Full dosage should be given to the pelvis, followed by hysterectomy after a few weeks.

Radiation technique. The principles are similar to those for the cervix—usually two insertions, and external radiation if indicated by evidence of spread.

The uterine cavity is usually on the large side, and will hold a longer tube than most cases of cervical cancer—e.g. 65 mg (25 above, 20 + 20 below) of total length 7.5 cm.

Vaginal applicators are of various types, as for the cervix; some workers apply vaginal radium at only one of the two insertions. The Manchester and Sheffield techniques already described (above) are applicable also to the corpus. Other techniques are shown in Figures 23.11, 23.12. In the Marie Curie method, two small 8 mg containers are positioned at the upper corners of the uterine cavity, and the central tube contains 8 + 25 + 25 mg. Two flat boxes, each holding 4 × 5 mg are placed in the lateral fornices; a third box can be placed centrally over the external os if growth involves the cervix. In the Stockholm method, the uterine cavity is packed with as many small radium sources as it will accommodate—Heyman's capsules, holding 8–10 mg each. Each capsule has an attached numbered thread, so that they can be removed in correct sequence.

After hysterectomy, external radiation should be given to the pelvis if the growth is anaplastic or if spread was found beyond the uterus. 4000 rad can be given in four weeks—a smaller dose than for an intact pelvis, because of the decreased post-operative vascularity and the risk of adherent bowel in the pelvis.

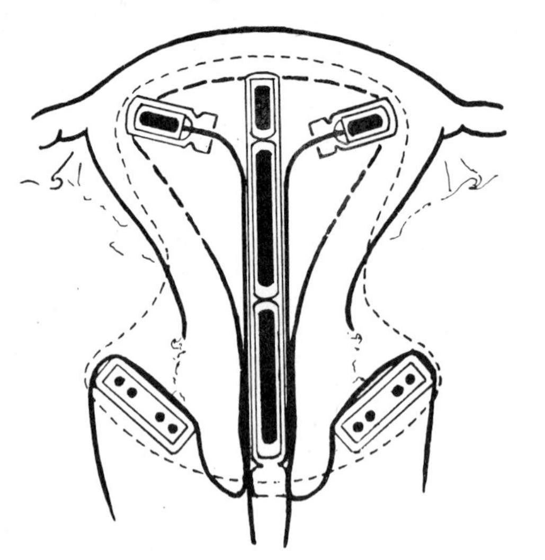

Fig. 23.11 A radium technique for body of uterus. Similar to treatment of cervix, but with two additional small containers inserted first at the upper end. Note the flat boxes, each holding four small radium tubes, inserted in the lateral fornices. *(Marie Curie Hospital)*

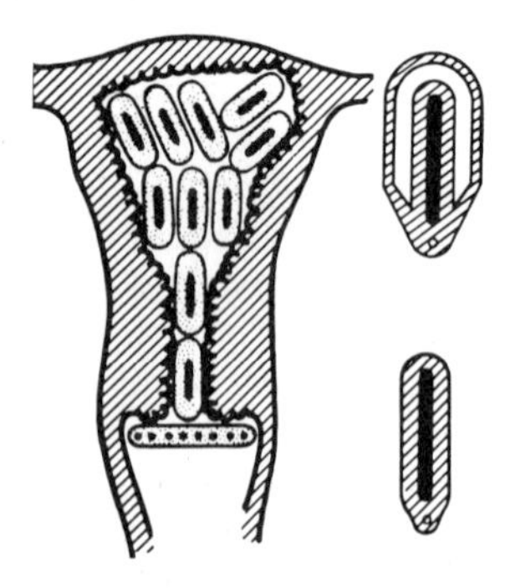

Fig. 23.12 Two types of Heyman applicators, and a uterine cavity packed with as many as it will hold. Note the flat box of radium against the cervix, and the eyelets in the applicators for threads. *(Hulbert, in 'Treatment of Cancer in Clinical Practice', Livingstone)*

If hysterectomy is contra-indicated for any reason, treatment is by radiation alone—e.g. two radium insertions using one of the above techniques at a week's interval, followed by supplementary external radiation as for cervical cancer. Doses similar to those of Table 23.1 are applicable. Supplementary external radiation is not used by all workers, and its desirability is a matter of individual judgment.

If the patient is unfit for a general anaesthetic or if there is any contra-indication to intracavitary insertion (see p. 227), external radiation alone is used, as for Stage 4 cervical cancer, with similar dosage.

If lymphography reveals paraortic node secondaries, it may be possible to extend the fields upward to cover this. But if the patient is obese or in poor condition, it is wise to be content with palliative radiation to pelvis only.

Results and prognosis. Cancer of the corpus is one of the more favourable cancers, as most are well-differentiated, grow slowly and metastasise late. For operable growths, five year survival rates are 60–70 per cent after hysterectomy; for patients under 60 it can be as high as 85 per cent.

Radiation is not generally used alone, unless surgery is contra-indicated or the growth appears too advanced. But it can achieve rates of 45–60 per cent, so that inoperability by no means connotes incurability. The Stockholm Centre, which has specialised in treating endometrial cancer by radium alone, has obtained survivals over 80 per cent; with some justice they can claim that the radiation response equals that of the cervix.

Adverse factors, such as anaplasia and extra-uterine spread, lower the survival rate to about 20 per cent.

Hormone therapy. In view of the hormonal influences on uterine growth in general, it is not surprising that hormones can play a (small) part in palliation of endometrial cancer—see p. 275.

Cancer of the ovary

The subject of ovarian tumours is large and complex. Most are benign, but one in five are malignant, especially after the menopause, when the proportion rises to one half. It may sometimes be difficult for the pathologist to distinguish benign from malignant lesions. Ovarian cancer comprises about 5 per cent of cancers in women.

There is a bewildering variety of histological types. The following classification is not comprehensive but will serve our needs:

1. *From germinal epithelium.* Most ovarian tumours arise from the epithelium which produces the germ cells (ova, egg cells), and occur during the reproductive years and up to age 65. The chances of malignancy increase with age.

   a. *Serous tumours*—often bilateral; so called because the fluid contents resemble serum* in chemical composition. This is the commonest type—serous cystadenocarcinoma; often with papillary projections on the inner surface, papillary cystadenocarcinoma.
   b. *Pseudomucinous* Tumours.* Mucin is the viscid secretion of intestinal and other glands. The secretory epithelium of these ovarian cysts produces a jelly-like substance with biochemical difference from ordinary mucin—hence the label. They are typically multicystic.
   c. Mixed and anaplastic growths—difficult to classify, and usually solid—as opposed to the common cystic types

2. *From germ cells*. These resemble tumours of comparable origin in the testis
   a. *Dysgerminoma* (more correctly disgerminoma*)—a rare malignant tumour of children and young women. It is analogous to seminoma testis in the male, and shows the same high radiosensitivity
   b. *Teratomas** arise from the various embryonic layers. Most are benign

3. *From specialised hormone-producing cells*
   a. *Granulosa* cell tumours* secrete oestrogenic* hormones and so cause precocious puberty in children or excess feminisation in adults, with menorrhagia* or post-menopausal endometrial hyperplasia and bleeding. Their malignancy is variable, but most are benign
   b. *Arrhenoblastomas** are very rare and usually benign. They may secrete androgenic* hormones and so produce signs of virilism (hair, voice etc.).

Spread. Most tumours are cystic, and malignancy begins on the inside. As the cancer progresses it may grow through to the outer (peritoneal) surface or the cyst may rupture and scatter cells into the peritoneal cavity. Cells become attached to, or invade, adjacent structures —uterine tubes, uterus, large and small bowel, bladder. 'Seeding' may deposit them far and wide on peritoneal surfaces, and multiple small nodules with some ascites* are commonly found at operation.

Lymphatic spread occurs to pelvic and paraortic nodes. Blood spread is to lungs, liver etc.

Natural history and symptoms. Ovarian tumours usually grow slowly and silently for some years. In the (uncommon) hormone-secreting types the first signal may be from the effects noted above. Benign tumours can attain huge size, resembling advanced pregnancy. Eventually pressure symptoms from the enlarging mass occur—pain, swelling of abdomen and lower limbs, dysuria from interference with the bladder. Malignant growths will cause loss of weight, ascites, multiple palpable pelvic and abdominal masses, maybe bowel obstruction, secondaries in nodes of neck etc., backache, vaginal bleeding (from uterine invasion).

Early detection is usually impossible. Growth is generally far advanced before symptoms take the patient to the doctor.

Treatment. This is primarily surgical, and only at laparotomy* can the full extent of spread be assessed, by direct inspection. Radical surgery—with removal of both ovaries, tubes and uterus—is the ideal, but practicable in only about 30 per cent of cases. If the growth appears localised and removal complete i.e. no obvious spread to the outside of the cyst or elsewhere and with a well-differentiated histology, no further treatment is indicated and the prognosis is good. But this is exceptional; as a rule, complete removal is impossible. Even so, it is advisable to remove as much of the growth as is practicable, to leave a smaller residue for further therapy to deal with. Not uncommonly the cyst may rupture during operative manipulation, with great risk of scattering seedling cells over the abdominal cavity. In all such cases—the majority—as well as for all anaplastic growths, additional treatment is indicated.

If the residue seems to be confined to the pelvis, megavoltage radiation may be given, by antero-posterior fields, up to 5000 rad in four weeks. The highest tolerable dose is needed, since the radiosensitivity of most growths is low. High doses carry some definite risk of morbidity and late complications (see p. 232), especially as there is likely to be some bowel adherent in the operation field, but this risk should be accepted if there seems to be no chance without further treatment.

Lymph-node metastases are common, and some therapists enlarge the field upwards to include the paraortic region (as in Fig. 24.7, p. 243).

If there are widespread abdominal deposits, external radiation would have to include the whole abdomen and pelvis. It is impossible to give worthwhile dosage to such a huge block of tissue, and radiation should not be attempted.

If there has been spill of potential seedlings in the peritoneal cavity, it is reasonable to instil a cytotoxic drug in the cavity at the end of the operation; an alternative is a radioactive isotope (gold, yttrium—see p. 98) at the time or post-operatively, to destroy microscopic deposits before they can take root.

Post-operative radiation or cytotoxic drugs?
The value of radiation is now increasingly questioned, and it seems probable that cytotoxic drugs usually give better results. The most commonly used include thiotepa, cyclophosphamide, chlorambucil; multi-drug regimes are better than single agents. See Chapter 31.

*Ascites.** Malignant peritoneal effusions are common and distressing. Repeated tappings are exhausting for the patient, and the loss of protein in the fluid is debilitating. External radiation is of no value, but useful results can be achieved in 50 per cent of cases by intraperitoneal instillation of a cytotoxic drug (or a radioactive isotope).

Dysgerminoma is a special case. Like seminoma testis it is highly radiosensitive and good results can be achieved even in advanced cases. It has the same mode of spread to paraortic nodes, and external radiation should be given to pelvis and nodes, to the same dose levels—see page 243.

Prognosis. Most ovarian cancers are already disseminated at diagnosis and the overall five-year survival is about 20 per cent. Anaplastic growths fare badly; the small minority of localised differentiated growths do well. Dysgerminoma being so radiosensitive has the best prognosis of all—about 80 per cent.

Cancer of vagina

Most cancers in the vagina are secondaries—from uterus, ovary, rectum etc. Primary cancer is nearly always squamous, and occurs in the 60+ age group. Local spread soon involves rectum, urethra, bladder etc. Lymphatic spread is early, to inguinal and pelvic nodes. Blood spread is late.

Treatment. Excision may be possible for small growths, but is usually ruled out by invasion or poor general condition. It is an awkward site for the radiotherapist; the technique will depend on the size and situation. If it is near the cervix, radium treatment on the same lines as for cervical cancer is usually practicable. Lower in the vagina, small radium needle implants may be possible. Otherwise intracavitary applicators should be considered, usually single line central sources, giving surface doses of 6000–7000 rad in a week.

More extensive growths are treated by external megavoltage/cobalt beam, usually a simple parallel pair. Dosage is limited by rectal tolerance, i.e. about 5000 rad in four weeks. If pelvic nodes are to be included, larger fields and lower doses are used.

The *prognosis* is generally poor, and five-year survival seldom exceeds 20 per cent.

Cancer of vulva—see page 190.

24. Kidney, bladder, testis, prostate

Cancer of the kidney

Tumours of the kidney are less common in adults than in children. The childhood nephroblastoma is dealt with on page 263.

Renal carcinoma. In adults almost all renal cancers are adenocarcinoma and arise from the epithelium of the renal tubules. Other names in use are—*hypernephroma* and Grawitz's* tumour.

(The word hypernephroma means 'above the kidney', in the same sense as suprarenal or adrenal. The name was given because the tumour cells resemble adrenal cells microscopically and most tumours do in fact arise at the upper pole of the kidney, near the adrenal body. It is a misnomer and should be dropped. Grawitz believed the tumour arose from misplaced—ectopic—adrenal tissue.)

Spread is locally to the kidney bed (perinephric fat) and neighbouring organs (bowel etc.). Regional nodes are soon invaded, along the renal blood vessels and paraortic group. Even more important is a tendency to invade the renal veins and so be carried widely to lung, liver, bone, brain etc.

Clinical features. A common presentation is painless haematuria. There may be pain in the loin, or a large mass may even be noticed by the patient. A distant metastasis may be the first sign—pathological bone fracture, pulmonary deposits with haemoptysis or cough, cerebral deposits with neurological symptoms etc., while the primary is still silent.

In investigation, the most important procedure is intravenous pyelography* (IVP), which will usually show distortion and filling defects of the normal shadows of the contrast medium. In doubtful case, retrograde pyelography may help, i.e. contrast medium injected at cystoscopy via the ureteric orifice, to outline the renal pelvis. Another radiological procedure is a renogram—a catheter is passed from the femoral artery in the thigh up the aorta to renal level and contrast medium injected to display the blood vessels; this shows the arteries feeding the tumour to be numerous and splayed out.

Treatment. Surgery is the method of choice, including any enlarged adjacent nodes. Some are clearly inoperable—a fixed growth, widespread metastases, poor general condition.

Pre-operative radiation has some advocates, who believe that survival figures are thereby improved, but its value is uncertain.

Post-operative radiation is more frequently practised, sometimes routinely in all cases, sometimes only if the surgeon has found local spread beyond the kidney itself, or if local nodes are invaded, or if the growth is anaplastic.

Radiation technique. The fields should cover the whole of the renal bed, and must extend across midline to include regional nodes on both sides of the aorta as in Figure 28.1, page 263. Care must be taken to exclude the other kidney, and the IVP films will help here. It is very difficult or impossible to avoid the spinal cord, and simple parallel antero-posterior fields of generous size are as good as any complicated technique, although the cord is in the treated volume. In obese patients a third field may be added from the side.

Tumour doses of about 4000 rad in four weeks are tolerable. Some sensitive abdominal organs are of necessity included in the high dose region—small bowel, colon, stomach (especially on left side) and a large fraction of the liver on right side. But serious radiation damage to these organs rarely occurs.

If the growth is inoperable, radical radiation is not possible, but useful palliation of symptoms can be secured, e.g. pain, haematuria.

Metastases can be treated for relief of pain etc., especially in bone (though the response is not as good as in breast cancer). Hormone therapy—see p. 275.

An interesting feature is the rare instance where an isolated secondary appears at a late date e.g. in lung or brain, and is removed surgically with long-term success.

Prognosis. The outlook for the patient depends on the stage at diagnosis. Extra-renal local invasion and metastasis, especially via the blood stream, are of very serious significance, and these are more liable to occur in anaplastic growths. The overall five-year survival is about 30 per cent. Whether post-operative radiation really improves this is debated, though it is justifiable in presence of known residues.

Cancer of the bladder

Bladder cancer is now increasing in frequency, and the cause probably lies in the urinary excretion of carcinogenic irritants derived from the environment. Workers in the aniline dye and rubber industries excrete carcinogenic chemical compounds in the urine which have been isolated. Apart from these specific industrial diseases, many other pollutants in air, food and drink have similar effects on the urinary tract, including products of the carcinogenic chemicals in absorbed tobacco-smoke.

There is an exceptionally high incidence in some (sub)tropical countries, notably Egypt, due to the frequency of bilharziasis* involving the bladder. This is a water-borne parasitic infection, carried by snails; the organisms penetrate the skin and eventually eggs are formed which have a characteristic terminal spine. These lodge in the bladder wall and cause intense inflammatory changes and ultimate malignancy.

Stones (calculi) in the bladder are not causative factors.

Pathology. The commonest site is on or near the base (trigone*) and adjacent side walls near the ureteric orificies.

Most are *transitional cell* carcinoma (80 per cent), 10 per cent are squamous, from epithelial metaplasia,* the rest are rare adenocarcinoma, sarcoma and lymphoma.

(Transitional—the epithelial lining of the urinary tract, from kidney to urethra, is of the histological type called transitional. It is a bad name, given originally in the mistaken belief that it somehow represented a transitional stage between a simpler and a more complex type of epithelium.)

The chief types on visual inspection are (a) papillary* or papilliferous* and (b) solid, with or without ulceration (Fig. 24.1). *Multiple growths are very common*.

Papilloma is a small growth on the surface lining, typically with a narrow base and numerous branching fronds (Fig. 24.1, mucosal T1). They tend to be multiple, to appear in crops, and ultimately to degenerate into carcinoma—i.e. they are *pre-malignant*.

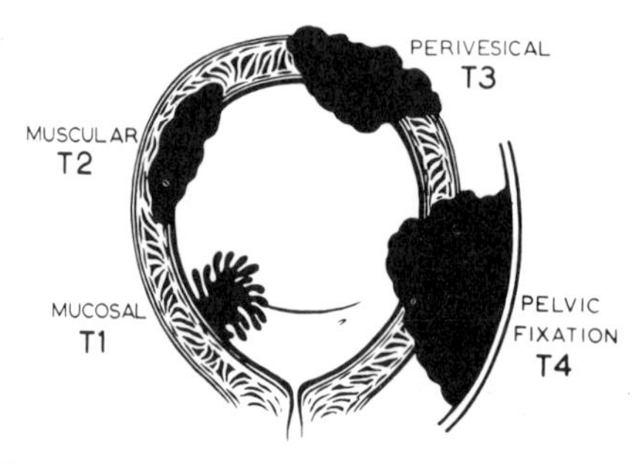

Fig. 24.1 Staging of bladder cancer. *(Wallace, Tumours of the Bladder, Livingstone)*

Papillary carcinoma has a broader base with shorter and thicker fronds. Confined at first to the mucosa and submucosa, they eventually invade the muscle coat and then the outside of the bladder.

Solid carcinoma is nodular, often ulcerated, grows more rapidly and infiltrates early. Squamous tumours are usually of this type.

As usual, the degree of differentiation is important, and anaplastic tumours grow faster and infiltrate sooner.

After muscle has been invaded, lymphatic spread is to pelvic and then paraortic nodes. Blood spread to liver, lung, bone etc. also occurs.

Natural history. The typical history is of *painless haematuria*. Papillomatous types grow slowly, and may cause no other symptoms for a long time. When anaplastic or later growths invade muscle, there will be frequency, dysuria and pain, especially when extravesical (i.e. outside the bladder) spread involves pelvic tissues. Bacterial infective cystitis is always liable, and aggravates symptoms.

Obstruction of one or both ureters can occur at any time, with no symptoms at first. Later there may be upper urinary tract infection, pain in the flank(s) and eventual renal failure from back pressure.

Investigations. Malignant cells can be detected in the urine, and this is a valuable screening method for industrial workers at risk. It is comparable with cervical cytological screening, and can give an early warning signal before symptoms arise.

Investigation of the symptomatic case includes:

1. Urine—for red blood cells, pus cells and bacteria, as well as cytology for malignant cells.
2. Blood chemistry, especially blood urea level for evidence of renal impairment.
3. *IVP* to assess renal function and reveal any abnormalities in the kidneys. Obstruction at the lower end of the ureter(s) will be shown by dilatation of ureter and renal pelvis (hydro-ureter and hydronephrosis). Filling defects in the bladder will also show up.
4. *Cystoscopy* is the most important of all. The whole of the bladder interior is inspected, the number, site, size and character of the tumours are noted and a biopsy taken. While the patient is relaxed under the anaesthetic, a bimanual examination of the pelvis is made, with a finger in rectum or vagina and the other hand on lower abdomen. In this way the tumour may be palpable and any extravesical spread assessed.

Clinical *staging* is shown in Figure 24.1.

Treatment will depend on the staging, histology, size and multiplicity of the tumours. The choice is from:

1. *Surgery*

a. T.U.R. (trans-urethral resection). Superficial growths can be removed, usually by diathermy, by instruments passed via an operating cystoscope (closed resection).
b. Open resection, by abdominal operation to expose the interior of the bladder (suprapubic cystotomy*).
c. Cystectomy—either partial or total. Total removal will involve diverting the ureters, e.g. to open on abdominal wall.

2. *Radiation*. By far the most important method is external megavoltage. Other possibilities are—interstitial solid isotopes, and intracavitary fluid isotopes.

To this list should be added—intracavitary cytotoxic drugs.

Stage 1. For small superficial lesions, single or multiple, TUR is the method of choice. All visible tumours are destroyed by diathermic fulguration. Papillomas, though technically benign, are undoubtedly premalignant. Typically, recurrent crops appear in the course of time, maybe for years, and the bladder can be kept clear by regular cystoscopic follow-up, with fulgurations as necessary.

Multiple papillomata (papillomatosis) involving most of the bladder lining are too extensive for TUR, but can be treated by an intracavitary fluid isotope or cytotoxic drug (see below). This is unlikely to be successful for long, and then the only adequate treatment is total cystectomy, since external radiation can only be palliative in this condition.

The transition from benign to malignant papilloma may be difficult or impossible to detect. The earliest stage of papillary carcinoma, i.e. before invasion of the muscular wall, can also be treated via the operating cystoscope, but careful cystoscopic follow-up will be essential. As a rule, it is difficult to exclude early invasion in depth, so that open operation may be preferred, with excision down to the base of the tumour plus interstitial implantation as for Stage 2.

For tumours well away from the bladder base, especially at the apex (vault), partial cystectomy is a possible alternative.

There is an important objection to opening the bladder in the presence of malignancy—i.e. cancer cells may be implanted in the incisional lines of the bladder or abdominal wall, to give rise to recurrent growths. This is a strong argument for external radiation as opposed to cystotomy (with or without implant). If the tumour is large, over 3 cm, external radiation is preferable, as large areas are difficult to implant properly.

Stage 2. If the growth is at the vault, partial cystectomy is feasible, but results are apt to be disappointing.

Interstitial methods are suitable only if muscle invasion is slight, since a single layer can cover a slab only 1 cm thick, which may not be adequate for the full microscopic extent of invasion.

Failing implant, the alternatives are—total cystectomy (with its operative hazards and obvious objections) or radical external radiation.

Stage 3. Lymph-node involvement is almost certain, and surgery therefore unjustifiable. External radiation is the best method, usually on a palliative basis. Occasionally it may be worth considering radical radiation if the patient's condition is good.

Stage 4. Palliative radiation is the only possibility, apart from cytotoxic drugs. For some advanced cases, particularly in the elderly, the best management is no active therapy beyond sedatives and nursing care.

Radiation techniques

Radical external radiation is mainly for growths thought to be still confined to the bladder, i.e. Stages 1 and 2. If any field dimension has to exceed 10 cm, the case is unsuitable for radical treatment. Maximum tolerable dosage is needed, and full-scale localisation and beam-direction therefore essential.

Localisation is by intravesical contrast medium (cystography) either in the diagnostic X-ray department or—preferably—on a simulator in the therapy department. A radio-opaque solution (similar to that used in IVP) is injected into the bladder via a catheter, and antero-posterior and lateral radiographs taken. Opaque markers on the skin give the relations of the bladder to the skin surface. The bladder contour is outlined on each film. If the patient is going to be treated both prone (lying on his face) and supine (on his back), films are taken in both positions, since there is always some shift in position of the bladder. The patient's body outline is obtained as described on page 68 and drawn on paper. The position of the rectum, just in front of the sacrum, is marked. Field planning with isodose contours then takes place, so as to include the bladder with a safe margin in the high dose region.

Useful results were obtained with multiple beams in the era of DXR at 250 kV, but there is no place today for anything less than megavoltage/cobalt beam. The critical organ of tolerance is the rectum where the dose should be kept as low as possible.

It cannot be completely avoided, but the posterior fields are angled so as to give a favourable distribution, with rapid fall-off towards the rectum, as shown.

1. A common arrangement is one anterior and two postero-lateral obliques (Fig. 24.2). If the patient turns over on the treatment couch for the posterior fields, beam direction may be by pin-and arc (see p. 71). If the bladder has been opened surgically, bolus should be used over the lower abdominal scar, to bring the high dose to the surface, to deal with any microscopic deposits in the wound.
2. Another method is rotation therapy, for which the central position of the bladder makes it a good subject (Fig. 24.3). Rectal dosage is decreased by omitting e.g. the posterior 60° of the full circle, as shown.
3. A further alternative, using anterior fields only, is shown in Figure 24.4.

The bladder should be emptied before each treatment session.

Dosage is typically 5500–6500 rad in five or six weeks, according to size of fields.

Palliative external radiation is very useful if the growth is unsuitable for radical treatment, or the patient too old or in too poor condition. Good symptomatic relief is generally obtained for haematuria, pain, dysuria, frequency of micturition. Technique can be simpler, and field size need not be so carefully restricted as in radical treatment. A pair of antero-posterior fields, e.g. 15 × 12 cm, usually suffice. If there is extravesical spread or an anaplastic growth, the whole of the pelvis should be included, to cover regional nodes, as in cancer of cervix. A depth-dose of 3000 rad in two weeks may be adequate, or up to 4500 rad in up to four weeks, depending on age etc.

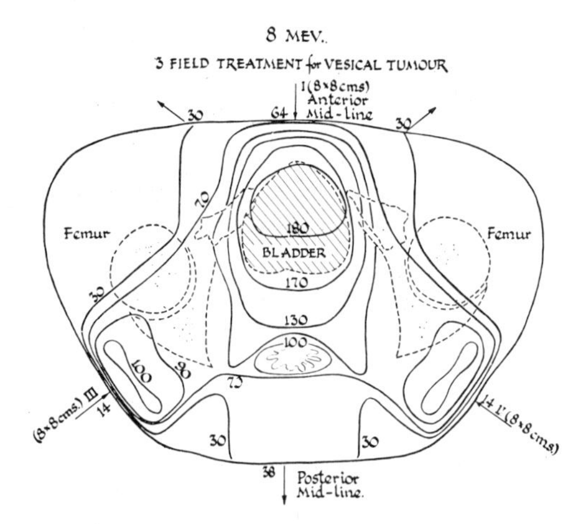

Fig. 24.2 Three-field megavoltage technique for bladder cancer.

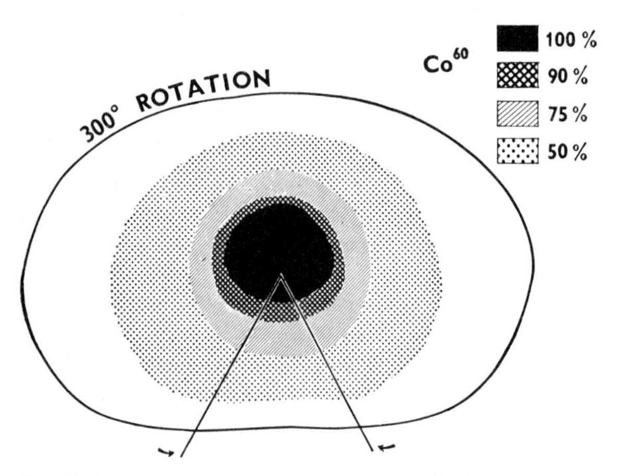

Fig. 24.3 Dose distribution for treatment of bladder by cobalt beam—rotation therapy. Note that the full circle is not used here; 60° at the back are omitted, to decrease dosage to the rectum. (*Paterson, Treatment of Malignant Disease, 2nd edition, Arnold*)

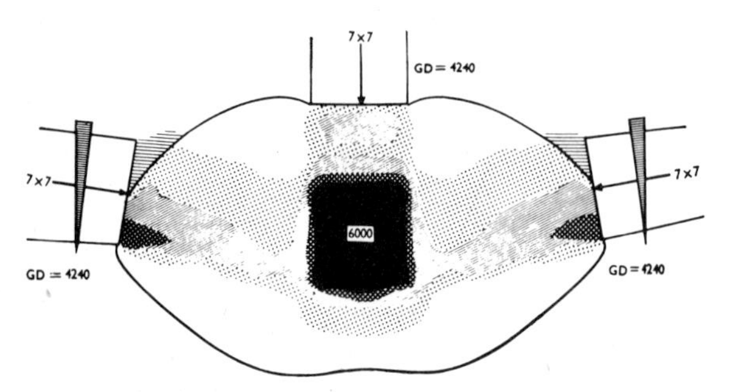

Fig. 24.4 Treatment of the bladder by three anterior fields at 4 MV (GD = given dose = incident dose). (Paterson, *Treatment of Malignant Disease*, 2nd edn., Arnold.)

Pre-operative radiation has some advocates, as improving results of cystectomy. The whole pelvis is covered, and 2000–3000 rad given in one to three weeks. Surgery follows in about a month.

Radiation reactions. Before radiation begins, attention is paid to the patient's general condition, nutrition etc. Anaemia after bleeding may need correction by blood transfusion. Infection of the bladder (cystitis) will be shown by the routine examination of the urine (as well as suspected from the symptoms), and an appropriate antibiotic is given. The urine should be made sterile if possible before radiation starts, since inflammation has adverse effects on radiation response, but with an ulcerated mass this may not be possible till the growth has shrunk under radiation. Treatment should not be held up in these circumstances, but dose-rate should be lowered at the start.

Early reactions.

1. Frequency and urgency, from radiation cystitis during and after the course, are common but not usually serious unless bacterial infection is gross. Painful spasm may require an antispasmodic drug. Fluid intake must be strongly encouraged—at least five pints daily. The patient should be warned that he may pass fragments in the urine (blood clot and tumour) and a little fresh blood.
2. Bowel reactions (rectum etc.) are also to be expected in almost every case—usually mild diarrhoea and tenesmus.* With sympathetic explanation and encouragement, they need not hold up treatment. If they are severe, dosage may have to be reduced or suspended for a few days.

Late reactions. Months after completion of radiation, symptoms may arise suggestive of recurrence, but due to post-radiation effects.

1. The bladder wall may be so contracted by fibrosis and bladder volume so reduced thereby that uncontrollable frequency may make life intolerable. Ureteric diversion may be required.
2. Telangiectases* on the bladder lining may develop, with repeated bleeding. It may be possible to seal them off with the diathermy point at cystoscopy. But they may become uncontrollable and require cystectomy.
3. Late bowel reactions (rectum etc.) are comparable to those after treatment of cancer of cervix (p. 232) though less common. Loops of bowel trapped in the

pelvis by adhesions after previous surgery or inflammatory disease are especially at risk. There may be telangiectatic bleeding, ulceration, even necrosis and perforation; if conservative treatment fails, surgical intervention will be required.

Other radiation methods

External megavoltage has proved to be the most valuable of all the radiation techniques available, and many departments use no other means. For completeness we shall describe some of the alternatives, but with the possible exception of gold grains they are obsolescent and were developed before the megavoltage era.

1. *Interstitial*. These methods, as noted above, are possible only as single-plane implants to irradiate a layer not more than 1 cm thick. Suprapubic cystotomy is carried out, the bulk of the tumour removed and the base diathermised. The radioactive sources are then implanted in the usual way, following the Paterson-Parker rules. Radiographs are taken to help in dosage calculations.

a. *Gold grains* (see p. 90) are inserted with the aid of a gun and magazine (Figs. 24.5, 24.6). The implant is permanent and adjustments of position and time are therefore not possible. One or two of the grains may later be sloughed off from the surface and be passed harmlessly in the urine. The tumour dose aimed at is 6000–7000 rad.

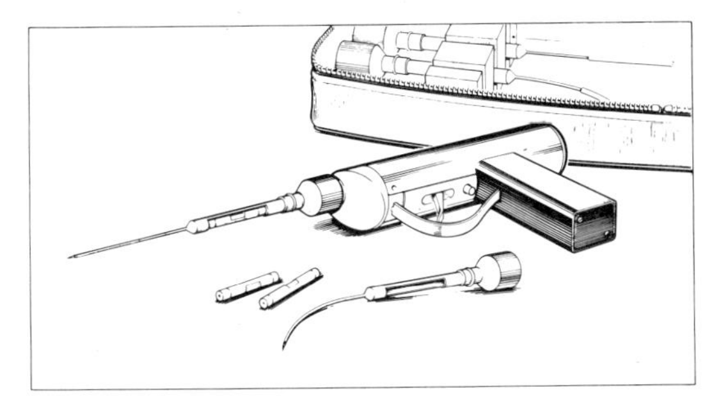

Fig. 24.5 Implantation gun and magazines for gold grains. (Medical Supply Association Ltd.)

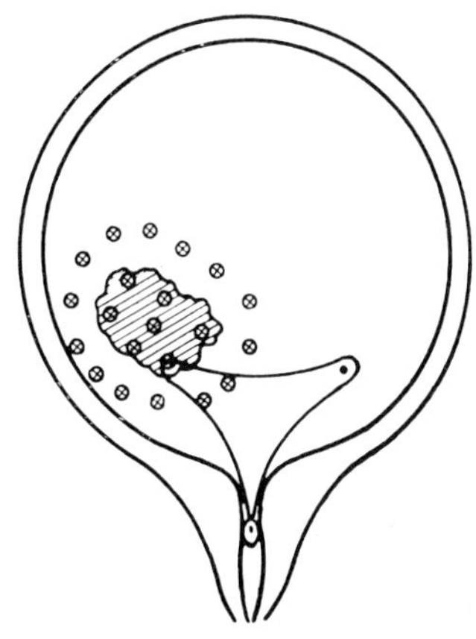

Fig. 24.6 A gold grain implant in the bladder. (*Paterson, Treatment of Malignant Disease by Radiotherapy 2nd edition, Arnold*)

b. *Needles*—radium, cobalt or caesium. A single plane of needles may be used, to give about 6500 rad in seven days—quite comparable with the tongue implant of Figure 19.1 page 193. But there are disadvantages—they are difficult to insert in the confined space and curved wall; they must be removed after a week, necessitating slight re-opening of the wound under another general anaesthetic; post-operative infection of bladder and wound is much more liable than with a 'closed' technique, and the stay in hospital is longer.

c. *Iridium wire* (see p. 89). This has been used as a form of implant. Small 'hairpin' loops are inserted after cystotomy, by special needles, into the bladder wall. Each loop has a thread attached at the bend, and the threads are all drawn through a urethral catheter to the exterior at the end of the operation. After study of the radiographs, the requisite length of time is determined—about 6–7 days for 6000—7000 rad—after which the catheter, together with the threads and iridium loops, is withdrawn via the urethra—no general anaesthetic is needed for this.

2. *Intracavitary*. As much as possible of the tumour mass is first removed by TUR (trans-urethral resection).

a. A liquid isotope in the bladder cavity directly, usually gold-198. About 3000 rad can be given to the surface in three hours, and the fluid then removed. A crop of papillomata can be removed in this way. Owing to the low penetration of the betas, the method is suitable for only very superficial tumours. Special facilities and equipment are needed and there is always some radiation hazard involved.

b. Intracavity *cytotoxic* therapy may be mentioned here. Intravesical application of a solution of a cytotoxic drug (e.g. thiotepa) can be used for multiple papillomata. It is left in the bladder for half an hour and then withdrawn. Treatment can be repeated.

Treatment of recurrences. After radical radiation, further radiation should never be given—the result would be painful necrosis. If total cystectomy is practicable, it should be considered. A second palliative course of external radiation may be considered. Cytotoxic therapy has little to offer so far.

Results and prognosis. As usual, these are governed by extent of invasion and degree of anaplasia. For early cases, Stage 1, five-year survival is about 75 per cent, for Stage 2—50 per cent, for Stage 3—20 per cent, for Stage 4—10 per cent. The overall figure is about 35 per cent.

Careful selection of cases for radiation to radical levels is important to ensure best results. For later stages, worthwhile palliation is usually achieved, often for prolonged periods.

Cancer of the testis

Tumours of the testis (the diminutive form 'testicle' is also in common use, with its adjective 'Testicular') are usually malignant. About 10 per cent arise in imperfectly descended testes (plural of 'testis'). In embryo, the testes arise on the posterior abdominal wall, and gradually migrate downwards through the inguinal canal into the scrotum. Descent may be arrested at any point on this path. Undescended testes are much more liable to malignancy than the fully descended.

Pathology. There are two main types—(1) Seminoma* and (2) Teratoma.* Much rarer is the Choriocarcinoma.*

Seminoma is the commonest. It derives from the precursor of the spermatozoa. Spread may be local, to epididymis* and spermatic cord, but lymphatic spread is more important. The first group of nodes to be invaded is the upper paraortic, at the level of the kidneys; this is due to embryological development from this site, as noted above. Further lymphatic spread may be upwards to mediastinum and even supraclavicular nodes via the thoracic duct, or downwards to lower paraortic and pelvic nodes. If extra-testicular tissues of the scrotum are invaded, including scrotal skin, their draining inguinal nodes may be invaded. Blood spread is far less common.

Teratoma derives from primitive tissue which can produce any elements from any of the three embryonic layers (ectoderm, mesoderm, endoderm). They may thus contain a wide variety of tissues—bone, gut, nerve, squamous epithelium etc. of all stages of differentiation, including highly anaplastic areas. Local and lymphatic spread occur, but the most important is early blood spread to lung, liver, bone etc.

Choriocarcinoma (also called chorionepithelioma) is a subvariety of teratoma, in some ways analogous to the chorionic tissue of the placenta. In the female, choriocarcinoma (see p. 283) is a malignant uterine tumour appearing after pregnancy and deriving from the placenta. In both females and males it secretes hormones, which can be detected and estimated quantitatively after renal excretion, in the urine. They are known as gonadotrophic* hormones and are the basis of urinary tests for pregnancy. Assays of the urine, in males or females, provide evidence of active growth of primary or secondary choriocarcinoma and are therefore useful in follow-up of patients as evidence of metastasis, after the primary has been dealt with.

Clinical features. A typical age for teratoma is 15–40, for seminoma, 30–50. The usual presentation is a gradually enlarging painless testicular swelling. The first sign, however, may come from a secondary deposit, e.g. pathological fracture in bone, haemoptysis from lung, or a large abdominal mass of secondary nodes, or nodes in the neck.

Choriocarcinoma may produce gynaecomastia* (breast enlargement) in males, due to the feminising hormones secreted.

Preliminary investigation includes a chest film for enlarged mediastinal nodes and lung deposits; lymphography for pelvic and paraortic nodes; IVP to outline kidneys and ureters, for possible displacement by tumour masses.

Treatment. The general rule is—*surgical removal of the primary mass (orchidectomy), followed by external radiation to lymphnode regions*.

Orchidectomy is a short and simple operation, and should be done even if secondaries are known to be present. It is much simpler and more satisfactory than radiation, and forestalls possible local recurrence and fungation through scrotal skin. The real problem is how best to deal with early lymph-node metastasis. Extensive surgical procedures to attempt removal of paraortic and other nodes were tried in the past, but carried high morbidity and mortality and proved unjustified by results. They have been abandoned in favour of modern radiotherapy which gives much superior results.

Post-operative radiation to regional nodes is nearly always indicated, except in advanced cases with widespread metastasis. Lymphography may have revealed silent paraortic node involvement, but even if there is no evidence of metastasis, radiation should still be carried out, in case of microscopic deposits (so-called 'prophylactic' radiation). Experience has shown the great value of this programme, especially for seminoma.

Radiation technique. Figure 24.7 shows the general principle—coverage of all regional nodes, pelvic and paraortic, by large specially shaped antero-posterior fields. In former pioneer days the whole abdomen was included, but serious and even fatal damage to kidneys occurred later (see p. 172). Today great care is taken to exclude all or most of both kidneys. An IVP should be carried out to check this, if not previously done. For these large fields, megavoltage is very desirable, at increased FSD. The shape is roughly an inverted T. The fields are best outlined on the simulator. If lymphography has been done, it is a great help in delimiting the field margins, since the opacified nodes are easily visible on the screen. On the treatment machine the requisite shape of fields is obtained by lead blocks.

Some therapists confine the pelvic part of the fields to the affected side only, using an L-shape (Fig. 24.8). In any technique the operation scar should be included and covered by bolus.

The fields extend from the upper border of the first lumbar vertebra to the lower border of the symphysis pubis. There is a difference of opinion about including the scrotum with the remaining testis in the field. Figure 24.7

shows it excluded, but some (a minority) include it, both to prevent local recurrence and to forestall the growth of another tumour in the second testis (this is uncommon—less than 1.5 per cent of cases—but much commoner than in normal men). If the testis is included, the patient will inevitably be sterilised, (but not become impotent) as the precursors of the spermatozoa will be destroyed. The consequences, and reasons for the treatment, must be fully explained to the patient beforehand. Even if the testis is not inside the field, it will receive some scattered radiation if megavoltage is used—about 150–200 rad–mainly from the posterior field. If the patient already has a family, it is a good alternative to sterilise him by vasectomy,* to exclude any genetic complications. If he is

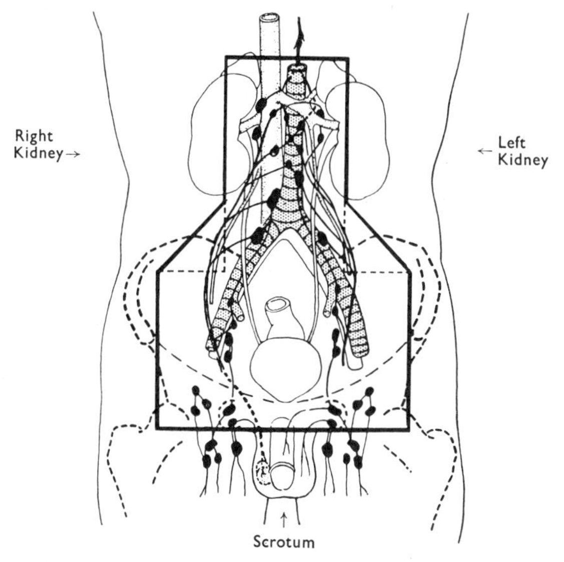

Fig. 24.7 Lymphatic drainage of testis and outline of polygonal megavoltage field to include nodes of groins, pelvis and paraortic area. Scrotum and other testis not included in the field. *(Raven, Cancer, Vol. 5, Butterworths)*

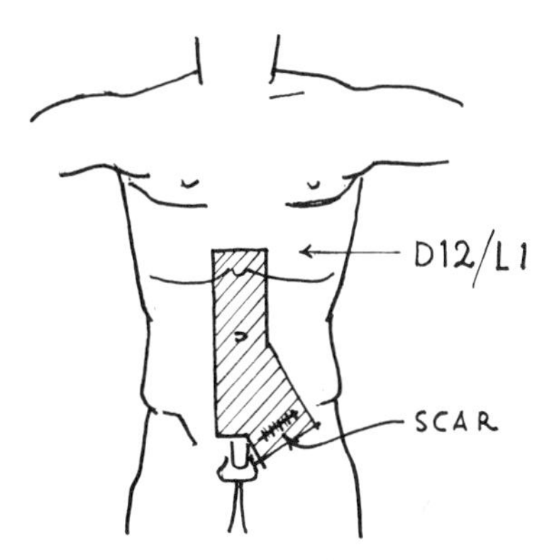

Fig. 24.8 Anterior L-shaped field for regional nodes (paraortic, pelvic and inguinal) after removal of (left) testis. Bolus is placed over the scar area, to bring the skin dose to the maximum, in case of implanted tumour cells. A matching posterior field completes the scheme.

young and wishes to retain the possibility of a family, the radiation should be at orthovoltage (250 kV) and the testicle enclosed during treatment in a lead chamber with walls 1 cm thick. This will prevent the testis receiving more than about 10 rad, an acceptable dose.

Seminoma is one of the most radiosensitive of all cancers; a tumour dose of 3500 rad in four weeks is adequate, and usually well tolerated. Teratoma, however, is much more resistant, and dosage must be pushed to the limit of tolerance—about 4500 rad in five weeks. A considerable amount of bone marrow (pelvis and vertebrae) lies within the treated volume, and blood counts should be made twice weekly. The white cell count falls progressively, but should not be allowed to go much below 2000—each case needs close individual attention. Radiation sickness and/or diarrhoea may occur in susceptible individuals, and it may be necessary to suspend treatment for a few days. Psychological support from the staff, and an anti-emetic drug, will be useful. Patients should be in hospital for all or most of the course.

If mediastinal nodes are known or suspected to be involved—e.g. if enlargement is visible on the chest film, or if the lymphogram shows invasion of nodes high in the paraortic region—the abdominal radiation should be followed by a further course, after a four-week interval, to include mediastinal and supraclavicular nodes, to similar dose level. If there are secondaries from seminoma in the lung fields, radiosensitivity is high enough to make radiation worthwhile. A chest 'bath' should be used, with large antero-posterior fields, giving 2000 rad to the whole chest in three weeks. Regression and disappearance of visible masses can usually be obtained.

For the earliest stage of teratoma—i.e. all investigations for spread negative—it is reasonable to withhold further therapy and leave on observation only, since over 75 per cent of these will never need further treatment. For widespread teratoma, radiation is inadequate, but useful results are now obtainable from cytotoxic drugs.

Cytotoxic therapy is now playing a major role in testicular tumours, especially disseminated teratoma. Combinations including Vinblastine, Actinomycin-D and Adriamycin are invaluable for teratoma. For disseminated seminoma, Melphalan is useful. Some workers now use cytotoxics 'prophylactically' after radiation, to deal with possible micro-metastases.

Prognosis. As usual, this varies with the histological type, and extent of spread at the time of diagnosis. Semonoma is the most favourable. In former days, when orchidectomy was virtually the only means, five-year survival was about 25 per cent. Today combined management with radiation has brought this up to about 75 per cent, and even higher if there are no demonstrable nodes at the start.

Teratoma is less favourable—about 50 per cent, but vigorous cytotoxic therapy has improved the outlook, and many patients with widespread metastatic disease can be kept going in reasonable comfort for two years or more. The outlook in the more anaplastic teratomatous growths is very poor, and hardly any survive two years. The same applies to choriocarcinoma, since the tumour is not 'foreign' as it is in the female, and so does not provoke any immunological rejection response (see p. 284).

Cancer of the prostate

Cancer of the prostate gland, situated just below the base of the bladder, is a common disease of elderly men. After lung and bowel it is the next most frequent fatal cancer, and accounts for 7 per cent of all male cancer deaths.

Pathology. Almost all are adenocarcinoma. Direct spread is to bladder and rectum, then to other pelvic tissues. Lymph spread is to pelvic and paraortic nodes, and sometimes to the inguinal group in the groin. It readily invades pelvic veins and thence to pelvic bones and vertebrae. Bone secondaries are often of sclerotic* type, showing as increased density in the radiograph instead of the usual bone erosion.

The prostate has analogies with the breast and is under hormonal control. Removal of male hormone by orchidectomy,* or administration of female hormones (oestrogens*), causes shrinkage of the normal gland, and often of cancerous prostatic growth also.

The gland secretes an enzyme—acid phosphatase—which can be estimated in the blood. Abnormally large amounts may be found in many cases of prostatic cancer, especially if there are widespread metastases. This is useful in diagnosis.

Early cancer is not uncommonly found unexpectedly on histological examination of prostates removed for benign enlargement.

Clinical features. The prostate often undergoes benign enlargement, and the early symptoms of cancer may be similar—i.e. increased frequency and difficulty of micturition. Urethral obstruction eventually causes chronic urinary retention, and back pressure may damage the kidneys. Later there may be sacral, sciatic or perineal pain from infiltration of nerves in the pelvis. Bone metastases cause pain, and pathological fracture may be the first symptom.

Investigations include: rectal examination to assess the size of the gland and local invasion, skeletal films and bone scan, IVP to assess renal function, blood chemistry including serum acid phosphatase.

Treatment. Radical prostatectomy is ideal, but only a very small fraction are suitable. Most patients are unfit for major surgery, and early diagnosis before dissemination is anyhow rare. Such patients are best managed by hormonal methods. Most tumours regress on administration of oestrogens (e.g. Stilboestrol*). Side effects may follow oestrogen therapy—nausea, enlargement of breasts, atrophy of testes, pigmentation of nipples.

(It is of historical interest that this was the first clinical application of hormonal cancer control—by Huggins of Chicago in 1941. He was later awarded a Nobel prize.)

The beneficial effects on primary and secondaries may be prolonged, even several years, but the growth eventually escapes from control. When this occurs, or if oestrogens fail, improvement may be obtainable by removing the source of male hormones, by operative removal of the testes (bilateral orchidectomy). Further details on hormones are in Chapter 30.

If urinary obstruction persists, TUR (trans-urethral resection), to remove the offending part of the prostate, is indicated. In this operation the instruments are passed by the surgeon along the urethra, and portions of the prostate are scooped out with a cutting edge.

Painful bone secondaries may be irradiated e.g. to 3000 rad in two weeks, but results are less reliable than in breast cancer. For widespread bone deposits, pain may be relieved for many months by intravenous injection of radioactive phosphorus (P-32). The tumour tissue causes a regenerative reaction in adjacent bone, with increased metabolic turnover—hence the increased selective uptake of phosphorus (the mechanism is the basis of isotope bone scans).

Radiation to the prostate itself gave such poor results in previous years (orthovoltage, radium needles, infiltration with radio-gold etc.) that it was largely abandoned, and prostatic cancer was considered insufficiently responsive to be worth attempts at radical radiotherapy.

Megavoltage is still not widely used with radical intent, but has proved capable of achieving local control in a proportion of inoperable cases, so that it is now coming into use and giving useful results with improved technique. Technique and dosage may be similar to that for bladder cancer (see above). Localisation is assisted by contrast medium in both bladder and rectum. The best results are claimed for small-volume technique with fields about 7 × 7 cm, delivering about 7500 rad in about $7\frac{1}{2}$ weeks. Since the rectum is immediately adjacent, it must share in the high dosage. Rectal reactions will inevitably occur, and the risk of occasional serious damage has to be accepted. See also page 275.

Prognosis. The natural history of prostatic cancer tends to be long. The average five-year survival is about 40 per cent. If there are no distant metastases, hormonal management alone is capable of achieving this. The rare cases of operability do better, but if there are secondaries at the time of diagnosis, survival is only about 20 per cent.

25. Lympho-reticular tissues and bone marrow

Malignant lymphomas

The term lymphoma is now used to include all the primary malignancies of lymph nodes. A simple classification, and in order of frequency, is as follows:

1. *Hodgkin's* disease*
2. *Lymphosarcoma*. This includes a wide spectrum of malignancy ranging from well-differentiated *lymphocytic lymphoma* to undifferentiated (or lymphoblastic) lymphoma
3. *Reticulum cell sarcoma*. The modern name for reticulum cell is 'histiocyte',* hence the alternative name *histiocytic lymphoma*.
4. *Follicular* lymphoma*. This used to be (and still may be) called *giant follicular lymphoma*.

Modern pathologists classify lymphomas not only according to the cell type but also by the tissue pattern. The cells may either form a diffuse sheet or be arranged in nodules which resemble normal lymphoid follicles. Most of the above types can therefore be subdivided into *diffuse* or *nodular* types. The diffuse is more anaplastic and of worse prognosis than the nodular.

The various types show clinical as well as histological differences, especially in regard to prognosis and survival. But they also have many similarities, so that some general statements can be made covering all or most of them:

a. A common presentation is node enlargement in a single focus, especially in the neck, and other groups appear to follow by spread from the first. Other cases seem to be of multifocal origin, i.e. to arise at several sites simultaneously
b. These foci almost always respond well to treatment at first, but a time may come (if primary treatment fails) when widespread deposits are present, including bone marrow. Lesions may eventually become refractory to treatment and the patient succumbs to exhaustion and infection
c. Progress of disease is often slow, and patients may do well for many years
d. The role of surgery was, till recently, restricted mainly to node biopsy, but is now assuming greater importance, especially in Hodgkin's disease, for laparotomy* and splenectomy etc.
e. Radiation is a major part of the management of most cases, especially in the earlier stages while disease is still confined to lymphoid tissue
f. Radiation is usually by large fields, in some form of regional therapy, to include large blocks of lymphatic tissue
g. Cytotoxic chemotherapy is now of major importance, and often the chief or even the sole weapon in later stages when disease has spread beyond lymphoid tissues

Hodgkin's disease

Clinical features. The first symptom is usually painless enlargement of a group of nodes, particularly in the neck, but sometimes in axilla or inguinal region. In some cases mediastinal or abdominal nodes may enlarge first, causing cough, dyspnoea or pain in chest or abdomen. The spleen may be palpably enlarged. Later spread to extra-lymphatic tissues may involve almost any organ—liver, bone, central nervous system, skin etc., with a variety of associated symptoms (jaundice, nerve palsies etc.). General symptoms include—malaise, loss of weight, tiredness, pruritus (itching—often wrongly spelled 'pruritis'), sweats (especially at night). There may be a characteristic type of fever (Pel-Ebstein fever, after the workers who first described it) in waves lasting a week or two, separated by afebrile intervals.

A peculiar symptom, of unknown cause, is severe pain in nodes etc. after taking even small quantities of alcohol.

Investigations. The first step is usually node biopsy, to establish the diagnosis and the histological type. To enable vital decisions on treatment to be made, it is essential to obtain as much information as possible on the extent of involvement (Staging). Some or all of the following investigations should be carried out:

Full blood count including E.S.R. (erythrocyte sedimentation rate)
Chest X-ray
Lymphangiography and IVP
Bone marrow biopsy
Isotope scan of liver and spleen
Laparotomy

Lymphangiography (or *lymphography*) came into use in the late 1950s. A blue dye is injected intra-dermally between the toes on the dorsum of each foot. Movement and massage assist its take-up by lymphatic vessels which are thus made visible. A lymphatic trunk is exposed by dissection (under local anaesthesia) and a radio-opaque contrast medium (similar to that used in pyelography) is injected.

It passes along the vessels of the limb, to successive groups of nodes. Enlarged nodes with irregular 'moth-eaten' patches due to malignant deposits can be demonstrated. But if the whole node has been replaced by malignant tissue, the lymphatic entry will be completely blocked and the node will not show up at all.

This technique has opened up the pelvic and paraortic nodes to inspection. Normally any enlargement cannot be detected by palpation till the node mass is huge. Now we can examine the state of the nodes in the early stages, and this information is of immense value in improving our assessment and management. It is unfortunately not possible to visualise mediastinal nodes in the same way. After a lymphogram, the contrast medium remains visible in the nodes for up to 18 months, and follow-up radiographs are valuable in detecting enlargement at an early stage.

IVP should follow lymphography. This will provide evidence of renal function, and the position of the ureters will be another indicator. One or both may be displaced and so provide indirect evidence of node masses which have not taken up the radio-opaque lymphographic contrast medium and so cannot be directly visualised.

Isotope scan of liver and spleen (see Ch. 10) will reveal the size of the spleen even if it is not palpable, and may show 'cold' areas in the liver suggestive of malignant deposits.

Surgical *laparotomy* is advisable in many cases, as shown in Table 25.1. This is a drastic step, involving some morbidity and even occasional mortality—a high price to pay. The justification for this lies in the vital importance of finding out the precise state of affairs inside the abdomen, since the optimum line of treatment (radiation or drugs) will depend on this information, and there is no other means of getting it. The surgeon removes the spleen (splenectomy) which is sent to the pathologist for microscopy. He also examines the paraortic and other node areas, and takes one or more nodes for biopsy; the lymphographic picture may help to guide him to suspicious nodes. The liver surface is also inspected and biopsies are taken of liver tissue and any other suspicious masses found. Before laparotomy a case might appear to be early—Stage 1 or 2 (see below) with nodes above the diaphragm. If abdominal involvement is discovered, the stage is changed to 3, or even 4 if liver biopsy is positive—and this will change the whole scheme of treatment. It is in any event advantageous to remove the spleen, since clinical assessment is grossly fallacious. If it is not palpable clinically, many are nevertheless found to be invaded; and surgery is a much more efficient way of dealing with an involved spleen than radiation.

Histological type. It is important to determine the histological type, from the node biopsy. Four groups are recognised, and they have great prognostic significance. From most favourable to least favourable they are:

Lymphocytic predominance (LP)
Nodular sclerosing (NS)
Mixed cellularity (MC)
Lymphocytic depletion (LD)

Staging is of fundamental importance.

Stage 1. Limited to one group of lymph nodes
Stage 2. Two or more groups of nodes, all on the same side of the diaphragm
Stage 3. Disease on both sides of the diaphragm, but not extending beyond the lymphoreticular system
Stage 4. Disease extending beyond the lymphoreticular system (e.g. liver, bone, skin etc.)

'Lymphoreticular system' includes lymph nodes, spleen and Waldeyer's* ring (described on p. 195). All stages are also subclassified into A or B:

A—general symptoms absent.
B—general symptoms present—especially fever and weight loss; pruritus and night sweats are less important and by themselves do not signify a B.

Treatment. Up to the 1960s textbooks of medicine stated that Hodgkin's disease was invariably fatal. Modern radiotherapy—and now chemotherapy—have revolutionised treatment and prognosis.

Table 25.1 summarises current methods and indications for the various histological types and clinical stages.

Radiation technique. The general rule is—*radiation for the early localised case, cytotoxics for the later generalised case*. For individual groups of nodes, or deposits in bone

Table 25.1 Management of Hodgkin's Disease

Clinical stage	Histological type	Laparotomy	Treatment
1A, 2A	Nodular sclerosis Lymphocyte predominance	No	Radiotherapy Mantle or inverted Y
1A, 2A	Mixed cell Lymphocyte depletion	Yes	Radiotherapy Mantle plus inverted Y (= TNI)
1B, 2B, 3A	All types	Yes	Radiotherapy TNI
3B, 4	All types	No	Combination chemotherapy

TNI = Total Nodal Irradiation

etc., simple techniques delivering 3000–4000 rad in two to four weeks by single or parallel fields (Fig. 25.1), used to give excellent local results. But nodes in adjacent or other regions sooner or later enlarged, and though successive local treatments were satisfactory for a time, malignancy eventually generalised beyond control. Hence the poor prognosis from this piecemeal approach—though a few cures were achieved. This conservative method has now been abandoned, and replaced by an aggressive attack on all the lymphatic tissue in the upper (or lower) half of the body, on one or other side of the diaphragm, in one large single block (Fig. 25.2). If necessary, treatment of one half can be followed by treatment of the other—i.e. *total nodal irradiation* (TNI). It is this new and radical approach that is responsible for the changed outlook, made possible by megavoltage equipment.

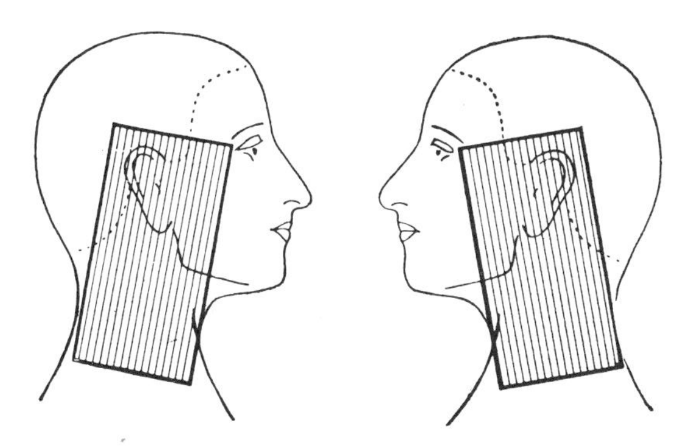

Fig. 25.1 Large-field therapy for reticulosis. Suitable e.g. for lesion confined to tonsil. Two fields, 25 × 15 cm, at 250 kV or 4 MV, to include whole of pharynx and adjacent lymph—node area. (Paterson, *Treatment of Malignant Disease*, 2nd edn., Arnold)

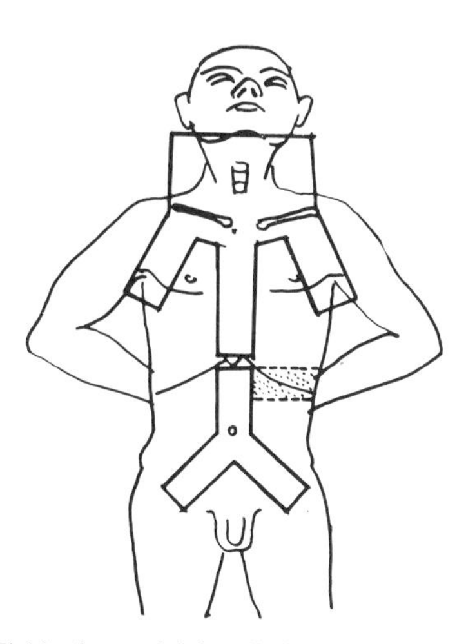

Fig. 25.2 Fields for nodal irradiation, e.g. Hodgkin's. The upper field covers the nodes above the diaphragm ('mantle' technique). The lower field covers the nodes below the diaphragm (inverted Y). Matching posterior fields are used. The upper part of the abdominal fields can be extended on left side to include the spleen or splenic pedicle.

In Stage 1 or 2 case, with nodes of neck, axilla or mediastinum involved, radiation is by the method shown in Figure 25.2. From its resemblance to a cloak it is known as the *'mantle' technique*. Large anterior and posterior fields cover the whole of the neck, from mastoid processes downwards, plus infraclavicular and pectoro-axillary areas, plus the whole of the mediastinum down to the 12th dorsal vertebra. An alternative is to take it down lower, to the lower border of the second lumbar vertebra, where the spinal cord ends, and so avoid the risk of dangerous overlap with a later abdominal field, which could damage the cord. This is the 'extended mantle' technique, and is advisable e.g. for Stage 3A i.e. as the first part of TNI. The lower border of the field is recorded on check films taken with marking wires in place, for later reference and dovetailing with abdominal fields.

The upper border is at the level of the mandible, and the mouth is excluded. It is not necessary to include the tonsillar region (Waldeyer's ring) unless there is evidence of involvement. If so, the fields must go up to the base of the skull, front and back.

To obtain these large fields, increased FSD will be necessary. At the level of the larynx, the decreased antero-posterior thickness of the neck must be allowed for in the dosage calculations. The larynx can be shielded for part of the course by lead on the anterior field. Some workers also shield the cervical part of the spinal cord by a narrow 2 centimetre strip on the posterior field. It is, of course, not possible to extend this downwards to the thoracic part of the field, otherwise mediastinal nodes would not receive full dosage. The thoracic section must be carefully shaped on the simulator if available, by means of lead blocks (p. 32). The mediastinum must be fully included. It is not possible to avoid the apical parts of the lungs, but the rest is shielded.

The tumour dose should be 3500 rad minimum in four weeks.

Treatment below the diaphragm is by large inverted Y fields (Fig. 25.2) covering paraortic and pelvic nodes. If total nodal irradiation is carried out, the second course should follow the first after about six weeks. To avoid overlap with the mantle fields, the check films mentioned above should be used, and further films taken to ensure that the field borders do not overlap. Because of the magnification effect due to the patient's thickness, this will mean a gap of about 2 cm between the edges on the skin. The previous lymphogram is invaluable in shaping the outlines of the fields, since the contrast medium is visible on fluoroscopic screen or film. This enables the therapist to use adequate but not excessive margins and so keep the volume dose (and pelvic bone marrow dose) to a minimum. The upper end of the abdominal field can be extended to the left to include the splenic area, whether the spleen has been removed or not. Radiation here will inevitably include part of the kidney; this must be accepted and should have no serious consequences.

Dosage is of the same order as for the mantle.

In females this technique involves ovarian sterilisation. Some workers have attempted, for younger women, to avoid this by operative moving of at least one ovary from its lateral position in the pelvis and anchoring it in the midline. It would thus receive only scattered radiation, about 100–200 rad, which would have no clinically significant effect (but would carry some risk of genetic damage to future offspring).

With these very large treated volumes—especially the inverted Y—reactions are to be expected, nausea, sore throat, diarrhoea etc., and are managed on the usual lines. Frequent blood counts, including platelets, are essential, and low counts may require suspension of treatment for a few days. The patient should be in hospital if possible, especially for the abdominal course.

For elderly or frail patients, or for treatment of isolated deposits, where aggressive therapy is not suitable, the old-fashioned local methods should be used.

Cytotoxic chemotherapy. For Stage 3B and 4, and for recurrences after radical radiation, chemotherapy is indicated. Formerly, single agents were used, with poor or modest results. Current regimes use multiple agents in combination, and results are outstandingly good. Remissions are nearly always achieved, and patients can be kept going for long years by judiciously spaced and repeated courses. Details are given on pages 282–3.

Prognosis. Modern radiotherapy, plus combination chemotherapy, has improved the overall five-year survival from under 40 per cent in the 1940s to over 60 per cent. For Stages 1 and 2, 90 per cent is now being achieved, and even 70 per cent for Stage 3 at major centres. Very careful staging and selection of appropriate therapy is essential, to obtain these excellent results, and this is the justification for all the elaborate measures described above. Even in late stages, the results of chemotherapy have been so good that it now seems likely that normal life expectancy is being achieved in some cases.

Non-Hodgkin's lymphoma

This is a group with a wide range of malignancy, from semi-benign to rapidly aggressive. They may be staged in a way similar to that for Hodgkin's. Management varies according to the histological type and degree of spread. Many of the principles outlined for Hodgkin's disease are applicable, with individual variation. Laparotomy is rarely indicated, unless required for initial diagnosis or removal of involved stomach or bowel.

Reticulum cell sarcoma. This is the least radiosensitive and carries the worst prognosis. Node enlargement is the commonest presentation, but there may be involvement of tonsil, bone, bowel, etc.

For Stages 1A and 2A, radical radiotherapy should be given as for Hodgkin's. Otherwise chemotherapy and palliative radiotherapy are used.

Lymphosarcoma. The radiosensitivity is as good as for Hodgkin's, but the prognosis is not so good. Involvement of stomach or bowel is relatively common and may be the first sign. Gastric lymphoma is discussed on page 211.

There may be an associated lymphatic leukaemia, or this may appear terminally. The general lines of management are: radiation for local masses and cytotoxics for generalised disease. Some cases with a solitary focus, e.g. tonsil, are easily curable locally, and may remain permanently free of disease.

Follicular lymphoma. This is the least aggressive of the group, and the most radiosensitive. It carries the best prognosis of all. There are often spontaneous remissions. Treatment can afford to be on non-aggressive lines. Doses of 3000 rad in three weeks are adequate for local control.

Mycosis Fungoides*

This is a neoplasm of the reticular* tissue of the skin, with some characteristic features. It runs a very chronic course, via a pre-mycotic phase of dermatitis for many years. There are itching erythematous patches, which later become thickened to form plaques, and finally fungating tumours, often likened to tomatoes. There may be terminal visceral involvement, of liver etc.

Individual lesions are very radiosensitive, and palliation can be given, often for years, by superficial radiation. Doses of a few hundred rads usually suffice. Superficial electron therapy is also useful, either to small patches or to large areas—even the whole trunk or body—from linear accelerators, betatrons or iridium units (p. 74). In the later stages, cytotoxics may help, combined with steroids.

Burkitt's lymphoma is another special member of the lymphoreticular group. It is described on p. 284.

Leukaemia

Leukaemia is a group of diseases characterised by cancerous proliferation of leukopoietic (white cell forming) tissue throughout the body, chiefly bone marrow and lymphoid tissue. The incidence is everywhere increasing and has more than doubled in the past 25 years. The cause is unknown, but there is an undoubted relation to exposure to ionising radiation:

1. Survivors of the atom bomb explosions at Hiroshima and Nagasaki showed a twenty-fold increase as compared with normals
2. Patients treated by X-rays for ankylosing spondylitis have a six-fold incidence (see p. 270)

An infective virus has been found to produce leukaemia experimentally in some animals, but there is no evidence of an infective origin in humans.

Pathology. White blood cells are of several types, and any type may undergo malignant change, so that we can classify the varieties of leukaemia according to the particular cell type. The normal process of transformation of primitive stem cells into mature adult cells is somehow disturbed and arrested, so that these primitive cells accumulate in the marrow and enter the circulation in large numbers, to give the typical peripheral blood picture. The marrow becomes filled with leukaemic cells which crowd out other components—red cells (with resultant anaemia) and platelets (with resultant impaired blood clotting and tendency to haemorrhage).

Resistance to infection is lowered because of the immaturity of the white cells. The end comes with failure of normal marrow function—severe anaemia and usually terminal infection or haemorrhage.

The chief varieties are—(a) *lymphatic* and (b) *myeloid** or *granulocytic** leukaemia—arising from the lymphocytic series or the polymorphonuclear leucocytes. Rarer types are the monocytic and eosinophil leukaemias.

Leukaemia is also subdivided into (a) *acute* and (b) *chronic* types, according to the rapidity of the clinical course, but the malignant process is essentially similar apart from the time factor.

Acute leukaemia may be of lymphatic, myeloid or other type, but the clinical picture is similar in all. It may occur at any age, but is most common in childhood (see p. 261). The onset is usually sudden, with fever, sore throat and bleeding from mouth or elsewhere. Typically there is a total white cell count of 20 000–50 000 per cubic mm, of which about 90 per cent are very immature or blast* cells (lymphoblasts, myeloblasts etc.). The diagnosis is obvious on the blood film, and is confirmed on examination of bone marrow, obtained from sternum or iliac crest, which shows heavy infiltration with blast cells. Often the cells are so primitive that it is impossible to say whether they are of lymphoid or myeloid type.

Some cases are at first aleukaemic—i.e. there is no excess of leucocytes in peripheral blood, but the marrow shows typical appearances.

Chronic leukaemia is of two main types—lymphatic or myeloid (granulocytic). The cells are more mature than in acute cases, and the clinical course is much more prolonged, with periods of spontaneous remission, before the fatal outcome. Diagnosis rests on the cells in peripheral blood and marrow.

1. *Chronic myeloid leukaemia* occurs mainly in early adult life and middle age. The typical picture is of progressive tiredness due to anaemia, with gross splenic enlargement which may cause a dragging sensation in the abdomen. The spleen may reach a huge size and appear to fill most of the abdomen. Lymph nodes are not usually enlarged. Occasionally there is leukaemic infiltration of the skin. The blood count shows a total white count between 100 000 and 500 000, of which 50–70 per cent are polymorphs and 10–20 per cent myelocytes (precursors of the polymorphs). There is always some degree of anaemia.

2. *Chronic lymphatic leukaemia* is usually seen in men of late middle age. The onset is insidious, with tiredness from anaemia. Enlargement of spleen, liver, lymph nodes of neck, axillae and elsewhere, is found. The spleen does not reach the huge size of the chronic myeloid variety. The blood shows a total white count up to 100 000, of which 80–90 per cent are small lymphocytes, and associated anaemia. Leukaemic infiltration of skin, bone etc. may occur.

Treatment

1. *Acute leukaemia* is treated mainly by *cytotoxic chemotherapy*—see p. 249. This has given remarkable and revolutionary results in the last decade, and the prognosis has been radically improved.

Radiation however has an essential supplementary role, to treat the central nervous system (CNS), which is usually involved by leukaemic deposits on the meninges. These cells are not effectively reached by the drugs given by mouth or intravenous injection, since they do not cross the 'blood-brain barrier'. If they are not specially dealt with, they will cause relapses, with headache, vomiting, etc. All patients should therefore be given 'prophylactic' therapy, early in the period of remission, 6–8 weeks after diagnosis. Its great value has been proved by experience.

There are two techniques—(a) irradiation of the entire cranio-spinal axis and (b) irradiation of the cranium combined with intrathecal* injection of a cytotoxic drug. In the first method, radiation is applied in the same manner as for medulloblastoma (see p. 255, but in this case the posterior part of the orbit should be included) giving 2400 rad by megavoltage in twelve fractions in three weeks. For children under two years, use 2000 rad, under one year, 1500. In the other method, a similar dose is given to cranium only, and the rest of the CNS treated by intrathecal Methotrexate, five doses given twice weekly. Temporary alopecia* is usual with either method, and wigs should be ordered beforehand, at least for the girls.

Local radiation is also valuable for relief of bone or joint pain. Single doses of 200–300 rad suffice.

Supportive therapy is important in all types and phases of leukaemia, especially the acute, and can be crucial—e.g. blood transfusion for severe anaemia, antibiotics for infection. Fluid intake must be kept high, to assist the kidneys to eliminate the breakdown products of destroyed leukaemic cells.

Immunotherapy is also proving helpful. The body's defence system can be stimulated by injections of vaccine (e.g. BCG)* given between the courses of chemotherapy.

Results in the acute myeloid type (AML) are definitely inferior to those of lymphatic (ALL). But remission rates of over 50 per cent can now be achieved in adults by intensive chemotherapy. Early relapse, however, is the rule, and the prognosis still remains extremely poor, in contrast to childhood.

2. *Chronic leukaemia* is also now treated mainly by drugs, notably Busulphan (see p. 280). Radiation used to be the mainstay of treatment, and still has its uses. The aim of treatment is to relieve symptoms and maintain the quality of life. No treatment is needed if the general condition is satisfactory and white cell count and haemoglobin levels reasonable. The best indicator for treatment is a fall in haemoglobin, which usually goes along with a rise in the white count. Treatment may also be required if an enlarged spleen causes symptoms; or node masses may be troublesome, even if only psychologically. Regular blood counts are essential, and should be charted graphically, so that the rate as well as the extent of change can be followed. A total white count of 10 000–15 000 is aimed at, and as the total falls the haemoglobin level usually rises, with improvement of symptoms. Too much attention should not be paid to the actual white total, and it is not essential to bring it down to these levels; some patients feel better with a higher count. The haemoglobin level is the more important indicator. Care must be taken not to overshoot the mark by excessive treatment which may inhibit the marrow so effectively as to lower the white count to dangerously low levels and expose the patient to devastating infection.

Radiation may be used for: (1) local treatment of the spleen, especially in the granulocytic type, (2) local treatment of lymphnode masses, (3) local deposits in bone, skin or elsewhere, (4) total body irradiation.

1. Radiation to the spleen used to be the standard form of therapy, and is usually quite effective in lowering the white count. It is also useful for decreasing the size of an uncomfortably large spleen, or controlling pain from perisplenitis (irritation from inflammatory involvement of the capsule of the spleen). To lower the white count it is not necessary to irradiate the whole of the spleen. Simple direct anterior or lateral fields, e.g. 15 × 10 cm, are adequate. Treatment should be gradual, beginning with 25 rad incident and increasing by 25 a day to about 150–200 rad. Daily blood counts are made, and the course continues till the level is judged low enough. The number of doses will depend on the rapidity of the response—bearing in mind that the drop will continue for several days after the last dose. Although the disease is widespread, satisfactory control for a long time is generally possible by splenic irradiation alone. The spleen is one of the chief sites of production of the abnormal white cells, but the effects of splenic irradiation are not confined to the spleen—there is also a distant effect on bone marrow etc. in a way not understood. Courses of radiation to the spleen can be repeated, but eventually resistance develops, just as it does to any form of therapy in leukaemia. Splenic irradiation is seldom indicated in lymphatic leukaemia.

2.—3. For node masses or local deposits, low doses of 500–1000 rad are usually adequate. Node irradiation is more often indicated in the lymphatic type, and the white count usually falls at the same time as a result.

4. Total body radiation was used mainly for the myeloid type. Either radio-phosphorus (P–32) was given (orally or intravenously) or external radiation encompassing the whole body at increased FSD. These methods have in most centres been superseded by multi-drug regimes, but may still be of value on occasion.

Eventually the time comes when therapy no longer avails, and marrow failure sets in. Symptomatic and supportive therapy alone are then indicated—transfusion, antibiotics, sedatives etc.

Prognosis. In acute leukaemia the picture has recently changed dramatically, because of cytotoxics. Life expectation used to be a matter of months, and early death inevitable. The greatest change has been in acute lymphoblastic leukaemia, which is the typical form in childhood. With modern intensive therapy, i.e. multi-drug regimes plus CNS radiation, patients are surviving ten years and more. We are now beginning to talk of 'cure', though the treatment is so recent that it will be many years before we can be sure. It is reasonable to believe that *at least 25 per cent can now be permanently controlled.* This is one of the most spectacular 'breakthroughs' of modern medicine.

In chronic leukaemia the picture is different, and there have been no major advances. Chronic lymphatic leukaemia is sometimes almost benign, with a slow course over 10–20 years and very little in the way of symptoms. But average survival in chronic types is only 2–4 years; current treatment probably has little effect on length of life, though it can improve the quality.

Polycythaemia (Polycythaemia* Rubra Vera, Primary Polycythaemia)

This is an uncommon disease of bone marrow, mostly in middle aged males. There is hyperplasia of all constituents—erythroblastic,* leucoblastic, megakaryocytic*—i.e. the precursors of red and white cells and platelets, which are all increased in the peripheral blood. The clinical picture is dominated by the greatly increased number of red cells in the peripheral circulation (hence the name), but there may eventually be complete marrow exhaustion (aplastic anaemia), or the erythroblastic tissue may be exhausted first and the leucoblastic proliferation continue, leading to leukaemia.

Clinical features. The patient complains of headache, dizziness and tiredness, and has a cyanosed* plethoric appearance. The spleen is palpably enlarged. The increase in the red cell volume decreases the fluidity of the blood i.e. there is increased viscosity leading to thromboses (in brain, limbs etc.) and haemorrhages.

The red cell count may be up to 10 million or more, with equivalent increase in haemoglobin. The red cell volume is increased; this is the haematocrit* or packed cell volume i.e. the volume occupied by the red cells in 100 ml of centrifuged blood, normally about 45 per cent.

Other causes of polycythaemia must be excluded before making the diagnosis. *Secondary polycythaemia* can occur from the stimulus of anoxia (decreased oxygen caused by deficient oxygenation of the blood) at high altitudes, in congenital heart disease and in certain lung diseases. It may also be found associated with renal lesions, especially cancer of the kidney. The bone marrow picture is also helpful in distinguishing the various types.

Treatment. The aim is to produce prolonged reduction of the red cell volume to a fairly normal level. This may be done by—(a) venesection (b) chemotherapy or (c) radiation.

Venesection (vein puncture) reduces blood volume rapidly and so gives quick symptomatic relief. It may be technically difficult because the high viscosity makes the blood flow very sluggishly. The effect is very temporary, since it does nothing to remove the cause in the marrow. It may be useful, e.g. before an operation, to decrease the danger of thrombosis.

Cytotoxic chemotherapy (usually busulphan) can be given in short courses, repeated as necessary. Prolonged treatment is needed, with close supervision and frequent blood counts. Most centres have given it up in favour of radiation.

Radiation used to be given either to the spleen or by trunk baths or whole body baths. These have now been almost entirely superseded by *radioactive phosphorus* (P–32)—see Table 10.1 page 115—which is much easier and more convenient. Phosphorus is an essential constituent of all cells, especially of nuclei, and is therefore taken up to a greater extent in rapidly dividing than in slowly dividing cells. It will thus tend to be concentrated more in certain types of neoplastic tissue and in those normal tissues where growth occurs at a relatively fast tempo such as the haemopoietic cells of bone marrow; here the uptake will be even higher in malignant marrow conditions. Dosage is empirical, usually 4–7 millicuries intravenously, according to the patient's weight and the severity of the case. It can also be given orally, 8–12 mCi to allow for incomplete absorption from the bowel. The patient need not be in hospital, since the radioactivity excreted via the bowel and kidney is not of a dangerous order. The red cell count begins to fall in 6–8 weeks, and a repeat dose is not given for at least three months. A decision to re-treat is based on clinical and haematological findings and the required dose is modified in the light of these.

About 85 per cent achieve excellent remission, and most of them need only a single injection for this. The very few who prove resistant are treated by chemotherapy or venesection every month or so. The length of remission varies considerably, averaging about two years.

One possible disadvantage of P–32 is that it may cause late leukaemia, but this can occur in any event. P–32 probably does raise the incidence slightly, but on balance the risk is quite acceptable, and it is the treatment of choice in almost all cases.

Prognosis. Without treatment the average survival is about 5 years, but modern therapy has more than doubled this. The end comes with marrow failure, leukaemia, thrombosis, haemorrhage or heart failure.

Myelomatosis (Multiple myeloma*)

This is a disease of bone marrow, occurring in late middle age.

Pathology. The cell concerned is the plasma* cell, which is believed to arise from the primitive reticulum cells of the marrow. Neoplastic proliferation leads to marrow destruction and failure, and also to local destruction of bone, so that this lesion is often classified with bone tumours.

A remarkable feature is that the plasma cells produce abnormal proteins, which appear in the blood and can easily be detected. Abnormal proteins may also be found in the urine (Bence-Jones* protein). Renal failure may occur from plugging of renal tubules with precipitated protein.

Clinical features. The commonest symptom is bone pain, e.g. from a vertebra or rib. Radiographs may show generalised bone rarefaction (osteoporosis) or very characteristic punched-out osteolytic* lesions, especially in skull, ribs and vertebrae. Pathological fractures are common, or collapse of vertebrae with serious neurological complications, including paraplegia* from compression of spinal cord.

Diagnosis is confirmed by examination of bone marrow, which reveals the typical plasma cells.

Treatment is unsatisfactory. Localised bone pain can usually be relieved by external radiation, e.g. 3000 rad incident over the affected part of the spine, by DXR or cobalt beam, in two weeks. Generalised radiation with P-32 (dosage and intervals as for polycythaemia—see above) may be helpful in widespread involvement.

Chemotherapy is also widely used, and can give useful results. The chief drugs are—melphalan, cyclophosphamide and steroids.

Supportive therapy is important, especially blood transfusion for anaemia. Pathological fracture may need operation and pinning (see p. 259).

Prognosis is very variable, but most die within two or three years. Intensive drug therapy may add another year or so to life expectation in those who respond.

Solitary myeloma (plasmacytoma) of bone

Myeloma confined to a single bone is unusual. When it does occur it can be successfully treated by local radiation—3500–4000 rad in three or four weeks. Most cases however, develop multiple lesions eventually, though it may not be till several years later.

Extramedullary plasmacytoma*

Plasma cell tumours may appear in organs other than bone marrow, and are sometimes benign, but may be the first manifestation of multiple myeloma. The commonest sites are in the upper air passages, e.g. larynx.

*Ewing's*Tumour*

This is an uncommon but characteristic tumour in bone, usually in a child or young adult. Its pathological nature is disputed, but it is probably a reticulum cell sarcoma of bone marrow, and not a primary bone tumour. Some apparent cases have proved to be secondary to a neuroblastoma discovered at post-mortem examination. It arises usually in the shaft of a long bone, causing pain. It may be mistaken at first for an inflammatory bone lesion (osteomyelitis).* Radiography shows a typical appearance—central erosion with surrounding subperiosteal new bone, giving an 'onion skin' picture. Early and wide spread is usual, especially to other bones and lungs. Secondaries in regional nodes occur less commonly.

Diagnosis must be confirmed by surgical biopsy.

Treatment. The tumour is highly radiosensitive, and local control is obtainable, but metastasis is very much the rule and hardly any cases survive five years. Amputation is scarcely ever justified, since external radiation usually achieves local control. The whole bone (not just the apparently affected part of the bone) should receive about 5000 rad in four or five weeks, which secures pain relief and local healing. Similar treatment may be given to metastases in other bones. Even lung deposits can be successfully treated with doses of 2000–3000 rad.

Cytotoxic drugs are now improving the outlook. The optimum regime is intensive radiation to heal the local lesion, with concurrent chemotherapy to deal with microscopic secondary deposits. Multiple agents should be used in combination—actinomycin-D, adriamycin, vincristine and cyclophosphamide—preferably all four together. Chemotherapy is continued for at least two years, with treatment-free intervals.

Prognosis. Surgery and/or radiation give cure rates under 10 per cent. Intensive chemotherapy is now achieving much improved survival.

26. Central nervous system

Tumours of the central nervous system

Tumours of the CNS are uncommon, but occur in both adults and children. Benign tumours can be as important as malignant, because, as space-occupying masses within the rigid skull, they can cause pressure symptoms and can be fatal if vital brain centres are compressed—e.g. meningioma.

Nearly 20 per cent of tumours in the cerebral hemispheres are blood-borne *secondary deposits* from primary growths elsewhere—especially lung, breast, prostate and kidney—and are often multiple.

Pathology. Table 26.1 gives details of the chief types. It is noteworthy that none are tumours of actual nerve cells themselves, but only of supporting tissues and glands.

Table 26.1 Tumours of the central nervous system

Tissue of origin	Histological type	Typical site and degree of malignancy	Relative incidence
	Glioma		
Neuroglia (connective tissue of CNS)	Astrocytoma	Commonest glioma. Ranges from benign to highly malignant. Adult—cerebrum. Child—cerebellum.	
	Oligodendroglioma	Adult—cerebral cortex.	50 per cent
	Ependymoma	Child—ventricles, esp. 4th. Wide range of malignancy. Seeds via CSF.	
	Medulloblastoma	Child—cerebellum. Anaplastic, highly malignant. Seeds via CSF	
Meninges	Meningioma	Adult—benign. Malignancy (sarcoma) possible but rare.	30 per cent
Pituitary gland	Chromophobe adenoma Eosinophil (acidophil) adenoma Basophil adenoma	Benign. Malignancy (carcinoma) very rare	12 per cent
Pineal gland	Pinealoma	Mostly malignant, especially children. Seeds via CSF	Less than 0.5 per cent
Developmental anomaly	Craniopharyngioma	Benign (but can be fatal by pressure)	Rare

Old names for anaplastic astrocytoma—spongioblastoma, glioblastoma multiforme.
N.B. For the derivation etc. of these exotic names see Glossary.

Spread is by local invasion, but metastasis outside the CNS almost never occurs. The absence of lymphatic vessels helps to account for this.

Local spread can occur by 'seeding' of cells via the cerebro-spinal fluid (CSF), giving rise to multiple deposits on the surface of brain and spinal cord. This is characteristic of medulloblastoma, ependymoma and pinealoma.

Clinical picture. Though the tumours listed are very diverse in nature and have individual life histories, symptoms and signs arise in two distinct ways: (1) general effects from increased intracranial pressure, and (2) local effects from local pressure and damage to nerve cells etc. These can arise from any space-occupying lesion, whether neoplastic, inflammatory, traumatic etc.

1. *General*. There is a classical triad—headache, vomiting and papilloedema (oedema of the optic disc, the white patch on the retina where the optic nerve emerges). Headache, often worse in the morning, is of a throbbing kind and gradually becomes worse. Papilloedema can be seen on direct inspection through the pupil of the eye with an ophthalmoscope, even before the patient notices any disturbance of vision. Drowsiness, mental deterioration and personality changes may occur; with a slow growth, the patient may be labelled neurotic, and end up in a mental hospital.

2. *Local*. Localising signs and symptoms depend very much on the situation of the tumour. Some typical examples:

Cerebellum—incoordination in walking, vertigo.

Frontal lobe—intellectual impairment, memory defects.

Temporal lobe—disorders of language, visual field defects.

Brain stem—cranial nerve palsies, involvement of motor and sensory tracts.

Epileptic attacks are common with tumours of cerebral hemispheres.

Diagnosis. A full history is taken including accounts from relatives, then complete clinical neurological examination made. Radiographs of the skull may show for example calcification in a tumour, separation of cranial sutures (between bones) from raised intracranial pressure in a child, enlargement of the sella* in pituary tumours, erosion of skull by meningioma.

Echo-encephalography by ultrasound is simple and will detect displacement of midline structures. Electro-encephalography (EEG) may also be of help by showing abnormal waves.

Cerebrospinal fluid is obtained by lumbar puncture, and may show a suggestive raised cell count or protein content.

Tests of hearing and vision are carried out, including charting of visual fields.

Contrast radiography is usually important and valuable. (1) Air introduced by lumbar puncture (air encephalography) or directly into the ventricles (ventriculography) may demonstrate distortion of the subarachnoid space or ventricles. (2) Angiography—a radio-opaque medium is injected into the carotid or vertebral artery, to reveal the cerebral vascular pattern. Displacement of the normal position of the vessels will localise the lesion, or an area of abnormally profuse vascularity may be shown (e.g. meningioma).

Radioisotope scanning (p. 112) is a more recent and effective method of diagnosis and localisation, involving no greater inconvenience to the patient than an intravenous injection, and does not require admission to hospital.

The most recent method is by the elaborate (and expensive) X-ray Computer-Assisted Tomography introduced originally as the EMI* scanner (see Ch. 12). This is easy and simple for the patient. It is very informative, and now replacing most of the older methods wherever it is available.

The above techniques may not always differentiate a benign from a malignant tumour—but surgery is usually needed in any case.

Treatment

Surgical exploration by opening the skull (craniotomy) is indicated as a rule, but the type and location of the tumour determine what is done. Complete removal is the theoretical ideal, but but not often possible in practice. Benign tumours may be dealt with radically, e.g. meningioma or benign astrocytoma. But because of rapid growth, infiltration or situation, many are only incompletely removable, if at all, and can be treated only palliatively. This applies to most gliomas. Partial excision may be worthwhile for temporary relief. Wider excision, though it might not be fatal, would often do unacceptable neurological damage and demean the quality of existence to subhuman level. In malignant astrocytoma, glioblastoma of adults and medulloblastoma of children, it is usually unwise to do more than take a biopsy.

In some cases surgery is ruled out because of poor general condition or the site of the tumour, e.g. in the brain stem. In others the surgeon must close the skull in the knowledge that all or part of the tumour is still present. To these should be added the cases of recurrence after attempted radical removal. The question of primary or post-operative radiation then arises.

Radiation tolerance of the CNS (brain and spinal cord) is of basic importance, in addition to the sensitivity of the actual tumour cells, which will vary with their degree of anaplasia. Nerve cells are among the most radioresistant cells in the body. This is associated with the fact that mature nerve cells never go into mitosis; once destroyed, they are never replaced. But they are vulnerable to damage to their supporting tissues (neuroglia*), and especially the small blood vessels on which they depend for oxygen and other nutrients. Obliterative endarteritis* induced by radiation is the limiting factor in nervous tissue tolerance. Danger to normal tissue therefore arises from: (1) Direct damage to nerve cells, (2) strangulation by fibrous scar tissue and (3) choking of blood vessels. Of these the last is the most important. The mechanisms are comparable to those seen in other organs, but the effects are more serious because of the irreplaceability of nerve cells and their fundamental importance in the body's economy.

The most radiosensitive parts are: motor areas of cerebral hemispheres, brainstem (in which vital areas such as the respiratory centre are located), spinal cord. More tolerant parts are: frontal, temporal and occipital lobes, to which higher dosage can be given. In addition, as usual, the treated volume and time factors (overall time, fractionation) must be taken into account.

If the whole brain is irradiated, dosage should not exceed 3500 rad in 10 treatments, or the equivalent in longer time. Smaller fields, e.g. to brain stem, can be given 5000 rad in four weeks. For spinal cord, if a length of 15 cm is treated, 3500 rad in two weeks should be the limit, though short lengths can be given up to 4500 in three weeks.

Heroic attempts at radical dosage are always unwise. Necrosis is a probable result, with consequences worse than the original disease. The symptoms and signs of high dose effects may be difficult or impossible to distinguish from those of tumour recurrence.

Radiosensitivity of brain tumours shows a wide spectrum. At one end, meningioma and differentiated astrocytoma are highly resistant; at the other, medulloblastoma and pinealoma are highly sensitive.

Radiation techniques. For a few lesions, e.g. small mid-brain tumours, small field arrangements are indicated. But as a rule the extent of invasion is uncertain, and generous margins should be used, e.g. quarter or half-brain volumes, or even the whole brain. For whole-brain treatment, a simple parallel pair is adequate (Fig. 26.1), e.g. 15 × 12 cm on cobalt beam or megavoltage, giving 3500 rad in 10 treatments. Simple wedge-pair arrangements are suitable for half or quarter-brain treatments.

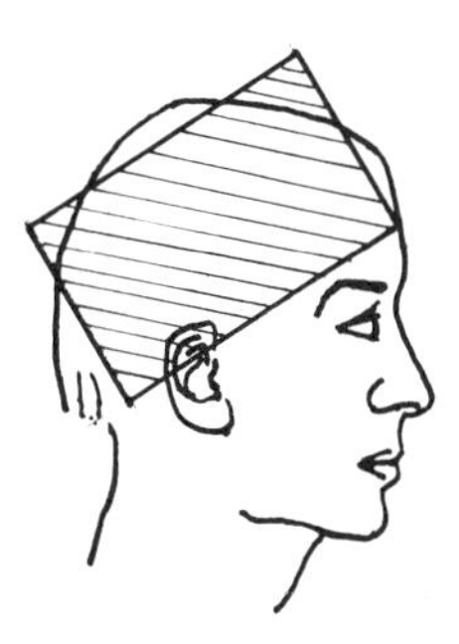

Fig. 26.1 Lateral field for parallel pair for whole-brain irradiation.

Raised intracranial pressure should first be relieved, by the preliminary surgical decompression, or failing this by medical means (diuretics* or steroids, usually dexamethasone). Radiation is liable to increase the pressure inside the skull, at least temporarily, and if there is clinical evidence of raised pressure and surgical decompression has not been done, radiation should begin slowly, with low daily dosage, to avoid exacerbation of symptoms.

A restless patient may need drug sedation to enable treatment to be carried out, and it may be difficult to immobolise the head. If there is any risk of violent movement (e.g. an epileptic attack), treatment may be possible on the patient's bed with restraining side-pieces in place. Alternatively he should be on a stretcher on the floor, to avoid any risk of falling from a height during treatment.

Whole CNS radiation. For those tumours which tend to spread widely via the CSF—medulloblastoma, pinealoma, ependymoma—special technique is needed, to cover the whole cerebrospinal axis. The commonest of these is *medulloblastoma*. This occurs in children, typically between ages four and eight, and arises in the cerebellum. Primary treatment is surgical with removal of part of the growth. If nothing further is done rapid recurrence follows. Spread may take place over the surface of the whole or part of the brain and spinal cord. It is therefore essential to treat the whole of this large area. Fortunately the growth is one of the most radiosensitive of all, and cases are now surviving many years, with permanent cures after adequate radiation.

Technique. The principles are shown in Figures 26.2, 26.3, 26.4 The aim is to irradiate the entire cranial and spinal cord cavities. Several basic techniques with minor modifications are available. One is a large posterior tennis-racket shaped field, at appropriately increased FSD using orthovoltage (DXR), wide at the top to cover the whole head, narrow below (4–5 cm) for the spine. A lead table, supported on pillars, with a central slit, outlines the spinal part of the field. The head is flexed downwards to flatten the posterior surface. The edge of the cranial part of the field lies along the orbito-meatal line, i.e. just above the orbit and through the external auditory meatus, to avoid the eyes but include the whole brain. This technique gives inadequate dosage to the anterior part of the brain, so supplementary fields are added, either two anterior obliques (DXR) or a parallel wedge pair (MXR)—Figure 26.3.

An alternative method (Fig. 26.4) is by separate fields for skull and spine. Two lateral megavoltage fields are applied to the skull, again bounded by the orbito-meatal line. This will include the spine down to second or third cervical vertebra. Dosage is 4000 rad in four weeks, supplemented by a small field to the posterior fossa (the site of the primary growth), giving an extra 500 rad (5 × 100 incident)—i.e. five weeks overall. The rest of the spine is

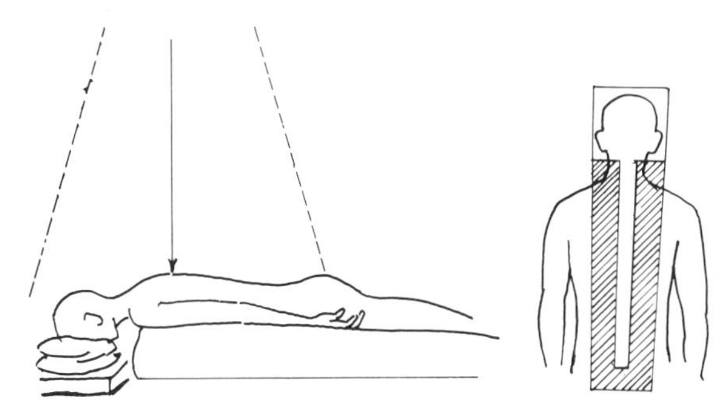

Fig. 26.2 Posterior field to include the whole cranio-spinal axis. The spinal part is demarcated by lead protection (cross-hatched).

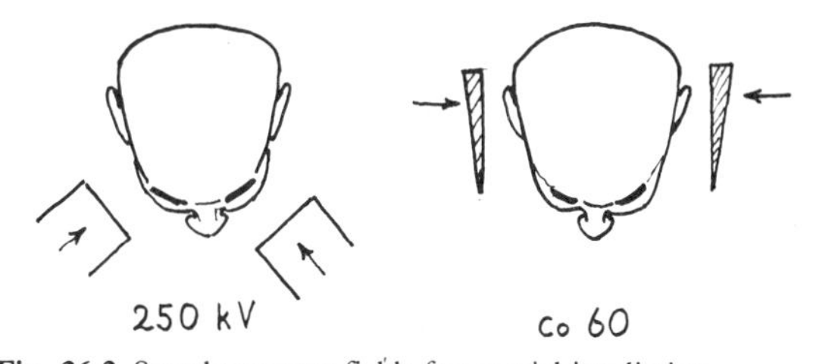

Fig. 26.3 Supplementary fields for cranial irradiation.

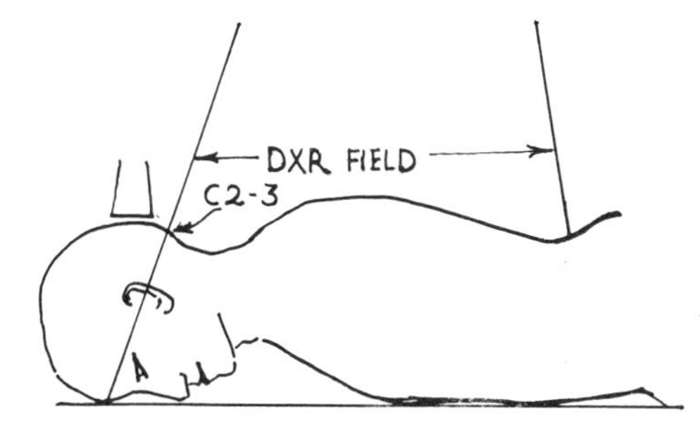

Fig. 26.4 Alternative scheme of cranio-spinal irradiation for medulloblastoma. Orthovoltage field for spine, megavoltage pair for cranium and supplementary field to posterior fossa.

covered by a DXR field at 250 kV (to avoid excessive dosage to abdominal contents), at 100 cm FSD measuring 30–35 cm long by 4–5 cm wide, according to the size of the child. It extends from the third sacral vertebra below (where the spinal dura* ends) to second or third cervical above, to match the border of the above cranial field. To avoid excessive dosage at the junction, a gap of 0.5 cm is left between the margins of the two fields. A lead table can be used over the spine, but it is preferable to make an individual plaster cast of each patient and mould a lead jacket to fit the whole of the back, with the central gap. The weight of the jacket is borne by the treatment couch at each side, not by the patient. The incident dose is 4250 rad in five weeks, which gives a tumour dose at the spinal cord (about 6 cm deep) of about 2750 rad. Very careful positioning and alignment of field margins are essential, to avoid under- or over-dosage. Some workers prefer two cobalt beam spinal fields of unequal length, at short SSD, alternating the fields every day, so that the effects of inhomogeneous dosage at field junctions are minimised.

In the few cases where a young child is too restless, general anaesthesia may be necessary before each treatment, by means of intravenous Ketamine.

A recent innovation is to add cytotoxic therapy—vincristine and CCNU (see Ch. 31)—during and after the radiation course.

Reactions. With proper shielding, constitutional reactions should be mild. Blood counts are taken weekly, but should cause no anxiety. There will be total epilation,* but the hair will grow again in three months. (It is a useful statagem to shave the head first, if there is no objection raised; it will lessen the reaction by taking advantage of build-up otherwise lost because of the presence of hair). The reaction over the spine will go to deep erythema and dry desquamation, followed by pigmentation. There will be some late effect on vertebral bone growth, with slight but insignificant shortening of height. No late mental impairment has been seen.

Intracranial metastases are very common. On rare occasions (e.g. renal cancer) a deposit has proved to be the only secondary in the body, and even been successfully removed surgically. But usually cerebral secondaries are seen in fairly advanced and disseminated malignancy.

Symptoms can in some cases be improved by medical means, especially decongestion by dexamethasone (see above). Good results may also follow radiation. Treatment is usually to the whole brain; anything less is seldom worthwhile, especially as deposits are so often multiple. A parallel pair, about 15 × 13 cm is used (Fig. 26.1) giving 3000 or 3500 rad in 10 treatments, and often gives good palliation. The hair will always be affected, and fall out soon after the course. This should be explained in advance to the patient (and/or relatives). It is a simple matter to provide a wig at an early stage, and this is a great comfort, especially to women. The hair grows again in a few months, but incompletely.

Prognosis. Since the natural history of the various types is so diverse, general statements are of no value. Most meningiomas and some astrocytomas are curable surgically by complete removal. A well-differentiated astrocytoma may take ten years or more after diagnosis to become dangerous, whereas the anaplastic type (glioblastoma) is usually fatal within months. Tumours in vital situations e.g. midbrain, can be rapidly fatal even if small.

Of the malignant growths, the best results are achieved in the medulloblastoma of childhood, where five-year survival is now 40 per cent, with results still improving. As noted above, our knowledge of the safe limits of dosage has virtually abolished the danger of damaging effects on nervous tissue or bone.

The pituitary inevitably receives fairly high dosage in most cerebral cases, but late endocrine damage is almost unknown.

Spinal cord compression and paraplegia

Pressure on the cord is often insidious, leading to paraplegia* and incontinence (of bladder and bowel) if not relieved. Symptoms begin with sensory and /or motor disturbances in lower limbs, or bladder upset. There may be deposits of growth in vertebrae (primary or secondary) encroaching on the spinal canal, or there may be no bone involvement, but deposits in the canal outside the coverings of the cord (extradural). Common lesions include—breast and lung secondaries, Hodgin's deposits, myeloma.

To be effective, treatment must be early, before neurological damage becomes irreversible. These cases are emergencies, and the quickest and most effective treatment is surgical (if the patient's condition permits), i.e. decompression by laminectomy* (removal of the lamina, the bony shelf forming the back or arch of the spinal canal). Removal will usually be necessary over several adjacent vertebrae; the offending mass is disclosed and partially excised. Post-operative radiation may then be advisable, on the same lines as for bone secondaries in the spine (p. 259).

In some cases, e.g. lymphoreticular, radiosensitivity may be high enough to justify preliminary treatment directed to the appropriate spinal level indicated by the neurological signs. If this does not lead to rapid improvement, laminectomy should be done. Cytotoxic drugs may also be useful in these emergencies, but local radiation, if available, is usually less upsetting and more reliable.

If diagnosed and treated in time, compression is usually relieved and the patient is spared the misery of terminal paraplegia and incontinence. Treatment is therefore justifiable in spite of the poor general prognosis.

Tumours of the pituitary gland

The pituitary* is the most important of the glands of internal secretion (endocrine*). Its own hormonal secretions control other glands—gonads, adrenals, thyroid etc.—and many important processes including growth and sexual development. It has been aptly called 'the leader of the endocrine orchestra' (Fig. 30.1, p. 273). To understand the types of tumour arising in the gland it is necessary to know something about the constituent cells. These, and the tumours derived from them, are classified by their staining reactions—i.e. their ability to take up certain dyes when sections are exposed to special stains for microscopic study. We thus have:

1. *Chromophobe** cells, difficult to stain at all, and producing *chromophobe adenoma*.
2. *Eosinophil** (acidophil) cells taking up acid dyes, especially a rose-coloured stain called 'eosin', and producing *eosinophil (acidophil) adenoma*.
3. *Basophil* cells, taking up basic (as opposed to acidic) dyes, and producing *basophil adenoma*.

Technically they are all benign tumours (carcinoma is extremely rare) and do not metastasise. However, some are dangerous because of their size, pressing on the brain or adjacent optic nerves, which can lead to partial or complete blindness. In addition to these local pressure symptoms, hormonal upsets usually occur, from underproduction or overproduction of hormones. Some typical clinical pictures will result, as follows:

1. *Chromophobe adenoma* is the commonest (about 66 per cent) and may grow to a large size, producing its effects mainly by pressure on surrounding structures and increase in the general pressure inside the skull. Radiographs may show enlargement of the pituitary fossa. Chromophobe cells do not produce any secretion, but the tumour compresses the rest of the gland and so causes endocrine upsets owing to hormonal deficiencies.

2. *Eosinophil adenoma* (30 per cent). Eosinophil cells produce a hormone controlling growth. An adenoma is an overgrowth of these cells, with excessive production of growth hormone. If it occurs in a growing child, a 'giant' may result owing chiefly to overgrowth of the skeleton. In an adult, where skeletal growth is already complete, the resultant disease is *acromegaly**—the extremities (hands and feet) are thickened, the nose broadens, the jaw appears large and heavy, etc. Local pressure effects may be serious—especially headache and interference with vision.

3. *Basophil adenoma* (4 per cent). This is always a tiny microscopic tumour, but its effects are profound, as basophil cells produce hormones governing many sexual and other processes. In a woman, the menstrual periods cease and she becomes obese, with raised blood pressure etc. The disease is called 'pituitary basophilism' and is commonly linked with the name of Cushing (Cushing's Syndrome), a famous American neurosurgeon who unravelled it in the present century.

Treatment is either by surgery (hypophysectomy)* or radiation or both. If there are troublesome pressure symptoms, and especially if vision is threatened, surgery is usually the first line. Complete removal by surgery, however, is almost impossible (without unacceptable damage to vital structures or vision) and post-operative radiation is therefore often indicated. If pressure symptoms are mild, radiation can be tried first, and surgery kept in reserve for failure, or deterioration of vision.

Radiation technique is shown in Figure 26.5. A head mould is made and four oblique megavoltage fields used, above the level of the orbits. The tumour dose is 4000 rad in four weeks.

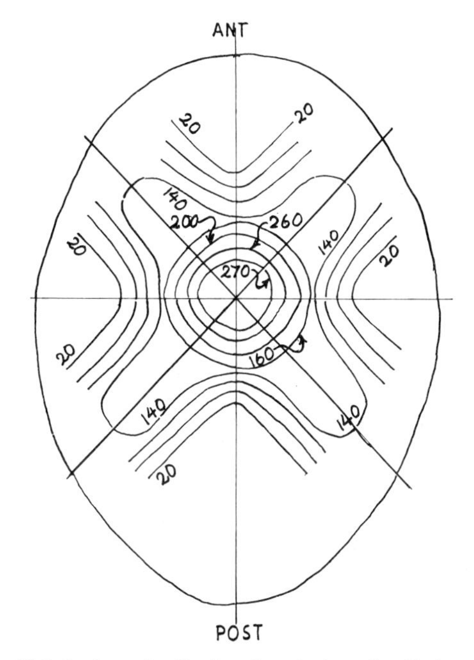

Fig. 26.5 Isodose distribution for pituitary irradiation. Four oblique megavoltage fields at 8 MV, each 4 × 4 cm.

Craniopharyngioma

This is also known as Rathke's* pouch cyst or suprasellar cyst and is a developmental anomaly. In the embryo the anterior part of the pituitary is developed from a narrow projection from the foregut, leading from what will later be the nasopharynx through the floor of the skull. Remnants of this may persist and form cystic tumours which may compress the pituitary. The best treatment is surgical removal or decompression with evacuation of the cyst. It is radio-resistant, but post-operative radiation may be added to prevent recurrence—5000 rad in four weeks by small fields.

27. Bone and soft tissues

Tumours of bone

The chief primary malignant tumours of bone are—*osteosarcoma, fibrosarcoma and chondrosarcoma*. Tumours of reticular tissue and bone marrow are often classified and discussed under 'bone tumours'. But they are really tumours *in* bone rather than tumours *of* bone. (The same holds for secondary cancers in bone). They include—Ewing's tumour, myeloma and various reticulosarcomas, dealt with in Chapter 25.

Osteosarcoma (osteogenic* sarcoma)

This is the commonest and most malignant primary bone tumour. It arises in bone-forming cells (osteoblasts). Local spread is along the marrow cavity, then the periosteum* is raised and soft tissues are invaded. Spread to lymph nodes is rare, but blood spread to lungs etc. is typically very early—usually before diagnosis, and it is this fact which accounts for the very serious prognosis.

It affects mainly children and young adults in the second decade. The typical site is near the ends of long bones, especially around the knee. There is another rise in incidence in the elderly, associated with Paget's* disease of bone (a skeletal disorder of unknown cause, leading to deformities—'osteitis deformans'—and thickening of bone) which carries an even worse prognosis.

There is a causative association with radiation. In the orthovoltage era, radiation was applied, e.g. for tuberculosis of bones and joints and other lesions, and late bone sarcoma occasionally followed. Another classical example is the development of bone sarcoma after deposition of radioactive isotopes (radium and thorium) after accidental ingestion (see p. 151).

Clinical features. The presentation is by pain, swelling and possibly pathological fracture after mild trauma. Radiographs show irregular bone destruction and some typical appearances of cortex, periosteum etc. A chest film may show secondary deposits even at the start. A surgical biopsy should be taken to confirm the diagnosis and exclude other tumours and inflammatory lesions.

Treatment. Bone tumours are notoriously radio-insensitive in general. The logical treatment is therefore by amputation. Till very recently, radical pre-operative radiation was used, followed by amputation in a few months if no lung secondaries had appeared. The best treatment now is primary amputation followed by cytotoxic therapy to deal with micrometastases. Combination chemotherapy is advisable e.g. vincristine and high-dose methotrexate followed by folinic acid rescue (see p. 280). Adriamycin and melphalan are also in use.

Radiation is still useful in the rare cases of inoperable sites e.g. a vertebra.

Prognosis. Four out of five victims (80 per cent) used to die within two years, and the five-year survival was about 15 per cent. Whether adjuvant (= helping, assisting) chemotherapy is going to fulfil its present early promise, remains to be seen.

Chondrosarcoma

Malignant tumours of cartilage may arise in pre-existing benign chondromas, or may be malignant from the start. Adults aged 30–50 are affected. Common sites are—flat bones (pelvis, scapula), ribs and long bones. Growth may be very slow indeed, with eventual swelling and pain. Spread is chiefly local, with terminal blood dissemination to lungs.

The only satisfactory treatment is surgical, if practicable. They are the most radioresistant of all tumours. Removal must be wide, and radical procedures such as forequarter or hindquarter amputations may be required.

If surgery is ruled out by the patient's condition or the tumour site, palliative high dose megavoltage, as for osteosarcoma, may offer temporary growth restraint. Chemotherapy is a reasonable supplement.

Fibrosarcoma

Fibrosarcoma of bone resembles osteosarcoma, but is less malignant and slower growing. It arises from the fibrous layer of the periosteum, often near the lower end of the femur.

It is radioresistant, and radical surgery is the only satisfactory treatment. Radiation and cytotoxics can have palliative value.

Osteoclastoma (Giant cell tumour of bone)

This is an uncommon but interesting bone tumour. The microscopic picture typically shows many large multinucleated 'giant' cells resembling osteoclasts,* hence the alternative names.

The pathology is disputed. Some regard it as a tumour of osteoclasts, but it more probably originates from connective tissue of bone marrow. Many bone lesions, neoplastic and otherwise, contain giant cells, and this has caused much confusion. It is usually classified as a benign tumour, but at least 15 per cent are malignant (sarcoma), either from the start or by later degeneration as revealed at follow-up.

Onset is usually between ages 20 and 35, with gradual swelling and pain, and possibly pathological fracture. Most occur in the region of the knee, in femur or tibia; also upper humerus and lower radius. Radiographs may show characteristic appearances, with bone destruction by a multilocular cystic lesion expanding the cortex. Diagnosis should be confirmed by open biopsy.

Treatment is by surgery, radiation or both. In general, surgery is best for accessible sites: (1) by thorough curettage, with or without post-operative radiation; this method is often followed by recurrence, or (2) complete local excision, which is the preferred surgical method, even though amputation may sometimes be involved. Radiation has often been used as primary treatment, and many lesions certainly respond, sometimes to low doses, e.g. 3000 rad in two weeks. But their radiosensitivity varies rather widely, and if radiation is used primarily (e.g. in inoperable sites such as a vertebra) dosage should be in the region of 5000 rad in five weeks.

Prognosis is difficult to assess, and varies with the histology and degree of differentiation. Late sarcomatous degeneration, even after 15 years, is now well recognised and has raised suspicion that it may sometimes be induced by previous radiation. Most lesions, however, can be controlled, and surgery offers the best long-term prognosis.

Secondary tumours in bone

The great majority of bone tumours are not primary, but of metastatic origin. The most frequent primary sites are—breast, lung, prostate, kidney and thyroid—i.e. mostly adenocarcinomas. The metastatic deposits come via the bloodstream, and this means that the growth is already disseminated, often widely, and the ultimate prognosis almost always very poor, though great relief is possible for months and even years.

The usual *clinical presentation* is pain, caused by tension of the periosteum or pressure on nerve roots as they emerge from the spinal canal (Fig. 27.1). *Early diagnosis* is important, before bone destruction has advanced far enough to cause pathological fracture or compression paraplegia*. X-ray films may show the deposits, usually as osteolytic (i.e. bone-destroying) rarefactions, but sometimes as patches of increased density or sclerosis (especially in prostatic cancer, also in breast cancer at times).

But a secondary deposit, e.g. in a vertebral body must be of some size—about 2 cm—before it will show up in a film. If we have a patient with a known primary liable to bony secondaries—e.g. breast cancer—complaining of low back pain or sciatica, our suspicions should be at once aroused. A film may show spinal secondaries, but even if it does not, clinical suspicion alone will usually justify local X-ray therapy on the reasonable assumption that secondaries are responsible.

In the *routine investigation* of many types of cancer, skeletal X-ray surveys may be made—either complete, or of spine, ribs and pelvis—but unsuspected findings of silent secondaries are very uncommon. A more searching method is now available in the *radioactive bone scan* (see p. 113) which can show up secondaries many months before the stage of possible radiological diagnosis. The finding of latent secondaries will obviously affect the whole treatment plan and prognosis.

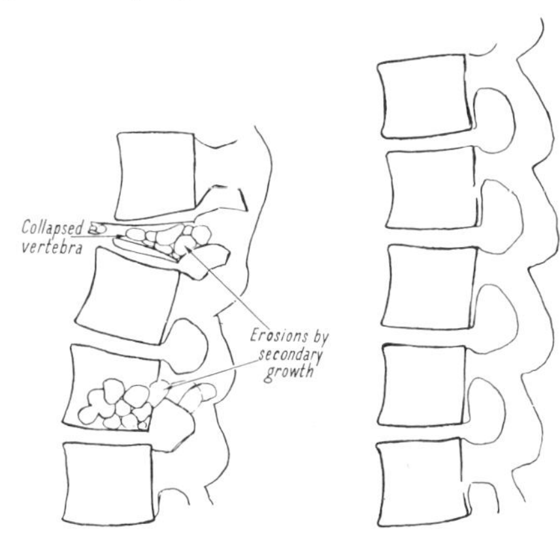

Fig. 27.1 Diagrammatic radiological appearances of normal spinal column and secondary deposits from cancer (e.g. breast).

Treatment of diagnosed deposits will depend on the total situation:

1. What is the local condition of the bone?—is it considerably eroded?—is there a pathological fracture present or impending?
2. Is it a solitary secondary (as far as can be determined) or are there widespread deposits, bony and/or visceral?
3. What is the patient's age, general condition, and reasonable expectation of remaining life?
4. Is the growth hormone-dependent, and if so, what part can hormone therapy play?

As a rule, the urgent primary objective is *relief of pain*, and for this radiation is usually the method of choice. In most cases the symptomatic results are most gratifying. This palliative role of radiotherapy is one of its great historic triumphs and would alone justify the installation of an expensive unit. For a case like that of Figure 27.1, a simple field, either deep therapy or high energy, suffices, giving about 2000 rad in 4–5 days (Fig. 27.2). For a limb a simple parallel pair of fields is used.

If *pathological fracture* of a long bone has occurred—e.g. femur or humerus—operative internal fixation of the fragments by an intra-medullary metal pin will give immediate relief and allow early mobilisation. The deposit can then be irradiated to destroy the growth locally.

For an elderly patient with widespread involvement from for example lung cancer, simple pain relief and general nursing care will contribute to a peaceful passage. But for a middle-aged woman with bone secondaries from breast cancer, even if multiple, good management, using a combination of local X-rays and general hormone therapy, may often keep her alive and fit for many months or even years, e.g. oophorectomy, adrenalectomy etc. (see page 273). Similar management for prostatic cancer with oestrogens, orchidectomy and local X-ray therapy also gives good results.

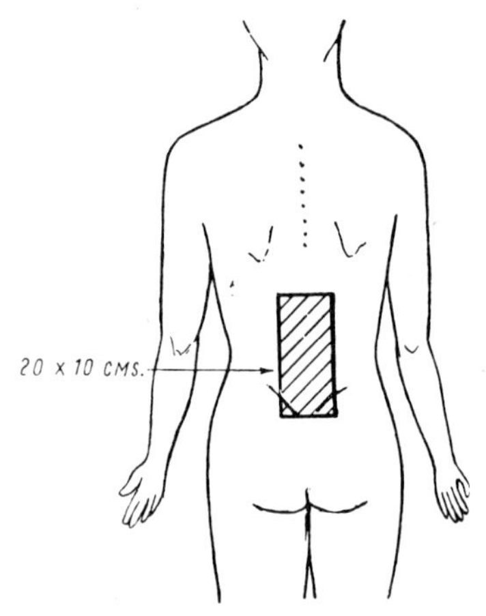

Fig. 27.2 X-ray field to lumbar spine, to treat secondary deposits.

If it involves a weight-bearing bone such as the femur, a longer course—e.g. 3500 rad in two weeks—may lead to good re-calcification and allow weight-bearing again in 2–3 months. But if bone deposits are widespread and hormonal improvement is not applicable, treatment should be confined to the painful areas only, because attempts to irradiate many bones will exhaust the limited marrow tolerance still left. For such cases *single shot therapy*, giving 1000–1500 rad in one session, is simple and easy for the patient, and can give good relief.

The nursing care of these patients is important and can be difficult. The risk of pathological fracture must always be kept in mind and forestalled—or dealt with if it has occurred—by use of plaster shells, fracture boards, arm slings and pillows, plastic neck collars etc. If plaster slabs or shells are used, it may be advisable to cut large holes in them for the X-ray treatment fields. If the ends of the fragments after pathological fracture override, orthopaedic traction with a Thomas's splint and Balkan frame may be needed. The splint may have to be removed at each treatment session, though it may be possible to wheel the whole bed to the treatment room without disturbing the patient.

Tumours of soft tissue

Sarcoma of soft tissue may occur almost anywhere in the body and at any age. They are classified by the tissue of origin—*Fibrosarcoma* (fibrous tissue), *liposarcoma* (fatty tissue), *myosarcoma* (muscle), *synovioma* (synovial* membrane of joints, tendon sheaths etc.)

They have many features in common—slow growth, late terminal spread to lungs, and high radioresistance. Local swelling and later pain are the presenting symptoms. Biopsy is usually necessary.

Treatment should be surgical if possible, even amputation if need be. Post-operative radiation and cytotoxic drugs may be helpful supplements. Local recurrence is common. In inoperable or recurrent cases, palliation is often possible, but high dosage is needed, e.g. 6000 rad or more, unless the mass is huge, when such doses would result in little except necrosis. Sometimes, however, the response is better than expected, and long-term survival can follow; so most cases are at least worth a trial.

Myosarcoma is of two varieties: (1) *leiomyosarcoma** which is managed as above and (2) *rhabdomyosarcoma*,* mostly in children, discussed on page 264.

Synovial sarcoma or malignant synovioma can form a large fleshy mass spreading widely and infiltrating bone and soft tissue. The lower limb is the usual site, but any joint can be involved. Post-operative local recurrence is common. If surgery is impracticable, long-term palliation is sometimes achieved by high dosage, e.g. 6500 rad in six weeks.

28. Cancer in children

Between the ages of one and fifteen—after accidents, poisoning and violence—malignant disease is the commonest cause of death in Britain. This is due to the control of infections in childhood diseases in the past quarter century (diphtheria, whooping cough, meningitis, measles etc. etc.). In the past decade or so there have been major advances in treatment and prognosis of several types of malignancy. Since the total number of cases is relatively small, it is best for treatment to be concentrated in a few special centres, where a specialist team can handle all the cases from a wide region and so accumulate expertise for the benefit of all patients—otherwise nobody would gain sufficient experience to be really competent. The team should include—paediatric physician, surgeon, radiotherapist, chemotherapist and/or haematologist.

To keep matters in perspective—a general practitioner is unlikely to see more than a couple of cases in a working lifetime. Diagnosis is in any case usually difficult, since the early signs tend to be vague—failure to thrive, fever, pallor, loss of appetite, headache—all of which are common to many other disorders.

In most types, the mode of origin of the growth differs from that in adults, being a developmental malformation (which may even be present at birth) rather than a result of 'chronic irritation'. Hence surface (epithelial) tissues, which are the usual sufferers from repeated trauma, are not the usual sites of origin in children.

Hereditary and genetic factors play a part. This is very obvious in retinoblastoma. Other congenital defects are liable to be associated—e.g. mongols have 30 times the normal risk of developing leukaemia.

Immunological factors are also probably important. The antibody recognition and defence systems are immature in the young child, and this may account for the frequency of acute lymphoblastic leukaemia. Whether virus infection plays any part in induction is still uncertain and unproved, though the idea is very tempting, especially as comparable leukaemias can certainly be induced by viruses in laboratory animals.

Just as normal growths in children is more pronounced than in adults, so malignant growth tends to be rapid and throw off early metastases. They are correspondingly *radiosensitive* but this unfortunately *does not mean radiocurable*.

Types The commonest cancers are:

Leukaemia—mainly acute	35 per cent
Central nervous system—glioma	30 per cent
Connective tissues—including bone	12 per cent
Neuroblastoma	8 per cent
Wilms's tumour or nephroblastoma	5 per cent

Acute leukaemia

Some details have already been given, pages 248–50.

This is the commonest cancer of all in childhood. It is of lymphocytic type (acute lymphoblastic leukaemia or ALL) in 80–90 per cent of cases. The clinical picture comprises fever, headache, ulceration of mucosal surfaces (mouth etc.), haemorrhages in skin and elsewhere, loss of weight, and maybe prostration. There may be enlarged lymph nodes, liver and spleen.

Until 1948, only supportive therapy was possible—blood transfusion for anaemia, antibiotics for infection, drugs for pain; few cases survived more than a few months. Cytotoxic drugs have now revolutionised treatment and prognosis. Control of the complications of treatment, especially infection, has also contributed greatly to the improved results. (For detail on drugs see Ch. 31.)

Almost all children with ALL obtain a *complete remission*—i.e. disappearance of all signs of disease, clinically and in peripheral blood and bone marrow—in about 4–6 weeks. The drugs used are—vincristine (intravenous) once a week, and prednisolone (a steroid, oral tablets) daily. After remission is induced, the improvement is consolidated with more aggressive agents such as adriamycin, cyclophosphamide and asparaginase. Prophylactic radiotherapy to the CNS is added soon after the end of the first course, as described on page 249.

Supportive therapy (as above) is important to cope with the two chief consequences of the disease or the marrow depression induced by the drugs—i.e. haemorrhage and infection.

Maintenance therapy is important, to ensure that the remission is as prolonged as possible. The best drugs are: 6-mercaptopurine daily and methotrexate weekly, with occasional vincristine. There are various regimes used by different workers, all giving good results. The total white cell count should be kept between 2000 and 3000 per mm^3, and the exact dosage is adjusted in the light of the blood and marrow picture. During maintenance therapy the child is at home leading a full and normal life.

How long should maintenance therapy be continued?

This is an obviously important, even vital, issue. Since most of the drugs are immunosuppressive (i.e. toxic to the body's defense mechanisms) there is always a definite hazard of serious and often fatal infection—'opportunistic' infection—from bacteria, viruses and fungi. To continue the drugs longer than really necessary is therefore worse than useless. The optimum length of time for drug maintenance appears to be about 2–3 years. When the drugs are stopped, a proportion of patients will relapse. A second remission may be obtainable by renewed treatment, but the chances now are much poorer, and as a rule only palliation will be possible. For relapses involving the CNS, the whole CNS (brain and spinal cord) should be re-treated by radiation.

Experience has shown that there may be a relapse from leukaemic involvment of the testicles, and some workers therefore advise 'prophylactic' testicular irradiation (even though this involves inevitable sterilisation).

In cases where ordinary treatment has failed, heroic attempts have been made by destroying all haemopoietic tissue by total body irradiation, and then re-populating the bone marrow by marrow grafts taken from a sibling.

Acute myeloblastic leukaemia (AML) is less responsive to chemotherapy than ALL. A high proportion fail to achieve complete remission, and the value of maintenance therapy is less certain. The value of prophylactic radiation to CNS is also doubtful.

Neuroblastoma

This is the commonest solid tumour in children. It is rare in adults, and the great majority occur before the age of five; over half are under two years, and it may even be present at birth.

Pathology. It arises in sympathetic nervous tissue, which includes medulla* of the adrenal gland and similar tissue at other sites (posterior abdomen, mediastinum, pelvis, neck). There is considerable variation in its biological behaviour. In a few rare cases, mostly under one year old, spontaneous regression occurs, with maturation into an adult non-malignant form (ganglioneuroma*)—even if disease is advanced and metastatic—but most spread locally and soon metastasise. *Most cases already have widespread secondaries when first seen.*

Local spread is to kidney, liver etc. Lymphatic spread is to paraortic nodes. Blood spread takes it to liver and bones—but rarely to lungs, in contrast with Wilms's tumour.

Clinical features. There may be a painless swelling in abdomen or loin, but often the first indication is due to metastases, e.g. bone pain or pathological fracture. Widespread bone secondaries with an occult primary is a common picture. The liver may be palpably enlarged. Skull metastases may cause unilateral exophthalmos* (protrusion of the eye). Intravenous pyelography (IVP) helps to distinguish it from Wilms's tumour, by showing it to be extrarenal. A full radiographic skeletal survey should be made, and bone marrow examined for malignant involvement. A total body isotope bone scan is better than radiography for revealing secondaries.

Catecholamines. The physiological activity of the sympathetic nervous system is mediated by complex organic substances called catecholamines. Metabolic breakdown products appear in the blood and are excreted in the urine. Most cases of neuroblastoma excrete abnormally large quantities, and the most important product is called vanillyl-mandelic acid (VMA). Biochemical tests of urine are very valuable both for diagnosis and to monitor progress under treatment and detect early recurrence.

Treatment. Surgery, radiation and chemotherapy all have important roles. Since most cases are generalised at the time of diagnosis, generalised (systematic) treatment is indicated in all but the early stages of development. Complete *surgical excision* is rarely possible, because of extensive local infiltration, but partial excision may be possible. Surgery therefore is seldom the main form of treatment.

Post-operative radiation is usually advisable. In abdominal cases (i.e. the majority), it will be necessary to include one or both kidneys, but the kidney dosage should not exceed 1500 rad in two weeks. Technique and dosage are similar to that for Wilms's tumour (see below). For an abdominal case, not more than 2500 rad should be given in four weeks, by a megavoltage pair, depending on the age and the treatment volume. It may be impossible to avoid irradiating the ovaries. In thoracic or pelvic cases, smaller volumes are involved, and 2500–3000 rad may be given in 3–4 weeks; humeral or femoral epiphyses* are shielded.

Radiation is good palliation for painful bone deposits—single doses of 750–1000 rad suffice.

For inoperable tumours, some use radiation in the hope of causing shrinkage and possibly making the mass operable. It is better to start with chemotherapy, and if shrinkage occurs, then proceed to surgery, maybe with the help of pre-operative radiation.

Cytotoxic therapy plays a major role and is finding increasing application now. It is the initial treatment for widespread metastatic disease, and the initial response is usually good, with pain relief and bone healing. Unfortunately it has not so far reduced the recurrence rate. The best drugs are: vincristine, cyclophosphamide and adriamycin together, intravenously, at fortnightly intervals. This can be done on an out-patient basis. They cause some sickness, for which an anti-emetic drug should be given, and also total alopecia* (baldness). The parents should be warned of this, and wigs provided for all but the youngest patients. The hair always grows again. Treatment should be continued at weekly or fortnightly intervals for up to two years.

Neuroblastoma is remarkable in that even cases with very widespread metastases—liver, bone etc.—are not hopeless, and are worth treating. The response to cytotoxics may be astonishingly good.

Prognosis. Neuroblastoma shows the highest rate of spontaneous regression of any tumour—up to 8 per cent. Survival depends more on the child's age than the extent of disease. Survival for two years usually means cure. For children below one year old, the two-year survival is 75 per cent; over one year, it drops to 20 per cent. Even with minimal treatment, response in the very young is so good, that there must be immunological factors at work to account for it; these factors are lost as the child ages. The outlook is particularly bad in the presence of bone involvement, except in the very young.

In practice, the response is poor in most cases. Intensive combined therapy has not much improved the survival rate (in contrast to Wilms's tumour and rhabdomyosarcoma). Palliation can usually be achieved, but overall survival has not changed.

Nephroblastoma (Wilms's* tumour)

This is one of the commonest solid tumours of childhood, and occurs mostly before the age of five.

Pathology. It arises from neoplastic change in a developmental malformation of the kidney, from ectodermal* and mesodermal* tissues. This accounts for the presence not only of primitive renal tubules, but also of muscle, cartilage etc. Although growth is rapid and the mass as a whole is anaplastic and radiosensitive, some of its elements are less responsive and give rise to local recurrence. The kidney may be enormously enlarged, with local adhesions and invasion. Spread is usually early. Lymphatic spread is to paraortic nodes, and blood spread to lungs, liver, bones etc. Bilateral growths occur in 10 per cent of cases.

Clinical features. The age range is a few months to eight years, but 90 per cent are under five. The typical presentation is a large swelling in the flank of a small pale infant. IVP will confirm its renal origin, and help to distinguish it from neuroblastoma (above). Haematuria may occur, but pain is not marked till late.

Full pre-treatment assessment should be made including chest film and skeletal survey (isotope whole body scan if available).

Treatment. Till recently, the standard treatment was immediate surgical nephrectomy plus post-operative radiation. Sometimes pre-operative radiation was practised if the mass was very large, in order to shrink it and improve the surgeon's chance. The results were poor, since more than 50 per cent proved to have silent lung metastases at the time, which were ultimately fatal. Improvement came only when cytotoxic chemotherapy was added to supplement surgery and radiation, to control these latent metastases, and when cases were concentrated and managed at specialist centres.

At operation, the surgeon removes the kidney with its associated vessels and nodes. If nodes appear to be involved, he also carries out a block dissection of paraortic nodes on the affected side, from the diaphragm above to the bifurcation of the aorta below. He examines the abdomen for malignant spread—liver, bowel, pelvis etc. It is very helpful if he outlines the limits of the tumour bed by radio-opaque clips for the guidance of the radiotherapist in laying out the radiation fields.

Radiation technique. Radiation is begun as soon as possible, and in children it is not necessary to await wound healing. Cobalt beam or megavoltage should be used. A pair of parallel fields, antero-posterior, covers the tumour bed and regional nodes (Fig. 28.1). They extend from the dome of the diaphragm to the sacral promontory, and cross the midline to include the whole width of the vertebrae. The other kidney is carefully shielded, with the aid of the IVP films. The typical dosage is 3250 rad in four weeks.

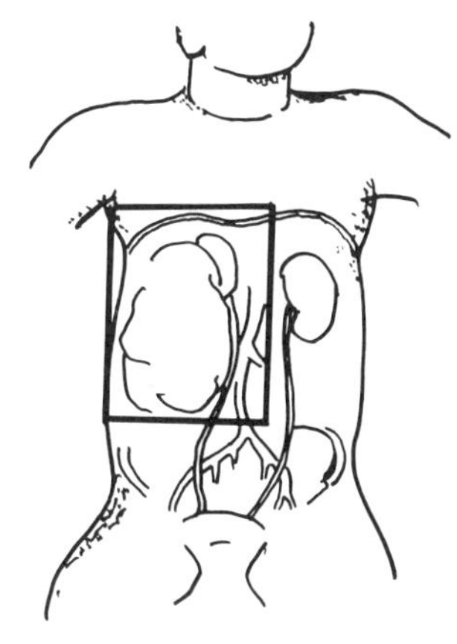

Fig. 28.1 Anterior field of megavoltage pair for nephroblastoma (Wilm's tumour)

An alternative technique, if there is evidence of spread, is an abdominal 'bath' treating the whole of the peritoneal cavity, from pelvic floor to diaphragm. The femoral heads are shielded, so as not to interfere with bone growth. The opposite kidney is shielded from the start—any involvement here is left to be dealt with by the subsequent chemotherapy. A depth dose of 3000 rad is given in five weeks.

Pulmonary secondaries should be treated by chest baths, irradiating all the lung tissue. The shoulder joints are excluded. A tumour dose of 2000 rad is given in three weeks. Even if lung secondaries appear localised, it is unsafe to treat less than the whole chest.

Radiation reactions must be watched—nausea, vomiting, white cell and platelets falls. The dosage and overall time may need adjustment accordingly. Impairment of growth of the vertebrae may result in slight but insignificant shortening of height. Formerly, when the border of the treatment field was taken only up to the midline, only half of any vertebral body was affected; this could lead to unbalanced growth and distortion, with resultant scoliosis.* If the whole vertebral width is included, as it should be, this inequality of bone growth is avoided; but scoliosis can still occur from fibrosis and contracture of the muscles of the back on the treated side, especially if additional supplementary dosage is given (as some advise) to the tumour bed locally.

If abdominal baths are applied in girls, amenorrhoea and sterility will follow. This should be explained to the family.

Cytotoxic chemotherapy. The chief purpose of *adjuvant cytotoxic therapy* is to destroy silent secondaries in their microscopic stage. There are various good regimes, and treatment can even begin on the day of operation. One regime is vincristine (intravenous) once a week for six doses, then once a fortnight for the next two years. It has little effect on bone marrow, but the chief toxic effects are on the nervous system—see page 280.

Another good drug is actinomycin-D. Adriamycin is now also under trial. Agents have usually been given singly, one or the other. The latest evidence goes to show that their effect is even better when combined, and multi-drug regimes will almost certainly be the rule in future.

Prognosis. Younger children tend to do better than older, but this is because the older child tends to have a more advanced lesion. Formerly, the average five-year survival was about 30 per cent, and about 50 per cent for children under two. Now, at the best specialist centres, rates of 80–90 per cent are being achieved, giving Wilms's tumour a better prognosis than any other childhood cancer.

Rhabdomyosarcoma*

This is the commonest of all the childhood soft tissue sarcomas. It occurs at two main sites: (1) head and neck, including orbit—80 per cent, and (2) lower urinary tract. It is highly malignant, grows rapidly, invades locally and spreads widely via lymph and blood.

Primary treatment is usually by *surgery*, but even radical removal often fails, because of local recurrence or distant metastasis. Post-operative *radiation* should be given—5000 rad in four weeks if possible.

Till recently, long-term survival was rare, but intensive *chemotherapy* is now improving the prognosis considerably. Surgery and post-operative radiation are used, plus a regime of triple chemotherapy using vincristine, actinomycin-D and cyclophosphamide in regular pulses. This begins immediately on diagnosis, continues throughout the course of radiation, and carries on for two years. Adriamycin is now also in use.

Complete regression is now achieved in most cases, and mutilating surgery is no longer justifiable. It is too early to assess the long-term prognosis with the new regime, but current results are very encouraging.

Retinoblastoma

This is a rare tumour, but of great interest as one of the very few hereditary growths and the only genetically determined neoplasm that is malignant from the start. It is caused by a gene mutation which is dominant, so that parents who have been cured of retinoblastoma are very likely to find the disease in their children. Most cases, however, appear in children with no such family history, i.e. as a result of new spontaneous mutations.

It arises from primitive retinal cells and forms flat or projecting masses which invade the retina and other tissues of the eye, eventually destroying vision and in extreme cases invading orbital soft tissue, bone etc. The chief route of spread is along the optic nerve, to the cranial cavity and brain. Distant metastasis is late, to nodes, lungs etc.

Clinical features. Most cases are unexpected, and discovered by the parents at an average age of two, rarely after six. There may be a squint, or an opaque light reflex ('cat's eye reflex') in the normally dark pupil, followed by loss of vision. These cases are always locally advanced. In 25–30 per cent of cases, growths are bilateral, and the other eye should be examined ophthalmoscopically at regular intervals after the first eye has been treated, to detect new lesions at an early stage. If there is a family history of retinoblastoma, all children should be examined from infancy regularly up to the age of five.

Treatment. For the typical new (i.e. advanced) case, there is no hope of useful vision, even though the growth is radiosensitive, and the eye should be removed surgically. As great a length as possible of the optic nerve is removed with the eye, and carefully examined microscopically. If the nerve is found to be invaded up to the cut end, there will clearly be residual tumour in the remaining part, and post-operative radiation should be given to the orbit and the region of the nerve stump. A cobalt beam wedge pair is suitable, giving 4500 rad in four weeks.

The situation is very different with a small growth discovered in the other eye, or in a child examined because of the family history. Here every effort should be made to preserve not merely the eye, but also useful vision. Radiation is here the treatment of choice, since the growth is radiosensitive. The most favourable situation is the posterior part of the retina, for the anterior part with the lens can then be spared. One technique is external radiation by a single lateral megavoltage field carefully aimed behind the lens, and directed 5° posteriorly to avoid the opposite eye. A tumour dose of the same order as above is given. Immobilisation of young children may be a problem; it may be necessary to give a soporific drug to put the child to sleep. If so, treatment is preferably three times a week instead of the usual five, and 10–15 per cent must be deducted from the total dosage to compensate for the larger fractions in the same overall time.

Another technique is shown in Figure 28.2. The external surface of the eyeball is exposed surgically, and a specially prepared disc holding radioactive cobalt is sutured in position, overlying the tumour mass. The exact position of the growth must be determined by previous ophthalmoscopy. The disc is removed at a second operation about a week later. There is a range of applicators,

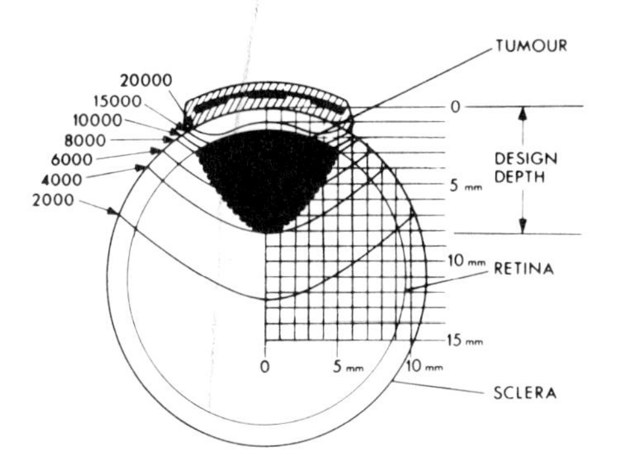

Fig. 28.2 A Cobalt-60 ophthalmic applicator in position at the back of the eye for treatment of a retinoblastoma (shown black). (Courtesy of the Radiochemical Centre, Amersham, England)

circular, semicircular and crescentic, made of platinum loaded with cobalt. They are calibrated to give a minimum tumour dose of 4000 rad; the highest dose, at the surface of the eyeball will be over 20 000 rad, but this is well tolerated. This can be very satisfactory treatment for a small solitary lesion, but since growths are commonly multicentric, i.e. more than one in the same eye, external radiation is more widely applicable.

Prognosis. For a rare tumour like this, and the need for specialised and meticulous technique, it is advisable for treatment to be concentrated at a very few centres in any country, so that adequate experience can be gained. Prognosis depends very much on the stage of advancement at the time of treatment. The aim of any treatment except surgery is preservation of useful vision—otherwise surgery would always be the method of choice. Postoperative radiation for known residues is certainly useful. Most small tumours at the back of the eye can be controlled, with useful vision—up to 90 per cent. This falls to 30 per cent for multiple growths or in the anterior part of the eye.

Medulloblastoma—this is dealt with on page 255.

29. Non-malignant disorders

Cancer is not the only disease process where radiation can be useful. It was inevitable that such a powerful weapon should be widely explored and applied in the early days, before its dangers and limitations were—or could be—realised. There is, in fact, scarcely a disorder in the textbooks for which radiotherapy has not been tried by some enthusiast in the pioneer days, and unfortunately a good deal of damage—even fatal damage—was done unwittingly before the serious consequences emerged.

A striking example came to light only a few years ago. Radiation of an enlarged thymus* in infancy used to be practised, and it caused gratifying shrinkage. Years later, however, carcinoma of the thyroid developed in some cases, undoubtedly induced by the effects of the rays on the immature thyroid cells.

Where treatment appeared to be useful, e.g. in many skin disorders, repeated dosage was commonly given for recurrent trouble, since the cumulative effects of even low doses were for long not appreciated.

Nowadays the field for non-malignant treatment has been severely restricted, both because of accumulated knowledge of the long-term effects on the various organs and the development of modern specific therapy such as antibiotics.

Mode of action. This is not fully understood, but the following mechanisms are probably at work:

1. In *inflammatory conditions*—breakdown of white blood cells and liberation of enzymes and antibodies; suppression of immune reactions (lymphocytes); promotion of more rapid 'ageing' of inflammatory cells, thereby accelerating the healing process.
2. *Reduction of excessive secretion* from glands—endocrine (thyroid, ovary, pituitary), sebaceous and sweat glands of skin.
3. *Reduction of cellular infiltration*—e.g. in eczema.
4. *Epilation**—destruction of hair follicles.
5. *Anti-pruritic*—the mode of action is quite obscure.
6. *Reduction of benign overgrowth*—e.g. keloids, warts, angioma, hyperkeratosis.

We will outline the chief situations where radiation may still be of value.

Dermatology

Twenty-five years ago it was possible to say that X-rays were the most important therapeutic weapon of the dermatologist. With the advent of antibiotics, steroids, etc., this is no longer true, but radiation is still useful and large skin departments commonly have their own X-ray machines; otherwise their patients are treated in the radiotherapy department.

Since great penetration is not required—and would in fact be a positive drawback—low-voltage superficial X-ray units are used. The most superficial type of all—*Grenz-ray* units (see Ch. 3)—are particularly useful, since the very low penetration of Grenz* rays gives them some notable advantages:

The epidermis can be treated with very little effect on deeper structures such as hair roots and sweat glands. It is difficult to cause serious skin damage, and late cancer production is unknown. There is no danger to underlying eyes, testes, bone etc., no risk of epilation, and treatments can be repeated many times.

If superficial X-ray machines are used, a range of voltages is available, from about 40 to 140 kV, and a change of filter is usually needed at each change of voltage. If changes of kV are to be made in the course of the day's work, great care is necessary to ensure that the correct filter is used on each occasion. Mistakes can very easily occur, leading to incorrect dosage. It is a matter of history that such accidents have led to serious trouble, damage to patients, and lawsuits. It is most desirable that some fool-proof warning system be built into the circuit, to make incorrect combinations of kV and filter impossible (see p. 21).

Figure 18.1, page 187 is a comparative summary of the depth doses achieved by the various qualities of radiation. The depths of the important structures are seen to be:

Basal cells of epidermis	1.0 mm
Sebaceous glands	1.5 mm
Hair roots	3.0 mm
Sweat glands	3.4 mm

When X-ray treatment is used in skin disorders, it is usually as a last resort, when other treatments have failed. Dosage is usually quite low—e.g. 50–100 rad at weekly, fortnightly or monthly intervals. Accurate records must be kept, since relapses are common. *It is a good rule never to exceed a total of 1000 rad at any site during a patient's lifetime*.

Gonads. When irradiating near ovary or testis, the possibility of damage, including gene mutations, must be appreciated. A 10 cm diameter field at 15 cm FSD, 100kV (HVL = 1 mm Al) over the pelvis will deliver 6 rad to the ovary at 8 cm depth (for 100 rad on the skin) or 8 rad at 140 kV—a most undesirable amount. The testis, lying just beneath the skin of the scrotum, is even more vulnerable. For such cases, Grenz-rays would be preferable.

Some examples in dermatology:

1. *Angioma*. The common *strawberry naevus*—a birthmark of infants—may grow rapidly and alarmingly for weeks. Almost all regress spontaneously within a few years and the *best policy is masterful inactivity*. The few indications for active treatment are: (a) medical—if it is causing trouble, such as feeding difficulty or respiratory obstruction, or threatens to ulcerate, and (b) psychological—to appease a worried mother who will not be reassured unless 'something is done'.

Possible treatments include—surgery, injection of coagulant chemicals, electro-coagulation, and freezing; or radiation, which is the simplest to apply. The speed of resolution can certainly be accelerated by treatment, and if the angioma is seen before its growth has reached its peak (which is not usually the case), a dose of superficial X-rays can be given, 400 rad. Care must be taken not to irradiate the gonads, epiphyses of bone, or the thyroid; if these are nearby, non-radiation methods are preferable.

Deep X-ray therapy may occasionally be needed, for deeper lesions, e.g. in the pharynx. Adequate sedation, even general anaesthesia, may be necessary to keep the child still on the treatment couch.

Port-wine stains (capillary angioma) differ in not disappearing spontaneously, and being usually unresponsive to X-rays.

2. *Keloid** is an overgrowth of fibrous tissue after healing of a wound, forming a thick ugly scar. It is common after burns. Apart from its appearance, there may be unpleasant itching—even pain. Modern therapy usually begins with injections of cortisone preparations locally. Alternatively, or if injections fail, radiation can be used —e.g. a single dose of 1000 rad carefully confined to the keloid. This can flatten an early vascular fleshy mass, but not after it has fibrosed. An old keloid can be excised, and the line of excision irradiated to prevent recurrence. The same warnings about proximity of immature thyroid and gonads should be remembered, and radiation to neck or pelvis is always best avoided in young people. Moreover, spontaneous regression is common.

3. *Warts*, especially on the sole of the foot, are common. They are epidermal overgrowths caused by an infective virus. If the usual dermatological treatment fails, X-rays are effective in single doses of 1000 rad. But necrosis, though very rare, can follow, and radiation is better avoided.

4. *Hyperkeratosis.** This is a little warty nodule on the skin and may be *premalignant*—e.g. pitch-warts on workers' hands or face (p. 186). Tiny nodules may be given a single dose of 2000 rad, larger lesions may be treated on the same lines as epithelioma, especially if there is any doubt about their malignancy.

5. *Psoriasis** is a common disease whose cause is still unknown. Small or large areas of skin may be involved by brownish patches covered with scales. X-rays in doses of 50–100 rad will usually cause the patches to disappear. But recurrence is the rule, and patients may go from place to place for repeated treatments and end up with radiation dermatitis and even malignancy. Radiation is frowned on, except for special circumstances, e.g. lesions of palms and soles, or the nail bed, where the usual treatments tend to fail.

There are many other skin conditions where radiation has been and sometimes still is of use—eczema, acne etc. etc.—but it would be pointless to attempt to go into detail. X-ray therapy is becoming of decreasing importance.

Endocrine* disorders

The excessive secretion of an overactive gland can always be repressed by adequate radiation. Some examples:

1. *Pituitary* (see also p. 257). In eosinophil adenoma, radiation usually reduces the pressure symptoms, but the hormonal effects (acromegaly) rarely improve, since the structural damage cannot be undone. In basophil adenoma, the tumour is usually microscopic, and commonly associated with an adrenal tumour. If there is no demonstrable adrenal tumour, radiation may be given to the pituitary, since it is simple and carries no operative risk. In case of failure, surgery is still available.

2. *Ovary*. The ovary, beside producing the ova or egg-cells which unite with the sperm to form the next generation, secretes female sex hormones (oestrogens and others) which control secondary sexual characters and the various cyclical changes in the female organism. Towards the age of the menopause this ovarian function may become irregular and cause intermittent excessive uterine bleeding (menorrhagia*). If medical treatment fails, surgical hysterectomy is often advised. An alternative is to suppress ovarian hormonal secretion by radiation. This can be applied in two ways:

(a) *Internal*—an intrauterine tube (radium or one of its substitutes), as used in treatment of uterine cancer, will irradiate and inhibit both the endometrium* and the ovaries. A 50 mg tube left in for 40–48 hours will deliver about 500–600 rad to the ovaries. General anaesthesia is needed, and this provides the essential opportunity for a thorough pelvic examination, uterine curettage* ('scrape') and biopsy, to exclude malignancy as the cause of bleeding. This technique is very effective and still used, though it can lead to some degree of endometritis and uterine discharge.

(b) *External*—by X-rays with simple antero-posterior

fields. This is a much simpler technique, but it is always important to exclude malignancy first. With megavoltage, fields about 14 × 12 cm are adequate, and a single depth dose of about 450 rad. The dose needed to suppress ovarian function is not a fixed quantity, but depends on age and individual variation. A younger patient will need a higher dose than an older. At age 35, it would be preferable to give 1250 rad in five fractions rather than a single dose.

Induction of an *artificial menopause* in this way involves *sterilisation* of the patient, since formation of ova is also inhibited. In almost all cases, this raises no problems and is in fact desirable. It is wise, however to obtain the written consent of both the patient and (if possible) her husband, who should be asked to sign a form stating that the effects of the treatment have been explained, understood and accepted.

In former days low-dose ovarian radiation was used for e.g. dysmenorrhoea (painful periods) and even sterility, as well as inflammatery pelvic disease. This cannot be too strongly condemned, in the light of modern knowledge of gene mutations and the potential danger to succeeding generations.

Radioactive iodine and the thyroid gland

For the physical and technical aspects relating to this section reference should be made to Chapter 10.

The thyroid gland synthesises the hormones thyroxine (T4) and tri-iodothyronine (T3). which are essential for the maintenance of normal metabolism, utilising iodine from the diet. As with other endocrine function, the activity of the thyroid is controlled by the pituitary through the thyroid stimulating hormone TSH (see p. 206). By substituting radioactive iodine, this unique affinity for the chemical provides a useful method of producing a high dose of radiation within a very small volume of tissue.

Thyrotoxicosis (hyperthyroidism)

This is a clinical state associated with raised levels of circulating thyroxine (T4) and tri-iodothyronine (T3). As a result there is an increase in the body metabolism manifested by agitation, palpitations, profuse sweating and weight loss. In some cases there is protrusion of the eyes (exophthalmos). The disease affects females more commonly than males and usually in the 25–45 age group. There is sometimes a significant enlargement of the thyroid gland (goitre*). Occasionally when a large gland extends behind the sternum (retrosternal extension) there may be evidence of obstruction of the trachea.

The high iodine affinity provides methods of diagnosis as well as treatment.

*Myxoedema.** (hypothyroidism) is a state of thyroid hormonal deficiency, causing sluggish metabolism, lethargy, loss of hair etc.

Investigation of thyroid function

The important *in-vivo* tests of the thyroid function are those associated with the uptake of radioactive iodine (^{131}I, ^{132}I) or other isotopes (technetium-99m). These may provide a measure of the physiological activity of the gland (tracer techniques) or provide a measure of size and position (imaging or scanning). Both of these tests require sophisticated apparatus and are *in-vivo* (i.e. performed on the patient) but recent tests—which are biochemical—can be performed in the laboratory (*in-vitro*) on samples of blood. In some cases it is necessary to perform both types of tests to establish a firm diagnosis.

The tracer investigation

As soon as radioactive isotopes became available, around 1950, this quickly became a routine procedure which was particularly useful in separating doubtful cases from, for example, anxiety neurosis. It can readily be done on an out-patient basis.

The patient attends in the isotope laboratory before breakfast and takes a very small dose of ^{131}I (15μCi) diluted in water and drunk through a straw. He may return home for breakfast and return back four hours later. He then lies on a couch, and a counter above the neck measures the radiation emitted from the gland. This gives a measure of the amount of ^{131}I which has been absorbed by the cells of the thyroid from the test drink, following its passage from the gut into the bloodstream. The count is repeated at 48 hours, when a blood sample may be taken, and the ^{131}I content of the blood proteins estimated.

Interpretation of the tracer test

The chief relevant details are given in Chapter 10, including Figures 10.1 and 10.2, pp. 110, 111.

1. *Gland*. Normally the measured radioactivity rises gently to a plateau, reached soon after 24 hours, but the toxic gland removes iodine from the blood at such a rapid rate that the ^{131}I curve rises to a sharp peak in a few hours and then falls. Thus the measured four-hour uptake will distinguish most thyrotoxic cases quite definitely.
2. *Plasma*. Normally the blood is cleared of iodine gradually, and at 48 hours very little is left, but in the toxic cases there is a secondary rise due to secretion of newly formed hormone, incorporating some of the ^{131}I. The 48-hour figures for protein bound ^{131}I give almost conclusive evidence in most cases.

The thyroid scan

An image of the gland may be obtained by giving the patient a small quantity of radioactive iodine or technetium (see above) and using a special detector.

The rectilinear scanner will 'look' at the activity within

the gland in automated cm steps and produce a composite colour picture of the size and shape of the gland with areas of high (red) and low (blue) activity clearly defined (see Fig. 29.1).

Variation in shape and activity are easily detected and the test is invaluable in revealing retrosternal goitre and particularly in investigation of thyroid cancer.

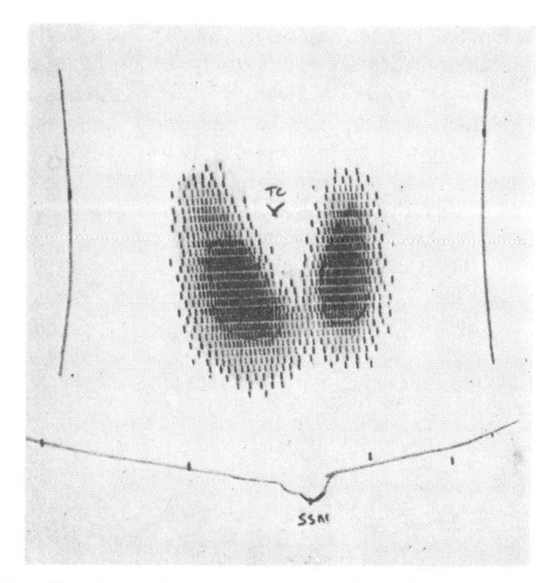

Fig. 29.1 A thyroid scintiscan. In a colour picture the darker central area would be red, the outer area blue.

'In-vitro' tests (see p. 113)

1. Measurement of circulating thyroxine and tri-iodothyronine by chemical means (as distinct from the radioactive test).
2. Measurement of TSH. The level is low in thyrotoxicosis, as other mechanisms of thyroid stimulation are in operation in this disease.
3. T3 Resin uptake—which reflects the amount of hormone binding to TBG (Thyroid Binding Globulin).

In-vitro tests are now more important than tracer studies, so that the administration of radioactive material to the patient can very often be avoided altogether.

The treatment of thyrotoxicosis

A small proportion of patients with thyrotoxicosis will improve spontaneously, but active treatment is required in most patients. Essentially, treatment is concerned with the reduction of circulating thyroxine, either by partial destruction of the gland or by interrupting the synthesis of the hormone.

Destructive therapy may be acheived either by surgery (removal of most of the thyroid gland) or by using the special radioactive iodine in carefully measured doses. Both methods have a place in the management of hyperthyroidism, but radioactive iodine therapy has the advantage of being relatively simple and devoid of serious side-effects and complications. Medical control is achieved by the use of drugs, such as Carbimazole and Propylthiouracil. Drug therapy may rarely be complicated by serious side-effects but is usually used as a preliminary method of treatment before surgery. It may be especially important where surgery or ^{131}I cannot be used because of other medical conditions, poor general condition or pregnancy.

Radioactive iodine therapy

The technique is simple. A dose calculated from the size of the gland and the uptake of radioactive iodine (in the tracer test) is given to the patient under the same conditions as for the diagnostic test. Unless the patient is ill, this can be arranged on an out-patient basis—although the patient should stay at the hospital for a few hours until the possibility of vomiting and other side-effects has subsided. The size of the therapeutic drink is of the order of 2–8 mCi which is usually sufficiently low to avoid the need for special precautions to prevent contamination and ensure the safety of staff. If the disease is not controlled by a single treatment, then more doses may be necessary, having allowed up to 12 months to elapse in order to assess the full effect of the first dose.

The drawback of this treatment is the progressive effect of the irradiation, which frequently leads to increasing atrophy of the gland with a slow decrease in the amount of circulating hormones, which can ultimately produce hypothyroidism—the clinical syndrome of myxoedema.

This treatment is usually restricted to patients over 40, because of the theoretical risk of inducing leukaemia or thyroid cancer—though this has never occurred after more than 20 years experience. There is also the possibility of gene mutations in the reproductive cells of ovary and testis.

'Rheumatic' disorders

This is a very mixed group, of obscure pathology—some inflammatory, some degenerative, some with an immunological basis. X-rays were once used extensively, but have now been largely given up.

Ankylosing spondylitis.** This is a disabling process, mostly in young men, with pain and stiffness, leading to 'poker back' rigidity. It usually begins in the sacro-iliac joints, spreading later to the small joints between the vertebrae, and the costo-vertebral joints (where the ribs are attached to the spine). The chronic inflammation leads eventually to ossification of ligaments, fusion of joint surfaces and bony ankylosis. Radiation usually gives dramatic pain relief and enables the patient to return to work. Until 1955, deep X-ray therapy was widely and successfully used as the mainstay of treatment directed either to sacro-iliac joints alone, or the whole spine. Doses up to 3000 rad were applied. Then it was discovered that a small proportion of treated patients developed *leukaemia*, undoubtedly a late radiation effect and comparable to the leukaemia induced by radiation from atom bombs etc. The risk is small, one in several thousand, but 5–10 times that of a normal population.

This led to the abandonment of the treatment by most radiotherapists, in favour of general measures, physiotherapy and drugs. But there are still a few cases where, if other methods fail, localised X-ray therapy can be quite justified for relief of a painful and crippling disorder. (And drug therapy also carries some risks of damage.) The slight risk should be fully explained to the patient, and the final choice left to him—more often than not, he will consider it an acceptable hazard.

Dosage should be of the order of 1000 rad in 7–10 fractions, daily or alternate days.

Peptic ulcers

The symptoms of gastric and duodenal ulcers are relieved by counteracting or abolishing the acid secreted by the gastric mucosa. Treatment is by medical and/or surgical measures. If conservative treatment fails and if surgery is contra-indicated by age etc., radiation to inhibit or reduce the acid secretion is worth a trial.

The technique is simple. The stomach is localised (on the simulator etc.) with a barium swallow, and antero-posterior fields about 10 × 10 cm outlined. A megavoltage course of 1500 rad to the stomach is given in 10 daily fractions.

Ophthalmology

The chief indications for radiation in benign eye conditions concern lesions of the cornea (the transparent window). The cornea itself is 1 mm thick and is about 3.5 mm distant from the front of the lens. Any radiation applied must be very superficial, and avoid damaging dosage to the lens. These conditions are fulfilled by the applicator shown in Figure 8.11. It is shaped like a contact lens, curved to fit the surface of the eyeball. The radiation source is *radioactive strontium* (Sr-90)—see page 91. The radiation consists of beta-rays alone. Various models are available. The active source may be circular and central, as illustrated, to treat the whole of the cornea, or eccentric to treat a segment only, or annular (ring-shaped) to treat the periphery of the cornea. Other models are of smaller dimensions, with long handles so that they can be held by hand on the part of the eye to be treated. A typical dose rate at the surface is about 100 rad per minute. The eye is anaesthetised by instilling a few drops of cocaine-like solution and the shell is slipped under the eyelid.

Examples of its use:

Corneal vascularisation. The normal cornea is completely devoid of blood vessels. Ulcers or infections may lead to abnormal extension of tiny vessels from the surrounding conjunctiva to heal the lesion (Fig. 29.2). But the new vessels are liable to persist, even after their purpose has been served, and with them any corneal scar is liable to persist also. If they can be obliterated, fading of the corneal opacity is encouraged. Localised radiation can produce a reaction that seals the vessels, and this is the object of therapy. Small leashes of vessels may be given 500 rad weekly for four or five treatments.

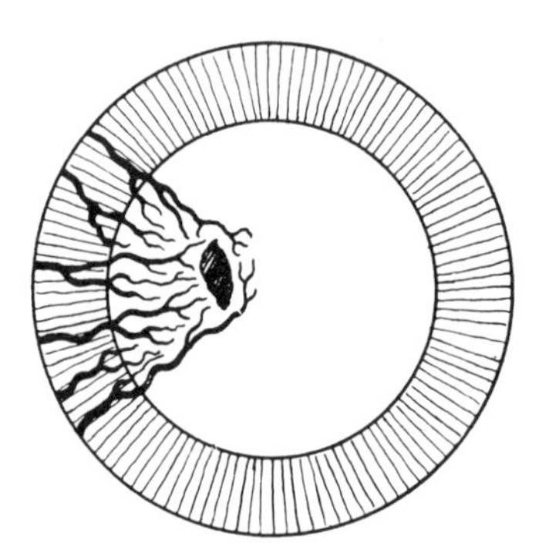

Fig. 29.2 Vascularisation in corneal ulcer.

Corneal grafts. An opaque cornea can be excised surgically and replaced by another cornea which is grafted on in its place. The success of the operation is, however, liable to be ruined by the associated inflammatory reaction, with vessels growing in from the margin, leading to opacification of the grafted cornea. Such vessels can be attacked in the way just described, with similar dosage. An annular source is most useful. Treatment may be pre-operative, post-operative, or both, and can make all the difference between success and failure. A typical course is of four weekly applications, giving 600 rad at the surface each time.

Neoplasms. It is occasionally possible to treat a small neoplasm on the conjunctiva—e.g. melanoma—by a strontium plaque, giving 2500 rad once a week, to a total of about 15 000 rad.

A good alternative source of superficial radiation on the eye is Grenz-ray therapy (see Dermatology above).

30. Hormones in oncology

Hormones in cancer have two distinct but related aspects: (1) their role in the causation of certain forms of cancer, and (2) their value in controlling already established cancers.

Historical development

The relationship between hormones and tumours is not a new concept. An early milestone was the work of a Scottish surgeon Beatson in 1896 (i.e. about the same time as the discovery of X-rays and radium). Basing himself on veterinary work on cows which demonstrated a relation between the ovary and milk production, he removed the ovaries of two young women with breast tumours and found the tumours then regressed. It was thought at the time that the mechanism involved removal of some 'ovarian irritation' responsible for the breast lesions, but recurrence soon took place, and though others repeated the procedure, the results were irregular and improvements only temporary, so that the operation fell into disuse and its full significance was not appreciated.

Later animal work drew attention again to hormonal factors—e.g. the incidence of spontaneous breast cancer in mice was reduced by ovariectomy. Other striking experiments produced breast tumours in mice by prolonged administration of ovarian hormones. Clinical interest revived when in 1941 Huggins in U.S.A. reported the favourable results of orchidectomy on cancer of the prostate (see p. 244), and a spate of work has followed ever since, involving the hormonal activity of the gonads, adrenals, pituitary, thyroid, etc.

Experimental induction of neoplasia by endocrine manipulation can easily be achieved in laboratory animals. The essential mechanism seems to be a *prolonged overstimulation of the target issue* by virtue of its physiological action, and not by acting as a crude irritant like a carcinogenic chemical. We cannot apply the results of these experiments directly to human cancer, since the experimental endocrine disturbances are more severe and prolonged than ever occurs in humans.

Of human cancers, hormones seem likely to be somehow involved in those of the prostate, breast, uterus, thyroid and possibly kidney. Many other organs are governed by hormones—e.g. ovary, testis, adrenals—but hormone therapy has so far not been found to influence any growths of these tissues.

Hormones as therapeutic agents

It is a most remarkable fact that hormones should be able to influence an established cancer at all. It seems to run counter to the conventional idea of cancer as a completely independent growth process, 'running wild' and submitting to no restraint. This is of the greatest interest and importance to modern theories of the essential nature of cancer, and we now regard the boundary between cancer and non-cancer as very much less clear-cut than we used to think.

There are only a few human cancers where hormone therapy is of practical value, and these are in organs where hormonal stimulation and control are known to be important—mainly prostate, breast, thyroid and body of uterus—and only a modest proportion of these do actually respond as a rule. Remissions, when they occur, may last for weeks, months or (exceptionally) years, and the relief in patients who do respond may be substantial and even dramatic.

Hormone therapy is superior to surgery, radiation or cytotoxic drugs, in that it does no significant damage to normal tissues and is capable of influencing widely disseminated growths.

Unfortunately, the effects are *temporary*, *never permanent*, the growth escapes eventually from control, and recurrence sooner or later is inevitable. Moreover, hormone therapy is almost always accompanied by unwanted *side-effects*, which are sometimes objectionable and may even be dangerous. Clearly we have to balance these against the potential benefits, and in practice it may be necessary to stop treatment because of this.

Hormone therapy involves changing the existing hormonal pattern or balance in a patient. There are, broadly, two main ways of achieving this:

1. *Addition*—administration of hormone preparations, by oral or intramuscular (IM) route
2. *Subtraction*—surgical removal or radiation inhibition of endocrine glands, to remove the supply of particular hormones

Addition of hormones

1. *Female hormones* of ovarian origin:

    a. *Oestrogen**—is responsible for enlargement of breasts at puberty and at menstrual cycles, and for maturation and maintenance of secondary sexual characters and sexual organs

b. *Progestogen**—its primary function is to maintain the uterus with its implanted fertilised egg during the early months of pregnancy

Oestrogens. Preparations include:

Stilboestrol*—the most frequently used of all, a synthetic oestrogen.
Ethinyloestradiol—one of several preparations alternative to stilboestrol.
Premarin—brand name of a preparation of natural oestrogens derived from pregnant mares' urine. Useful if others cause sickness.

Side-effects of oestrogens occur in a third of patients:

a. Nausea and vomiting—especially from stilboestrol
b. Fluid retention—which can precipitate pulmonary oedema and congestive heart failure; a low-salt diet helps to combat this
c. Enlargement of breasts
d. Deepening of pigmentation of nipple and areola.*
e. Uterine bleeding, from stimulation of the endometrium, even long after the menopause. This is also liable to occur if the oestrogen is stopped abruptly; dosage should be tapered off gradually. The patient should always be warned of possible bleeding to avoid alarm
f. Hypercalcaemia (in excess of the normal 9–11 mg of calcium per 100 ml), due to increased mobilisation of calcium from bone. This can also occur apart from oestrogen therapy. It causes decreased neuromuscular excitability, with weakness, anorexia, constipation and generalised pain. The pathology is obscure, there may or may not be osteolytic secondaries, and it is possible that the growth itself produces a substance which acts like parathyroid hormone in releasing calcium from bone. At high levels (16–20 mg), fatal coma and anuria can result.

Progestogens. Preparations include:

Hydroxyprogesterone (Primolut-Depot).
Medroxyprogesterone (Provera).
Norethisterone (SH420).

Side-effects: fluid retention is possible.

2. *Male hormones or androgens** which cause masculinisation. Preparations include:

Testosterone propionate
Fluoxymesterone (Ultandren)

Side-effects: fluid retention; virilisation after three months on treatment, especially hirsutism (growth of hair on face etc.), deepening of voice, increased libido.
Androgens also have a protein-anabolic action and the allied anabolic steroids also have some androgenic effects:

Nandrolone phenylpropionate (Durabolin).
Nandrolone decanoate (Deca-durabolin).
Drostanolone (Masteril).

3. *Adrenal hormones* have complicated effects on electrolyte and water balance, carbohydrate and protein metabolism. They are called *corticosteroids* (from adrenal cortex).
In cancer therapy the most commonly used preparation are:

Prednisone and Prednisolone.
Dexamethasone.

Side-effects are important and potentially serious, especially after prolonged dosage:

a. Exaggeration of normal physiological action, causing hypertension, muscle weakness, diabetes, fluid retention and a Cushing-type syndrome (see p. 257) including 'mooning' of the face due to local oedema.
b. Euphoria—a mild degree of this is valuable, but there may be more serious mental disturbance, especially with a past history of mental disorder
c. Modification of tissue reactions, resulting in spread of infections, or failure of wound or ulcer healing—e.g. haemorrhage or perforation may occur from a peptic ulcer
d. Osteoporosis, especially in the elderly, which may cause for example vertebral collapse

Note—patients on steroids should carry cards giving details of drugs and dosage and possible complications. Long-term therapy leads to atrophy of the adrenal cortex. This is potentially dangerous, since the adrenals can no longer react to stress by increased hormone secretion, and any acute infection, injury etc. can lead to a state of collapse like an Addisonian crisis. Increased doses must be given in these circumstances, including operative procedures.

Subtraction of hormones

1. *Ovaries.* The chief source of female hormone may be removed surgically—*oophorectomy* (ovariectomy) which is usually a minor procedure, necessitating only a few days in hospital. Alternatively, the secretory activity of the ovaries can be suppressed by a sterilisation dose of X-rays (as for menorrhagia—p. 267); this is simpler, but has the drawback that it takes about three months to become effective
2. *Testes.* The male hormone can be simply abolished by bilateral *orchidectomy**
3. *Adrenals.* The glands may be removed by surgical *adrenalectomy*. If large enough doses of corticosteroids are given (cortisone, prednisone etc.), the glands will atrophy from disuse and this is tantamount to a *'medical adrenalectomy'*
4. *Pituitary.* This is the master endocrine gland, as it controls all the others by its various secretions (Fig. 30.1). Its removal leads to atrophy of ovaries, adrenals, thyroid etc. This may be achieved (though not perfectly) by surgical hypophysectomy,* or the gland may be destroyed by radiation. High dosage is required, which can be attained only by interstitial methods—e.g. gold grains, or an yttrium rod.

   These can be inserted transnasally, with the help of radiographic control with a fluorescent screen. It is a much less formidable procedure than resection, and can be tolerated by a patient unfit for surgery (Fig. 30.2)

   When the gland is destroyed, some of the missing hormones—especially cortisone and thyroxine—must be supplied artificially for the rest of the patient's life.

Hormone therapy in breast cancer

The normal breast is subject to cyclical growth changes at every monthly cycle and every pregnancy. These are responses to controlled stimulation by hormones of ovary, pituitary and adrenals (Fig. 30.1). Whatever the role of hormones may be in the causation of breast cancer—and we know very little about this—there is no doubt about their effectiveness in treatment in many cases.

Hormone therapy is generally reserved for treatment of the late or recurrent case, especially where dissemination has taken place. In practice, *two-thirds of all cases require palliative treatment for the later stages*.

For local palliation of metastases causing symptoms—especially secondaries in bone—radiation is more effective and should be used. Palliative surgery may also have a place—e.g. for recurrent nodules in the skin flaps, or operable masses of cancer in the breast itself. We *resort to hormones when disease becomes too extensive or generalised.* There are exceptions to this rule—e.g. in an elderly patient we may prefer to begin with hormone treatment, even for a localised growth, since many growths at this age respond well, and it may be less of a strain than surgery or radiation.

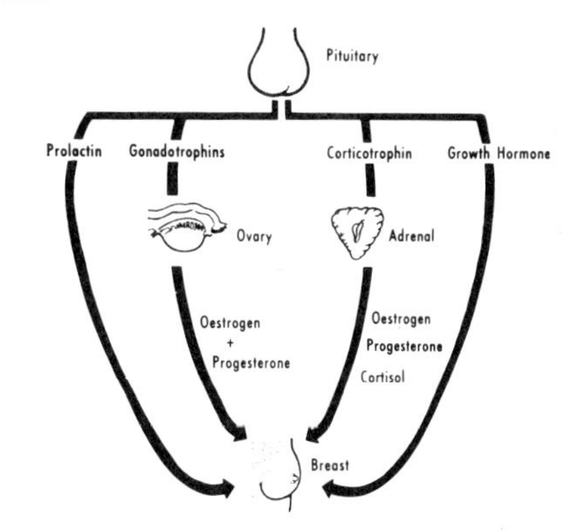

Fig. 30.1 The hormonal control of the normal breast *(Courtesy Prof. Forrest, in Pye's Surgical Handicraft, Wright)*

Fig. 30.2 An yttrium rod implanted in the pituitary through a screw-type trocar. *(Courtesy Prof. Forrest in Pye's Surgical Handicraft, Wright)*

Choice of hormonal method

In general, we begin with the simpler methods first and, if the treatment causes improvement, use it until the growth escapes from control—as it always does eventually. Then we turn to another method in the light of the previous response (or lack of response) and such factors as age, general condition, etc. When the time comes to consider the more drastic methods of operative ablation, we must carefully weigh the reasonable expectation of life against the inevitable discomforts of treatment and the price will often be judged too high.

Unfortunately there is no simple test available to indicate the probable response of a particular patient to treatment, or to pick out the best hormone to use. It is therefore usually a matter of trial and error. Various biochemical tests have been investigated, but none is of great value.

First-line treatment

In the first instance, the best guide is the patient's ovarian hormone status:

1. Pre-menopausal
2. Late post-menopausal, i.e. 10 years or more after the menopause
3. Early post-menopausal

1. In pre-menopausal patients, *therapeutic castration* is the first line in hormonal management. Surgical *oophorectomy* is best, especially if there is any urgency (e.g. multiple bone secondaries, involvement of brain or liver) or if growth is rapid. *Radiation castration* (as described on pp. 267–8) is an effective alternative but takes 2–3 months to work, and this delay is usually undesirable.

   This method succeeds in relieving 20–30 per cent of cases and the remission may last for up to two years. Bone pain is usually relieved, often quite dramatically, and osteolytic lesions may recalcify. Soft tissue lesions usually respond less well.

   '*Medical castration*' by administration of androgenic hormones is a possible alternative but gives inferior results.
2. In late post-menopausal patients, *oestrogens* are first choice—usually stilboestrol if tolerated. In the older patients (60+) the response is often very gratifying, and a large breast mass may melt away and ulceration heal. Bone secondaries respond less well. Good response occurs in a third of cases; the longer the time since the menopause, the more likely is oestrogen to succeed.

   Remission may last up to 18 months.
3. In early post-menopausal patients, the problem is more difficult, since the hormone status is more uncertain. Up to two years after the menopause, *oophorectomy* should still be advised, since some oestrogen continues to be produced by the ovaries. An indication of this can be obtained by cytological examination of a vaginal smear.

   Between two and five years after the menopause, *androgen therapy* should be used. After the fifth year, oestrogens should be given first.

Second-line treatment

When the effect of the above first-line treatment has worn off, and reactivation of growth occurs—or if the treatment fails completely after a trial of 2–3 months—we can fall back on second lines:

1. Administration of hormones—androgens, corticosteroids, progestogens
2. Endocrine ablation—adrenalectomy, hypophysectomy, pituitary implantation

Androgens. If there has already been a good response to castration, androgen therapy will yield a useful improvement in 30 per cent of cases. But if castration has failed, it is unlikely to help.

Where there has previously been a good response to oestrogens, androgen therapy will yield a useful improvement in 20 per cent of cases.

Corticosteroids are valuable even if sex hormone therapy fails. They can give striking relief of bone pain, and are indicated for hypercalcaemia and secondaries in lung, liver and brain. The increased feeling of well-being is also useful. If the patient is too ill for endocrine surgery, steroids are generally indicated.

Progestogens may be useful for soft tissue deposits, but rarely improve bone or visceral deposits. If sex hormone therapy has failed, they are a possible alternative to steroids.

In general, the important thing seems to be to *change the hormonal environment* of the growth and this is probably more important than the precise nature of the hormone used. If a growth responds to, say, oestrogen and eventually begins to grow again, further improvement may sometimes be obtained simply by withdrawing the oestrogen. When further activation of growth occurs later, another useful response may follow administration of oestrogen again.

Anti-oestrogens. Compounds showing antioestrogenic activity have been used to inhibit breast cancer. The most used is *tamoxifen* (proprietary name—Nolvadex). It avoids some of the unwanted side-effects of oestrogens and androgens.

Endocrine ablation

Adrenalectomy (bilateral) is particularly useful if there has been a good previous response to castration, and is more effective than medical suppression with steroids.

Hypophysectomy is also likely to be useful if castration has been effective. The choice between adrenalectomy and hypophysectomy will depend on local experience and the availability of interested surgeons.

Successful remissions are obtained in 30 per cent of treated cases. But the mortality rate is high, 10–15 per cent.

Pituitary implantation, e.g. by radioactive yttrium (Fig. 30.2) is a less severe procedure and may succeed even if there has been no useful response to previous hormone therapy.

Major endocrine surgery has had a considerable vogue in various places, but enthusiasm is declining. Useful results are certainly obtainable in a modest proportion of cases, but the price is high and the quality of remaining life is always doubtful.

Cancer of the male breast

The best form of hormone therapy is surgical castration, the relatively simple procedure of bilateral *orchidectomy*. This is effective in two-thirds of cases, a remarkably high proportion, and double the success rate of oestrogen therapy or oophorectomy in females.

Oestrogen therapy succeeds in only about 15 per cent and is therefore not the method of first choice.

Radiation castration is not practicable in the male. When tumour control ceases after orchidectomy, steroid therapy is more likely to help than adrenalectomy or hypophysectomy.

Summary conclusion

Hormone therapy is a palliative for advanced and disseminated breast cancer, and is effective only in a minority of cases (20–30 per cent). The average period of control in successful cases is 1–2 years and the quality of life can be good. More than one method can be used in sequence.

The more drastic methods of endocrine ablation should be applied only to very carefully selected cases, after weighing up all the factors.

Cancer of the prostate

This is one of the commonest cancers in men (Table 13.2, p. 148). Radical prostatectomy is seldom possible, as few patients are seen early enough. Hormone therapy is therefore the treatment of most cases, and fortunately 90 per cent of prostatic cancers are dependent on hormone stimulation, i.e. androgens, for the gland fails to develop in the absence of androgens. Growth can thus be slowed by reducing the patient's supply of androgens.

This can be achieved either by bilateral *orchidectomy* or administration of *oestrogens*—or both. Most workers begin with oestrogens and proceed to orchidectomy if the response is not satisfactory.

There are many oestrogen preparations; the most often used is Stilboestrol, but if it proves nauseating, one of the others may be better tolerated (see above). Hormone treatment usually results in considerable softening and shrinkage of the growth, with relief of symptoms.

When primary hormone therapy fails, relief can often be obtained by corticosteroids, and this is in practice likely to be more effective than adrenalectomy. Apart from hormone therapy, good palliative relief of obstructive symptoms is available from transurethral resection. Painful secondary deposits are common, especially in pelvic and vertebral bones. These may respond to localised external radiation, but not as reliably as breast cancer secondaries.

Since megavoltage is now widely available, there is a revival of interest in high-dose irradiation of the primary growth, and good palliation of bladder-neck obstruction can be obtained by small-field beam-directed external radiation—see page 244.

Cancer of the thyroid

Hormone therapy has been discussed on p. 206.

Cancer of body of uterus

The endometrium is another tissue subject to hormonal control, and some cases of uncontrolled cancer of the uterine body may benefit from *progestogen* therapy (see p. 272).

Cancer of the kidney

The kidney and uterus are related embryologically, both being derived from the 'urogenital ridge'. This helps to explain why, in a small proportion of cases *progestogens* can help in advanced renal cancer.

31. Medical oncology—Cytotoxic chemotherapy

The subject of hormone therapy might well be included under 'Medical Oncology',* but the heading is widely used to connote the specialised use of cytotoxic (cell-poisoning) drugs.

Chemical methods are as old as medicine itself and nearly as old as mankind. Medical history is littered with a wonderful assortment of anti-cancer agents—extracts of plant and animal tissues, herbs and potions, inorganic and organic substances of diverse kind. They are now mainly of historical interest only, though some slight success was achieved by compounds, especially of arsenic and lead. Even in the 17th and 18th centuries, agents such as belladonna, mercury, antimony and arsenic were used, and in the 19th century, a little over 100 years ago, some definite success was achieved with preparations of arsenic in chronic leukaemias and lymphomas.

The modern story goes back no further than the 1930s, with the chemical identification of carcinogenic substances present in tar. Soon after came the discovery that these substances could also *inhibit* the growth of normal and cancer tissues—just as X-rays and radium can be both destructive of cell growth and carcinogenic. This chemical clue was followed up by testing allied chemical compounds on animal tumours in the laboratory. The turning point came during the Second World War, from the investigation of poison gases. (Even during the First World War sulphur 'mustard' gas, was seen to have, apart from its vesicant action on the skin, severe depressant effect on bone marrow, and leucopenia* was noted in those dying of mustard gas.) War gases were studied again, and in the 1940s the nitrogen 'mustards' were found to cause gene mutations and chromosome damage leading to cell death at subsequent mitosis—just like radiation. These agents, are therefore *radio-mimetic* (i.e. radiation-imitating). Profound effects were observed on the marrow, and cautious trials were begun on cancers involving marrow, with some encouraging results. The stage was now clearly set, and research has since proceeded with gathering momentum. Many thousands of complex chemical compounds have been synthesised and tested on animals. Only a few have proved of value in human cancer, but a new chapter has been opened.

Medical oncology is virtually a new department, and cytotoxic chemotherapy now takes its place alongside surgery and radiotherapy as one of the three major agents in cancer therapy. Some doctors now specialise in medical oncology, just as others do in surgical or radiation oncology. The best results are achieved in centres where there are teams of specialists in close cooperation, working out treatment policies for particular types of cancer and holding joint clinics for assessing patients. In this way research and development are most efficiently pursued, new techniques and drugs evaluated, and the benefits passed on quickly to new patients. In many centres, chemotherapy is practised also by radiotherapists, haematologists, surgeons, physicians and others, depending on local interests and resources. Technicians in radiotherapy departments will not usually be personally concerned with this type of treatment, but should have a general appreciation of the subject, since many of the patients they treat will be actual or potential candidates for cytotoxic therapy as part of the total treatment plan. A course of cytotoxic therapy may even be carried out at the same time as the radiation course, or just before or after.

The subject is new, and current developments rapid. Inevitably and properly, progress had to be cautious and tentative, after experimental work on animals. Trials could be justified only on patients who had failed to respond to established methods (surgery and/or radiation). With increased experience, it became justifiable to use these drugs at earlier stages, and in a few types it is now the treatment of choice, not only for the later stages (e.g. Hodgkin's) but from the start (e.g. acute leukaemia, choriocarcinoma). Examples have already been described where cytotoxic therapy is now an essential part of the total treatment plan—e.g. many tumours of childhood—and more and more cancers will undoubtedly be managed in this way in the future.

In some departments it is—or was—true that cytotoxic therapy tends to be reserved only—or mainly—for the late and hopeless case, sometimes as a gesture to solace the patient or relatives that 'everything possible is being done'. This may have been reasonable years ago when the subject was in its infancy, but we now have enough knowledge to make the therapy more rational and scientific.

Any drug can be misused, and until experience is gained it is virtually inevitable that it will be wrongly used at the start. Dosage and timing are of critical importance, just as in radiotherapy. The 'harmless' aspirin tablets that the public consumes by the ton, can be lethal when a child finds a bottle and helps himself too liberally. Most modern and many other drugs are potent and dangerous and have unpleasant side-effects; they must be strictly controlled if they are to do more good than harm. Cytotoxic drugs are also potent and dangerous, and we are only slowly learning how to use them to achieve the maximum useful effect with the minimum harm and discomfort. But we *are* learning, and our patients are reaping the benefit. Until recent years, cytotoxics were really useful only in leukaemia and some lymphomas, i.e. growths of scattered or disseminated type. It still remains true that most of the 'solid' common tumours (e.g. lung, colon, stomach, cervix) are not responsive to current chemotherapy. But we are now seeing increasing success in prolonging life and comfort and improving the prognosis generally, in many cancers where formerly little could be offered.

Mode of action. The ideal drug would be one which killed only cancer cells and spared normal tissues completely. This is no more possible (at least yet) than it is with radiation. The most important effect of the drugs in use is on cell mitosis, by interference with the synthesis or functioning of vital nuclear constituents—DNA, RNA, nucleic acids, proteins. They are therefore toxic to all cells, malignant or normal—just like radiation. Their use in cancer depends on the fact that conditions of growth in a cancer are abnormal. The cells of a particular cancer *may* have some inherent chemosensitivity, or they *may* go into mitosis more frequently than their normal counterparts. However, many (even most) cancers do not proliferate faster than normal tissues—many in fact grow more slowly. *But they are less capable than normal tissues of repairing cell damage, so that normal cells can recover while cancer cells do not; this is probably the most important factor of all*. Their effects will thus be comparable to Figure 17.1 page 178.

As with radiation, a given dose kills a certain fixed proportion of cells (compare Figs. 17.5, 17.6, pp. 181–2) irrespective of the actual number of cells present. Just as much of the drug will therefore be needed to kill a few cells in a small tumour as to kill many cells in a large mass. Treatment must therefore be continued long after a growth is no longer detectable clinically. At the present time it is not possible to eradicate most 'solid' cancers by drugs, for several reasons—(a) inherent resistance of the cells, (b) failure of the drugs to reach all the cancer cells, (c) our ignorance of just how long treatment should be continued,(d) toxic side-effects which set a limit to dosage.

Dosage and toxicity. Since the effects are indiscriminate, all tissues are bound to be more or less affected. But damage to normal tissues can be minimised, just as we have learned to do in radiotherapy. Intermittent administration of high doses of drugs will result in stepwise permanent losses of tumour cells, while normal tissues still retain their powers of recovery. This is better than continuous therapy with small doses of drugs, which used to be practised. Low doses are harmful as they accelerate drug resistance by eliminating the more sensitive cells and leaving the most resistant cells to flourish. The cell damage shows itself only when the cell goes into mitosis—or attempts to—just as with radiation. *Rapidly growing normal tissues are therefore affected as much as—maybe more than—the cancer itself, and it is this fact that usually sets the limit to practical therapy*. The tissues most affected are those undergoing rapid division:

1. Haemopoietic tissue especially bone marrow
2. Mucous membranes, especially bowel

(These are the chief limiting factors in actual therapy)

3. Hair and skin
4. Gonads—ovary and testis
5. The reticulo-endothelial cells responsible for the immune response

When high-dose treatment is given, patients should be in hospital, and examined daily for toxic effects. Frequent blood counts—haemoglobin, total red and white cells, differential count if indicated, and platelets—should be charted graphically. High fluid intake should be maintained, to avoid bladder damage from products of the drugs excreted in the urine, and to remove the excess uric acid produced by rapid cell breakdown in the tumour, which can be toxic to the kidney.

Bone marrow. Most cytotoxics have profound and critical effects on the rapidly dividing normal marrow, and the resultant drop in red and white cells and platelets can rapidly lead to anaemia, haemorrhage, infection and death. Polymorph levels of 1000 or less can be tolerated for a week or so, but in general the total should not be allowed to fall much below 2000. Platelet counts should not sink from the norm of 200 000 to much below 80 000. Severe anaemia and thrombocytopenia (platelet deficiency) should be corrected by blood transfusion. Antibiotics may be indicated for infection, actual or threatened, but there is always some risk of 'opportunistic' secondary fungal infection, which can be devastating.

Sterile isolation units are not necessary for most regimes, but may be justifiable for special cases, e.g. choriocarcinoma or leukaemia.

Mucous membranes. The alimentary tract, including the mouth, is vulnerable. *Ulceration in the mouth* is often the first sign of toxicity and may occur the next day. The ulcers are typically on the lip margin, shallow and often painful. There may be colicky abdominal pain and *diarrhoea*—especially with 5-fluorouracil. *Bladder ulceration*, with dysuria and haematuria, may occur, especially with cyclophosphamide, from urinary excretion and concentration in the bladder; this is usually due to inadequate fluid intake.

Skin and hair. Alopecia* is to be expected after, for example, cyclophosphamide, methotrexate, adriamycin. Patients must be told what to expect, and their agreement must be obtained. Hair regrows after the end of treatment, or even if treatment continues at lower dosage. A wig should be provided early in the course, especially for women. Some of the older men may be unconcerned, but for younger men a wig is equally good for morale.

Skin rashes occasionally appear, possibly because of hypersensitivity. Erythema may develop, e.g. round a skin metastasis.

Gonads. Spermatogenesis is repressed in the male. In women, prolonged treatment may cause amenorrhoea. Cytotoxics should be avoided in pregnancy—most certainly in the first four months, when there is a high risk of fetal abnormalities and abortion. Gene mutations, inherited by offspring, are a potential hazard in both sexes.

Nervous System. Nausea and vomiting are common. Anti-emetic drugs given an hour beforehand, e.g. prochlorperazine (Stemetil) or metoclopramide (Maxolon) are helpful. Vincristine is particularly liable to cause neurotoxicity—see below.

Immunosuppression. Cytotoxics are used for deliberate suppression of the immune response on special occasions, e.g. surgical transplants of kidneys etc. They can diminish patients' ability to react to infections. The significance of this in cancer is obscure. Some workers are worried that they may damage the body's defence mechanisms against cancer, but the whole subject of the immunology of cancer is so complicated and doubtful that it is impossible to be dogmatic. All treatments carry potential dangers, and risks have always to be carefully balanced.

Other toxic effects include

Daunorubicin is toxic to the heart beyond a certain level of total dosage.

Bleomycin can cause dangerous lung fibrosis, and skin changes.

Busulphan can cause lung fibrosis and skin pigmentation.

Steroids lead to salt and water retention, and can cause diabetes etc.

Resistance to cytotoxics develops eventually—just as it does to antibiotics—and when any member of a group of drugs ceases to be effective, resistance extends to all the other members of the same group. But there may still be a good response to another group. To minimise the effects of resistance, modern *combination therapy* has been developed, using two or more drugs together or in sequence. Recent revolutionary advances have depended largely on this kind of combined attack, and the use of single agents is now not usually justifiable.

The best indication and justification for cytotoxic therapy is the presence of *widespread malignancy. Since so many cases, though clinically localised are actually widely disseminated at the time of diagnosis, it is not surprising that these drugs are assuming ever increasing importance in oncological management.* The correct combination of surgery, radiotherapy and chemotherapy is of literally vital importance. We have still a lot to learn, but progress is definite and encouraging.

Classification of cytotoxic drugs

The available agents may be divided into a few broad groups:

1. Alkylating agents
2. Antimetabolites
3. Plant products } Chromosome inhibitors
4. Antibiotics }
5. Miscellaneous

Table 31.1 gives details of the most frequently used drugs.

1. Aklylating agents

'Alkyl' is the name of the chemical structure common to the alcohols—hence the name. Alkylation is the chemical process of combination of this structure with other chemical groups, including structures of biological importance. These include protein constituents and precursors, especially of the cell nucleus, and the result is gross interference with their functioning in mitosis etc.

These agents are all chemically related to mustard gas—hence the name of the first useful substance developed, *Mustine*. *Mustine* is still one of the most effective of all cytotoxics. It is unstable, and should be dissolved immediately before injection. Like mustard gas, it is intensely vesicant and irritant to skin and other tissues. If injected outside the vein, local pain and possibly necrosis result. Phlebitis and thrombosis may be caused, and the drug is therefore often injected into the tubing of a saline infusion, though this is not essential for the experienced doctor. Nausea and vomiting usually follow within a few hours. To counter this, drugs such as chlorpromazine (Largactil) may be given both before and after the injection. Though injection is best given in the evening, followed by a sedative for the night, most patients receive injections as out-patients during the day.

Side-effects include epilation and temporary amenorrhoea. Excessive dosage, as for most cytotoxics, leads to anaemia, thrombocytopenia, fever and bleeding, which can be fatal. The white cell count begins to fall almost from the start, is at its lowest after a week, and takes about six weeks to recover.

Chief uses:

a. Malignant lymphomas—especially Hodgkin's in later stages. Rapid relief may be obtained in acute toxic cases, and severe symptoms may be controlled much earlier than by radiation. Fever and weakness may subside in a few days, but pruritus is less often relieved.
b. Malignant pleural and peritoneal effusions—may be diminished by intracavitary injection of mustine in low concentration. Some will be absorbed into the

circulation and this must be borne in mind if other therapy is combined.

It is particularly useful where speed of action is important—e.g. mediastinal obstruction or spinal cord compression—especially if radiation is not available; or the two agencies may be combined.

Chlorambucil has the advantage of oral administration, though its action is slower than that of mustine. Its effects are similar, but the lymphocyte count is more depressed.

Chief uses:

a. Malignant lymphomas—especially chronic lymphatic leukaemia and lymphosarcoma (which have many similarities and merge into each other).
b. Cancer of ovary—at least as good as thiotepa or cyclophosphamide.

Cyclophosphamide also has the great advantage of effectiveness by mouth. It is particularly liable to cause *epilation* of the scalp, and somewhat liable to produce sterile cystitis and *bladder* ulceration with bleeding. These bladder complications can be avoided or minimised by ensuring a high fluid intake. It is the least toxic of the group, and is relatively kind to platelets. It is useful for a wide variety of cancers.

Chief uses:

a. Lymphomas—Hodgkin's: chronic leukaemias, etc.
b. Myelomatosis—especially combined with melphalan
c. Carcinoma—ovary, breast, lung etc.
d. Burkitt's lymphoma

Thiotepa is usually given by intramuscular injection; also into serous cavities for effusions.

Chief uses. Cancer of breast and ovary.

Melphalan is commonly used for multiple myeloma. It

Table 31.1 Cytotoxic agents used in cancer therapy

Drug	Proprietary name	Administration route	Main uses
Mustine (Nitrogen Mustard)		Intravenous, Intracavitary.	Reticuloses, especially Hodgkin's. Malignant effusions. Other cancers, e.g. lung.
Chlorambucil	Leukeran	Oral	Chronic lymphatic leukaemia. Other reticuloses, especially lymphosarcoma, Hodgkin's. Other cancers, e.g. ovary.
Cyclophosphamide	Endoxan(a) Cytoxan	Intravenous. Oral (especially maintenance). Intracavitary	Reticuloses—most types, including Burkitt's. Myelomatosis. Other cancers, e.g. lung, ovary, breast.
Thiophosphoramide	Thiotepa	Intravenous or intramuscular. Intracavitary.	Ovary; breast; malignant effusions.
Melphalan	Alkeran	Oral. Intravenous	Myelomatosis; malignant melanoma (limb perfusion). Other cancers, e.g. ovary.
Busulphan	Myleran	Oral.	Chronic myeloid leukaemia.
6–Mercaptopurine	Purinethol	Oral.	Acute leukaemia; Choriocarcinoma.
Methotrexate		Oral. Intravenous. Intra-arterial infusion, plus citrovorum factor (intramuscular).	Acute childhood leukaemia. Choriocarcinoma. Burkitt's lymphoma. Infusions—cancers of head and neck, pelvis etc.
5–Fluorouracil (5 FU)		Intravenous.	Cancers of gastro-intestinal tract.
Vinblastine	Velbe Velban	Intravenous.	Hodgkin's. Choriocarcinoma.
Vincristine	Oncovin	Intravenous.	Acute childhood leukaemia. Some other reticuloses and cancers.
Actinomycin-D		Intravenous.	Wilms's tumour: Teratoma testis: Choriocarcinoma.
Adriamycin		Intravenous.	Acute leukaemia. Childhood tumours. Breast, ovary, teratoma.
Daunorubicin		Intravenous.	Leukaemia. Childhood tumours.
Bleomycin		Intravenous, Intracavitary. Intramuscular.	Squamous carcinoma. Late lymphomas
L-Asparaginase		Intravenous.	Acute leukaemia.
Cytosine Arabinoside	Cytarabine	Intravenous.	Acute leukaemia.
Procarbazine	Natulan	Oral.	Hodgkin's.

has also proved useful in perfusion of upper and lower limbs for malignant melanoma.

Busulphan is the best example of preferential action on a particular tissue, since it destroys abnormal stem blood cells in *chronic myeloid leukaemia* at doses too low to damage normal cells. It is now the treatment of choice and preferable to radiation. Toxic effects—p. 278.

Blood counts should be at least weekly, and the drug continued till the white count is about 20 000 and then stopped—or earlier if platelets fall below 100 000. The count will continue to fall for a time, and the aim should be to stabilise it at around 10 000–15 000. Maintenance dosage may then be started and adjusted as the blood picture dictates—including the all-important haemoglobin level.

Whether this treatment actually prolongs life is uncertain, but it does usually keep the patient symptom-free until near the end and enable him to lead a worthwhile existence.

2. Antimetabolites

These are chemical variations of important constituents of cell nuclear metabolism. They act by competing with the normal chemical constituents in building up nuclear proteins etc. The normal components are 'elbowed out' and the cell forms an abnormal product which is useless for mitosis and so leads to cell death.

Methotrexate is usually given orally or intravenously. It has also been used by the intra-arterial route. The typical sign of toxicity is soreness or ulceration in the mouth; alopecia is common. The side-effects can be minimised if the drug is given as an infusion over a period not exceeding 24–36 hours. Toxic effects are more liable after prolonged oral administration.

The toxic effects can be countered by an antidote given intramuscularly or orally—*Folinic acid* (also called *citrovorum factor or leucovorin*). *Folinic acid 'rescue'* reverses the toxic effect on cells—not on cells which have already been damaged, but it counters the further action of the remainder of the methotrexate which has not yet been excreted. High dose *pulses* of methotrexate, with subsequent 'rescue' are now being used.

Chief uses:

a. Lymphomas, especially acute leukaemia
b. Cancers of head and neck, by infusions
c. Choriocarcinoma
d. Burkitt's tumour

6-Mercaptopurine (6 MP) has similarities to methotrexate. It is usually given orally in daily doses. The chief toxic effect is on the marrow, less often on the liver.

Chief uses: Similar to methotrexate, especially for leukaemias. It is of little or no value in other malignancies.

5-Fluorouracil (5 FU) has been used particularly for *gastrointestinal adenocarcinoma* and is the only drug that has proved of any value for these. It can be very toxic, with diarrhoea, mouth ulcers, vomiting, alopecia, marrow depression and dermatitis, especially if misused. It is usually given intravenously and is widely used in multi-drug schedules (breast, ovary etc.).

It has been given as a hepatic arterial infusion for liver secondaries from the bowel, and some useful results claimed.

Another use is as a local application in *ointment* form on the skin, to treat superficial hyperkeratoses and basal cell carcinoma recurrent after radiotherapy.

Cytosine Arabinoside (Cytarabine) is given intravenously and is valuable in acute lymphoblastic leukaemia in children.

Other antimetabolites include—*thioguanine*, *hydroxyurea*, *BCNU* (bis-chloroethyl nitrosourea) and CCNU (chloroethyl-cyclohexyl nitrosourea) which is useful for tumours of central nervous system.

3. Plant products

There are two valuable alkaloids obtained from the periwinkle plant (vinca rosea). Their mode of action is unknown.

Vinblastine is probably the least toxic agent of all, though great care must be taken to avoid leakage outside the vein, where local necrosis may result.

It is widely used in the later stages of Hodgkin's and in combination therapy (see below).

Vincristine is also widely used in many combinations, for leukaemia (especially acute lymphoblastic), Hodgkin's and other lymphomas, many carcinomas and malignant melanoma, and most childhood tumours.

It is a relatively safe drug, though locally irritant. But it is distinguished by its *toxic effect on nerve tissue*, which can be very serious. It causes pain, muscle weakness and atrophy, beginning with foot-drop; also severe constipation from its effect on the bowel musculature. These effects must be watched for, and the drug must be stopped at the first sign of weakness, or the nerves may not recover.

4. Antibiotics

All antibacterial agents obtained from fungi etc. are routinely screened for anti-tumour activity in animals. This has led to several valuable products. They act by binding to DNA and so inhibiting it.

Actinomycin-D is locally destructive if leaked outside a vein. It can cause nausea and vomiting, marrow depression, sore mouth and alopecia.

Chief uses:

a. Childhood tumours—Wilms's tumour, Ewing's tumour, soft tissue sarcomas
b. Teratoma of testis
c. Choriocarcinoma

Adriamycin is given intravenously. Toxic effects

are—marrow depression, nausea and vomiting, sore mouth, alopecia. A special danger is to the heart—*cardiotoxicity*—so that the total dosage must be strictly limited.

It is proving useful in leukaemias, childhood tumours (especially neuroblastoma) and in combinations for many cancers (breast, ovary etc.).

Daunorubicin (Daunomycin, Rubidomycin) is given intravenously. Its toxic effects are very similar to those of adriamycin, including cardiotoxicity. It is of use in leukaemias, and some childhood tumours.

Bleomycin differs from other antibiotics as it is taken up selectively in relatively high concentration by *squamous tissues*, and also by the lungs. Its main use is for squamous carcinoma of skin and other structures, e.g. head and neck cancers. Another unusual feature is the *absence of marrow toxicity*, so that the blood picture is not disturbed. It can therefore be used in Hodgkin's disease etc. when other drugs have failed. It may be given by intramuscular or intravenous route, or into cavities for effusions.

Toxic effects are seen on the skin and mouth—swelling, discoloration and hyperkeratosis of hands and fingers, changes over pressure points (knees and elbows), soreness and ulcers in the mouth. A sinister complication is pneumonitis leading to lung fibrosis, which can be fatal. Chest films should be taken at the start and at intervals, and treatment stopped at the first radiological sign of this toxic effect.

5. Miscellaneous

There are many other compounds, of which we will describe only a few:

Corticosteroids have already been discussed under Hormones (p. 272). In addition to their hormonal effects, they have other actions, including lymphocyte destruction, which are not well understood, but make them of considerable use in some disorders. The best example of this use is in acute leukaemia, especially the lymphoblastic type. They are important constituents of most drug regimes for Hodgkin's and other lymphomas (e.g. MOPP—see below) and are given orally.

Procarbazine is also given by mouth. It can cause nausea and vomiting and some marrow depression and neurological changes. Its action is obscure, and it is of no value in carcinoma. Its chief use is in combination therapy for the lymphomas (e.g. MOPP).

L—Asparaginase is an enzyme, given intravenously. It breaks down asparagine, an amino-acid essential to cell metabolism and growth. Normal cells can synthesise asparagine, but some tumours, especially leukaemias, cannot. The enzyme destroys the amino-acid and so denies it to the cancer cells and causes their death indirectly by deprivation. Its main use is in acute lymphoblastic leukaemia, also in some combination regimes.

Localised treatment

When a drug is given by mouth or injection, the whole of the body is exposed to its effects. It would clearly be advantageous to limit its effects to the tumour area if possible, and so spare bone marrow etc. This can be achieved, to some extent at least, in suitable cases by using appropriate routes of administration, in the following ways:

1. *In serous cavities.* Recurrent malignant exudates in the pleural or peritoneal cavity can be treated by exposing the involved walls of the cavity to the drug. The treatment is exactly comparable to the use of radioactive isotopes and since it is a simpler method, involving no radiation hazard, and can give comparable results, it is now generally used as the *first choice in treatment.*

It is injected by needle and syringe into the cavity, after most of the fluid has been removed by aspiration. If the method fails, isotopes can still be tried. Malignant deposits lining the cavity will be attacked directly; some of the drug will be absorbed through the serous* membrane and reach the general circulation, but rarely in sufficient quantity to cause serious marrow depression.

The drugs most frequently used are mustine, thiotepa and cyclophosphamide.

2. *In the bladder.* In a comparable way, thiotepa may be injected into the bladder via a catheter, for treatment of multiple papillomatosis. This is again similar to isotope therapy with radio-gold. The solution is left in the bladder for half-an-hour and then withdrawn. The superficial cytotoxic effect can destroy small growths on the bladder lining.

3. *By perfusion.* This is illustrated in Figure 31.1. If a growth is situated on an extremity (leg, arm or head), the drug can be largely confined to the tumour area by isolating the blood circulation temporarily from the rest of the body. The main artery and vein are exposed surgically and catheterised. The catheters are connected to a motorised pump, and in this way a solution of the drug is pumped through the artery into the tumour area and out again through the vein into the external part of the circuit. This continuous circuit is maintained for about an hour. The growth is thus exposed to a high concentration of the drug

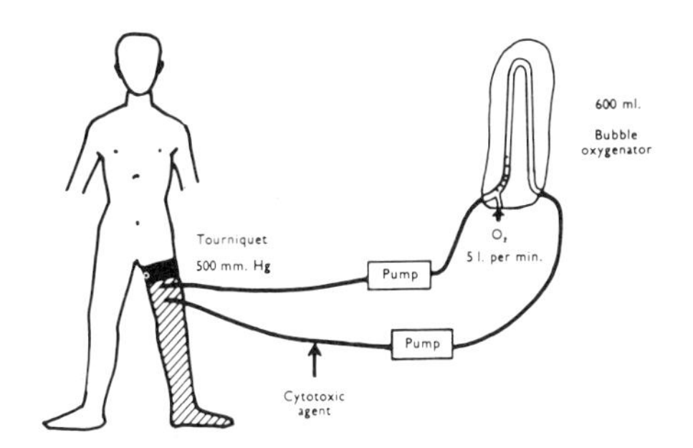

Fig. 31.1 Regional perfusion of a peripheral growth. *(Courtesy Prof. Forrest in Pye's Surgical Handicraft, Wright)*

for a short time, much higher (up to 20 times as high) than can be achieved by oral or intravenous methods, while the rest of the body is largely protected. In practice it is impossible to isolate a part perfectly and there is always a little seepage of the drug into the general circulation, but this is of no serious import.

If the tumour is on a limb, a tourniquet is fastened round the proximal end until the perfusion is over, to achieve more complete isolation from the rest of the body. This technique has proved useful for peripheral *melanoma*, and the best drug here is *melphalan*.

It is a specialised method, and carries appreciable dangers such as bleeding and sepsis post-operatively.

4. *By infusion*. If the tumour area cannot be easily isolated for a perfusion circuit, simple *intra-arterial infusion* can be carried out. A catheter is inserted into the main artery only, either by open exposure or percutaneous puncture. The exact position of the catheter tip and the vascular area served can be ascertained by injecting a radio-opaque contrast medium; adjustments can be made, if necessary, by advancing or retracting the catheter. The extent of skin surface affected can be made visible by injecting a dye (e.g. sulphan blue). The drug is injected through the catheter, either by gravity feed or pump, and almost immediately reaches the tumour region in high concentration. A proportion will, of course, enter the general circulation and produce the usual effects on bone marrow and elsewhere.

The catheter can be maintained in position for many days, and injections made at frequent intervals. Antiseptic precautions are important, as septicaemia is a real threat in a patient whose resistance is likely to be low—as well as thrombosis, peripheral neuritis and haemorrhage.

A recent innovation is to provide a small portable clockwork pump, so that the patient can be ambulant and outside the hospital, and inject the daily dose himself.

The most frequently used drug is *methotrexate*, since its undesirable effects can be countered by the antidote folinic acid.

Unlike perfusion, infusion can be used for e.g. pelvic growths (bladder, cervix, rectum), and even liver and stomach.

Effect on pain. Intra-arterial injections often relieve pain, even in the absence of obvious effect on tumour growth. This is due to a direct toxic effect on nerve endings and can be of real clinical value. The action can at times be too pronounced and cause unwanted paraesthesia (abnormal sensations) and even temporary paralysis.

5. *Direct application*. High concentrations have been used with some success in such instances as:

a. Direct surface application to superficial basal cell lesions on the skin, especially 5 FU
b. Direct injection infiltration of growths, e.g. subcutaneous secondary nodules of breast or lung cancer

Combination chemotherapy

Useful results came from agents used singly, but there was always an obvious possibility that a *combination* of two or more drugs might be superior to a single agent, just as several antibiotics together may be more potent in eradicating bacterial infection. One drug in the combination may kill tumour cells resistant to the other drugs, and vice versa. There will also be less opportunity for resistant cell lines to be left.

Another variation is to use them in *sequence—sequential therapy*—over a definite cycle. Examples:

Choriocarcinoma—see below
Acute leukaemia—page 261

Hodgkin's disease. At first, single agents were used continuously for the late stages—mustine, vinca alkaloids (vincristine or vinblastine), procarbazine, and corticosteroids (prednisolone etc.). When one drug ceased to be effective, another was tried, in the above order. The results were certainly encouraging, with complete remission in 25 per cent, and 25 per cent five-year survival. A major breakthrough was the introduction of *quadruple therapy*, four drugs used together—the now famous *MOPP regime*. (M = Mustine, O = Oncovin = vincristine, P = procarbazine, P = prednisone or prednisolone). Complete remission now came in 80 per cent, with 75 per cent five-year survival—a revolutionary achievement on any reckoning. *This regime is now the standard treatment* in all but the early stages of Hodgkin's and many non-Hodgkin's lymphomas. Minor variations are in use—e.g. some prefer vinblastine to vincristine (Oncovin).

Each course takes a fortnight. The prednisolone and procarbazine are given daily by mouth; the mustine and Oncovin intravenously on the first day and again a week later. Patients need not be in hospital, except perhaps overnight and on the first course, to observe reactions. After two weeks' rest the next course is begun. Courses are repeated at monthly intervals to a total of 6–10. If remission is complete, further courses are given at three-monthly intervals for a year, and at four-monthly intervals for the next year.

The *general principles of combination therapy are:*

1. The drugs should be effective as single agents, but have different modes of action
2. Treatment is *pulsed*—i.e. a short intensive course is followed by a free interval before the next. This gives time for recovery of bone marrow, bowel mucosa, nervous system and immune mechanisms. High doses given over short periods are generally better than low doses given over long periods.
3. Each drug should have a different major toxic effect, so that, while the therapeutic effect is cumulative, specific toxicity is minimised. Drugs of the same group, e.g. alkylating agents, should not usually be combined

4. There is less chance of resistance to any one agent showing itself, since potentially resistant cells are likely to be vulnerable to one of the other agents
5. Different drugs are believed to exert their effects at different points in the cell's developmental cycle. E.g. mustine acts at any stage of the cycle, whether the cell is resting or in mitosis (like radiation). Vinblastine, vincristine and methotrexate act in the phase of DNA synthesis, before mitosis. Cyclophosphamide, 5 FU and actinomycin-D are inactive in the resting stage, but active at any stage of the mitotic cycle. Combinations are therefore calculated to achieve a higher percentage kill.

There is now a long list of multiple drug regimes. Some have been mentioned under acute leukaemia (p. 261). Others in use include—COPAD (cyclophosphamide, Oncovin, prednisone, adriamycin), VAC (vincristine, actinomycin-D, cyclophosphamide), MOB (methotrexate, Oncovin, bleomycin) and many similar combinations. Experience is steadily accumulating and this is perhaps the most actively growing edge of the oncological attack. The shape of things to come is emerging from the shadows, but it will be many years before experience can be sifted to give us the most successful regimes.

Combined radiotherapy and chemotherapy

This is another attractive possibility. Examples:

1. Acute lymphoblastic leukaemia in children, where results have been dramatically improved by irradiating the central nervous system (skull and spinal cord) after remission is achieved by drugs—see page 249
2. Ewing's tumour—it is always wise to assume that there is disseminated occult disease even if only one bone is clinically affected. Results are now markedly improving by irradiation of the bone first, then intermittent multi-drug therapy—see p. 252.

Combined surgery and adjuvant chemotherapy

This is another obvious field to explore. Drugs are given after primary surgery, to destroy silent microscopic metastases and so increase the proportion of long-term survivors. So far, this combination has been applied mainly in children's tumours, with considerable success—e.g. Wilms's embryoma of kidney (see p. 264).

In theory this combination ought to be effective, and many attempts were made to improve results in the common adult cancers—lung, breast etc.—using single agents, but with little or no success. Warning voices have been raised that such therapy might even be worse than useless, by depressing immunological defense mechanisms. With increased experience of combination therapy, the problems are now being tackled again. Promising examples are:

1. Bone cancer. Most cases (80 per cent) are doomed because of early silent secondaries before operation (p. 258). Recent adjuvant chemotherapy immediately after amputation has shown improved early results which will, we hope, be maintained with increased permanent salvage
2. Breast cancer. Early spread is again the menace. Trials are now in progress using adjuvant chemotherapy in patients with proved axillary secondary nodes. It is a considerable burden on the patient to undergo such long-term chemotherapy, which carries its own hazards. But since we know that the presence of axillary secondaries lowers the five-year survival from 70 to 30 per cent, it seems a risk worthwhile. We have the lessons learned from the disseminated lymphomas to guide us, and we know that useful improvement can often be obtained even in advanced breast cancer, from multi-drug therapy. The hope is that long-term freedom and normal life expectancy will be achieved.

   Early results of current trials already show definite lengthening of the recurrence-free interval, but longer time is needed for full assessment.

Cancers cured by chemotherapy

It is now possible to talk of radical curative cytotoxic therapy, where drugs are the sole—or the chief—means of treatment. Some examples have already been given—acute leukaemia (p. 249) advanced Hodgkin's (p. 248). Two further examples will now be described, for their special interest and as signposts to future possibilities.

1. *Choriocarcinoma** (or Chorionepithelioma). This is the malignant variety of *gestational trophoblastic disease*. The fertilised ovum gives rise to the trophoblast* which burrows into the endometrial wall to implant and nourish the ovum. Maternal tissue is destroyed till a balance is struck between fetal and maternal elements with eventual formation of the placenta. This equilibrium may be disturbed, by factors unknown, with production of disorders ranging from the simple to the highly malignant. It is unique among tumours, as it is derived not from tissues of the host, but from a 'parasite' (the fetus).

The simple variety is known as *hydatidiform*mole*, where maternal tissue is invaded somewhat more than normally. The intensely malignant type is the *choriocarcinoma*. It is almost always a result of pregnancy, usually soon after the start, and half of them are preceded by moles. The uterus becomes enlarged, with ulceration and haemorrhage. Vaginal bleeding may be very heavy. Early, even the first, signs may come from secondaries, e.g. in lungs, with cough and dyspnoea. Diagnosis is confirmed by microscopy of fragments passed via the vagina or after uterine curettage.*

Hormone tests are important. The tissue produces large amounts of *human chorionic gonadotrophin (HCG)*—the basis of pregnancy tests. This is very valuable both as an aid to diagnosis and to monitor progress after treatment, since persistently positive tests point to the presence of metastases.

Treatment used to be by hysterectomy and/or radiotherapy. Surgery is now rarely practised, and though the growth is radiosensitive, modern treatment now consists essentially of chemotherapy.

The most valuable drug is methotrexate in fairly high dosage, with folinic acid 'rescue' (see above). Severe toxicity may result and special isolation units are desirable (though not essential) to counter the complications. A good response is achieved in over 80 per cent. In refractory cases, combinations including 6-mercaptopurine and actinomycin-D may be successful.

If the growth is still confined to the uterus, the long-term success rate is 90 per cent, but if there are distant metastases, this drops to 25 per cent. After successful chemotherapy, further pregnancies are possible—though the high risk of gene mutations in the ovarian germ cells may make this a doubtful blessing.

This is a remarkable result by any standard, and represents perhaps the greatest triumph yet of chemotherapy in cancer. But choriocarcinoma is not a typical cancer. It arises in fetal tissue which is genetically 'foreign' to the host (the patient) and is therefore likely to give rise to antigens. Immunological factors almost certainly play a considerable part in the response, and this would account for the spontaneous regression (with no treatment at all) even of metastases, which certainly can occur on rare occasions (see p. 156).

It does not follow therefore that other tumours will respond as well, on the basis of our present knowledge. Still, this does not lessen the immense practical value of the treatment, or its theoretical importance in pointing the way to potential similar triumphs in other lesions.

2. *Burkitt's Tumour.* In 1958 a British surgeon Burkitt described a peculiar tumour affecting the jaws and abdominal viscera in African children. Histologically it was a type of lymphoma, probably lymphosarcomatous. Sporadic cases occur outside Africa—in Europe, U.S.A. etc. Growth is very rapid, and without treatment most die within six months. Surgery is difficult, mutilating and usually ineffectual, since in clinical jaw cases there is usually undetected abdominal involvement also.

The treatment of choice is cytotoxic—methotrexate or cyclophosphamide—better still, both together. Severe marrow depression follows, the hair falls out and cystitis is frequent. But complete regression occurs in over 50 per cent of cases, with apparent cure in 15 per cent. In case of recurrence, re-treatment may succeed.

There are also other intriguing aspects. The geographical area of distribution is exactly the same as that of disease-carrying insects, e.g. mosquitoes, and a virus is now believed to be the causative agent. Just as with choriocarcinoma, immunological factors probably play a part—even a predominant part—in the success of treatment. Complete regression can occur after even a single dose of the drug—as though some mechanism had been triggered into action. Even spontaneous remissions are known to happen. Clearly it is a very untypical cancer, but at least an interesting and exciting pointer to possibilities of wider application in other cancers.

Summary conclusion

Cytotoxic chemotherapy is a recent and revolutionary development and has a long way still to go. In the last edition of this book, only a few years ago, we said 'We cannot yet claim any permanently curative results'—but the situation has already changed. Experience was inevitably slow in accumulating, mistakes were made, toxicity was alarming, the use of single agents was the rule, and initial optimism was succeeded by a wave of pessimism. This is a common sequence of events when new modes of treatment are discovered—just as in the case of ionising radiation.

So far, the greatest successes have been in the field of disseminated 'soft' tumours—such as leukaemias and lymphomas. In children's tumours the impact has also been revolutionary. In the common adult 'solid' cancers, the contribution has so far been disappointing, but the field is being worked over again, in the light of our newer knowledge of combination therapy and a more scientific appreciation of what is called 'cell population kinetics' (see Ch. 17) as a basis for correct dosage, timing etc. The role of pre-operative and, post-operative chemotherapy, and the optimum combinations of surgery, radiation and drugs for each particular type of cancer—not forgetting the potential critical importance of immunology—have still to be worked out. We are really only at the beginning of the road.

The brilliant results e.g. in choriocarcinoma and Burkitt's tumour give us inspiration and hope. The recent spectacular advances in molecular biology, and knowledge of the key biochemical structures of the cell, especially the nucleus and mitosis, could well enable us, before the end of the century, to pin-point the initial steps in malignancy. If so, we could—one hopes—devise rational means of either forestalling or correcting it.

Glossary

L. = Latin G. = Greek
The ending 'oma' signifies tumour or neoplasm
The ending 'itis' signifies inflammation
The ending 'osis' signifies state or condition

Ablate. (Verb), *Ablation* (noun). L. ab = from, away; latus = part of the verb 'to bear, take, carry'. Take away, remove.

Acromegaly. G. acro = tip, extremity; mega(1) = large. In the metric system mega = a million times e.g. megavoltage.

Adenocarcinoma. G. aden = gland; karkinos = crab. A cancer of glandular tissue.

Ala nasi. L. = wing (of the nose or nostril). The curved outer wall of each nostril.

Alopecia. G. alopex = fox. A disease like fox mange, causing loss of hair. Baldness.

Alveolar (adjective). *Alveolus* (noun). L. alvus = belly; alveus is the diminutive = trough or cavity. Used for the tooth sockets in the jaw bones; also for the surface tissue of upper and lower jaws. Other uses include—the terminal ducts of the small bronchi and the sac-like dilatations where gas exchange occurs; also the terminal secretory parts of glands, (Fig. 14.4, p. 158); hence 'alveolar carcinoma'.

Amenorrhoea G. a = not; men = month; rhoea = flow. Absence of normal menstruation, cessation of periods.

Anaplasia (noun), *Anaplastic* (adjective). G. ana = again, back; plasia = forming, moulding. Reversion of cells to a more primitive, embryonic-like type with increased reproductive activity, including malignancy.

Androgen (noun), *Androgenic* (adjective). G. andro = man, male; gen = produce. An agent, especially a hormone, with masculinising effects from stimulation of accessory sex organs and secondary sexual characters (e.g. deepened voice, moustache).

Angioma. G. angeion = vessel. Tumour usually of blood vessels (haemangioma); also of lymphatic vessels (lymphangioma).

Ankylosing Spondylitis. G. ankylos = bent; spondylos = vertebra. Stiffening or fixation of intervertebral joints; in advanced cases causes gross forward bending of spine.

Aplastic. G. a = not; plastic = forming. A. anaemia—due to lack of formation of blood cells, especially red.

Areola. Diminutive of area (L) i.e. any small area. Used of the pigmented ring round the nipple.

Arrhenoblastoma. G. arrhen = male; blast = germ. Tumour of gonadal cells of ovary, liable to produce androgenic hormones and masculinisation.

Aryepiglottic. One of the laryngeal cartilages is called 'arytenoid' from a G. word meaning 'ladle', after its shape. Epi = on, upon; glottis = aperture of larynx. A fold of mucous membrane goes from arytenoid to epiglottis (Fig. 20.1 p. 201).

Ascites. G. ascos = bag, wine-skin. Accumulation of fluid in peritoneal cavity.

Astrocytoma. G. astron = star; cytos = receptacle, cell. The microscopic appearance of the cell (one of the neuroglial cells) is star-shaped.

BCG. Bacillus of Calmette (1863–1933) and Guérin (1816–1895), two French bacteriologists. A variety of tubercle bacillus, used as a vaccine for immunisation against pulmonary tuberculosis.

Bence-Jones. London physician and chemical pathologist, 1814–1873.

Bilharziasis. Bilharz, German parasitologist, 1825–1862, gave his name to a genus of worms—bilharzia (now called schistosoma). B. is the resultant disease after infestation.

Biopsy. G. bios = life, opsis = vision. Examination of tissue taken from a patient during life.

Blast. G. = bud, offspring. Used of immature or primitive cells e.g. precursors of adult blood cells.

Bowen. American dermatologist, 1857–1941.

Bremsstrahlung. German bremse = brake; strahlung = radiation. 'Braking' radiation, produced by the sudden slowing down of electrons by atomic collisions. They are exactly analogous to ordinary X-rays generated by the impact of electrons in the tube when they are suddenly halted by the target.

Bronchogenic. G. bronchos = windpipe; gen = produce, beget. Arising in a bronchus.
(Compare 'osteogenic'—arising in bone).

Buccal. L. bucca = cheek. Used mostly of the oral aspect.

Cachexia (noun), *Cachectic* (adjective). G. kakos = bad; hexis = state or condition. Wasting in chronic disease.

Cataract. From G. for 'to dash down'. Usually refers to a large waterfall, but there is an obsolete meaning of 'portcullis' (sliding gate). The medical use, applied to the lens of the eye, is a figurative extension of this.

Choriocarcinoma. G. chorion = the outermost membrane enclosing the fetus before birth. It consists of several layers, of which the outermost is the *trophoblast* (G. trophe = nourishment; blastos = germ) with projections that become embedded in the uterine mucosa. The metabolic exchanges between mother and fetus take place through these.

Chromophobe. G. chroma = colour; phobos = fear. Resistant to histological stains.

Cicatrising. L. cicatrix = scar, the fibrous tissue which replaces destroyed normal tissue.

Colloid (al) G. colla = glue. A colloidal solution differs from an ordinary solution. The molecules are so large that they do not pass freely through the body membranes. A colloid injected at a given site therefore does not move around freely, but tends to stay where it is deposited.

Colostomy. G. kolon = large intestine; stoma = mouth. (Making of) an artificial anus by an opening into the colon, brought onto the abdominal wall.

Cuirasse French = breastplate (armour). *Cancer-en-cuirasse*—extensive malignant invasion of the skin of the chest wall.

Curettage. French 'curer' in the sense of 'to clear, cleanse'. A *curette* is a small instrument in the form of a loop, ring or scoop, with sharpened edge and long handle, to scrape the interior of a cavity in order to remove tissue for biopsy or treatment.

Cyanosed. G. kuanos = dark blue. With livid skin due to deficient oxygenation of blood.
Cystoscopy. G. kystis = bladder; skopos = watcher. Inspection of the interior of the bladder by a miniature sort of telescope passed along the urethra.
Cystotomy. G. kystis = bladder; tome = cutting. Cutting into and opening of the bladder through the abdominal wall.
Cytology. G. kytos = receptacle, cell; logos = discourse, study. The science of the structure and function of cells.

Desquamation. L. de = from, away; squama = scale. Shedding of the epidermis in scales or flakes.
Diuretic. G. and L. dia = through, in sense of 'thorough'. L. urina = urine. Diuresis is excretion of urine in excess of normal, induced e.g. by drugs. Such a drug is a 'diuretic'.
Dura (mater). L. = hard (mother). The tough outer membrane enclosing brain and spinal cord.
Dysgerminoma. (Better 'disgerminoma'). 'Dis' is from L. and G. for 'two' in sense of 'twain, asunder'. G. dys = bad, difficult, the opposite of 'eu' (eugenic, euthyroid etc.). L. germen = bud, sprout. A rare tumour of ovary, of gonadal germinal cells, corresponding to seminoma of testis.
Dysmenorrhoea. G. dys (see above); men = month; rhoia = flow. Difficult or painful menstruation.
Dyspnoea. G. dys (see above); pnoea = breathing. Difficult breathing, shortness of breath.

Ectoderm(al). G. ecto = outside; derma = skin. The outer of the three primary layers of cells in the embryo. (The others are mesoderm and endoderm).
EMI. Initials of Electronic and Musical Industries Ltd., original manufacturers of the equipment for CAT (computerised axial tomography).
Endarteritis. G. endo = within. Inflammation of the inner lining of an artery. May lead to arteriosclerosis and obstruction of the vessel.
Endocrine. G. endo = within; krino = sift, separate. Applied to glands of internal secretion.
Endometrium. G. endo = within; metra = womb (meter = mother, same root as L. mater). The internal lining membrane of the uterus.
Endoscopy. G. endo = within; skopeo = examine. Inspection of the interior of a channel or hollow organ.
Eosinophil. G. eos = dawn; philos = loving. Eosin is an (acid) rose-coloured dye. 'Acidophil' is an alternative name. Used of cells that take up acid (in contrast with basic) stains.
Ependymoma. G. ependyma = garment. The name was given by the great 19th century German pathologist Virchow to the lining membrane of the cerebral ventricles and central canal of the spinal cord.
Epidermis. G. epi = on; derma = skin. The outer (epithelial) layer of the skin.
Epididymis G. epi = on; didymos = twin (used in the plural to denote 'testicles'). A structure attached to the back of the testis, consisting mainly of coils of the excretory spermatic duct.
Epilation L. e = out; pilus = hair. Removal of hair.
Epiphysis. G. epi = upon; physis = growth. An extremity of a long bone, originating in a centre of ossification separate from that of the shaft.
Epithelium. G. epi = upon; thele = nipple. Applied originally to the thin skin covering the nipples. Now used of the non-vascular layer covering all free surfaces, internal and external—skin, mucous and serous membranes and all the glands etc. derived from them.
Erythoroblast(ic). G. erythros = red; see Blast. Primitive immature red blood cell.

Eustachius. Italian 16th century anatomist. The Eustachian tube joins middle ear and nasopharynx.
Ewing (James). American pathologist, 1866–1943. Director of the Memorial Hospital, New York, and author of a famous textbook on Pathology of Tumours.
Exenteration. G. ex = out; enteron = bowel. Removal of viscera.
Exophthalmos. G. ex = out; ophthalmos = eye. Protrusion of the eyeball. See Proptosis.
Extramedullary. L. extra = outside; see 'medullary'. In oncological contexts the medulla usually means the spinal cord, originally regarded as the 'marrow' of the spine.

Fallopius. Italian 16th century anatomist.
Fistula. (plural—fistulae or fistulas). L. = pipe, tube. An abnormal passage leading from a cavity (e.g. abscess) or hollow organ to the surface, or from one cavity to another.
Follicular. L. follis = bellows; the diminutive folliculus = little bag or sac. Microscopically the thyroid shows glandular tubes lined with cells. On cross-section they form little cavities, vesicles or follicles, with the secretory cells lining the walls and the secretion in the lumen. Follicular carcinoma may closely resemble normal thyroid tissue and be difficult to distinguish; or the follicles may be large and irregular.
Fornix, plural—fornices. L. = arch, vault. E.g. at the sides of the uterine cervix.
Fundus. L. = bottom. The base of an organ i.e. the part remote from the external opening.

Ganglioneuroma. G. ganglion = tumour under skin near tendon; still used in this sense. Another modern sense is enlargement or knot on nerve, from which nerve fibres radiate; also—aggregation of nerve cells. G. neuron = nerve. A tumour of adult sympathetic nerve cells.
Glioma. G. glia = glue. Neuroglia ('nerve glue') is the specialised connective tissue or binding cells of the central nervous system.
Glossitis. G. glossa = tongue. Inflammation of the tongue.
Goitre. L. guttur = throat. A chronic enlargement of the thyroid gland. (Compare 'guttural').
Gonadotrophic. G. gone = seed; hence gonad = germ or sexual gland. Trophe = nourishment.
Granulation. L. granulum, diminitive of granum = grain. Formation of tiny fleshy projections on the surface of a wound in the process of healing.
Granulocyte. As above + cytos = cell. Mature polymorphonuclear leucocyte containing granules in the cytoplasm which take up acid (eosinophil or acidophil) basic (basophil) or neutral (neutrophil) stains.
Granulosa. The inner lining of the ovarian follicle*, from which the ovum develops, is called 'membrana granulosa'—granular membrane, and the constituent cells 'granulosa cells'. They do not show granules microscopically and the origin of the name is obscure.
Grawitz. German pathologist 1850–1932.
Grenz. German = boundary. Grenz rays lie on the border between ultra-violet and X-rays.
Gynaecomastia. G. gynaeco = woman, mastos = breast. Excessive development of male mammary tissue, maybe even with secretion of milk.

Haematocrit. G. haem = blood; krino = sift, separate (compare 'endocrine'). Centrifuge for separating cells of the blood from plasma. Also used for 'haematocrit value', i.e. the percentage figure of the relative cell volume in the blood.
Haemolytic. G. haem = blood; lytic = dissolving, destroying.

Haemopoietic. G. haem = blood; poiesis = making. (Same root as poet).

Haemoptysis. G. haem = blood; ptysis = spitting. Spitting of blood from lungs or air-passages.

Histiocyte. G. histion, diminutive of histos = web, tissue; cytos = cell. One of the cell types of the reticulo-endothelial system. Also called 'macrophage'.

Histology. G. histos = web, tissue. Microscopic study of tissues and organs, and of cells arranged in tissues. Contrast 'cytology'—study of individual cells.

Hodgkin, Thomas. 1798–1866. Physician-pathologist at Guy's Hospital, London. The first to describe the disease as a separate entity.

Hydatidiform. i.e. with the form or appearance of a hydatid. G. hydatis = drop of water (hydro = water). A cyst containing watery fluid, especially that formed by the larva of a tapeworm. *Hydatidiform mole* (L. and G. mole = meal)—a polycystic mass resulting from proliferation of the trophoblast (see Choriocarcinoma) and cystic degeneration of its projections into the uterine wall.

Hydronephrosis. G. hydro = water; nephros = kidney. Dilation of the pelvis etc. of the kidney due to obstruction to the flow of urine at some lower level.

Hydroureter. G. hydro = water. Distension of the ureter with urine due to obstruction e.g. by a tumour.

Hyperbaric. G. hyper = over, in excess; baros = weight. At a pressure higher than that of the atmosphere.

Hyperkeratosis. G. hyper = over, in excess; keras = horn. Overgrowth of the horny (outer) layer of the epidermis.

Hypophysectomy. G. hypo = under; physis = growth; ectome = excision. *Hypophysis* thus = undergrowth or downgrowth, applied to the pituitary which appears to grow from the base of the brain.

Intrathecal. L. intra = within; G. thece = box, receptacle. Refers e.g. to the loose sheath covering spinal cord and its surrounding cerebrospinal fluid.

Ipsilateral. L. ipse = self, same; later = side. On the same side.

Ischaemia. G. ischo = check, restrain; haem = blood. Local anaemia due especially to narrowing or obliteration of blood vessels.

Keloid (or *Cheloid*) G. chele = claw; eidos = form, appearance. Hyperplastic fibrous scar tissue, especially after burns or wounds, with claw-like processes radiating from its extremities.

Keratinised. G. keras = horn. Hornified or cornified (L. cornu = horn); made horny. Compare Hyperkeratosis.

Kerato-acanthoma. G. keras = horn; acantha = thorn. Non-malignant keratinising (see above) tumour of epidermis. One of the normal epidermal layers is composed of 'prickle cells' with tiny microscopic processes looking like thorns—hence 'acanthoma', a tumour of these cells.

Keratosis. See Hyperkeratosis.

Koilonychia. G. koilos = hollow; onych = nail. Malformation of the nails, in which the surface is spoon-shaped i.e. concave instead of the normal convexity.

Laminectomy. L. lamina = thin piece or plate; G. ectome = excision. Removal of the posterior part of one or more vertebral arches, to expose the spinal canal and gain access to the spinal cord etc.

Laparotomy. G. lapara = flank; tomia = incision. Opening the abdominal cavity in order to inspect the contents or perform an operation on a viscus etc.

Leiomyosarcoma. G. leio = smooth; myo = muscle. Sarcoma of smooth muscle.

Lesion. L. laesio = hurt, injury. Any abnormal change in the texture or functioning of an organ. E.g. any inflammation, wound, infection, tumour or other more or less localised abnormality.

Leucopenia. G. leuco = white; penia = poverty, shortage. Diminution of the normal number of white blood cells; the opposite of leucocytosis.

Leucoplakia. G. leuco = white; plak = plate, plaque. Disturbance of maturation of epithelium causing heaped up whitish patches; usually in the mouth, but also vulva etc.

Lumen. L. = light or opening. A cavity or space enclosed by the walls of a tube, cell etc.

Medullary. L. medulla = marrow, from medius = middle. 'Medulla' is used of any soft marrow-like element, especially in the centre of a part.

Medulloblastoma. See Medullary. G. blastos = sprout, shoot, germ. Medulloblast is a primitive type of glial cell (see Glioma). In this case it refers to a particular part of the cerebellar region.

Megakaryocyte. G. mega = big; karyon = nut or nucleus; cyt = cell. An exceptionally large cell in bone marrow, from which blood platelets derive.

Melanoma. G. melan = black. Tumour derived from cells capable of forming the pigment melanin (seen in freckles etc.). Pigment may or may not actually be formed; if not, the tumour is 'amelanotic' (G. a = not).

Menorrhagia. G. men = month; rhag = break, burst. Abnormally profuse or prolonged menstruation.

Mesodermal. G. meso = middle. See Ectoderm.

Metaplasia. G. prefix 'meta' refers especially to change (of place, condition etc.). Plasia = formation. Transformation; formation of a type of tissue by cells normally producing another type. E.g. transitional epithelium of the bladder may be changed to squamous.

Metastasis. See Metaplasia. G. stasis = standing, position. Change of place; especially, production of secondary tumours remote from the primary.

Mucosa. L. = mucous. Short for 'membrana mucosa'—mucous membrane, an epithelial surface lubricated by secreted mucus.

Mycosis Fungoides. G. myke = mushroom, fungus; eidos = shape, form. A malignant lymphoma of skin producing fungating tumour masses.

Myelitis. G. myelo = marrow. Inflammation of (1) the spinal cord, once regarded as spinal 'marrow' or (2) bone marrow.

Myeloma. G. myelo = marrow. Usually refers to plasma cell myeloma (or plasmacytoma), a tumour of plasmacytes (plasma cells) of bone marrow. Plasma cells are concerned in antibody production. When the lesions are diffuse the disease is 'myelomatosis' or 'multiple myeloma'.

Myxoedema. G. myxa = mucus; oidema = swelling. In advanced thyroid hormone deficiency there is infiltration of gelatinous fluid into the tissues, giving a feel of hard oedema. The name was given in the mistaken belief that the infiltrate was mucus.

Naevus (Nevus), plural = naevi. L. = mole, spot, blemish. A general term applying to any congenital lesion, often pigmented, especially on the skin. The chief types are (1) angioma and (2) melanoma.

Neoplasia, *Neoplasm*. G. neo = new; plasia = formation. New growth, tumour (in the oncological sense).

Neuroglia. See Glioma.

Oestrogen. G. oestros = gad-fly, sting, frenzy; gen = beget, produce. Oestrus (animal heat or rut) is that part of the female sexual cycle during which mating is accepted. The hormone that brings on oestrus is 'oestrogenic'.

Oligodendroglioma. G. oligos = few or small; dendron = tree. Oligodendroglia refers to cells of the neuroglia which have relatively few and short branches.

Oncology. G. oncos = mass, bulk, tumour. The science of tumours in all their aspects.

Oophorectomy. G. öon = egg, ovum; phoros = bearing, carrying; ectome = excision. Removal of the organ that bears the eggs (oophoron = ovary).

Orchidectomy. G. orchis = testicle; ectomy = excision. The orchis family of plants is so named after the resemblance of the tubers to the shape of a testicle. Surgical removal of testicle(s).

Os. L. = mouth. Plural = ora (hence-oral). NB: another word with same spelling = bone (plural = ossa—hence ossicle, osseous).

Osteo G. = bone.

Osteoclast G. clast = broken. A large many-nucleated cell in bone marrow, concerned in absorption and removal of bone tissue.

Osteogenic G. gen = produce. Means either formed from bone or forming bone.

Osteolytic G. lytic = dissolving, destroying. Destructive of bone.

Osteomyelitis. G. myelo = marrow. Inflammation of bone marrow.

Paget, Sir James. 1814–1899. Famous London surgeon. Several conditions are named after him, including (1) Paget's disease of bone (see p. 258) a skeletal disease leading to thickening and softening of bone and bending of weight-bearing bones, and (2) Paget's disease of nipple, see page 216.

Pancoast 1875–1939. American radiologist. Pancoast's syndrome = pain in arm etc. due to nerve involvement by a tumour at the apex of the lung.

Papilla (plural = papillae). L. = nipple. Diminutive of papula = pustule or pimple, probably from an older root 'pap' = swell. Any small nipple-like projection. Hence *papillary*, *papilloma* (plural = papillomas or papillomata), *papilliferous* (bearing papillae), *papillomatosis*.

Para G. = by the side of. The primary meaning refers to position e.g. parathyroid (alongside the thyroid).

Paracentesis G. centesis = puncture. The perforation of a cavity by a hollow instrument to remove fluid.

Parametrium (plural = parametria). G. metra = uterus (see Endometrium). The connective tissue etc. between the lateral border of the uterus and the pelvic wall.

Paraplegia G. plegia = blow or stroke. The name was originally applied to a stroke involving one side, now called hemiplegia (hemi = half) but was later used to refer to involvement of the lower limbs.

Peau d'orange. French = peel of orange. In locally advanced breast cancer lymphatic obstruction causes oedema of the skin while the sweat ducts are tethered so that their orifices become noticeable. The resultant pitted appearance is likened to orange peel.

Peptic. G. pepto = cook, digest. Relating to the stomach or gastric digestion.

Periosteum G. peri = around; osteo = bone. The thick fibrous sheath adherent to and surrounding a bone.

Petrous G. petra = rock. Applied to the hard part of the temporal bone protecting the inner ear. Temporal is from L. for 'temple', that part on each side of the head being metaphorically regarded as the 'temple of the head'.

Phimosis. G. = muzzling. Contraction of the orifice of the prepuce (foreskin) so that it cannot be drawn back over the glans penis.

Piezo-electric. G. piezo = press, squeeze. Refers to electricity generated by pressure on certain crystals e.g. quartz.

Pinealoma. L. pinus = pine-cone. The pineal body is a cone-shaped structure behind the third cerebral ventricle.

Piriform. L. pirum = pear. Pear-shaped. The spelling 'pyriform' is incorrect, but sanctified by usage.

Pituitary. L. pituita = mucus or phlegm. Nasal secretion was thought by the ancients to come through the skull floor from the base of the brain.

Plasma cell. See Myeloma

Plasma. See Serum.

Polycythaemia rubra vera. G. poly = much, many; cyt = cell; haem = blood. An increase of blood cells in number, especially red cells. L. rubra = red; vera = true, genuine—i.e. in distinction from secondary polycythaemia, see page 251.

Proctitis. G. proctos = anus. Medically, used to refer mainly to the rectum. Inflammation of rectal mucosa.

Progestogen. L. pro = in favour of; gesto = bear; G. gen = produce. Applied to hormones causing changes in the endometrium to prepare it for the reception of the fertilised ovum.

Proptosis. G. pro = forward; ptosis = falling. Forward displacement of a part, especially protrusion of the eyeball (= exophthalmos).

Pseudomucinous. G. pseudo = false. Pseudomucin is a gelatinous material resembling mucin. See page 234.

Psoriasis. G. psora = the itch. A skin eruption with reddish-brown scaly patches. Itching is actually not at all characteristic.

Pyelography. G. pyelo (= L. pelvis) = trough, basin; refers to the pelvis (collecting basin) of the kidney. G. grapho = write. Radiography of the urinary tract.

Pyometra. G. pyo = pus; metra = uterus. Accumulation of pus in the uterine cavity.

Pyriform. See Piriform.

Pyrogen. G. pyr = fire; gen = produce. An agent causing a rise in temperature. Especially, a substance of unknown nature, but probably (foreign) protein, liable to be present in solutions injected intravenously.

Rathke, German anatomist, 1793–1860. Rathke's pouch or diverticulum—see page 257.

Reticulum, *Reticulosis.* L. diminutive of rete = net. Net-like structure. (compare reticule—a network bag).

Rhabdomyosarcoma G. rhabdo = rod, strip. Malignant tumour of striated (striped) muscle. Compare Leiomyosarcoma.

Salpingitis. G. salpinx = trumpet. Used to denote a tube with flared end, in particular the Fallopian* tube along which the ovum passes to the uterus.

Scirrhous (adjective). G. skirrhos = hard. Overgrowth of tough fibrous tissue in a tumour gives it a hard feel. The noun *scirrhus* is also used for a hard tumour.

Sclerotic. G. scleros = hard. Sclerosis refers to hardening of chronic inflammatory origin e.g. in the walls of arteries. The sclera (or sclerotic) is the white of the eye, a fibrous coat forming the outer surface of the eyeball, except for the cornea in front.

Scoliosis. G. scolio = bent, curved. Lateral curvature of the spine.

Sella. L. = seat, saddle. The saddle-shaped part of the upper surface of the sphenoid (G. = wedge-shaped) bone which houses the pituitary. Also called 'sella turcica' = Turkish saddle, after its shape.

Seminoma. L. semen = seed (from the verb 'to sow'). Tumour of male germ cells, precursors of the spermatozoa.

Serous, *Serum.* L. = whey, watery fluid. Serum is the fluid part of the blood which separates from the clot after coagulation.

Distinguish from *plasma*, the fluid part after removal of blood cells, which is still coagulable. Serous membranes line the closed cavities of the body, especially pleural and peritoneal, and are moistened by exuded fluid similar to serum.

Situ. L. situs = site. *Carcinoma-in-situ* = cancer cells 'in position', still in their tissue of origin (epidermis etc.) before breaking through the basement membrane.

Spondylitis.—see Ankylosing

Spongioblastoma. G. spongia = sponge; blastos = germ. The spongioblast is a primitive neuroglial cell, precursor of the astrocyte. Multiple cavities in the cytoplasm in microscopic preparations give it a sponge-like appearance.

Squamous. L. squama = scale. Refers to the scaly part of epidermal and other surfaces.

Stenosis. G. stenos = narrow. Narrowing of any channel.

Stilboestrol. G. stilbo = glisten. Stilbene is the chemical name of a complex hydrocarbon (used in dyes etc.). For oestrol see Oestrogen.

Stroma. G. = something spread out. The supporting framework, generally connective tissue, of an organ or structure, as opposed to the specific cells of the organ or neoplasm etc.

Submental. L. sub = beneath; mentum = chin. Below the chin and floor of mouth.

Synovial. The word *synovia* was coined by Paracelsus (physician and chemist, died 1541), applied to nutritive fluid peculiar to various parts of the body. Now refers to the viscid lubricating fluid secreted in joints and tendon sheaths from their lining membranes.

Telangiectasia. G. telos = end; angion = vessel; ectasis = extension, dilatation. Dilatation of small or terminal blood vessels, especially in the skin, like tiny varicose veins, producing a purplish, blotchy, spidery appearance.

Tenesmus G. = straining, from the verb 'to stretch'. A continual inclination to empty the bowel (or bladder), with painful spasm, but little or no discharge.

Teratoma. G. teras = monster. A tumour composed of various tissues (e.g. bone, teeth etc. etc.) not normally existent at the site.

Thrombopenia or *thrombocytopenia.* G. thrombos = clot; penia = poverty, scarcity. Diminution in number of blood platelets.

Thymus. G. = warty excrescence. Refers to the lymphoid organ in the superior mediastinum and lower neck, present in childhood. It is of great importance in immunological development.

Tracheostomy. G. trachea = rough—i.e. the rough artery; the ancients believed it was an artery, and the cartilaginous rings give it a rough feel. G. stoma = mouth. Formation of an opening into the trachea, to relieve obstruction of the airway.

Transurethral. L. trans = across, through; urethra from same root as 'urine'. Refers to a procedure such as partial prostatectomy carried out by instruments passed along the urethral channel.

Trigone. G. tri = three; gon = angle. Triangle, especially the triangular area at the base of the bladder, between the openings of the two ureters and the urethra.

Trophoblast. See Choriocarcinoma

Uraemia. G. ouron = urine; (h)aem = blood. Toxic excess of urea and other waste products in the blood, due to impaired excretion.

Urea. Same root as 'urine'. The chief end-product of nitrogen metabolism in mammals, excreted in the urine.

Vasectomy. L. vas = vessel; G. ectomy = excision. In this context the vas is the duct conveying the semen (spermatic fluid) from the testicle towards the urethra. Surgical removal of a segment of the spermatic duct for the purpose of sterilisation.

Waldeyer. German anatomist-pathologist, 1836–1921.

Wilms. German surgeon, 1867–1918.

Bibliography

PART I

General radiation physics

Alexander P. '*Atomic Radiation and Life*'. London: Penguin.

Coggle, J. E. (1971) '*Biological Effects of Radiation*'. London: Wykeham Publications.

Hay, G. A. & Hughes, D. (1972) '*First Year Physics for Radiographers*'. London: Baillière Tindall.

Johns, H. E. & Cunningham, J. R. (1969) '*The Physics of Radiology*'. 3rd edn, Thomas.

Meredith, W. J. & Massey, J. B. (1977) '*Fundamental Physics of Radiology*', 3rd edn, Bristol: Wright.

Radiotherapy physics

Catterall, M., Rogers C., Thomlinson, R. H. & Field, S. B. (1971) 'An investigation into the clinical effects of fast neutrons'. *British Journal of Radiology*, **44**, 603–611.

Catterall, M. (1976) 'Fast neutrons—clinical requirements' *British Journal of Radiology*, **49**, 203–205.

Cohen, M., Jones, D. E. A. & Greene D. (1972) 'Central axis depth dose data for use in radiotherapy'. *British Journal of Radiology*, Suppl. 11.

Fredrickson, D. H., Whitton, J. B. & Karzmark, C. J. (1977) The possibility of monitoring patient position during treatment using the transmitted beam. *British Journal of Radiology*, **50**, 289–290.

Greene, D. & Stewart, J. G. (1965) Isodose curves in non-uniform phantoms. *British Journal of Radiology*, **38**, 378–385.

Hall, E. J. & Rossi H. (1975) The potential of californium–252 in radiotherapy. *British Journal of Radiology*, **48**, 777–790.

Hans, A. G. & Mark, J. E. (1973) Detection and evaluation of localising errors in patient radiation therapy. *Investigative Radiology*, **8**, 384–391.

Haybittle, J. L. (1964) A 24 Ci Strontium–90 unit for whole body superficial irradiation with beta rays. *British Journal of Radiology*, **37**, 297–301.

Hospital Physicists Association (1970) Guidance on the testing of sealed sources of radioisotopes for leakage and surface contamination. *Scientific Report Series*, **1**.

Hospital Physicists Association (1971) A practical guide to Electron Dosimetry (5–35 MeV). *Scientific Report Series*, **4**.

Hospital Physicists Association (1975) A practical guide to Electron Dosimetry below 5 MeV for radiotherapy purposes. *Scientific Report Series*, **13**.

Hospital Physicists Association (1977) Phantom materials for Photons and Electrons. *Scientific Report Series*, **20**.

Jolles, B. (1950) The reciprocal vicinity effect of irradiated tissues on a 'diffusable substance' in irradiated tissues. *British Journal of Radiology*, **23**, 18–24.

Massey, J. B. (1970) Manual of dosimetry in radiotherapy. I.A.E.A. *Technical Report Series*, **110**.

Tsien, K. C., Cunningham, J. R., Wright, D. J., Jones, D. E. A. & Pfalzner, P. M. (1967) *Atlas of Radiation Dose Distributions* I.A.E.A.

Radioisotopes

Beierwaites, W. H., Keyes, J. W., Carey, J. E. (1971) *Manual of Nuclear Medicine Procedures*. C.R.C. Press, Chemical Rubber Company.

Belcher, E. H. & Vetter, H. (1971) *Radioisotopes in Medical Diagnosis*. London: Butterworths.

Early, P. J., Razzak, M. A., Sodee, D. B. (1975) *Textbook of Nuclear Medicine Technology*. 2nd edn. The C V Mosby Company.

Oliver, R. (1971) *Principles of the Use of Radioisotopes Tracers in Clinical and Research Investigations*. Oxford: Pergamon.

Wagner, H. N. (ed) (1975) *Nuclear Medicine* New York Hospital Practice.

Radiation protection

Abbatt, J. D., Lakey J. R. A. & Mathias, D. J. (1960) *Protection Against Radiation*. London: Cassell.

H.M.S.O. (1972) *Code of Practice for the Protection of Persons against Ionising Radiations arising from Medical and Dental Use*.

Rees, D. J. (1967) *Health Physics*. London: Butterworth.

Ultrasonics

Hospital Physicists Association (1977) *A Guide to Medical Ultrasonics and Acoustics. Scientific Report Series*, **10**.

McDicken, W. M. (1976) *Diagnostic Ultrasonics—Principles and Use of Instruments. St Albans: Lockwood.*

Wells, P. N. T. (1977) *Ultrasonics in Clinical Diagnosis*. Edinburgh: Churchill Livingstone.

Wells, P. N. T. (1969) *Physical Principles of Ultrasonic Diagnosis*. London: Academic Press.

PART II

General and historical

Chesney, D. N. & Chesney, M. O. (1973) *Care of the Patient in Diagnostic Radiology*. Oxford: Blackwell. (Relevant also to radiotherapy.)

Curie, E. *Madame Curie*. Various editions e.g. (1962) London: Heinemann. (1959) New York: Pocket Books. (Biography of the discoverer of radium, by her daughter).

Davies, P. M. (1974) *Medical Terminology in Hospital Practice*. (A guide for all those engaged in professions allied to medicine.) 2nd edn. London: Heinemann.

Donizetti, P. (1967) *Shadow and Substance. The Story of Medical Radiography*. Oxford: Pergamon Press. (Includes radiotherapy).

Reid, R. (1974) *Marie Curie*. London: Collins.

Roberts, ff. (1972) *Medical Terms: Their Origin and Construction*, 5th edn. London: Heinemann.

Cancer

Ashley, D. J. B. (1972) *An Introduction to the General Pathology of Tumours*. Bristol: Wright. (Good short survey).

Capra, C. G. (1972) *The Care of the Cancer Patient.* London: Heinemann.
Harris, R. J. C. (1976) *Cancer.* 3rd edn. London: Penguin.
Harris, R. J. C. (ed.) (1970) *What We Know about Cancer.* London: Allen and Unwin.

More specialised

Ambrose, E. J. & Roe, F. J. C. (ed.) (1975) *The Biology of Cancer.* 2nd edn. Chichester: Ellis Horwood.
Del Regato, J. A. & Spjut, H. J. (1977) *Cancer: Diagnosis, Treatment, Prognosis.* 5th edn. St. Louis, Mo.: Mosby Co.; London: Kimpton.
Symington, T. & Carter, R. L. (ed.) (1976) *Scientific Foundations of Oncology.* London: Heinemann.

Radiotherapy

Barnes, P. A. & Rees, D. J. (1972) *A Concise Textbook of Radiotherapy.* (Does not include medical oncology.)
Deeley, T. J. (1975) *Principles of Radiation Therapy.* London: Butterworth.
Lowry, S. (1974) *Fundamentals of Radiation Therapy.* London: English Universities Press.
Walter, J. (1977) *Cancer and Radiotherapy* (a short guide for nurses and medical students). 2nd edn. Edinburgh: Churchill Livingstone.

More advanced

Fletcher, G. H. (ed.) (1973) *Textbook of Radiotherapy.* 2nd edn. Philadelphia: Lea and Febiger; London: Kimpton.
Moss, W. T., Brand, W. N. and Battifora, H. (1973) *Radiation Oncology.* 4th edn. St. Louis: C. V. Mosby Co.
Paterson, R. (1963) *Treatment of Malignant Disease by Radiotherapy.* 2nd edn. London: Arnold.

Radiation biology and hazards

Alexander, P. (1965) *Atomic Radiation and Life.* 2nd edn. Penguin Books. (Recommended.)
Coggle, J. E. (1971) *Biological Effects of Radiation.* London and Winchester: Wykeham Publications.
Hersey, J. *Hiroshima.* Various editions e.g. (1966) New York: Bantam Books. School edition, London: Hamilton. (Graphic account of the effects of the atom bomb.)
Mayneord, W. V. (1964) *Radiation and Health.* Nuffield Provincial Hospitals Trust. (Good short survey of the chief problems.)
The Safe Use of Ionising Radiations. A Handbook for Nurses. London: H.M.S.O.

More advanced

Duncan, W. & Nias, A. H. W. (1977) *Clinical Radiobiology.* Edinburgh: Churchill Livingstone.

Medical Oncology—*Cytotoxic Drugs.*

Greenwald, E. S. (1973) *Cancer Chemotherapy.* 2nd edn. London: Kimpton.
Priestman, T. J. (1977) *Cancer Chemotherapy: an Introduction.* Barnet, Herts: Montedison Pharmaceuticals Ltd.

For reference

Bagshawe, K. D. (ed.) (1975) *Medical Oncology.* Oxford: Blackwell.
Holland, J. F. & Frei, E. (ed.) (1973) *Cancer Medicine.* Philadelphia: Lea and Febiger.

Index